THE ART OF DIGITAL DESIGN

An Introduction to Top-Down Design
SECOND EDITION

Franklin P. Prosser
David E. Winkel
Indiana University

 Prentice-Hall International, Inc.

Library of Congress Cataloging-in-Publication Data

WINKEL, DAVID E. (date)
    The art of digital design.

    Order of authors' names reversed on previous ed.
    Includes bibliographies and index.
    1. Electronic digital computers—Design and
construction. 2. Minicomputers—Design and construction
I. Prosser, Franklin P. II. Title.
TK788.3.W56 1986   004.2'1   86-5042
ISBN 0-13-046673-5

*This edition may be sold only in those countries to which it is
consigned by Prentice-Hall International. It is not to be re-exported
and it is not for sale in the U.S.A., Mexico or Canada.*

Editorial/production supervision and
    interior design: Diana Drew
Manufacturing buyer: Gordon Osbourne

Printed in the United States of America

10 9 8 7 6 5 4

ISBN 0-13-046673-5 025

Prentice-Hall International (UK) Limited, *London*
Prentice-Hall of Australia Pty. Limited, *Sydney*
Prentice-Hall Canada Inc., *Toronto*
Prentice-Hall Hispanoamericana, S.A., *Mexico*
Prentice-Hall of India Private Limited, *New Delhi*
Prentice-Hall of Japan, Inc., *Tokyo*
Prentice-Hall of Southeast Asia Pte. Ltd., *Singapore*
Editora Prentice-Hall do Brasil, Ltda., *Rio de Janeiro*
Prentice-Hall, Inc., *Englewood Cliffs, New Jersey*

Contents

PREFACE TO THE SECOND EDITION   **xiii**

Part I   Tools for Digital Design

1   **DESCRIBING LOGIC EQUATIONS**   **1**

     The Need for Abstraction, Formalism, and Style,   1
         Style,   *1*
         Abstraction,   *2*
         Formalism,   *3*
     Logic in Digital Design,   3
         Logical Constants,   *4*
         Logical Variables,   *4*
         Truth Tables,   *5*
         Logical Operators and Truth Tables,   *6*
     Elements of Boolean Algebra,   9
         Basic Manipulations,   *9*
         Equations from Truth Tables,   *11*
         Truth Tables from Equations,   *15*
         Condensing Truth Tables,   *17*
         Don't-Care Outputs in Truth Tables,   *18*
     Karnaugh Maps,   19
         Building K-maps,   *21*
         Simplifying with K-maps,   *22*
         K-map Simplification Blunders,   *24*
         Other Ways of Reading K-maps,   *25*

Conclusion, 26
Readings and Sources, 26
Exercises, 27

2   REALIZING LOGIC IN HARDWARE   33

Representing TRUE and FALSE with Physical Devices, 33
Mixed Logic: Representing AND, OR, and NOT, 35
    Mixed Logic, 36
    Mixed-Logic Theory, 42
Building and Reading Mixed-Logic Circuits, 49
    Analyzing Mixed Logic Circuits, 49
    Synthesizing Mixed-Logic Circuits, 50
    Other Mixed-Logic Notations, 55
The Positive-Logic Convention, 55
    Reading Positive-Logic Circuit Diagrams, 58
Mixed Logic for Other Logic Functions, 61
    EXCLUSIVE OR and COINCIDENCE, 61
    Open-Collector Gates, 63
    Implementing the Less-Used Logic Functions, 66
Readings and Sources, 68
Exercises, 69

3   BUILDING BLOCKS FOR DIGITAL DESIGN   75

Integrated Circuit Complexity, 76
Combinational Building Blocks, 76
    Combinational and Sequential Circuits, 76
    The Multiplexer, 77
    The Demultiplexer, 80
    The Decoder, 83
    The Encoder, 86
    The Comparator, 87
    A Universal Logic Circuit, 90
    Binary Addition, 92
    The Arithmetic Logic Unit, 94
    Speeding Up the Addition, 95
Data Movement, 99
    The Bus, 100
Readings and Sources, 104
Exercises, 105

4   BUILDING BLOCKS WITH MEMORY   108

The Time Element, 108
    Hazards, 108
    Circuits with Feedback, 111

Sequential Circuits, 112
    Unclocked Sequential Circuits, 112
    Clocked Sequential Circuits, 118
Clocked Building Blocks, 120
    The JK Flip-Flop, 120
    The D Flip-Flop, 120
Register Building Blocks, 126
    Data Storage, 126
    Counters, 126
    Shift Registers, 131
    Three-State Outputs, 133
    Bit Slices, 133
Large Memory Arrays, 134
    Random Access Memory, 135
    Read-Only Memory, 139
    Field-Programmable Read-Only Memories, 142
Programmable Logic, 143
    The PLA, 144
    The PROM as a Programmable Logic Device, 145
    The PAL, 147
    PAL, PLA, and PLE Programming, 150
Timing Devices, 152
    The Single Shot, 152
    The Delay Line, 153
The Metastability Problem, 154
Conclusion, 155
Readings and Sources, 156
Exercises, 157

Part II The Art of Digital Design

5 DESIGN METHODS 165

Elements of Design Style, 166
    Top-Down Design, 166
    Separation of Controller and Architecture, 166
    Refining the Architecture and Control Algorithm, 168
Algorithmic State Machines, 170
    States and Clocks, 171
    ASM Chart Notations, 172
Realizing Algorithmic State Machines, 176
    Traditional Synthesis from an ASM Chart, 176
    The Multiplexer Controller Method of ASM Synthesis, 178
    The One-Hot Method of ASM Synthesis, 183
    The ROM-Based Method of ASM Synthesis, 186
Design Pitfalls, 188
    Clock Skew, 188

Contents

      Asynchronous Inputs and Races, 190
      Asynchronous ASMs, 190
      Sidestepping the Pitfalls, 193
      Debugging Synchronous Systems, 193
Conclusion, 194
      Summary of Design Guidelines, 195
Readings and Sources, 195
Exercises, 196

⑥   PRACTICING DESIGN   203

Design Example 1: Single Pulser, 203
      Algorithmic Solution of the Single Pulser, 204
      A Combined Architecture-Algorithm Solution, 206
      A Single-Pulser Building Block, 207
      Generalizing the Single Pulser, 207
Design Example 2: A System Clock, 208
      Statement of the Problem, 208
      Digesting the Problem, 209
      Algorithm for the System Clock, 209
      Implementing the Circuit, 210
      Critique, 211
Design Example 3: A Serial Bit Clock, 212
      Approaching the Problem, 212
      The Initial Architecture, 213
      The Control Algorithm, 213
      Implementing the ASM, 215
Design Example 4: Serial-Parallel Data Conversions, 217
      Specifying the Problem, 218
      Building the P → S Converter, 219
      Building the S → P Converter, 221
      Critique, 224
Design Example 5: Traffic Light Controller, 224
      Statement of the Problem, 224
      Preliminary Considerations, 225
      The Control Algorithm, 225
      Realizing the ASM, 226
      Choosing a Particular Traffic Signal, 229
Design Example 6: Simple Combination Lock, 229
      Stating the Problem, 230
      Digesting the Problem, 230
      Developing the Algorithm, 231
      An Elegant Combination Lock ASM, 231
      Architecture, 235
      Implementing the Control, 236
Design Example 7: The Black Jack Dealer, 237
      The Rules of Play for the Dealer, 238
      Stating the Problem, 239

Digesting the Problem, 239
Initial ASM for the Black Jack Dealer, 241
Reducing the Number of States, 241
Errors in the Algorithm, 244
Races, Again, 244
Process Synchronization, 246
The Single-Pulser Revisited, 247
The Final ASM for the Black Jack Dealer, 247
The Final Architecture of the Black Jack Dealer, 249
Implementing the Control Algorithm, 250
Summing Up, 253
Readings and Sources, 253
Exercises, 254

7   DESIGNING A MINICOMPUTER   260

PDP-8I Specifications, 261
PDP-8 Memory Addressing, 262
PDP-8I Instructions, 266
Architecture of the LD20, 272
Principal Elements of the Architecture, 272
Data Paths, 274
Preliminary Sketch of the LD20's Control, 279
Fetch and Execute, 281
Investigating the Fetch Phase, 281
Developing an ASM Chart for the LD20, 286
Conventions, 286
Memory Control, 287
State F1, 291
State F2, 293
State F3, 294
State F4, 296
State F5, 297
State F6, 297
State F7, 297
State IDLE, 298
Execute-Phase Processing, 300
State E0: The Single-Instruction Switch, 300
State E0: Detection of Manual Operation, 302
Execute Phase: Manual Commands, 303
Execute Phase: Memory-Accessing Instructions, 304
Execute Phase: The IOT Instruction, 304
State E0: Operate Microinstructions, 309
Conclusion, 315
Readings and Sources, 319
Exercises, 319

Contents

8 BUILDING THE MINICOMPUTER
324

Preliminaries, 324
    Auxiliary Variables, 324
    Labels for ASM Chart Location, 326
The Data-Routing System, 327
    Inputs to the Data Multiplexer, 327
    The Select Signals of the Data Multiplexer, 329
    ALU Operations, 329
    Register-Load Signals, 331
    Architecture and Control of the Link Bit, 333
State Sequencing System, 334
    The State Generator, 334
Special Systems, 337
    Priority Control in the Operate Instruction, 337
    The IOP Signal Enabler, 339
    Control of the Interrupt System, 340
    The Manual System, 341
    Logic Equations, 341
The Memory System, 342
    Memory Access to the LD20, 342
    The Memory Unit, 342
Finished!, 347
Readings and Sources, 347
Exercises, 348

9 INTERFACING TO THE MINICOMPUTER
352

Terminal Communications, 352
The PDP-8's Input-Output Protocol, 353
    From Terminal to PDP-8 (DA03), 354
    From PDP-8 to Terminal (DA04), 354
Requirements of the LD20-Terminal Interface, 355
    Interface Flags and Interrupts, 355
    The Receive Section (DA03), 355
    The Transmit Section (DA04), 356
Preliminary Architecture of the Interface, 356
    The UART, 357
    Incorporating the UART into the Design, 358
The Interface Control Algorithm, 358
    Clocking Events in the Interface, 358
    Synchronizing Signals to the LD20 and the UART, 360
    Algorithm for the Receive Circuit, 360
    Circuit Algorithm for the Transmit, 362
    Interrupt Generation, 364
Implementing the Terminal Interface, 364
Polishing the LD20's Input-Output Processes, 366
    Maintaining the LD20's Control of the Status Inputs, 366
    Attaching Several Devices, 367

A Final Word about Part II, 368
Readings and Sources, 368
Exercises, 369

Part III   Bridging the Hardware-Software Gap

10   MICROPROGRAMMED DESIGN   371

Classical Microprogramming, 372
Classical Microprogramming with Modern Technology, 375
   Microprogramming with Multiple Qualifiers per State, 376
   One Qualifier per State, 379
   Single-Qualifier, Single-Address Microcode, 382
   Comparison of the Microprogramming Approaches, 385
Moving toward Programming, 385
   Cleaning Up the Outputs, 387
   Enhancing the Control Unit, 388
   The 2910 Microprogram Sequencer, 390
   Choosing a Microprogram Memory, 393
The Logic Engine—A Development System for Microprogramming, 393
   The Base Unit, 396
   The Backpanel, 397
   The Supporting Software, 397
Designing and Debugging with the Logic Engine, 398
   The Initial Design, 399
   The Initial Testing of the Architecture, 401
   Developing the Control Program, 401
   Testing the System, 405
Designing a Microprogrammed Minicomputer, 406
   Developing a Microprogram, 406
   The Architecture of the LD30, 406
   A First Approximation of the Command Bits, 407
   Writing the High-Level Microcode, 408
   The Idle Phase of the LD30, 409
   The Manual Phase of the LD30, 411
   The Fetch Phase of the LD30, 412
   The Execute Phase of the LD30, 413
   Interrupt Processing in the LD30, 416
   Instruction Decoding in the LD30, 417
   Declarations for the LD30 Control Microprogram, 419
   Our Design of the LD30 is Complete, 423
Summing Up, 423
Readings and Sources, 425
Exercises, 426

Contents   **ix**

11   MICROCOMPUTERS IN DIGITAL DESIGN   430

The Computer as a Device Controller,   431
     Enter the Microcomputer,   *432*
The Microcomputer in Digital Design,   432
Data Flow in a Microcomputer,   434
     Memory-Mapped Input-Output,   *435*
     Separate Input-Output,   *435*
     Hardware for the Bus Interface,   *436*
     Bus Protocols,   *437*
Microcomputer Input-Output,   438
     Programmed Input-Output,   *438*
     Programmed Input-Output with Interrupts,   *440*
Design Example 1: A Wire-Wrap Controller—Pure Software Control,   442
     The Architecture,   *443*
     Specifications for the Controller,   *443*
     The Control Algorithm,   *445*
     Wrapping Up,   *449*
Design Example 2: A Terminal Multiplexer—Hybrid Hardware-Software Control,   449
     Specifications,   *451*
     The Structure of the Terminal Multiplexer,   *452*
     The Control Algorithm for the Microcomputer,   *455*
     Checking Out the System,   *458*
     Alternative Approaches,   *460*
Conclusion,   461
Readings and Sources,   462
Exercises,   462

Part IV   Digital Technology

12   MEETING THE REAL WORLD   465

On Transistors and Gates,   465
     The Flow of Electrical Charge,   *466*
     Metallic Conduction,   *466*
     Insulators,   *466*
     Semiconductors,   *466*
     Diodes,   *467*
     Bipolar Transistors,   *467*
     MOS Transistors,   *469*
     Gates from Transistors,   *471*
Bipolar Logic Families,   472
     RTL: Resistor-Transistor Logic,   *472*
     TTL: Transistor-Transistor Logic,   *473*
     ECL: Emitter-Coupled Logic,   *474*

Unipolar Logic Families, 475
    p-MOS and n-MOS Logic, 476
    CMOS: Complementary MOS Logic, 476
Three-State and Open-Collector Outputs, 477
    Open-Collector Outputs, 477
    Three-State Outputs, 479
Integrated Circuit Data Sheets, 480
    Electrical Data, 481
Performance Parameters of Integrated Circuit Families, 483
    Input and Output Loadings, 483
    Noise Margins, 485
Unused Gate Inputs, 485
The Schmitt Trigger, 486
A Power-On Reset Circuit, 488
Oscillators, 488
    The Schmitt Trigger Oscillator, 488
    The 555 Oscillator, 489
    Crystal-Controlled Oscillators, 490
    The System Clock, 490
Switch Debouncing, 491
Lamp Drivers, 492
Driving Inductive Loads, 493
    Protecting Switches with Diodes, 494
    Solid-State Relays, 494
The Optical Coupler, 494
Power Supplies, 495
    Remote Sensing, 496
    Integrated-Circuit Voltage Regulators, 496
Power Distribution, 497
    Losses in Power Distribution Systems, 497
    Combatting Inductance in Power Distribution Systems, 498
    Bypassing the Power Supply, 499
Noise, 500
    Crosstalk, 501
    Reflections, 502
    Line Drivers and Line Receivers, 504
    Summary, 505
Metastability in Sequential Circuits, 505
    What should you do?, 511
A Selection of Integrated Circuits, 512
Readings and Sources, 515

INDEX   **517**

Preface
to the Second Edition

This book is an introduction to the art of designing hardware for digital circuits. The design of computer hardware, once the exclusive province of the electrical engineer, is now of vital interest to the computer scientist, and techniques for the systematic solution of complex problems—the computer scientist's specialty—are of increasing importance to the electrical engineer.

Traditional approaches to design, which evolved prior to the integrated circuit, place great emphasis on the devices themselves. In earlier times, this was natural, since the devices were so expensive that the cost of the hardware controlled the design. This led to the development of many complex methods of logic minimization, state assignment, handling asynchronous circuits, and so on, that are now little used because of the smaller cost and greater power and flexibility of digital components. The complexity of traditional design methods can actually interfere with the designer's ability to create a straightforward, understandable, and correct design.

When we first studied digital design, we consulted traditional textbooks and practiced traditional design methods, but we found that these methods did not really help us much to solve complex digital hardware problems. A vast body of vital knowledge of design was missing from the books, and there was a great lack of systematic methodology for dealing with a digital problem as a system. Eventually, we realized that the traditional emphasis was misplaced. The difficult part of digital design is not choosing or assembling the hardware, but rather is understanding the problem and developing a systematic solution for the system's architecture and its control. For this book, we have looked closely at each design technique and have been ruthless in eliminating methods that do not contribute substantially to the goal of clear and correct design.

Here is our thesis. We must approach hardware design from the top, remaining aloof from the actual components as long as we can. We must understand the problem thoroughly and must let its requirements guide us to suitable hardware, rather than allow premature selections of hardware to force us to make inappropriate design decisions.

With the realization that the human cost of designing and maintaining a digital system far exceeds the cost of the materials, digital designers are developing a new approach to design. This new approach, which this book exploits, means that our tools must significantly assist us to understand, solve, and document complex hardware and software problems—if necessary, at the cost of additional hardware. The designer's mind must be uncluttered by unnecessary detail. To one experienced in computer programming, this has a familiar sound. Methods of designing software have improved drastically. Software writers accept as valid and powerful such concepts as structured programming and top-down design. This book is a contribution, in the same spirit, to the field of digital hardware design.

COURSE LEVEL

This book is for self-study and for classroom use. It should benefit the student or the professional unsophisticated about digital hardware, yet it provides an opportunity for old hands to come to grips with some newer trends in basic design. For background, we assume that the reader has an elementary knowledge of computer problem solving in a high-level language and some elementary exposure to the structure of computers and the use of assembly language. The student should be familiar with such number systems as binary, octal, and hexadecimal, and with number representations, particularly two's-complement. We assume no prior knowledge of electronics or hardware other than Ohm's law and simple formulas for series and parallel resistances.

In a college curriculum, this book is suitable for a first course in digital design. The course may be at the middle or upper undergraduate level in electrical engineering or computer science. The text is a modernization of the traditional first course in digital design, emphasizing the solution of design problems rather than the study of hardware. Students of computer science should feel at home with the structured methods and will be delighted to find that they can understand and design complex hardware. Electrical engineering students will gain insight into modern, systematic design principles and will develop an understanding of the computer scientist's emphasis on structure.

PLAN OF THE BOOK

This book first provides a foundation for digital design, emphasizing basic hard-wired design. It then introduces microprogramming and microprocessor-based design and prepares the student for further study of these topics. Throughout, we apply the principles of top-down design.

This book has four parts. Three of the parts form a sequence leading systematically from the fundamentals of logic through digital design with micro-computers, with an emphasis on solving digital problems using hardwired structures of the complexity of medium- and large-scale integration (MSI and LSI). The ordering of the topics is from primitive to complex, the natural way. The fourth part is a collection of information on digital technology that the student may read whenever appropriate. Part IV does not depend on the previous parts of the book.

In Part I, we develop the tools of digital design. In these four chapters, we present the theory and formalism required for systematic digital design. In Chapter 1 we cover the theory of logical expressions—Boolean algebra, truth tables, and useful techniques of simplification. In Chapter 2 we present the realization of logic in hardware, using small-scale integration circuits. In this chapter the crucial distinction between voltage and logic is introduced and the mixed-logic method of drafting physical circuits from logic expressions is developed. In Chapter 3, we develop a collection of basic design tools: useful combinational building blocks such as the multiplexer, decoder, and the arithmetic logic unit. The treatment emphasizes the systematic uses of the building blocks. In Chapter 4 we introduce the theory of circuits with memory—the sequential circuits. After establishing the theory and use of basic sequential elements such as flip-flops, we describe standard sequential building blocks such as registers, bit-slices, random-access and read-only memories, and programmable logic. In Chapter 4, the student is introduced to the real-world pitfalls awaiting the unwary designer; these topics are elaborated in Chapter 12.

Part II is an exploration of the art of digital design at the hard wired MSI and LSI levels. The theme of these five chapters is that a designer must understand the problem before becoming committed to specific chips and wires—top-down design. In Chapter 5 we introduce the structure of the solution to a digital hardware problem: the architecture and the control. We present the ASM method of expressing control algorithms. Chapter 6 consists of a series of digital design examples that illustrate the systematic solutions of common design problems at the MSI level. In Chapter 7, we execute a large-scale design—a complete minicomputer—using top-down style—and in Chapter 8 we translate that design into hardware, using the principle of deferring the decisions about hardware. Chapter 9 is an introduction to asynchronous design through the design of a terminal interface for the minicomputer.

Part III is a bridge between hardware and software. Here we consider the control of hardware using microprogramming and microprocessors. In Chapter 10 we discuss microprogramming and its impact on computer design. As an illustration, we develop a microprogrammed version of the minicomputer designed in Part II. Chapter 11 is an introduction to the use of conventional microprocessors and microcomputers in digital control.

In its single chapter, Part IV contains material on digital technology. The topics in Chapter 12 are independent of the previous material; students may explore these topics whenever they need the information. Representative topics are transistor technology, reading integrated-circuit data sheets, handling pull-up resistors, clocks, power systems and power distribution, noise problems, line driving, and metastability.

SUGGESTIONS FOR COURSE CONTENT

This book provides abundant material for a one-semester or two-quarter course at the undergraduate level, and adequate material for a year's study. Unless the student will study microprogramming in a subsequent course, we feel that Chapter 10 should be included. In a compressed schedule, an instructor might use Chapters 1–5, several of the Design Examples in Chapter 6, Chapters 7 and 8 (possibly excluding the treatment of the OPERATE instruction), and Chapter 10.

Chapter 12 (Part IV) is a buffer in that the instructor can adjust the degree of coverage of hardware technology without disturbing the main theme of the course.

LABORATORY

Laboratory experience is a vital part of any study of digital design. In our courses, taught for over a decade at Indiana University, the students construct, study, debug, and extend the complete minicomputer designed in Chapters 7, 8, and 9. This is a major project, occupying about a semester of laboratory work. The students experience a great sense of achievement when their mini-computer—a fully operational PDP-8 built with MSI technology—actually runs sophisticated PDP-8 software.

We have found that lengthy experiments with individual gates, flip-flops, and registers are unnecessary; the laboratory time is better spent working on the design and construction of more complex systems. Students master individual components in a natural way while studying the structure of larger systems. In any event, we encourage instructors to provide a digital laboratory to accompany the course for, after all, no design works until it is built, and executing designs on paper is only a part of the art of digital design.

With the installation in 1984 of our Logic Engine Development Systems as the basis for supporting the laboratory work, our students have been able to build a microprogrammed version of the large minicomputer lab project. They are able to compare firsthand and in detail the characteristics of hard wired control and microprogrammed control. Such lab experience is impossible without sophisticated microprogramming support. A laboratory manual for the mini-computer project, including both hard wired and microprogrammed versions of the control, is available from Franklin Prosser.

THE SECOND EDITION

In the time since our first edition appeared, we have observed increased acceptance of the principles of structured design. ASM charts now appear in many textbooks. Mixed logic, although not yet as extensively used as ASMs, has gained a proper central place in several new texts and manufacturers' handbooks.

For our second edition, we have extensively revised many sections while retaining the first edition's basic format. We feel that the following enhancements are of particular importance:

In Chapter 2, the theory of mixed logic is formalized and extended, and the treatment of positive logic is enlarged.

In Chapter 3, digital arithmetic is given more thorough treatment.

In Chapter 4, read-only memory and programmable logic are substantially emphasized, and the problem of metastability is addressed.

In Chapter 5, ROM-based controller synthesis, omitted in the first edition, is given its proper treatment.

Chapter 10 has been rewritten, and examples of top-down microprogramming based on our Logic Engine microassembly language LEASMB are included. A microprogrammed version of the minicomputer project of Chapters 7, 8, and 9 is fully expounded, and serves as an important vehicle to display top-down microprogramming techniques.

Chapter 12 contains new treatments of transistor electronics and bipolar and unipolar logic families, and several additional topics. Metastability, widely recognized in the design community at last, is thoroughly discussed, and guidelines for dealing with the problem are presented.

In all chapters, exercises have been revised and new exercises added. Each chapter ends with suggested additional readings and sources.

Franklin Prosser
David Winkel

Describing Logic Equations

Before you begin this journey into digital design, it is important that you understand the philosophy that will guide your study. If you have not read the Preface, do so now before you go on. There we discuss the issues that give rise to the need for good style and structure in digital design. Also, the Preface contains an outline of the book, which will give you a view of where you are heading and how you will get there. It is particularly important that you understand our approach to the details of digital hardware. The overriding emphasis is to let the problem solution dictate the hardware, rather than allowing premature commitments to hardware to coerce the solution. This conscious suppression of hardware detail during most of the design pays big dividends. Chips and wires and power supplies are still important—they are vital to success, and you will need a good background in many areas of digital technology in order to become an accomplished designer—but too often in digital design the hardware has dominated the solution to the problem. To head off this common malady, we have deferred to the end of the book almost all of the technology needed for your introduction to digital design. This information, in Part IV, does not depend on the material in Parts I through III. Read the topics in Part IV as you require or desire them while you are progressing through the first three parts of the text.

THE NEED FOR ABSTRACTION, FORMALISM, AND STYLE

Style

The human mind needs help when it tackles complex tasks. As we use the term, style is a method of partitioning a large problem into manageable subunits in a systematic and understandable way. The need for good style is more apparent as problems become larger. The most complex projects ever attempted by human beings have been computer programs; some, exceeding 500,000 lines of code, are so large that no one person is able to encompass the entire program, or even a significant part of it. The study of programming style was forced upon practitioners of the art as a way of gaining control over their projects. Programming style has blossomed into a rather well-defined set of techniques, bearing such names as "top-down" and "structured." The hardware of a large computer involves complexity on the same scale as these giant computer programs. The study of style in digital system design is not as well developed as its programming counterpart but is nonetheless essential to success. In this book, we emphasize style.

Here are some rules of good style in digital design:

(a) Design from the top down. The design starts with the specification of the complete system in a form compact enough that one person can readily comprehend it. The design proceeds by sectioning the digital system into subunits such as memory, arithmetic elements, and control, with well-defined interrelationships. You may then describe each unit in more detail and still retain the ability to comprehend both the whole structure and the details of the units. This process continues until you have completely specified the system in detail, at which time construction may begin.

(b) Use only foolproof techniques that will keep you on the narrow path of safety in the design process. Digital hardware allows a high degree of flexibility in design—so much flexibility that designers can bury themselves in clever and unusual circuits. Uncontrolled use of such flexibility promotes undisciplined, unintelligible, and incorrect design of products. This phenomenon has its counterpart (to a less severe degree) in computer software because assembly language programming offers access to the full power of a digital computer. Experience in the solution of hardware and software problems has shown that we must restrict our design tools and techniques to those that can be shown to work reliably and understandably under a variety of circumstances.

(c) Use documentation techniques, at both the system level and the detailed circuit level, that clearly portray what you, the designer, were thinking when you reduced your problem first to an abstract solution and then to hardware. This often-violated precept boils down to common courtesy. Put yourself in the position of a user or a maintainer of your hardware design; in such a position you would be grateful for clear, complete documentation.

Abstraction

In our context, abstraction means dealing with digital design at the conceptual level. The concept of a memory, a central theme in every computer, is an abstraction. When starting a design we need to deal with conceptual elements and their interrelationships: it is only later in the design process that we need to worry about the realization of the concepts in hardware. This freedom from concern about the details of various hardware devices is absolutely essential if we are to get a good start on a new design of any complexity. Start at the top and begin reducing the problem to its natural conceptual elements. For example, a computer will need a memory, an input–output system, an arithmetic unit, and so on. We ordinarily begin a design at this highly abstract level, and carry the conceptual process down, level by level. Thus, at the next level we draw a block diagram of an arithmetic unit by interconnecting functional units such as registers, control units, and data paths. The initial abstraction is a critical part of any design, since bad early planning will inevitably lead to bad implementations. There is no way to rescue a bad design with clever tricks of Boolean algebra or exotic integrated circuits.

Formalism

Formalism is the theory of the behavior of a system. In a digital design, formalisms help us to establish systematic rules and procedures with known characteristics. Formalisms are important at all levels of design. In the traditional study of digital systems, one of the principal formalisms is Boolean algebra, the theory of binary logic, named after George Boole, who studied it long before the advent of digital computers. Boolean algebra is an essential tool for describing and simplifying the detailed logical processes at the root of digital design. Powerful and well developed as Boolean algebra is, it nevertheless becomes of less benefit as our level of abstraction increases. For instance, at the top ("systems") level of abstraction, where we are thinking in terms of the movement, storage, and manipulation of data, Boolean algebra is of little use. As we move closer to the detailed implementation—as our design becomes less abstract and more concrete—Boolean algebra begins to be a useful tool.

At the systems level, the formalisms are less well developed, appearing as structures and rules rather than mathematical constructs. High-level formalisms are nevertheless of great importance to good design, for only by adopting systematic methods at all levels can we hope to transform correct concepts reliably into correct hardware.

LOGIC IN DIGITAL DESIGN

Imagine trying to speak without the words *and*, *or*, and *not*. Discarding these little words would severely handicap our ability to express complex thoughts. Although *and*, *or*, and *not* each have several meanings in English, the most profound uses describe logical combinations of thoughts: "I have money for gas *and* my car is running"; "There is a paper jam *or* the printer is out of paper"; "She will *not* fail." Our thought processes are molded by our language, and when we design digital systems we will use *and*, *or*, and *not* in the same sense as above. In this book, we denote the specific logical uses of *and*, *or*, and *not* by the symbols AND, OR, and NOT.

It is nearly always useful to formalize heavily used concepts; by so doing we achieve compactness and are able to handle more complexity than if we wrestle with informal concepts such as the normal English language uses of *and*, *or*, and *not*. To pave the way for a Boolean algebraic treatment of digital logic, we will formalize the concepts of logical constants, variables, and operators.

Logical Constants

The statement "There is a photodiode error" is either true or false. The operators AND, OR, and NOT are likewise concerned with two logical values, true and false.

We will concern ourselves only with logic systems that can be formalized with a binary set of logical constants, TRUE and FALSE. Since we use these concepts so heavily, it is worthwhile seeking abbreviations. We will represent TRUE by T or 1, and FALSE by F or 0. In this context 1 and 0 are not decimal numerical values; they are abbreviations for TRUE and FALSE, and nothing more. We will use 1 and T, 0 and F interchangeably in the text. Each abbreviation has its value, and both are widely used in digital design. In the study of hardware implementation of logic in Chapter 2, we will show a strong bias toward the T,F notation, to avoid a common point of confusion. On the other hand, in much of this chapter the 1,0 form for TRUE and FALSE is convenient. Be prepared to accept and use either form.

Logical Variables

Consider the declaration

$$A = \text{photodiode error}$$

We use the *logical variable* A as an abbreviation for the cumbersome phrase "photodiode error" to achieve compactness. The variable A can have two values, T or F. If we do not have a photodiode error, then $A = F$; if we do, then $A = T$. Although it is possible to use single letters to represent logical elements, as we have above, it is usually better to use a more recognizable abbreviation, such as

$$PDE = \text{photodiode error}$$

In general, the abbreviations should be a compromise between clarity and brevity. It is not really necessary to abbreviate at all. We could use *PHOTODIODE.ERROR* as the name of the logical variable, but it is too long to be convenient. *A* is short but conveys no meaning; *PDE* is a good compromise. We often use numbers in logical variables to indicate a particular member of a set of variables. A common unit of computer information is the 8-bit byte. If we needed to examine the individual bits of the byte, we might choose to assign distinct names to the bit values:

$$B0 = 1 \text{ in leftmost bit}$$

.

.

.

$$B7 = 1 \text{ in rightmost bit}$$

Other notations for sets of variables will suggest themselves. Instead of the distinct names for bits in the byte, we might choose to use a subscripted variable $B_0 \ldots B_7$, with equivalent meanings.

In this book we will capitalize the letters in the names of logical variables and will always start each name with a letter. To preserve the mnemonic value, names of variables may include periods as separators; *GO.ON* is an example.

Truth Tables

Consider a set of logical variables, each variable of which may have one of two values, T or F. In digital design we are interested in combining logical variables (e.g., using AND, OR, and NOT) to produce new variables that again have only two possible values, T or F, for any combination of given variable values. In other words, we wish to study binary functions of binary variables.

For a set of logical variables, we may define any desired function by giving the function value for each possible set of variable values. A tabular form with input variables on the left and the function on the right is useful for this display. For example, here is a logical function X of three variables A, B, and C, shown in both the 1,0 and the T,F notations.

A	C	B	X		A	C	B	X
0	0	0	0		F	F	F	F
0	0	1	1		F	F	T	T
0	1	0	1		F	T	F	T
0	1	1	0		F	T	T	F
1	0	0	1		T	F	F	T
1	0	1	0		T	F	T	F
1	1	0	0		T	T	F	F
1	1	1	0		T	T	T	F

Such a display is called a *truth table*. Having chosen an ordering of the input variables (A, C, B in this case), we list all possible combinations of the variables' values, in binary numeric order. A tabulation in this standard form is called a *canonical truth table*. "Canonical" means standard. For three variables, the canonical truth table has $2^3 = 8$ rows, arranged from binary 000 through binary 111. Since each binary bit pattern corresponds to a decimal number, we may describe a row of a canonical truth table by its decimal number equivalent. For example, the row corresponding to the variable values 0110 for a four-variable function may be called row 6. When convenient, you may write the row numbers on the left of the canonical truth table.

For canonical truth tables, we may compactly describe the function by a vector of function values. For example, the three-variable truth table for X above yields an eight-element vector

$$X(A,C,B) = (0,1,1,0,1,0,0,0)$$

Although we usually choose to list the values of variables in canonical order, any other order of rows displays the same information. The following two truth tables are equivalent, but the right-hand table lacks the useful uniformity of the canonical form on the left:

Row number	D	E	Y
0	0	0	0
1	0	1	0
2	1	0	1
3	1	1	0

Row number	D	E	Y
2	1	0	1
0	0	0	0
1	0	1	0
3	1	1	0

Logical Operators and Truth Tables

We will now give a precise definition of the three logical operators AND, OR, and NOT.

NOT. We represent logical NOT by the overscore. Thus, if PDE is a logical variable, then

$$\text{NOT } PDE = \overline{PDE}$$

In words, we would describe the notation $\overline{PDE}$ as "PDE not." We may define NOT by listing in a truth table all possible values for an arbitrary logical variable, and the corresponding values of the logical NOT of that variable. Since a logical variable A can have only two values, 1 and 0, the following list is exhaustive:

A	$\overline{A}$
0	1
1	0

We regard the formal definition of logical NOT to be given by its truth table. Remember that 1 and 0 represent TRUE and FALSE.

AND. We represent logical AND by a dot separating two logical variables— we write B AND C as $B \cdot C$. We shall faithfully use the "$\cdot$" symbol to represent the AND operator even though some authors omit it when dealing with single-letter logical variables. Thus, if you insist upon single-letter names, you might interpret *BC* as *B•C*. This is dangerous because we may want to name a single logical variable with the two-letter name *BC*. In real-world logic design, single-letter names are not descriptive enough to be of use. There are only 26 possible single-letter names and a typical design may require many more than 26 names. We therefore give up the dubious advantage of having an implied AND for the real advantage of multiletter names for logical variables.

We will define AND with a truth table. There are two independent variables in the logical AND, each of which can assume either of the two values, 1 and 0. Therefore, specifying the function value for each of the four combinations of inputs completely defines AND:

B	*C*	*B•C*
0	0	0
0	1	0
1	0	0
1	1	1

The table corresponds to our intuitive notion of AND in that *B•C* is true only if both *B* and *C* are simultaneously true.

Just as we may generalize the English use of *and* to encompass more than two variables, we can do so for the formal logical AND. The truth table for *A•B•C* is

A	*B*	*C*	*A•B•C*
0	0	0	0
0	0	1	0
0	1	0	0
0	1	1	0
1	0	0	0
1	0	1	0
1	1	0	0
1	1	1	1

For more than three variables, the truth table becomes unwieldy, and we revert to a verbal definition of the logical AND:

The logical AND of several variables is true only when all the variables are simultaneously true.

OR. The symbol for logical OR is the + sign. Do not confuse this with the use of + in other contexts to represent arithmetic addition. Since in logic design the uses of logical OR will vastly outnumber the uses of an arithmetic plus, we choose a convenient single symbol for the OR operator and we use the more cumbersome word "plus" or the symbol "(+)" for an arithmetic plus. Here are two word statements translated into their corresponding logic design notations

A is true if B OR C is true $\qquad A = B + C$

2 added to 3 is 5 $\qquad\qquad$ $5 = 2$ plus 3, or $5 = 2\,(+)\,3$

The defining truth table for a two-input logical OR is

B	C	$B + C$
0	0	0
0	1	1
1	0	1
1	1	1

As with the AND operator, we may generalize the definition of the logical OR to more than two input variables. In words, the output is true if at least one of the inputs is true. For instance,

AY	PDE	X	$AY + PDE + X$
0	0	0	0
0	0	1	1
0	1	0	1
0	1	1	1
1	0	0	1
1	0	1	1
1	1	0	1
1	1	1	1

This completes our definition of AND, OR, and NOT. These logical operators operate on input variables to yield a single output. The NOT operates only on single variables, while AND and OR fundamentally operate on two inputs. For our convenience we may also think of AND and OR as multi-input operators. Truth tables are useful in many designs, but we need a more compact and powerful tool for representing logical manipulations. An algebra of logical operators, called *Boolean algebra* in honor of George Boole, who first explored the properties of the logical operators, is analogous to the familiar algebra of arithmetic operators. In the next section, we present some simple but important Boolean algebraic results.

ELEMENTS OF BOOLEAN ALGEBRA

Basic Manipulations

Boolean algebra is important to hardware designers because it allows the compact specification and simplification of logic formulas. Physical devices can perform the AND and OR functions, and it is this fact that raises Boolean algebra from the realm of interesting theory to the role of a vital design tool. The algebra may be developed from any of several starting points. Modern mathematicians derive Boolean algebra from a compact set of abstract postulates, producing an elegant and rigorous theory; however, in building a useful tool to assist structured digital design, we best achieve our goals by emphasizing the relationship of truth tables to logic equations. Truth tables and allied tabular displays play an important role in digital design and implementation. Therefore, we will assume as our starting point the existence of the two binary values TRUE (T or 1) and FALSE (F or 0), and the three operators AND, OR, and NOT, with behavior described by their truth tables.

Boolean algebraic formulas follow certain conventions. Our intuition tells us that we want the operators AND and OR to *commute* (e.g., $A + B \equiv B + A$) and *associate* [e.g., $A + (B + C) \equiv (A + B) + C$]. The operators also *distribute* according to the relations

$$A \cdot (B + C) \equiv (A \cdot B) + (A \cdot C)$$
$$A + (B \cdot C) \equiv (A + B) \cdot (A + C)$$

The conventional hierarchy of operator action in complex expressions is

First: NOT
then: AND
last: OR

Our notation for logical NOT (the overscore) explicitly shows the scope of action of the NOT operation, so the only possible confusion in evaluating expressions would occur with AND and OR. As the hierarchy shows, AND takes precedence. As an example, consider

$$X = \overline{A + B \cdot C}$$

Evaluation is in the order specified below by the parentheses, innermost parenthesized expressions being evaluated first. Thus

$$X = (\overline{(A + (B \cdot (\overline{C})))})$$

Parentheses are useful in Boolean equations to override the normal hierarchy, just as we use them for similar purposes in conventional algebra.

Below are some fundamental relations of Boolean algebra; memorize these results.

$$\overline{\overline{A}} \equiv A \tag{1-1}$$

$$A \cdot T \equiv A \qquad\qquad A + F \equiv A \tag{1-2}$$

$$A \cdot F \equiv F \qquad\qquad A + T \equiv T \tag{1-3}$$

$$A \cdot A \equiv A \qquad\qquad A + A \equiv A \tag{1-4}$$

$$A \cdot \overline{A} \equiv F \qquad\qquad A + \overline{A} \equiv T \tag{1-5}$$

$$\overline{A \cdot B} \equiv \overline{A} + \overline{B} \qquad\qquad \overline{A + B} \equiv \overline{A} \cdot \overline{B} \tag{1-6}$$

Each identity involving AND or OR operators comes in two forms, one emphasizing AND, the other emphasizing OR. This *principle of duality* is a characteristic of Boolean algebra, and it has important applications in the study of logic and in the implementation of logic functions with physical devices. The dual identities are related by this rule:

Change each AND to OR, and each OR to AND, and change each T to F, and each F to T.

Equation (1–6) is the well-known *De Morgan's law*. It is of special importance because it allows us to convert Boolean operators from AND to OR, and vice versa.

You may prove each of the foregoing identities by using the truth-table definitions of the logical operators. To illustrate the art of proving theorems with truth tables, we will prove the validity of De Morgan's law. Start with the form $\overline{A \cdot B} \equiv \overline{A} + \overline{B}$. Develop truth tables for the left-hand side and for the right-hand side of the identity. (When convenient, we may show several functions [outputs] in the same table: we write two or more truth tables in one package.)

A	B	$A \cdot B$	$\overline{A \cdot B}$	A	B	$\overline{A}$	$\overline{B}$	$\overline{A} + \overline{B}$
F	F	F	T	F	F	T	T	T
F	T	F	T	F	T	T	F	T
T	F	F	T	T	F	F	T	T
T	T	T	F	T	T	F	F	F

A truth table is an exhaustive list of function values for each possible combination of inputs; therefore, if two truth tables have identical rows, the functions behave identically. You see that the truth table for $\overline{A \cdot B}$ is the same as that for $\overline{A} + \overline{B}$; this proves Eq. (1–6).

De Morgan's law extends to more than two variables. For example, the following identities are valid.

$$\overline{A + B + C} \equiv \overline{A} \cdot \overline{B} \cdot \overline{C} \qquad\qquad \overline{A \cdot B \cdot C} \equiv \overline{A} + \overline{B} + \overline{C}$$

Several other Boolean identities find frequent use in our design work. You may demonstrate each relationship using truth tables or using the previous Boolean identities.

$$A + A \cdot B \equiv A \qquad\qquad A \cdot (A + B) \equiv A \qquad\qquad (1-7)$$

$$A + \overline{A} \cdot B \equiv A + B \qquad A \cdot (\overline{A} + B) \equiv A \cdot B \qquad (1-8)$$

$$A \cdot B + \overline{A} \cdot B \equiv B \qquad\qquad\qquad\qquad (1-9)$$

The left-hand form of Eq. (1–8) is not immediately obvious, but it is of great help in reducing the complexity of commonly occurring Boolean expressions. After De Morgan's law, Eq. (1–9) is perhaps the most widely used relation. It is the basis for several systematic Boolean simplification procedures. Presently, we will develop the one simplification method, Karnaugh maps, that will be of most benefit to our design work.

Truth tables serve as an easy means of verifying the validity of small Boolean equations, whereas the Boolean identities presented above are useful in manipulating both large and small Boolean equations. Here is an example of Boolean algebraic manipulations.

$$
\begin{aligned}
\overline{A \cdot (B + C \cdot (B + \overline{A}))} &= \overline{A \cdot (B + C \cdot B + C \cdot \overline{A})} && \text{(distribution law)} \\
&= \overline{A \cdot (B + C \cdot \overline{A})} && \text{[Eq. (1--7)]} \\
&= \overline{A \cdot B + A \cdot (C \cdot \overline{A})} && \text{(distribution law)} \\
&= \overline{A \cdot B + A \cdot (\overline{A} \cdot C)} && \text{(commutation law)} \\
&= \overline{A \cdot B + (A \cdot \overline{A}) \cdot C} && \text{(association law)} \\
&= \overline{A \cdot B + F \cdot C} && \text{[Eq. (1--5)]} \\
&= \overline{A \cdot B} && \text{[Eqs. (1--3) and (1--2)]} \\
&= \overline{A} + \overline{B} && \text{(De Morgan's law)}
\end{aligned}
$$

Equations from Truth Tables

If the truth table and logic equation are to work hand in hand as design aids, we must be able to derive a logic equation for a function from its truth table. Consider this example:

Row number	A	B	W
0	F	F	F
1	F	T	F
2	T	F	T
3	T	T	F

In words, W is true only if A is true and B is false (from row 2). More formally, $W = A \cdot \overline{B}$. Another example is

Row number	A	B	Y
0	F	F	T
1	F	T	F
2	T	F	T
3	T	T	T

Here, our intuitive understanding of the truth table is that Y is true whenever A is false and B is false (row 0), or whenever A is true and B is false (row 2), or whenever A is true and B is true (row 3). Thus

$$Y = \overline{A} \cdot \overline{B} + A \cdot \overline{B} + A \cdot B \qquad (1\text{--}10)$$

Incidentally, we may simplify this equation:

$$
\begin{aligned}
Y &= \overline{A} \cdot \overline{B} + A \cdot \overline{B} + A \cdot B & & \\
&= \overline{B} + A \cdot B & \text{[by Eq. (1--9)]} & \qquad (1\text{--}11) \\
&= \overline{B} + A & \text{[by Eq. (1--8)]} &
\end{aligned}
$$

Viewing this truth table another way, we may say that Y is *false* only when A is false and B is true (row 1):

$$\overline{Y} = \overline{A} \cdot B \qquad (1\text{--}12)$$

We have two equations for Y derived from the same truth table—Eqs. (1–10) and (1–12). Can you show that the expressions for Y are equivalent?

Which way is best for writing equations from truth tables—reading true conditions for the function, or reading false conditions? Both ways result in equivalent expressions, usually in somewhat different form. For equations of more than two variables, when there are many rows in the truth table, you will usually wish to use the method that involves the fewer AND terms. In this example, Eq. (1–12), derived from the false function values, yields the more direct and simple result.

Sum-of-products form. Equation (1–10) (and also Eq. [1–11]) expresses the function Y in the *sum-of-products form*. This is the most common form for deriving Boolean equations from truth tables, and in this context the form fits nicely with our thought processes. The name "sum of products" comes from an analogy of the Boolean operator symbols with those of arithmetic: the expression is a sum (OR) of product (AND) terms.

In sum-of-products form, a product term consists of the logical AND of a set of operands, each operand being a logic variable or its negation. (A trivial form of product term consists of a single variable or its negation.) A variable's name must appear at most once in a product term. For example, $\overline{A}$, $A \cdot \overline{B}$, and $\overline{A} \cdot B \cdot \overline{C}$ are valid product terms, whereas $A \cdot \overline{A}$ and $\overline{A} \cdot B \cdot B \cdot C$, although valid Boolean expressions, are not proper product terms.

A product term containing exactly one occurrence of every variable (either asserted or negated) is called a *minterm* or a *canonical product term*. A function expressed as a logical OR of distinct minterms is in *canonical sum-of-products form* or *disjunctive normal form*. Our intuitive method for deriving equations from truth tables yields the canonical sum-of-products form, as in Eq. (1–10). An expression may be in sum-of-products form yet not be canonical. Equation (1–11) is a noncanonical sum-of-products expression.

We may now state formal prescriptions for deriving sum-of-products logic equations from canonical truth tables:

> To derive a sum-of-products form for a function from a canonical truth table, write the OR (sum) of the minterms for which the function is true. Similarly, to derive a sum-of-products form for the complement of a function from a canonical truth table, write the OR of the minterms for which the function is false.

Applying these rules to the previous truth table yields Eqs. (1–10) and (1–12).

For n-variable functions, there are 2^n possible minterms. A minterm is sometimes designated by $\mathbf{m}_i$, where i is the number of the single canonical truth-table row for which the minterm yields truth. For example, $\mathbf{m}_4 = A \cdot \overline{B} \cdot \overline{C}$ yields truth only for variable values $A = 1$, $B = 0$, $C = 0$; this corresponds to row 4, since binary 100 is decimal 4. Again, $\mathbf{m}_2 = \overline{A} \cdot B \cdot \overline{C}$ produces truth only for row 2 (binary 010). (The name *minterm* denotes that the term is true for only one row of the table.) With this notation, we may describe canonical sum-of-products equations as sums of minterms. Equation (1–10) becomes $Y = \mathbf{m}_0 + \mathbf{m}_2 + \mathbf{m}_3$. Although this notation is important in certain developments of Boolean algebra, we will not use it frequently in this book.

Equation (1–12) is a canonical sum-of-products equation for $\overline{Y}$, albeit a somewhat trivial form containing only one term. We may use the minterm notation for this form also:

$$\overline{Y} = \mathbf{m}_1$$

Product-of-sums form. There is another formulation of logic expressions from truth tables: the *product-of-sums form*. This form consists of the AND (the product) of a set of OR terms (the sums), such as $(\overline{A} + B) \cdot (A + B) \cdot (\overline{B})$. In a product-of-sums expression, a sum term consists of the logical OR of a set of operands, each operand being a logic variable or its negation, and each variable appearing at most once.

A sum term that contains exactly one occurrence of every variable (either asserted or negated) is called a *maxterm* or *canonical sum term*. A function expressed as a logical AND of distinct maxterms is in *canonical product-of-sums form* or *conjunctive normal form*. The product-of-sums notation is not much used in practical design, since it lacks the sum-of-products' easy kinship with our thought processes. An expression may be in product-of-sums form yet not be canonical, if one or more of the sum terms is not a maxterm. For a three-variable function, $A + \overline{B} + C$ is a maxterm; $\overline{B} + \overline{C}$ is not. $W = (P + \overline{Q} + \overline{R}) \cdot (P + Q + R)$ is in canonical product-of-sums form; $W = (P + Q) \cdot (P + Q + R)$ is not canonical.

The prescriptions for forming product-of-sums logic equations from truth tables are:

To derive a product-of-sums form of a function from a canonical truth table, write the AND (product) of each maxterm for which the function is false. Similarly, to derive a product-of-sums form for the complement of a function, write the AND of each maxterm for which the function is true.

Applying these rules to the previous truth table, we have

$$Y = (A + \overline{B})$$
$$\overline{Y} = (A + B) \cdot (\overline{A} + B) \cdot (\overline{A} + \overline{B}) \tag{1-13}$$

The first equation agrees with Eq. (1–11), obtained by simplifying the sum-of-products form in Eq. (1–10). With the aid of the distributive law, we may simplify Eq. (1–13) to yield

$$\overline{Y} = \overline{A} \cdot B$$

in agreement with Eq. (1–12).

A maxterm is true for every row of the canonical truth table except one; we sometimes specify the maxterm by $\mathbf{M}_i$, where i is the row number for which the maxterm is false. For example, $(\overline{A} + B + \overline{C})$ is true for every combination of values of variable except $A = 1$, $B = 0$, $C = 1$, which designates row 5 (binary 101 = decimal 5). Maxterm $\mathbf{M}_0$ is $(A + B + C)$, since only for variable values 000 does the term produce a false value. The name *maxterm* connotes that the term is true for all but one set of variable values. We occasionally write canonical product-of-sums expressions, analogous to the sum-of-products formalism, as products of maxterms. For instance, the products of sums above become

$$Y = (A + \overline{B}) = \mathbf{M}_1$$
$$\overline{Y} = (A + B) \cdot (\overline{A} + B) \cdot (\overline{A} + \overline{B}) = \mathbf{M}_0 \cdot \mathbf{M}_2 \cdot \mathbf{M}_3$$

To illustrate the four rules for producing canonical equations, we derive the equations for a function W of three variables:

Row number	J	K	L	W
0	0	0	0	0
1	0	0	1	1
2	0	1	0	1
3	0	1	1	1
4	1	0	0	0
5	1	0	1	0
6	1	1	0	0
7	1	1	1	0

Sum of products on true outputs:

$$W = \overline{J} \cdot \overline{K} \cdot L + \overline{J} \cdot K \cdot \overline{L} + \overline{J} \cdot K \cdot L \tag{1-14}$$

Sum of products on false outputs:

$$\overline{W} = \overline{J} \cdot \overline{K} \cdot \overline{L} + J \cdot \overline{K} \cdot \overline{L} + J \cdot \overline{K} \cdot L \\ + J \cdot K \cdot \overline{L} + J \cdot K \cdot L \tag{1-15}$$

Product of sums on false outputs:

$$W = (J + K + L) \cdot (\overline{J} + K + L) \cdot (\overline{J} + K + \overline{L}) \\ \cdot (\overline{J} + \overline{K} + L) \cdot (\overline{J} + \overline{K} + \overline{L}) \tag{1-16}$$

Product of sums on true outputs:

$$\overline{W} = (J + K + \overline{L}) \cdot (J + \overline{K} + L) \cdot (J + \overline{K} + \overline{L}) \tag{1-17}$$

You should simplify each equation, using algebraic identities, and verify that the equations are equivalent.

Truth Tables from Equations

Sometimes you will wish to convert a logic expression into its truth-table form. If the expression is in sum-of-products form, the conversion is easy. Each product term will form one or more truth-table rows having a true function value. A canonical product term (one with all the variables in it; a minterm) produces one row with an output of TRUE. A product term with fewer variables yields more rows, since such a term is true for *any* values of the missing variables. Often more than one product term in the sum will contribute truth for a given row of the truth table. This is fine—double-truth is still truth!

As an example, consider this equation of three variables:

$$Y = J \cdot \overline{K} + \overline{J} \cdot K \cdot L + J \cdot K \cdot \overline{L} + K \cdot L \tag{1-18}$$

Term 1   Term 2   Term 3   Term 4

The truth table for this equation is

J	K	L	Y	
0	0	0	0	
0	0	1	0	
0	1	0	0	
0	1	1	1	(Terms 2 and 4)
1	0	0	1	(Term 1)
1	0	1	1	(Term 1)
1	1	0	1	(Term 3)
1	1	1	1	(Term 4)

$$(1-19)$$

Terms 2 and 3 are canonical; each contributes a true output to one row of the table. Terms 1 and 4, having a missing variable, contribute true outputs to two rows each.

Equations in product-of-sums form are most easily translated into truth tables by focusing on the conditions for having false function values. Each sum term in the product will assure a false expression value whenever *all* its variables are the opposite of the form in the term. For example, consider the following equation of three variables:

$$G = (\overline{A} + B + C) \cdot (\overline{A} + B) \cdot (\overline{A} + \overline{B} + \overline{C})$$

Term 1   Term 2   Term 3

Term 1 makes *G* false for row 4 (100) of the truth table; term 3 produces a result of false for row 7 (111). Term 2, with its missing variable *C*, produces a result of false for two rows, 4 (100) and 5 (101). Terms 1 and 3 are canonical: each contributes a false function value for one row; term 2 is not canonical. Here is the truth table:

Row number	A	B	C	G	
0	0	0	0	1	
1	0	0	1	1	
2	0	1	0	1	
3	0	1	1	1	
4	1	0	0	0	(Terms 1 and 2)
5	1	0	1	0	(Term 2)
6	1	1	0	1	
7	1	1	1	0	(Term 3)

For logic expressions of more general form than the sum of products or the product of sums, we fall back on the ultimate method of deriving truth tables: evaluating the function for every combination of values of the input variables. This means that we explicitly determine each function value. The process is often laborious, but (barring error!) is foolproof. For example, from the equation

$$K = (\overline{L + G}) \cdot (\overline{L} + G)$$

we get the following truth table by computing the value of *K* for each of the four sets of values of *L* and *G*:

L	G	K
F	F	T
F	T	F
T	F	F
T	T	F

Another approach for forming a truth table from a general Boolean equation is to manipulate the expression (usually using De Morgan's law) until it becomes a sum-of-products or product-of-sums form, and then use the methods presented earlier in this section.

Condensing Truth Tables

A canonical truth table of n input variables has 2^n rows arranged in binary numerical order on its inputs, corresponding to all the possible values of the input variables. The canonical form explicitly displays the function's value for every possible set of input conditions. This form of truth table—the only one we have used so far—is the counterpart of the canonical forms of sum-of-products and product-of-sums equations. Just as we frequently use Boolean equations in a simplified form, we also sometimes wish to deal with a simplified or collapsed truth-table notation.

Consider Eq. (1–18) again:

$$Y = J \cdot \overline{K} + \overline{J} \cdot K \cdot L + J \cdot K \cdot \overline{L} + K \cdot L$$

Term 1   Term 2   Term 3   Term 4

The canonical terms (2 and 3) each contribute one row to the canonical truth table in Eq. (1–19). Term 1, however, is independent of the value of variable L (L does not appear), so term 1 contributes two rows with true output to the canonical truth table, one for each value of the missing variable. If we are willing to abandon the canonical form for the truth table, we may introduce a shorthand notation for this situation. We collapse these two rows for term 1 into a single row and place an X for the value of the missing variable L. The X means "both values" and implies that the function value is independent of that variable whenever the other inputs are in their stated conditions. Similar arguments apply to term 4, which lacks the variable J.

Applying this concept to the expression, we may derive a shortened truth table

J	K	L	Y	
0	0	0	0	
0	0	1	0	
0	1	0	0	
X	1	1	1	(Terms 2 and 4)
1	0	X	1	(Term 1)
1	1	0	1	(Term 3)

(1–20)

Note how this truth table yields the original equation in a direct manner.

The X is the truth-table equivalent of the important Boolean algebraic identity of Eq. (1–9)

$$A \cdot B + \overline{A} \cdot B \equiv B$$

There are various applications of this identity that we could introduce directly into the truth table of Eq. (1–20) if we desired. For instance, here are two more stages in the condensation of this table:

J	K	L	Y		J	K	L	Y
0	0	X	0		0	0	X	0
0	1	0	0		0	X	0	0
X	1	1	1		X	1	1	1
1	0	X	1		1	X	X	1
1	1	0	1					

(1–21)

If we are presented with a truth table containing X's, the derived Boolean equation will not be canonical, but will contain some simplified terms. A sum-of-products equation for the right-hand truth table in Eq. (1–21) is

$$Y = K \cdot L + J$$

Satisfy yourself that the original Eq. (1–18) is equivalent to this, and that all these truth tables based on Eq. (1–18) are equivalent.

Condensed tables are convenient, since the original statement of a problem will often lead in a natural way to the condensing of rows. The main virtue of the notation is in allowing the truth table to reflect its origins more faithfully. A secondary virtue is in the resultant shortening. Attempts as in Eq. (1–21) to simplify truth tables by collapsing rows of a less simplified form are tricky and can lead the inexperienced into errors. Soon we will discuss a graphical method for simplifying logic functions that is easier for people to use.

Don't-Care Outputs in Truth Tables

Truth tables have a useful property that a logic equation cannot express. We often know from the nature of the problem that the function's value is irrelevant for certain combinations of the input variables. This situation usually arises when we know that the inputs will never legitimately assume certain sets of values. We may use a small dash "–" for the truth-table function value in such cases. The – means "don't care." In deriving a logic equation from the truth table, we are free to use either a T or F value for the don't-care dash, whichever will yield the more useful form. For instance, look at the condensed truth table below:

A	B	Y
F	X	T
T	F	–
T	T	F

Of the various equations that we may derive from this truth table, a choice of T for the don't-care output might yield the sum-of-products form

$$Y = \overline{A} + A \cdot \overline{B}$$

which can be simplified to

$$Y = \overline{A} + \overline{B}$$

whereas a choice of F for the don't-care gives the quite different equation

$$Y = \overline{A}$$

You may use either equation; your choice may depend on other factors in the problem design.

KARNAUGH MAPS

In the early years of digital computers, each logic element in a circuit was large and cumbersome by modern standards, and consumed considerable power. There was a natural emphasis on reducing the number of circuit elements to the bare minimum so as to cut the total cost. Designers developed many elaborate techniques for simplifying logic expressions, and much effort went into perfecting these tools. Today, hardware for digital logic is inexpensive, and in modern design work the emphasis on circuit minimization has given way to a concern for modularity and clarity in the design process.

One result of this shift in technology and design style is that circuit building blocks have become larger and more powerful, while the "glue" that holds them together (the logic equations) has become simpler and less voluminous. Although the minimization of complex Boolean equations is no longer of paramount importance, simplification of small and manageable equations remains a routine task that we should make as easy and mechanical as practical. Manipulating equations through the Boolean algebraic identities, as we have done in the previous sections, is an arcane art. There are few guidelines to follow other than our intuition (based on experience) and trial and error. You have seen that sum of products is a common form of Boolean expression. This form results from truth-table derivations and occurs in other steps of our design process. The Boolean identity that is of most frequent use in simplifying sum-of-products forms is Eq. (1–9), which allows the elimination of a variable and its complement when these have a common factor:

$$A \cdot B + \overline{A} \cdot B = (A + \overline{A}) \cdot B = T \cdot B = B$$

The Karnaugh map (K-map) is a graphic display whose visual impact assists us in the systematic application of this identity.* A Karnaugh map is a canonical

* The Karnaugh map is also known as a Veitch diagram.

truth table rearranged in form. Figure 1–1 is a truth table and its K-map for an arbitrary function of two variables. The Karnaugh map has a square for each truth-table row; each combination of variables identifies a square in the map. In the two-variable case, the values of one variable appear as the labels for the vertical edges of the squares ($B = 0$ and $B = 1$ in Fig. 1–1), and the other variable's values mark the horizontal edges ($A = 0$ and $A = 1$). Each square contains the value of the given function (Y_i) corresponding to the appropriate truth-table row, as specified by the labels on the edges of the K-map. For instance, the lower left square in the K-map of Fig. 1–1, corresponding to $A = 0$, $B = 1$, has the function value Y_1.

Row number	A	B	Y
0	0	0	Y_0
1	0	1	Y_1
2	1	0	Y_2
3	1	1	Y_3

Y:

B \ A	0	1
0	Y_0	Y_2
1	Y_1	Y_3

Figure 1–1. A truth table and its Karnaugh map.

Functions of more than two variables also have K-map representations. Three- and four-variable functions are easy to manage; with more than four variables, the K-map technique becomes unwieldy, but fortunately most of the simplifications of design equations that we encounter in practice involve no more than four variables. The three-variable map contains eight squares, corresponding to the eight rows in the canonical truth table. Figure 1–2 shows two notations for K-maps of three variables. Both forms are equivalent, and both are in common use. Our mild preference is for the left-hand form, but you should be familiar with each. We like the left-hand form because it has an explicit display of the values of the variables on each edge of the map. The right-hand form explicitly labels only the location of *true* values for each variable. Some people prefer this form; take your pick. Conventionally, the third variable (rightmost in the truth table) appears on the vertical edge, while the horizontal edge displays the first and second variables.

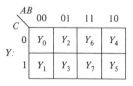

Y:

C \ AB	00	01	11	10
0	Y_0	Y_2	Y_6	Y_4
1	Y_1	Y_3	Y_7	Y_5

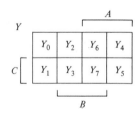

Figure 1–2. Two forms of a K-map of three variables.

Note carefully the order of the labels on the top edge. In moving from square to square across a row (and around the corner, also), the value of only one variable changes at a time. As you will see, this *unit distance* property gives the K-map its virtue in simplifying logic expressions.

Extending the K-map to four variables adds an additional variable to the vertical edge, resulting in 16 squares. Figure 1–3 illustrates both notations for Karnaugh maps of four variables.

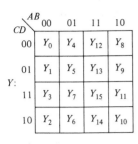

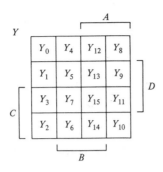

Figure 1–3. Two forms of a K-map of four variables.

Building K-Maps

There is a one-to-one correspondence between Karnaugh map squares and canonical truth-table rows. Deriving either the map or the table from the other is just a matter of rearranging the information. Don't-care outputs from truth tables are directly transferable to the appropriate K-map squares. Condensed truth-table rows yield values for more than one K-map square, in an obvious way.

We may view the K-map as a representation of the canonical sum-of-products form of a Boolean expression. Just as we may derive a truth table from a logic equation, we may move directly to a K-map from an equation, when this is appropriate. Figure 1–4 shows the three forms for a function of two variables.

The techniques for creating a truth table from a logic equation will also yield the K-map for the equation. The expression in Fig. 1–4 is already in canonical form, so the transformations among table, K-map, and equation are easy. Consider now the three-variable equation

$$V = A \cdot \overline{B} + B + A \cdot \overline{B} \cdot C$$

The first term in the sum yields 1's in the K-map when $A,B = 10$ and $C =$ anything. To give a true value, the second term requires $B = 1$, but A and C may be anything. The third term yields 1 for $A,B,C = 101$, which already appears in the map because of the $A \cdot \overline{B}$ term. The resulting map is Fig. 1–5.

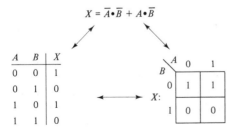

Figure 1–4. Three representations of a Boolean function.

Figure 1–5. K-map for $V = A \cdot \overline{B} + B + A \cdot \overline{B} \cdot C$.

Simplifying with K-Maps

Why bother with these maps? You may get a clue from Fig. 1–4. You have probably noticed that the expression for X can be simplified with Eq. (1–9):

$$X = A \cdot \overline{B} + \overline{A} \cdot \overline{B}$$
$$= (A + \overline{A}) \cdot \overline{B}$$
$$= T \cdot \overline{B}$$
$$= \overline{B}$$

How does the K-map display this simplification? The key point is that a certain term ($\overline{B}$ in this case) is ANDed with both A and $\overline{A}$. On the K-map this results in 1's in both the $A = 0$ and $A = 1$ squares for $B = 0$. Let's circle these adjacent 1's to remind us that the A variable disappears from the simplified expression because it appears as both A and $\overline{A}$ (see Fig. 1–6). Note how the simplified form $A = \overline{B}$ stands out more clearly: the condition for X to be true is that B is false (or $\overline{B}$ is true). Thus $X = \overline{B}$.

The drawing of circles (really ovals) among adjacent 1's is the basis for using K-maps in Boolean simplification. On K-maps of two variables, there are five ways to display applications of the basic identity of Eq. (1–9) with circles. Figure 1–7 shows these forms.

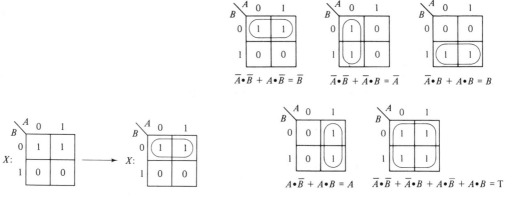

Figure 1–6. Circling adjacent 1's.

Figure 1–7. Simplifications of K-maps of two variables.

In using the K-map for simplification, we look for applications of the rules for circling. Depending on the position of the 1's, we may have several circles, each spanning a grouping of one, two, four, eight, . . . 1's. The K-map method requires that each 1 in the map appear in at least one circle, even if it is by itself. Circling two 1's causes two canonical terms to collapse into one term; one variable drops out. Four circled 1's bring four terms into one term, eliminating two variables. A proper group of eight circled 1's drops three variables, and so on.

On K-maps of three or more variables, some applications of the simplifying identity do not involve physically adjacent 1's. In these cases, we must draw "around-the-corner" circles. Figure 1–8a shows some typical ordinary circle patterns for a K-map of three variables; Fig. 1–8b gives all the around-the-corner patterns of two 1's; and Fig. 1–8c shows the only around-the-corner pattern involving four 1's.

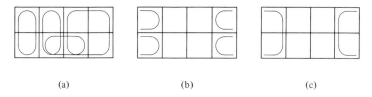

(a)   (b)   (c)

Figure 1–8. Typical circlings of K-maps of three variables.

Figure 1–9 shows three improper circlings. Diagonal or L-shaped arrangements do not correspond to applications of Eq. (1–9), nor does the circling of three 1's, since 3 is not a power of 2.

Figure 1–9. Improper K-map circling patterns.

K-maps of four variables are similar to the three-variable variety. Figure 1–10 shows some forms involving correct circlings of four and eight 1's. You should inspect these patterns until you are comfortable with their meaning.

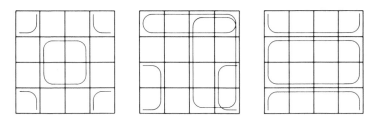

Figure 1–10. Typical circlings of four 1's and eight 1's on K-maps of four variables.

Here is the prescription for circling 1's in a K-map: Draw circles (ovals or around-the-corner patterns) around properly positioned collections of 1's, starting with the largest possible circles, and working toward smaller circles. Overlapping circles are appropriate when they allow a larger circle to appear. (Do not draw a circle that is completely within a larger circle; this would result in a redundant term and an incompletely simplified function.) Drawing the largest circles possible, cover all the 1's on the map. Use don't-care dashes "–" as either a 1 or a 0, as convenient. The point of using K-maps is to let the drawing display the simplified result in a systematic and mechanical fashion. When you have finished drawing circles, read off the simplified function as a sum of products, in which each circle contributes one product term to the sum.

Figure 1–11 shows two examples of functions of two variables, derived from their K-maps. Figure 1–11b yields no simplification.

The K-map method allows a simple derivation of the important identity of Eq. (1–8); Fig. 1–12 shows the process.

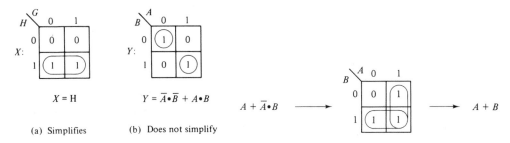

$$X = H$$

(a) Simplifies

$$Y = \overline{A} \cdot \overline{B} + A \cdot B$$

(b) Does not simplify

Figure 1–11. Simplifiable and unsimplifiable functions of two variables.

$$A + \overline{A} \cdot B \longrightarrow \qquad \longrightarrow A + B$$

Figure 1–12. Proof of the identity $A + \overline{A} \cdot B = A + B$.

Figure 1–13 shows the simplifications of two functions of four variables. Notice the use of overlapping circles to achieve the largest circles. Some 1's must be circled by themselves, yielding unsimplified four-variable terms.

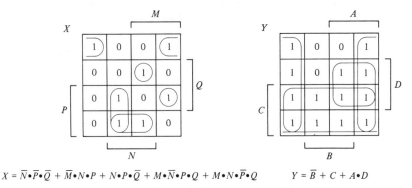

$$X = \overline{N} \cdot \overline{P} \cdot \overline{Q} + \overline{M} \cdot N \cdot P + N \cdot P \cdot \overline{Q} + M \cdot \overline{N} \cdot P \cdot Q + M \cdot N \cdot \overline{P} \cdot Q \qquad Y = \overline{B} + C + A \cdot D$$

Figure 1–13. Simplification of two functions of four variables.

K-Map Simplification Blunders

The most common error in simplifying expressions with K-maps is to fail to circle the largest possible groupings of 1's. A less common error is to introduce a redundant smaller circle within a larger one. Figure 1–14 shows some typical blunders and the correct forms. Although Figs. 1–14a and 1–14c produce correct Boolean expressions, these expressions are not as simple as the K-map method allows. Design criteria may sometimes require the use of an incompletely simplified

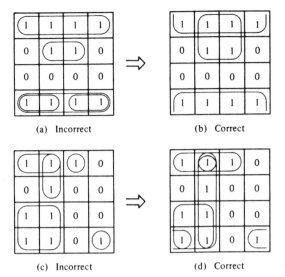

(a) Incorrect

(b) Correct

(c) Incorrect

(d) Correct

Figure 1–14. Common blunders in circling K-maps.

form, but these occasions are rare and we will not consider them here. Make all your circles as large as possible. Figure 1–14a has four circling errors: two incomplete circlings and two redundant circles. Figure 1–14c has three errors, all incomplete circlings.

Other Ways of Reading K-Maps

We have stressed the method of circling 1's and reading a sum of products for the function. If you are interested in developing the best facility with K-maps, you will want to investigate three other interpretations. These are analogous to the forms for reading expressions from truth tables shown earlier in this chapter. We give the three additional K-map methods below, with one example, leaving a comprehensive study of these techniques as grist for your mental mill.

> Method 1: Circle 1's and read a sum of products for the function (the normal method).
>
> Method 2: Circle 0's and read a sum of products for the inverse of the function (also a frequently used method).
>
> Method 3: Circle 0's and read a product of sums for the function.
>
> Method 4: Circle 1's and read a product of sums for the inverse of the function.

Figure 1–15a shows a K-map with 1's circled; Fig. 1–15b is the same map with 0's circled. The resulting equivalent simplified forms of the function are:

From method 1: $S = \overline{A} \cdot B + \overline{A} \cdot C$
From method 2: $\overline{S} = A + \overline{B} \cdot \overline{C}$
From method 3: $S = \overline{A} \cdot (B + C)$
From method 4: $\overline{S} = (A + \overline{B}) \cdot (A + \overline{C})$

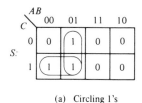

(a) Circling 1's

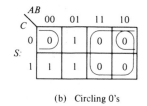

(b) Circling 0's

Figure 1–15. Two ways of circling a K-map.

CONCLUSION

You now have the knowledge of the foundations of digital logic that you need to continue your introduction to digital design. Boolean algebra and its allied techniques are a fascinating field of study, and you may wish to pursue these topics in more depth as your skill as a designer grows. We have just scratched the surface of the field, but the information in this chapter is sufficient for our purposes. Going deeper would deflect us from our goal, which is to build up the necessary tools as rapidly as possible in Part I so you may quickly reach the study of the design process in later parts of the book.

Now it is time to develop tools for systematically translating logic expressions into hardware.

READINGS AND SOURCES

BLAKESLEE, THOMAS R., *Digital Design with Standard MSI and LSI*, 2nd ed. John Wiley & Sons, New York, 1979.

BOOLE, GEORGE, *An Investigation of the Laws of Thought, on which are Founded the Mathematical Theories of Logic and Probability*, 1849. Dover Publications, 31 East 2nd Street, Mineola, N.Y. 11501, 1954. Where it all started. A Dover reprint of the classic work.

DIETMEYER, DONALD L., *Logic Design of Digital Systems*, 2nd ed. Allyn & Bacon, Boston, 1978. Good exposition of traditional switching theory and minimizations.

FLETCHER, WILLIAM I., *An Engineering Approach to Digital Design*. Prentice–Hall, Englewood Cliffs, N.J., 1980. Chapter 3 contains a good exposition of Karnaugh maps and minimization.

HILL, FREDERICK J., and GERALD R. PETERSON, *Introduction to Switching Theory and Logical Design*, 3rd ed. John Wiley & Sons, New York, 1981.

KARNAUGH, M., "The map method for synthesis of combinational logic circuits," *Communications and Electronics*, No. 9, November 1953.

MILLER, RAYMOND E., *Switching Theory, Vol. 1: Combinational Circuits*. John Wiley & Sons, New York, 1966. An important early work.

SHANNON, C. E., "Symbolic analysis of relay and switching circuits," *Transactions AIEE*, Vol. 57, 1938, p. 713.

VIETCH, E. W., "A chart method for simplifying truth functions," *Proceedings ACM*, Pittsburgh, 1952, p. 127.

EXERCISES

1–1. Have you read the Preface?

1–2. What is the goal of Part I of this book? Part II? Part III? When may you read the material in Part IV?

1–3. What are the basic logical operators in digital design? What are the constants? What does the numeral 1 mean in digital design?

1–4. How many different Boolean functions of two variables are there? Of three variables? Derive an expression for the number of different Boolean functions of n variables.

1–5. What is a canonical truth table? Give examples of a canonical truth table and a noncanonical truth table of three variables.

1–6. Give the operator hierarchies for AND, OR, and NOT. By inserting full parentheses, show the order of evaluation of these functions:
(a) $B \cdot \overline{A} \cdot C + D + \overline{E}$
(b) $\overline{A + B} \cdot C + D$
(c) $\overline{A + B} \cdot (\overline{C} + D)$

1–7. By using the operator hierarchies, write the following expressions with as few parentheses as possible:
(a) $(\overline{(Q + (R \cdot S))} + U \cdot V)$
(b) $((Q \cdot (\overline{R} + S)) \cdot (U + V))$

1–8. Prove the following identities by writing the truth tables for both sides.
(a) $A \cdot (B + C) \equiv (A \cdot B) + (A \cdot C)$
(b) $A + (B \cdot C) \equiv (A + B) \cdot (A + C)$
(c) $\overline{\overline{A}} \equiv A$
(d) $\overline{A \cdot B \cdot C} \equiv \overline{A} + \overline{B} + \overline{C}$
(e) $\overline{A + B + C} \equiv \overline{A} \cdot \overline{B} \cdot \overline{C}$
(f) $A + \overline{A} \cdot B \equiv A + B$
(g) $A \cdot (A + B) \equiv A$

1–9. The cancellation law of regular algebra states that

$$\text{If } X \; (+) \; Y = X \; (+) \; Z, \text{ then } Y = Z$$

Show by giving counterexamples that Boolean algebra has no equivalent cancellation law. In other words, show that the following statements are false:

$$\text{If } X + Y = X + Z, \text{ then } Y = Z$$
$$\text{If } X \cdot Y = X \cdot Z, \text{ then } Y = Z$$

1–10. NAND and NOR are Boolean functions sometimes used in design. NAND (NOT AND) is defined as AND followed by NOT; NOR (NOT OR) is defined as OR followed by NOT. Write the defining truth tables for A NAND B and A NOR B.

1–11. (a) Write each of the following expressions in a form that has no AND operators:

$$\overline{(A + B)} \cdot C \qquad A \cdot B \cdot \overline{C} + \overline{B} \cdot D$$

(b) Write each of the following expressions in a form that has no OR operators:

$$\overline{A} + \overline{B} + \overline{C} \qquad \overline{A} + \overline{B} + \overline{C \cdot D} + \overline{E}$$

1–12. Define the following terms:

(a) Canonical.
(b) Minterm.
(c) Maxterm.
(d) Sum-of-products form.
(e) Product-of-sums form.
(f) Canonical sum-of-products form.
(g) Canonical product-of-sums form.

1–13. Which of the following expressions is in sum-of-products form? Which is in product-of-sums form?

(a) $A + \overline{B \cdot D}$

(b) $C \cdot \overline{D} \cdot E + \overline{F} + D$

(c) $(A + B) \cdot C$

1–14. (a) Write a four–element vector describing the function $X(A,B)$:

A	B	X
F	F	F
T	F	T
F	T	F
T	T	F

(b) Derive a logic equation for X directly from the truth table.

(c) Derive a logic equation for X directly from your vector expression.

(d) Show that the results from parts (b) and (c) are equivalent.

1–15. Without formally deriving any logic expressions, deduce the value of each function W, X, Y, and Z:

A	B	C	W	X	Y	Z
0	0	0	0	1	0	1
0	0	1	0	1	0	0
0	1	0	0	1	0	1
0	1	1	0	1	0	0
1	0	0	0	1	1	1
1	0	1	0	1	1	0
1	1	0	0	1	1	1
1	1	1	0	1	1	0

1–16. Write the logic equations corresponding to the following:

(a) $X(A,B,C) = m_0 + m_2 + m_5$

(b) $\overline{X}(P,Q) = M_1 \cdot M_3$

1–17. Write the canonical truth tables for each of the following:

(a) $Y(V,W,X) = M_2 \cdot M_3 \cdot M_5 \cdot M_6$

(b) $Y(C,B,G) = m_1 + m_2 + m_7$

1–18. Show that Eq. (1–13) reduces to $\overline{Y} = \overline{A} \cdot B$:

(a) By using Boolean algebraic reductions.

(b) By developing the canonical truth table for each sum term in Eq. (1–13), then performing the AND of the truth-table function values to produce a truth table for $\overline{Y}$, and finally reading the sum-of-products logic equation for $\overline{Y}$ from the truth table.

1-19. Consider the following truth table:

A	X	YZ	G
0	0	0	0
0	0	1	0
0	1	0	0
0	1	1	1
1	0	0	0
1	0	1	1
1	1	0	1
1	1	1	0

Derive canonical equations for G or $\overline{G}$ in the following forms:
   (a) Sum of products on true outputs.
   (b) Sum of products on false outputs.
   (c) Product of sums on true outputs.
   (d) Product of sums on false outputs.

1-20. Prove the correctness of your answers to Exercise 1-19 by reconstructing a truth table for G from each equation.

1-21. Consider the following canonical truth table for two functions S and C:

P	Q	R	S	C
0	0	0	0	0
0	0	1	1	0
0	1	0	1	0
0	1	1	0	1
1	0	0	1	0
1	0	1	0	1
1	1	0	0	1
1	1	1	1	1

   (a) Express S and C as eight-element vectors.
   (b) Working directly from the vectors, write a canonical sum-of-products equation for S and a product-of-sums equation for $\overline{S}$. Show the equivalence of the equations by Boolean algebraic manipulations.
   (c) Repeat part (b) for functions C and $\overline{C}$.
   (d) Directly from the truth table, write a vector for the function $S \cdot C$. Working from the vector, give a logic equation for $S \cdot C$.
   (e) Repeat part (d) for the function $S + C$.

1-22. By Boolean algebraic transformations, show the equivalence of the forms in Eqs. (1-14) through (1-17).

1-23. Express Eqs. (1-14) through (1-17) using the minterm and maxterm notations.

1-24. Explain the use of X for "both values" and – for "don't care" in truth tables.

1–25. Derive the canonical truth tables that correspond to each of the following K-maps:

0	0	0	0
0	0	1	0

0	1	0	0
0	1	–	0
–	1	1	0
0	1	1	0

1–26. Plot the function in Exercise 1–19 on two K-maps, one map labeled as in Fig. 1–2a and the other as in Fig. 1–2b. Simplify the function if possible.

1–27. Draw a K-map for each of the truth tables below. Derive a simplified logic equation from each K-map.

A	B	C	M		A	B	C	M		A	B	C	D	M
0	0	0	1		0	0	0	1		0	X	0	X	1
0	0	1	1		0	0	1	1		0	X	1	1	0
0	1	0	1		0	1	0	0		0	0	1	0	–
0	1	1	1		0	1	1	0		X	1	1	0	0
1	0	0	1		1	0	0	–		1	0	0	0	1
1	0	1	–		1	0	1	1		1	X	X	1	–
1	1	0	0		1	1	0	0		1	0	1	0	1
1	1	1	0		1	1	1	0		1	1	0	0	0

1–28. Here is a K-map for a function S:

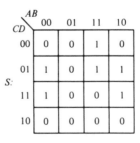

By circling zeros, give a logic equation for $\overline{S}$ as a sum of products with each product term containing two variables.

1–29. By circling zeros, simplify the functions in Fig. 1–13.

1–30. Assuming that there are three inputs A, B, and C, write a truth table to describe each of these ideas:

(a) The output should be true only when two or more of the input variables are true.

(b) The output should be true only when the number of true input variables is odd.

(c) The output should be true only when the number of false input variables is even.

Tools for Digital Design   Part I

(d) The output should be false only when exactly two of the input variables are true.

1–31. Simplify the functions derived in Exercise 1–30, using K-maps.

1–32. Often the natural formulation of a logic function is not in perfect sum-of-products or product-of-sums form. For example, consider the equation $M = \overline{A} \cdot (B + C) + A \cdot \overline{C}$. Simplify this equation using two different K-map circlings. Are the resulting sum-of-products forms less compact or more compact than the original? (A criterion for compactness is the number of binary AND and OR operators in the expression.) Can you perform elementary factorings on the K-map results that make the results more compact? Compare the original equation with each of the final equations.

1–33. Consider two 2-bit binary numbers, say, A,B and C,D. A function X is true only when the two numbers are different.
(a) Construct a truth table for X.
(b) Construct a four-variable K-map for X, *directly from the word definition of X*.
(c) Derive a simplified logical expression for X.

1–34. You are installing an alarm bell to help protect a room at a museum from unauthorized entry. Sensor devices provide the following logic signals:

$ARMED$ = The control system is active

$DOOR$ = The room door is closed

$OPEN$ = The museum is open to the public

$MOTION$ = There is motion in the room

Devise a sensible logic expression for ringing the alarm bell.

1–35. A large room has three doors, A, B, and C, each with a light switch that can turn the room light on or off. Flipping any switch will *change* the condition of the light.
(a) Assuming that the light is off when the switch variables have the values 0, 0, 0, write a truth table for a function $LIGHT$ that can be used to direct the behavior of the light.
(b) Derive a logic equation for $LIGHT$.
(c) Can you simplify this equation?
(d) How is this exercise related to Exercise 1–30?

1–36. Electronic watches display time by turning on a certain combination of seven light-bar segments to yield approximations of the shape of the decimal digits. For each digit position, the segments are labeled as follows:

The decimal digit displays have the form

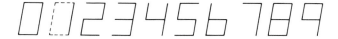

For example, the digit 4 has segments b, c, f, and g lighted. Internally, the watch represents a decimal digit by a 4-bit binary code, say, D,C,B,A. For example

$$
\begin{array}{cccc}
D & C & B & A \\
7 = 0 & 1 & 1 & 1
\end{array}
$$

(a) Develop a multi-output truth table for lighting the segments. The truth table will have inputs D, C, B, and A, and outputs a, b, c, d, e, f, and g. Notice that don't-care conditions arise naturally, since 4 bits can encode 16 combinations, whereas the decimal digits use only 10 of them. Binary codes above 1001 will never occur.

(b) Plot the light segment outputs a through g on four-variable K-maps, and derive a simplified equation for each segment.

1–37. The university pool room has four pool tables lined up in a row. Although each table is far enough from the walls of the room, students have found that the tables are too close together for best play. The experts are willing to wait until they can reserve enough adjacent tables so that one game can proceed unencumbered by nearby tables. A light board visible outside the poolroom shows vacant tables. The manager has developed a digital circuit that will display an additional light whenever the experts' desired conditions arise. Give a logic equation for the assertion of the new light signal. Simplify the equation, using a K-map.

2

Realizing Logic
in Hardware

REPRESENTING TRUE AND FALSE WITH PHYSICAL DEVICES

Boolean algebra and truth tables are our tools for expressing logical relationships. To use these tools in the real world, we must have some physical way to represent TRUE and FALSE, the fundamental constants of logic. Digital systems record T and F in several ways:

(a) *Punched cards.* A hole punched at a given spot in the card might represent logic 1; no hole at that spot would then represent 0.

(b) *Magnetic tapes or disks.* Magnetic tapes or disks represent logic data with magnetized areas on the recording surface. The designer might choose a south pole sticking out of the surface to be a 1, in which case a north pole would be 0.

(c) *Switches.* A switch has two states, closed and open. The digital designer may choose either state (but not both!) to represent logic truth; for example, an open switch may represent 1.

(d) *Voltages.* In digital electronic circuits, T and F are represented by voltage. For instance, the popular transistor–transistor logic (TTL) 74LS family of digital circuits produces two voltage levels: <0.5 V and >2.7 V.

Each of these four examples has only two states—in a punched card either there is a hole or there is not a hole. This two-valued, or binary, characteristic of the digital world makes Boolean algebra the natural way to formalize the behavior of these physical devices. Conversely, the need to implement logical constructs in physical devices makes these binary devices useful. If more than two values existed, more complex algebras would be needed to handle the multiplicity of values. Perhaps fortunately, engineers have had only limited success in designing reliable nonbinary devices. It is questionable whether multi-valued devices would be desirable, since the resulting systems would be harder to troubleshoot. In practice, it is important to know that any signal can have only two values, TRUE or FALSE.

The designer may select the physical representation for T and F. In a punched card, representing 1 with a hole seems natural, but such an assignment is not a logical necessity. With equal validity, we could let the absence of a hole represent 1. On magnetic tapes either a north or a south pole can be a 1; both conventions exist. The same is true of switches.

In this book, we will use electronic logic circuits extensively. We let H stand for the high-output voltage level of a digital device and let L stand for the low-output voltage level. Each family of devices has its own H and L output voltage ranges. (Chapter 12 contains information on the performance of the various families of devices. Remember, you may read the material on digital technology in Chapter 12 whenever you desire, without mastering the intervening material.) In this book, we use the 74LS TTL (low-power Schottky transistor–transistor logic) family as our main source for digital electronic devices.

A particular device is called an *integrated circuit* or a *chip*. Each type of integrated circuit has an identifying number and a descriptive name. In this book, most of the chip numbers are of the form 74LSxx, where xx is two or more digits specifying the particular member of the low-power Schottky TTL line. The chip has a set of pins that provides connections to the elements within the chip. Each pin has a pin number that identifies its location on the chip. The pin numbers are usually not printed on the chip itself, but they occur in a standard arrangement, with pin 1 marked. The pins are mounted in a *dual in-line* package (DIP).

Many chips of interest in the next few chapters have 14 or 16 pins. Here is a drawing of a 14-pin chip, about twice actual size:

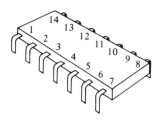

Operating power enters the chip through two pins, V_{CC} and GND (ground), usually at opposite corners. In the drawing, pin 7 is GND and pin 14 is V_{CC}. A chip may contain one or more independent logic elements.

The manufacturers' data books are authoritative sources of information on available integrated circuit chips. Learning to use a data book is a necessary part of building digital circuits. (Refer to Chapter 12; the bibliography lists some of the more comprehensive books.)

In this chapter, we will use simple devices to build the basic logic functions described in Chapter 1; historically, each such device is called a *gate*. Here we concentrate on chips with gates—the basic tools for realizing logic equations. In Chapters 3 and 4, you will meet more complex chips.

Like other forms of digital devices, electronic logic circuits offer a choice for representing T and F. Be flexible in your choice. Sometimes it is advantageous to let H represent T and to let L represent F; at other times the converse is more convenient. Either will do, if you let the rest of the world know your choice.

MIXED LOGIC: REPRESENTING AND, OR, AND NOT

We may represent logic truth by either of the two voltage levels in a digital electronic device. If we apply this notion faithfully, a powerful and beautiful design tool emerges as we represent logic equations with physical hardware. We now undertake the development of clear and systematic ways of building and describing hardware circuits for logic expressions. Efforts at solving digital problems yield logic equations and logical structures; the hardware must faithfully embody these equations and structures. Furthermore, we certainly wish the documentation of our hardware (our *circuit diagrams*) to convey the spirit of the solution to the original problem. Documenting the hardware for a logic circuit is called *digital drafting*.

The foregoing thoughts suggest some criteria for drafting methods:

(a) We wish to synthesize (create) a physical realization of any logic expression directly from the logic, in a straightforward, natural, and rigorous manner.

(b) We wish to be able to analyze (pick apart) a physical realization and directly recover the original logic expressions.

These are strong conditions; many digital drafting and construction techniques in use today do not meet them. The conditions require that the circuit diagram clearly and fully display both the logic and hardware. We can identify several implications of these requirements:

(1) The drafting notation should represent the Boolean expressions in AND, OR, and NOT form—the natural way we develop our logic.

(2) The correspondence between a logical value (T or F) and its voltage counterpart (H or L) should be evident everywhere in the circuit diagram.

(3) The notation should clearly identify each physical device in the circuit.

The key to satisfying these requirements is a representation called *mixed logic*. This notation was first published in a coherent form in 1971,* although

---

* P. M. Kintner, "Mixed logic: a tool for design simplification," *Computer Design*, August 1971, pp. 55–60; and F. Prosser and D. Winkel, "Mixed logic leads to maximum clarity with minimum hardware," *Computer Design*, May 1977, pp. 111–117.

mixed logic was used in the Philco TRANSAC computers in 1957 and the technique is probably even older than that. We will develop mixed-logic methodology carefully, since the principle is vital to clear, top-down design.

Mixed Logic

Showing the logic. We choose a unique symbol for each of the natural logical operators. The following shapes represent the AND and OR operators:

Whenever we see these shapes, we know that we are representing a logical AND or a logical OR function. Furthermore, every logical AND and OR in our original expression will appear in the circuit diagram as the corresponding shape. Inputs to the symbols enter from the left, and the output leaves from the right. The notations

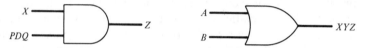

specify that $Z = X \cdot PDQ$ and $XYZ = A + B$. The graphic symbols for AND and OR can have more than two inputs. For example, Fig. 2–1 shows AND gates with 4 and 8 inputs. Figure 2–2 shows OR gates with 4 and 8 inputs.

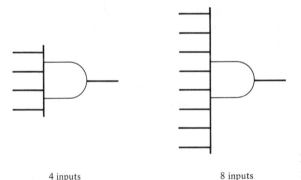

4 inputs   8 inputs

Figure 2–1. Symbols for AND with multiple inputs.

In circuit diagrams, the graphic symbols imply a physical device that performs the logic operation. In our applications, the physical device is an integrated circuit (usually of the 74LS family), and logic values appear as voltages on copper wires. Let's see how to record the voltage information on the diagram without altering the logic.

Logic conventions. How does a device represent T and F? There are two logic levels (T and F) and two voltage levels (H and L). Two useful possibilities exist:

(a) T is represented by H (and F is represented by L).

(b) T is represented by L (and F is represented by H).

The first form, with T = H, is called *positive logic;* the form with T = L is *negative logic.* When one of these relationships of truth and voltage is used consistently throughout a design, we refer to a *positive-logic convention* or a *negative-logic convention* for the design. The *mixed-logic convention*—our convention—allows us to use positive or negative logic at any point in our design, as we desire. The clarity gained by this innocent-sounding step is enormous.

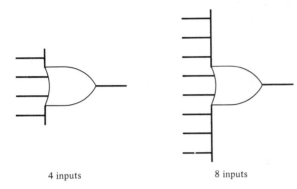

4 inputs   8 inputs

Figure 2–2. Symbols for OR with multiple inputs.

Showing the device. In our exposition of mixed logic, we represent T = L by a small circle on the corresponding terminal of the logic symbol. The absence of a small circle means that T = H at that point. The circles do not change the logic operation. For example, each of the symbols in Fig. 2–3 is a mixed-logic implementation of a logical AND function of two variables. We emphasize this point again: each of these symbols (and there are four more) represents a physical realization of the same truth table:

Logic		
Inputs		Output
F	F	F
F	T	F
T	F	F
T	T	T

Each symbol defines a particular type of physical device. Since we know the truth table (because of the symbol's shape) and the voltage representation of truth on each input and output (by the presence or absence of circles), we can

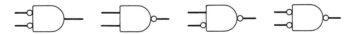

Figure 2–3. Symbols for logical AND.

immediately write down the voltage table for any symbol. Then, referring to a data book for integrated circuits, we can identify the device. For example:

	Logic				Voltage		
A	B	M		A	B	M
F	F	F		L	L	H
F	T	F	↔	L	H	H
T	F	F		H	L	H
T	T	T		H	H	L

Here, the symbol completely defines the behavior of both the logic and the voltage. Can we find a physical device with this voltage table? If so, we can use this mixed-logic symbol to perform AND with a real device. A look at the first chip in a TTL data book shows that the 74LS00 Quad Two-Input Nand Gate behaves this way. Now we have a symbol that describes both a logic operation and a piece of hardware.

The term *quad* in the chip's name signifies that the chip contains four similar gates. "Triple" means three gates on a chip; "dual," two gates, and so on. The number of gates per chip is determined by the number of inputs and outputs per gate and the number of pins on the chip.

The term *Nand* in the name of the 74LS00 chip arose historically from the view that this chip implemented the logical NAND (NOT AND) function in positive logic. The user of positive logic is forced into this interpretation, but since NAND is not a familiar and intuitive logic function, mixed logicians are not much interested in expressing logic in terms of NANDs. We discuss the positive-logic convention later in this chapter.

Another common TTL device is the 74LS02 Quad Two-Input Nor Gate. This device can perform the logical OR function when T = H at the inputs and T = L at the output:

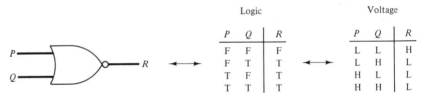

	Logic				Voltage		
P	Q	R		P	Q	R
F	F	F		L	L	H
F	T	T	↔	L	H	L
T	F	T		H	L	L
T	T	T		H	H	L

The name *Nor* derives historically from conventional positive logic, in which this chip performs the logical NOR (NOT OR) function.

Signal names in mixed logic. In a physical circuit, voltages represent values of the logic variables. We will refer to the actual voltage representation of a logic variable as a *signal*. When we label a circuit's inputs and outputs with the names of logic variables, we need a rule for also describing the voltage polarity of that variable at that point. Naming the logic variable is not enough; we must also name the signal. Here is the convention used in this book for creating names of signals that correspond to names of logic variables:

If a signal has T = L, append a terminal .L to the logic variable's name.
If a signal has T = H, append a terminal .H to the logic variable's name.

For example, the logic variable *PICK* might appear in a circuit as the signal *PICK*.L or as *PICK*.H, depending on its particular voltage representation at that point in the circuit. The terminal .L is always associated with a small circle on a line in the circuit diagram; the .H form is always associated with a line having no circle:

Understand two points thoroughly:

(a) The logic variable is the same in both representations. *PICK*.H and *PICK*.L are two different ways of physically forming the logic variable *PICK*.
(b) .L does *not* mean that the voltage is low. Rather, it means that *if* the voltage is low, *then* the value of the logic variable is *true*. .H is interpreted analogously.

Frequently, a circuit diagram contains both signals for a logic variable (on different wires!). We display the voltage convention on the wires of the circuit diagram and also in the signals' names. You may think this is redundant, but it is an important aid to clarity. If you show line and signal notations rigorously, your diagrams will show *all* the logic and *all* the voltage information in the circuit, clearly and conveniently.

Whenever it is necessary to write the name of a logic variable or signal more than once in a design, the need for the signal notation arises. This happens when signals appear on more than one page of the circuit diagram, when the designer prepares a master list of signal names for the circuit, and in other circumstances. Figure 2–4 illustrates a common digital drafting situation. In the figure, a signal generated on one page serves as an input elsewhere. With proper notation, there is never any doubt as to which signal is meant.

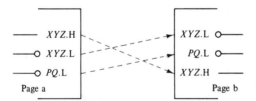

Figure 2–4. Illustrating the need for a convention for naming signals.

Naming signals is important. The particular notation for distinguishing the two signals for a variable is not crucial as long as you are consistent. Some people prefer to append a + and −, or ↑ and ↓, to a variable's name; others use a terminal / to indicate that T = L. There is no widely accepted standard convention. We use .H and .L in this book because of their strength in displaying voltage assignments.

Using this signal convention, we reach the final form of the earlier AND implementation:

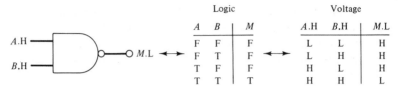

Logic				Voltage		
A	B	M		A.H	B.H	M.L
F	F	F		L	L	H
F	T	F		L	H	H
T	F	F		H	L	H
T	T	T		H	H	L

AND/OR duals. Here is another example, this time for an OR operation:

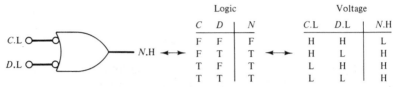

Logic				Voltage		
C	D	N		C.L	D.L	N.H
F	F	F		H	H	L
F	T	T		H	L	H
T	F	T		L	H	H
T	T	T		L	L	H

Inspect the voltage table. You can see that it is equivalent to the one in the previous example, and therefore again corresponds to the 74LS00 Nand gate. But this time the logic operation is OR. In this particular use of OR, T = L at both inputs and T = H at the output. We have identified two uses for the 74LS00 Nand gate—to implement AND and to implement OR. The AND–OR duality of gate usage is always present, and is related to the principle of duality in Boolean algebra. On a given gate, the dual symbols for AND and OR have reversed circles.

Similarly, you may derive mixed-logic notations for other devices. In Fig. 2–5, we have drawn some common chips and have shown their mixed-logic symbols and their standard chip names. In practice, it is simple to find the mixed-logic uses of any gate chip, since the data book gives one symbol directly, and the other one follows immediately by swapping AND–OR shapes and reversing the circles. Also, the conventional chip names are usually sufficient for writing down the mixed-logic symbols. Many integrated circuit chips embody AND and OR, giving us powerful and flexible tools for implementing the logic.

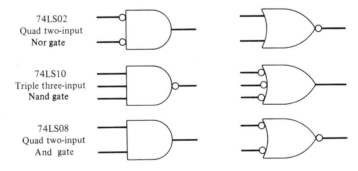

Figure 2–5. Mixed-logic symbols for some TTL gates.

Some detailed circuit examples. To bring home the exact implications of the mixed-logic notations, let's study the circuit in Fig. 2–6. Figure 2–6 means:

(a) There are two physical devices (both 74LS00 gates) that function as logical ANDs when T = H at the inputs and T = L at the outputs.

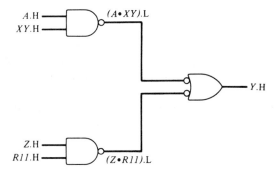

Figure 2–6. Schematic of
$Y = A \cdot XY + Z \cdot R11$.

(b) There is another physical device (again a 74LS00 gate) that functions as the logical OR when T = L at its inputs and T = H at its output.

(c) There are seven copper wires carrying voltages for the logic variables A, XY, Z, $R11$, $(A \cdot XY)$, $(Z \cdot R11)$, and Y.

(d) Truth (T) is represented by a high voltage (H) on signal wires A.H, XY.H, Z.H, $R11$.H, and Y.H.

(e) Truth (T) is represented by a low voltage (L) on signal wires $(A \cdot XY)$.L and $(Z \cdot R11)$.L.

(f) The circuit implements the logic equation $Y = A \cdot XY + Z \cdot R11$.

Consider another example, Fig. 2–7. This drawing tells us:

(a) There is one physical device (a 74LS00 gate) that functions as the logical OR when T = L at its inputs and T = H at its output.

(b) There is another physical device (again a 74LS00 gate) that functions as the logical AND when T = H at its inputs and T = L at its output.

(c) There are five copper wires carrying voltages for the logic variables P, $R2$, $(P + R2)$, AB, and $A3$.

(d) Truth (T) is represented by a high voltage (H) on signal wires $(P + R2)$.H and AB.H.

(e) Truth (T) is represented by a low voltage (L) on signal wires P.L, $R2$.L, and $A3$.L.

(f) This circuit implements the logic equation $A3 = AB \cdot (P + R2)$.

Figures 2–6 and 2–7 are straightforward illustrations of the mixed-logic drafting conventions.

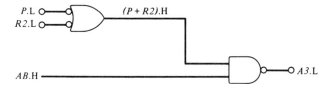

Figure 2–7. Schematic of $A3 = (P + R2) \cdot AB$.

Mixed-Logic Theory

With this introduction to basic notations, let's examine the theoretical basis of mixed logic. We will first consider the logical identity operation—one that we usually take for granted, but which provides important insight into the properties of mixed logic. Look at Fig. 2–8.

A logic variable A is the input to a box that performs the logical identity operation and produces Y as output: $Y = A$. The truth table for the behavior of the box is just that of the identity function. If we consider the possible voltage implementations of this box, we have four choices: $A.H$ and $Y.H$; $A.H$ and $Y.L$; $A.L$ and $Y.H$; and $A.L$ and $Y.L$. Each of these choices of voltage representation yields a voltage table, and each choice results in a different mixed-logic realization. As usual, we use mixed-logic circles to indicate that T = L.

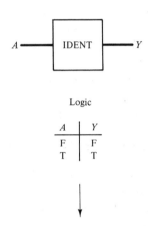

Logic

A	Y
F	F
T	T

Four choices of voltage

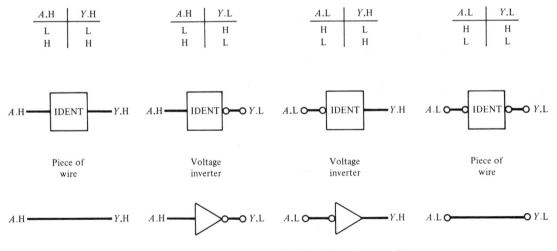

$A.H$	$Y.H$
L	L
H	H

$A.H$	$Y.L$
L	H
H	L

$A.L$	$Y.H$
H	L
L	H

$A.L$	$Y.L$
H	H
L	L

Piece of wire

Voltage inverter

Voltage inverter

Piece of wire

Figure 2–8. Realizations of the logical identity operation.

Tools for Digital Design   Part I

Contemplate the devices required to realize the four voltage tables that correspond to our four diagrams. When $A.H$ yields $Y.H$, we need only a piece of wire, since the voltage does not change in passing through the identity box. Similarly, when $A.L$ yields $Y.L$, we need only a piece of wire. On the other hand, the voltage table for $A.H$ and $Y.L$ specifies a voltage inverter, for instance the 74LS04 Inverter Gate. The voltage table for $A.L$ and $Y.H$ also specifies a voltage inverter. The triangle with a single circle on the input or output is the customary symbol for the voltage inverter.

We have four ways of performing the identity operation, two of which require pieces of wire and two of which require a voltage-inverting device. The mixed logician may use any of the four, as required in the circuit. The obvious choice is to use minimal hardware, and in most instances the wire will suffice. However, if the design calls for a change in the voltage representation with no change in logic, then the mixed logician has convenient ways to change the voltage representation. The use of a real device—the 74LS04—to accomplish the logical identity operation is an immediate consequence of mixed-logic theory.

Now consider the logical AND operation, as developed in Fig. 2–9. Two logic variables A and B are inputs to a box that must provide the logical AND of A and B as its output Y: $Y = A \cdot B$. When there are two inputs and one output, there are $2^3 = 8$ ways to select the voltage conventions. Eight voltage tables result from the eight assignments. Each voltage assignment produces its own mixed-logic diagram: the square box labeled "AND" shows the logic, and the circles show signals in which $T = L$. Since logical AND is of great importance to designers, we use the special AND shape instead of the box labeled "AND."

We may ask which of these eight realizations of logical AND correspond to readily available integrated circuits. A scan of a TTL data book discloses chips that embody four of the voltage tables. As shown in Fig. 2–9, these chips are the 74LS08 Quad Two-Input And Gate, which performs AND when $T = H$ everywhere; the 74LS00 Quad Two-Input Nand Gate, for AND when $T = L$ at the output; the 74LS02 Quad Two-Input Nor Gate, for AND when $T = L$ at its inputs; and the 74LS32 Quad Two-Input Or Gate, for AND when $T = L$ everywhere. Mixed logic gives us four useful building blocks for realizing logical AND. The other four voltage tables in Fig. 2–9 offer perfectly valid ways of performing logical AND, but we are unable to find readily available integrated circuits that conform to these voltage tables; we shall avoid using them in circuit diagrams meant for actual construction.

The positive-logic convention allows only one choice for logical AND—the 74LS08. Similarly, the negative-logic convention allows only one choice, the 74LS32. Mixed logic offers four.

The logical OR operation gives analogous results. Figure 2–10 shows the treatment. Again, we have eight ways of representing voltage at the inputs and output, and each choice transforms the truth table for logical OR into a voltage table. A tour through a TTL data book discloses that the same four integrated circuits that perform logical AND will also perform logical OR, but the voltages representing truth are different. The 74LS32 Or Gate, which performs logical AND when $T = L$ at inputs and output, will perform logical OR when

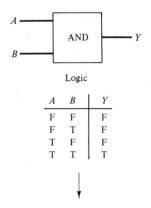

Logic

A	B	Y
F	F	F
F	T	F
T	F	F
T	T	T

Eight choices of voltage

A.H	B.H	Y.H
L	L	L
L	H	L
H	L	L
H	H	H

A.H	B.H	Y.L
L	L	H
L	H	H
H	L	H
H	H	L

A.H	B.L	Y.H
L	H	L
L	L	L
H	H	L
H	L	H

A.H	B.L	Y.L
L	H	H
L	L	H
H	H	H
H	L	L

AND   'LS08

AND○   'LS00○

○AND   (AND symbol)

○AND○   (NAND symbol)

A.L	B.H	Y.H
H	L	L
H	H	L
L	L	L
L	H	H

A.L	B.H	Y.L
H	L	H
H	H	H
L	L	H
L	H	L

A.L	B.L	Y.H
H	H	L
H	L	L
L	H	L
L	L	H

A.L	B.L	Y.L
H	H	H
H	L	H
L	H	H
L	L	L

○AND   (AND symbol)

○AND○   (NAND symbol)

AND   'LS02

○AND○   'LS32○

Figure 2–9. Realizations of logical AND.

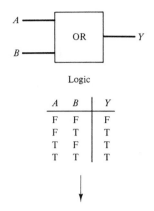

Logic

A	B	Y
F	F	F
F	T	T
T	F	T
T	T	T

Eight choices of voltage

A.H	B.H	Y.H
L	L	L
L	H	H
H	L	H
H	H	H

A.H	B.H	Y.L
L	L	H
L	H	L
H	L	L
H	H	L

A.H	B.L	Y.H
L	H	L
L	L	H
H	H	H
H	L	H

A.H	B.L	Y.L
L	H	H
L	L	L
H	H	L
H	L	L

OR 'LS32   OR 'LS02   OR   OR

A.L	B.H	Y.H
H	L	L
H	H	H
L	L	H
L	H	H

A.L	B.H	Y.L
H	L	H
H	H	L
L	L	L
L	H	L

A.L	B.L	Y.H
H	H	L
H	L	H
L	H	H
L	L	H

A.L	B.L	Y.L
H	H	H
H	L	L
L	H	L
L	L	L

OR   OR   OR 'LS00   OR 'LS08

Figure 2–10. Realizations of logical OR.

T = H everywhere. Similarly, the 74LS02 Nor Gate performs logical OR when T = L on the output only; the 74LS00 Nand Gate, when T = L at both inputs; and the 74LS08 And Gate, when T = L at inputs and output.

Mixed logic gives us four readily available implementations of logical OR. The other four representations, although valid, do not correspond to readily available integrated circuits. Positive logic permits only one choice for logical OR—the 74LS32; negative logic permits only the 74LS08.

Last, let's look at logic inversion. In Fig. 2–11 the development is similar to that of the logical identity operator, but *Y* must equal NOT *A*. The truth table shows that for logical NOT the logical value of the variable must be inverted. There are four voltage realizations of this truth table: *A*.H and *Y*.H, *A*.H and *Y*.L, *A*.L and *Y*.H, and *A*.L and *Y*.L. Each gives rise to its own mixed-logic diagram, with the square box labeled "NOT" surrounded by appropriate circles to display the voltage representation of truth at the input and output.

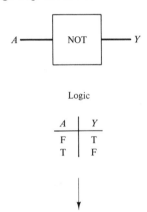

Logic

A	*Y*
F	T
T	F

Four choices of voltage

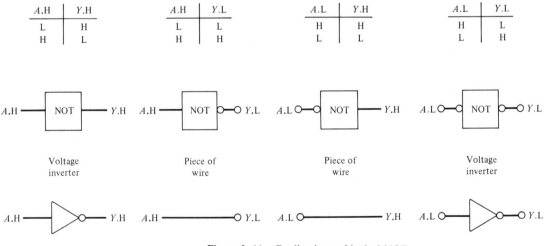

A.H	*Y*.H
L	H
H	L

A.H	*Y*.L
L	L
H	H

A.L	*Y*.H
H	H
L	L

A.L	*Y*.L
H	L
L	H

Voltage inverter   Piece of wire   Piece of wire   Voltage inverter

Figure 2–11. Realizations of logical NOT.

The four voltage tables correspond, respectively, to a voltage-inverting device (for instance, the 74LS04), a piece of wire, a piece of wire, and a voltage-inverting device. This is interesting: the mixed logician may implement logical NOT with a piece of wire! You can see what is required to achieve this design: the voltage representing truth is different at the two ends of the wire.

The positive logician and the negative logician each have only one way to implement logical NOT—with a voltage inverter. The mixed logician potentially has four ways, two using a piece of hardware and two using just the wire connecting the input and output.

We get the logical identity "for free" along a wire, as long as the voltage representations are the same at each end of the wire. We get logical inversion "for free" along a wire if the voltage representations for truth are different at the two ends. Occasions arise when we wish to change the voltage representation of a signal without performing any logic or to perform logical NOT while maintaining the same voltage representation at the input and output. In digital design, we do not consciously perform logical identity operations; we assume that we have identity logic at any point in our circuit unless otherwise specified. On the other hand, we do indeed frequently perform logical NOT; it is one of our three important logical operators. Now we make an important choice: the mixed logician chooses to use the voltage inverter solely to generate the logical identity. This means that we will never insert a voltage inverter—a real device—into our circuit unless absolutely necessary. This choice also assures that the voltage inverter is never used for logical NOT. A corollary is that logical NOT is performed along a wire if and only if the voltages representing truth are different at the two ends.

We already have useful symbols for AND and OR; these correspond to real devices in a circuit. Since we generate logical NOT without a real device, we need a notation to display logical inversion. In a diagram, logical NOT appears as a slash along the line. In a logic diagram, the slash is necessary to display the logic. In a circuit diagram that shows both voltage and logic, the slash is not strictly necessary, since we may infer logic inversion from the difference in voltage representations at the two ends of the wire. However, we always include the slash in mixed-logic circuit diagrams. Formally, the slash shows the point at which the voltage representation changes, and so one side of the slash should have a circle representing T = L. On either side of the slash we have our usual identity operation; logical inversion occurs at the slash:

The exact point along the wire at which we put the slash is arbitrary; the designer chooses a convenient place.

When we need the same voltage representation for the input signal and its inverted output, we may insert a voltage inverter to the left or to the right of the slash:

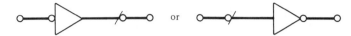

In practice, we usually omit the T = L circle on the logic-inversion slash, since it conveys no information that cannot be gained by following the wires from the slash to the ends. In this book, we will use the slash without the circle except, in a few cases, for emphasis.

Once again, note carefully that the voltage-inverting device performs no change in logic; the same logic variable is implemented on both sides of the voltage inverter, but with different voltage polarities. The device inverts voltage, not logic. To avoid using the term "inverter," which smacks of logic inversion, we sometimes refer to the voltage inverter as the "oops" function, implying that we are satisfied with our logic variable, but, "oops," we need to switch voltage levels.

Let's implement $Z = X \cdot Y$, making input X available as signal X.L (i.e., T = L), and Y available with T = H. We would prefer to have a physical device that performs like this:

No commonly available integrated circuit gate behaves this way. If instead we choose to use the 74LS02 Nor Gate, the circuit is:

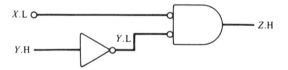

We inserted a voltage inverter to change signal Y.H into Y.L, the input required for our AND gate. You will find that adding voltage inverters where necessary requires no active thought; the little mixed-logic voltage-polarity circles direct you to do the right thing.

The 74LS04 Inverter is not the only way to invert voltage. Spare Nand and Nor gates provide several ways, such as those shown in Fig. 2–12. You may verify the validity of these methods by recalling the logical behavior of AND and OR. (Integrated circuit data books describe physical considerations that determine whether you should tie the input signal to both input terminals of the gate or should fix one input to H or L.)

This is the entire mixed-logic theory. We have laid a firm foundation for achieving our original circuit-design goals. We have tools to realize logical AND, OR, and NOT. We have notations that show the voltage representing truth at every point in the circuit and that show each physical device. We are able to keep a strict separation of logic and voltage in our diagram—all the logic is there and all the voltage behavior is there. These mixed-logic notations and theory can be applied immediately to more complex digital building blocks. As

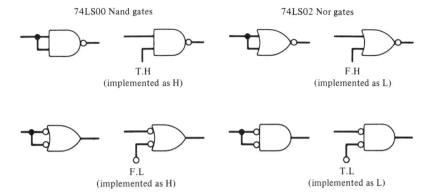

74LS00 Nand gates   74LS02 Nor gates

T.H   F.H
(implemented as H)   (implemented as L)

F.L   T.L
(implemented as H)   (implemented as L)

Figure 2–12. Nand and Nor gates used as voltage inverters.

we now turn to the analysis and synthesis of mixed-logic circuits, you will see that the mixed-logic methods allow the creation of a circuit from the original logic equation (synthesis) as well as the recovery of the original logic equation from the circuit (analysis), an advantage shared by no other method.

BUILDING AND READING MIXED-LOGIC CIRCUITS

Analyzing Mixed-Logic Circuits

In setting the goals of our circuit design methods, we said that a good circuit diagram should present the logic in a way that allows the reader to retrieve the designer's original expression. Mixed logic fulfills this condition; analyzing a mixed-logic circuit is simple. Here is the prescription: ignore circles and inverters (since they perform no logic in themselves), interpret the slash as logical NOT and the AND and OR symbols as logical AND and OR, and read the original logic equation from the diagram. For instance, read the circuit in Fig. 2–13 to recover the logic equation

$$Y = A \cdot \overline{B} + (\overline{C + D})$$

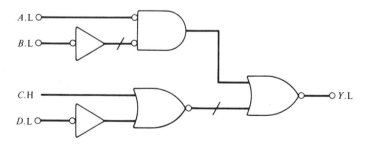

Figure 2–13. Implementation of $Y = A \cdot \overline{B} + (\overline{C + D})$.

In words, the thought process is

$$Y \text{ equals } A \text{ and } B\text{-NOT or } (C \text{ or } D)\text{-NOT}$$

If you feel the need for additional information on the diagram, insert the intermediate signal names and the missing circles on the slashes. You will quickly develop the skill to read a mixed-logic circuit without these extra legends.

Synthesizing Mixed-Logic Circuits

To implement a given logic equation, begin by sketching a picture of the equation, using the AND and OR logic symbols, and using the slash for logical NOT. Label the inputs and outputs with appropriate names of logic variables. At this point, the emphasis is on the logic, and it is this initial logic framework that allows us to analyze completed circuits with relative ease. Next, if the input or output variables have fixed voltage representations, add the appropriate .H, or .L and its companion circle. This step converts some of the logic variables into signals and fixes the voltage representation on some of the lines. You may assume that logic variables without a stated representation are available in either signal form, and you may use this flexibility in synthesizing the circuit. Usually, the devices available for implementing the AND and OR operations will be restricted. The tighter the restrictions, the easier the synthesis. For instance, if you are given only 74LS00 Nand gates, the synthesis is straightforward however complex the equations. There is no flexibility, since the only symbols available are:

This one chip is sufficient: we have a device for AND, one for OR, we can obtain voltage inversion with the 74LS00 if necessary, and we need no device for logical NOT.

With a wide choice of building blocks, we have more flexibility, and therefore more decisions. We wish to optimize the circuit. An obvious criterion for optimization is to minimize the number of voltage inverters, but other factors are often relevant, such as using up surplus gates on chips. When a variety of chips is available, the designer may display virtuosity in developing a satisfying implementation of the original equation.

Producing a valid circuit design is easy; producing an aesthetic design is an art. Nevertheless, the general process is straightforward. Select a likely chip for a gate, insert the corresponding mixed-logic symbol by adding required circles to the logic symbol, add voltage inverters where needed to make the circles in the design conform to the requirements of the logic, and then move to a neighboring element. As the synthesis proceeds, you may notice that a different choice of chip in a previous step results in fewer inverters. Within reason, you would probably wish to backtrack and introduce the change. In a short time, the circuit will converge to an acceptable solution.

AND and OR functions of more than two variables can be handled either with multi-input gates or with gates with fewer inputs. For instance, either of the forms in Fig. 2–14 describes a three-variable AND function. Design considerations dictate the choice; the original logic is not affected.

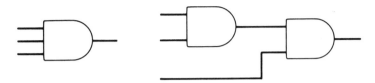

Figure 2–14. Realizing a 3-input AND.

Circuit drafting conventions. Circuit diagrams show the logic of the digital system; they frequently also specify the hardware. The information on the diagram varies with its intended use. On circuit diagrams that document actual constructions, each gate must display four hardware items:

(a) An arbitrary label for the specific chip containing the gate, for example, A13.
(b) The integrated circuit device type, for example, 74LS04.
(c) The pin numbers for each gate input and output.
(d) The function of each pin.

The labels for the chip and the type of device appear either within the symbol for the device or close by. The pin numbers appear outside the device symbol, close to each input and output line; and pin-function labels go within the symbol, adjacent to the corresponding input or output.

The mixed-logic symbols for realizations of AND and OR are usually sufficient to identify the type of device. Our habit is to omit explicit device labels where no confusion exists, in order to unclutter the diagrams, but specifying device types is always proper.

Pin-function labels are not necessary for most gates, since the symbol itself specifies the function of each input and output. On more complex integrated circuits, function labels are helpful and often necessary for clarity. We omit pin-function labels on gates, but include the labels on other circuit symbols.

On circuit diagrams intended for actual construction, we must include chip labels and pin numbers even if we choose to delete device-type notations. Most of the diagrams in this book deal with the logical properties of circuits rather than with the detailed hardware; usually, we omit pin numbers and chip labels to improve the readability of the diagrams.

A chip usually contains several gates; for example, the 74LS02 has four Nor gates. A common drafting convention is to label each gate as a fraction of a whole chip; for instance, each 74LS02 Nor gate is labeled ''¼ 74LS02,'' or ''¼ 'LS02.'' Circuit diagrams easily become top-heavy with notation, and in this book we omit the fraction unless there is a real possibility of confusion.

You may adopt our conventions or develop your own. In the next section, to illustrate the full notations, we display each gate with a chip label, device type (with fraction), and pin numbers. Thereafter, we preserve only information useful to the context.

Mixed-logic style. Experience in synthesizing mixed-logic circuits leads to concern about several points of style. We present a few points here and you will discover others as your mixed-logic skills grow.

Our habit of dropping the circle adjacent to the / operator is a matter of convention; you are free to accept or reject this convention. The slash is itself a convention, since we may infer the existence of logical NOT from a diagram of a voltage circuit without the slash. Nevertheless, we strongly recommend that you use the slash to indicate logical NOT; it adds greatly to the clarity of the diagram.

In dealing with voltage inverters and voltage polarities in mixed-logic syntheses, opportunities for decisions occur. Consider a design for $K = A + B$, with input signals $A.L$ and $B.H$. Any implementation with readily available gates will have an inverter on one input. Figure 2–15 shows two designs. Which is better? The designer who will need signal $K.L$ later would probably choose Fig. 2–15b. In other situations, Fig. 2–15a would be appropriate.

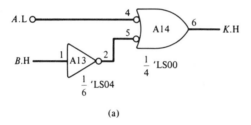

(a)

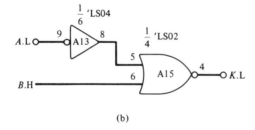

(b)

Figure 2–15. Two circuits for $K = A + B$ with signals $A.L$ and $B.H$. The diagrams in this section are fully labeled. The choice of a particular gate within a chip is arbitrary.

The next illustration is more clear-cut. To realize $G = A + \overline{B}$, again with inputs $A.L$ and $B.H$, a good circuit is

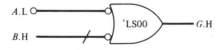

The circuit has no inverters. Designs using more gates are inferior, even if they perform the same logic; for example, the following circuit has two inverters:

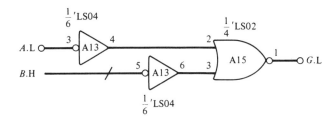

Even if we desire *G.L* as the output in preference to *G.H*, the second circuit is still bad, since for any voltage signal we are always only *one* inverter away from its other voltage counterpart: we could produce *G.L* from the first circuit by adding an inverter to the output. We might find some merit in the second circuit if, encountering it at the end of the entire design, we find a 74LS02 gate, some 74LS04 inverters, and *no* 74LS00 gate left over. Then the poor design might use up spare gates without the need for introducing a new 74LS00 chip. Don't look too hard for this kind of circuit optimization. Gates are cheap, and your brain is usually better used in thinking about other aspects of the design.

Be wary of subtly changing the logic with your drafting notations. Figure 2–16 shows three partial circuits, each an implementation of the equations $M = J \cdot \overline{X}$ and $N = G + X$. These circuits have identical wires and gates, and therefore must perform equivalent logic. But remember the guideline for digital drafting: the circuit must display the *original* logic. Only Fig. 2–16a is a proper implementation of the original logic equations. It is unlikely that the double logic inversion of *X* in Figs. 2–16b and 2–16c would have survived earlier simplifications of the logic. We would not expect to see such a form presented for construction.

We have found the following rule to be helpful in drafting logical NOT in complex logic circuits:

Place the slash as far to the right as practical on its signal line.

Frequently, an input to a device will always be fixed at F or at T. This often occurs with the complex chips described in Chapters 3 and 4, where you may wish never to enable a certain chip feature, or always to enable it. Suppose that we wish to say "never do it"—we wish to make a signal permanently FALSE. Here are two mixed-logic representations for the F operation:

In this unchanging operation, the voltage on the wire never varies. For convenience, we sometimes show only the voltage level (H or L) on the diagram:

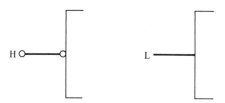

A similar analysis yields two obvious forms for the "always do it" operation. Part II contains examples of these shorthand notations.

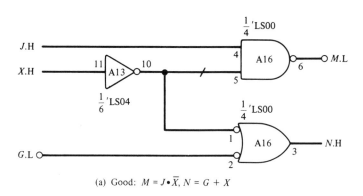

(a) Good: $M = J \cdot \overline{X}, N = G + X$

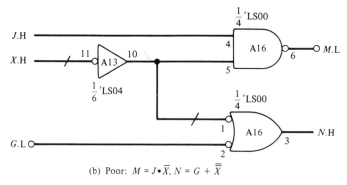

(b) Poor: $M = J \cdot \overline{X}, N = G + \overline{\overline{X}}$

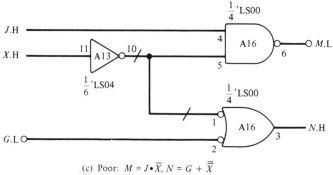

(c) Poor: $M = J \cdot \overline{X}, N = G + \overline{\overline{\overline{X}}}$

Figure 2–16. A good circuit and two poor circuits for $M = J \cdot \overline{X}, N = G + X$.

Other Mixed-Logic Notations

The presence or absence of the little circle in mixed logic shows the *voltage polarity* at each node in the diagram. Another popular mixed-logic notation for circuit diagrams is a small triangle to show the polarity of the voltage. A triangle lying above the line means T = H at that point; a triangle below the line means T = L. Figure 2–17 illustrates this notation.

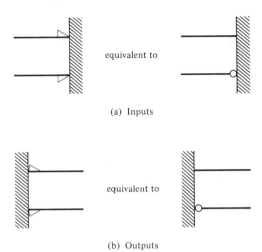

(a) Inputs

(b) Outputs

Figure 2–17. An alternative mixed-logic notation.

THE POSITIVE-LOGIC CONVENTION

The positive-logic convention has been widely used in engineering practice. The convention is easy to learn, but this ease of learning is deceptive, because designing real circuits with positive logic requires clumsy, constricting rules and transformations.

When the positive-logic convention is used, logical TRUE is represented everywhere by a high voltage, and FALSE is represented by a low voltage. Under these conditions, the voltage table for the 74LS02 Nor Gate can be transformed into the truth table for the logical NOR function:

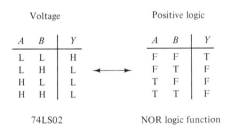

Voltage				Positive logic		
A	*B*	*Y*		*A*	*B*	*Y*
L	L	H		F	F	T
L	H	L		F	T	F
H	L	L		T	F	F
H	H	L		T	T	F
74LS02				NOR logic function		

Similarly, the voltage tables for the 74LS00 Nand Gate, the 74LS08 And Gate, and the 74LS32 Or Gate can be translated into truth tables for logical NAND, AND, and OR, respectively. A voltage inverter, for instance the 74LS04, performs the logical inversion function. In positive logic, we also may achieve logic inversion with Nand and Nor gates by feeding the same signal into both gate inputs; therefore in what follows we are free to use Nand or Nor gates or inverters for logic inversion. An application of De Morgan's law shows that the NAND function may be transformed into the inverted-input OR function:

$$A \text{ NAND } B = \overline{A \cdot B} = \overline{A} + \overline{B}$$

Similar transformations of NOR, AND, and OR produce inverted-input AND, inverted-input NOR, and inverted-input NAND logic functions, respectively.

In positive logic, since logic and voltage are tightly bound together, picking a physical gate is equivalent to picking a particular logic function. Nand and Nor gates arise naturally in most transistor-based technologies; fabricating And and Or gates requires the insertion of voltage inverters within the chips. The positive logician, in dealing with Nand and Nor gates, implements NAND and NOR logic or their De Morgan counterparts.

In the customary notation for circuits based on positive logic, a small circle represents the *logical inversion* operation. Figure 2–18 shows positive-logic interpretations of some common gates.

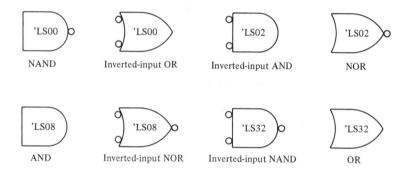

Figure 2–18. Interpretation of gates in the positive-logic convention.

To realize a logic equation, the positive-logic designer must transform the logic, either algebraically or graphically, into a form that corresponds to the chosen gates; in the process, the flavor of the original logic equation vanishes. An algebraic approach involves repeated applications of De Morgan's law to recast the original logic expression into one using the positive logic supported by the chips. Consider the following equation, to be rewritten so as to accommodate two-input Nor gates:

$$Y = A + B \cdot C \cdot \overline{D}$$

We may transform this equation into one involving only NOR operators (or NOR and inversion operators, if desired), by any of several sequences of tedious steps. Here is one such sequence:

Tools for Digital Design   Part I

$$Y = A + B \cdot \overline{C} + \overline{\overline{D}}$$

$$= A + \overline{\overline{B} + \overline{\overline{C}} + D}$$

$$= A + \overline{\overline{B} + \overline{\overline{C}} + D}$$

Such transformations are prone to error. Figure 2–19 is a positive-logic circuit diagram of the final form.

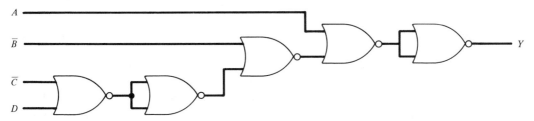

Figure 2–19. Positive-logic implementation of $Y = A + B \cdot C \cdot \overline{D}$, using De Morgan's law and Nor gates.

Digital-design textbooks in which positive-logic techniques are used contain prescriptions for drafting circuit diagrams using Nand and Nor gates without performing the algebraic transformations. There are separate sets of rules for using Nand gates alone, Nor gates alone, and certain combinations of Nand, Nor, And, and Or gates. The prescriptions require starting with a sum-of-products or product-of-sums logic expression, thereby forcing upon the designer a particular form of logic expression *solely to achieve a hardware circuit*.

We will illustrate circuits based on positive logic using the simplest set of rules—those for pure Nand-gate synthesis of two-level circuits. (The term *level* refers to the maximum number of gates encountered from an input to the output.) It is assumed that multiple-input Nand gates are available. The rules are:

1. Simplify the function as a sum of products.
2. Draw a Nand gate for each product that has at least two variables. The variables form the inputs to the Nand gate. These Nand gates are the first level of gates.
3. Draw a single Nand gate, using either the NAND or inverted-input OR graphic symbol at the second level. The inputs come from the first-level outputs.
4. For a product consisting of a single variable, insert an inverter at the first level; alternatively, use the complement of the variable as an input to the second-level Nand gate.

Applying these rules to the preceding example produces the two-level result shown in Fig. 2–20.

Rules for other two-level syntheses are more complex in application. Positive-logic textbooks do not contain rules for building a circuit for a general multi-

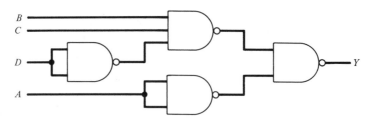

Figure 2–20. Positive-logic implementation of $Y = A + B \cdot C \cdot \overline{D}$, using Nand-gate rules.

level logic expression using arbitrary choices of gates—the rules would be too cumbersome.

To the mixed logician, these techniques are obnoxious and unnecessary. With mixed logic, we can readily develop a circuit for any logic equation, of any complexity, using any desired chips, while preserving the original structure of the logic equation. In mixed logic, there are no special cases for different types of gates.

In subsequent chapters, where we encounter complex integrated circuits with various true-high and true-low inputs and outputs, the benefits of the mixed-logic notation are even more pronounced.

Reading Positive-Logic Circuit Diagrams

The mixed-logic concept of performing a logic operation (NOT) without a physical device, and the related concept of a physical device (the inverter) that performs no logic, evolve directly from our insistence that the original logic be visible in the circuit diagram. In the positive-logic and negative-logic conventions, where logic and voltage are rigidly tied together, the inverter performs *logic* inversion. We sometimes wish to extract a logic equation from a fixed-convention circuit. As we have shown, it is not possible to recover the original logic; the circuit shows only a transformed version of the original. Can a mixed logician read a positive logic circuit? Certainly; there are several approaches. One method is to read the positive-logic diagram directly, interpreting Nand gates as logical NAND (NOT AND), Nor gates as logical NOR (NOT OR), inverters as logical NOT, and so on. Then write a logic equation from the circuit. Using this approach, we write A NAND B as $\overline{A \cdot B}$, and A NOR B becomes $\overline{A + B}$. If you are unhappy with the number of logic inversions in the result, then apply De Morgan's law to try to eliminate some of them. Similar approaches allow us to read negative-logic diagrams.

As an example, consider Fig. 2–21, a positive-logic-convention circuit. Following the prescription, we have

$$K = \overline{\overline{(\overline{A} \cdot B)} \cdot \overline{(B + \overline{C})}} \tag{2-1}$$

$$= \overline{(A + \overline{B}) \cdot (B + \overline{C})}$$

$$= \overline{(A + \overline{B})} + \overline{(B + \overline{C})}$$

$$= \overline{A} \cdot B + \overline{B} \cdot C \tag{2-2}$$

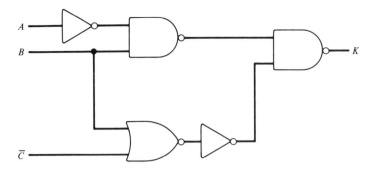

Figure 2–21. A circuit in the positive-logic convention. The circles represent logic inversions.

When do we stop manipulating the expressions for K? From the circuit in Fig. 2–21, we don't know. All the above Boolean algebraic forms of K are legitimate, but we have no way to tell what the designer originally had in mind. We can be fairly sure it was *not* Eq. (2–1)!

This method of reading positive-logic-convention diagrams produces a Boolean equation from the unmodified circuit diagram and transforms the equation into a more tractable form. Another method is to transform the positive-logic-convention circuit diagram into mixed logic, from which we read off an equation. This is a graphical method, whereas the first method is algebraic. For the graphical method, take the following steps. First, append .H to the positive logic inputs and outputs. If you wish, replace a negated input or output by the non-negated mixed-logic form. For example, an input $\overline{G}$ becomes a mixed logic $\overline{G}$.H, and you may express this as G.L with a circle on the input line. Next, wherever the circles do not match at the ends of a line, insert a slash to emphasize the implied logical NOT. Where a gate is surrounded by slashes, you may simplify the solution by altering the AND or OR gate symbol to its mixed-logic OR or AND counterpart. This is an application of De Morgan's law, and on the diagram the result is an inversion of circles and a change of the logic symbol to its dual. The circle inversions require rectification of the slashes on the gate input and output lines, leading to a simpler circuit. The process is shown in Fig. 2–22. After this conversion to a mixed-logic circuit, reading the logic is simple.

In Fig. 2–23 we depict a conversion of the previous example to mixed logic. The result is the same as Eq. (2–2). Again, we cannot be sure if this is the designer's original equation, since Fig. 2–21 does not preserve the original equation.

We might investigate how a mixed logician would handle the original synthesis problem. Equation (2–2) represents one of the possible starting points. In a mixed-logic synthesis for this equation, one choice for inputs is A.H, B.H, and C.L, in accordance with Fig. 2–21, and we might assume that the form of output K is unspecified. Figure 2–24 shows a mixed-logic circuit for Eq. (2–2). Because we permit K to appear in either signal form, Fig. 2–24 contains one less inverter than Fig. 2–21. Whether this ends up saving a gate will be determined by the needs of the larger design. We may require only signal K.H later, and thus the

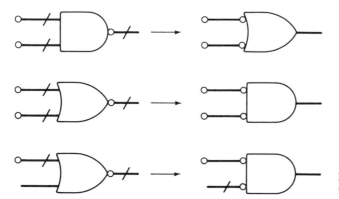

Figure 2–22. Some logic transformations on gates.

saved 74LS04 would reappear to convert $K.L$ to $K.H$. However, the mixed logician is not preoccupied with voltages; the drafting conventions handle voltages automatically. The mixed-logic method *creates the opportunity* for saving gates.

In our experience, for a given logic expression, mixed logic always produces a result that is at least as economical of hardware as other systems. Mixed-logic circuits often save hardware while preserving the original logic in the diagram.

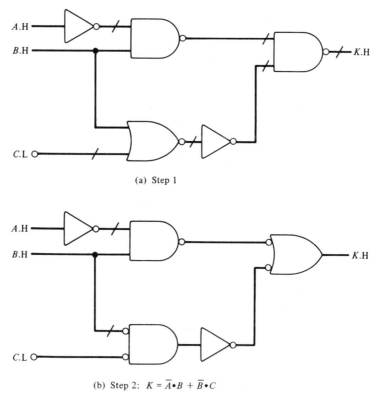

(a) Step 1

(b) Step 2: $K = \overline{A} \cdot B + \overline{B} \cdot C$

Figure 2–23. Converting Fig. 2–21 to mixed logic.

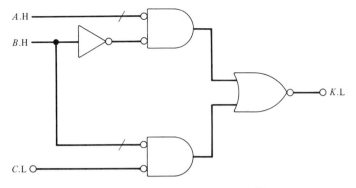

Figure 2–24. A mixed-logic circuit for $K = \overline{A} \cdot B + \overline{B} \cdot C$.

We close the discussion of other drafting conventions with a warning. In some parts of the electrical engineering community, "logic 1" and "logic 0" refer to the high-*voltage* level and the low-*voltage* level, respectively. This is an old use of the word *logic,* originating in the early days of digital circuits to distinguish a voltage level (a range of voltage) from an exact value of voltage. The terminology illustrates the confusion that arises when logic values and voltage levels are not kept as separate concepts. This jargon is giving way to more modern concepts of separate logic and voltage, but you will sometimes encounter the old usage, so be alert when you deal with existing documentation and when you talk with other hardware designers.

MIXED LOGIC FOR OTHER LOGIC FUNCTIONS

EXCLUSIVE OR and COINCIDENCE

We have stressed AND, OR, and NOT as natural logic elements for designers. Are there other Boolean functions of two variables that are used intuitively in designs? Two more functions are of sufficient value to be included in our set of simple logic building blocks. These correspond to our concepts of "different" and "same." The EXCLUSIVE OR (EOR) logic function is true only if its two inputs have different logical values; one input is true while the other is false. The COINCIDENCE (or EQUIVALENCE) function is true only when both of its inputs are the same—both true or both false. Logic operator notations and drafting symbols for these logic operations are

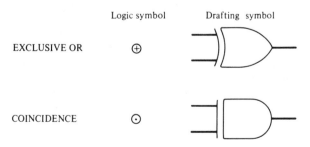

These operators have the following truth tables:

		Logic	
A	B	$A \oplus B$	$A \odot B$
F	F	F	T
F	T	T	F
T	F	T	F
T	T	F	T

Observe that $\odot$ is the inverse of $\oplus$. The methods used in Chapter 1 yield Boolean equations for these functions in terms of the familiar AND, OR, and NOT operators:

$$A \oplus B = \overline{A} \cdot B + A \cdot \overline{B}$$
$$A \odot B = \overline{A} \cdot \overline{B} + A \cdot B$$

These are well-known and useful expansions of EOR and COINCIDENCE. Memorize them. In the evaluation hierarchy for logical operators, EOR and COINCIDENCE fall below NOT and above AND.

A glance through a TTL data book reveals a chip with a promising name, the 74LS86 Quad Two-Input Exclusive Or Gate, whose voltage table is

74LS86 Voltage		
Inputs		Output
L	L	L
L	H	H
H	L	H
H	H	L

Our usual method of comparing a logic truth table with a device's voltage table produces an amazing result: the 74LS86 yields four mixed-logic realizations for the EXCLUSIVE OR logic function! The drafting symbols are:

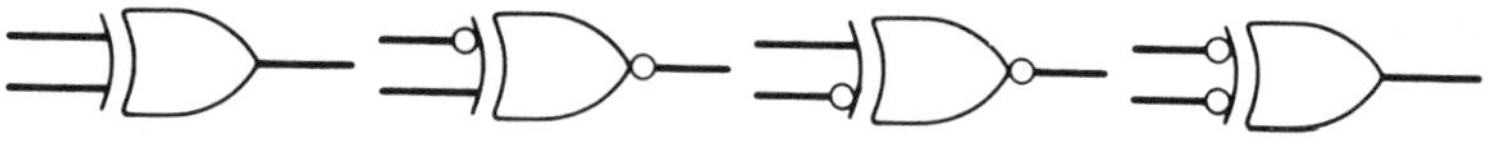

You should verify that substituting the indicated voltage values for the T and F in the truth table will in each case give a voltage table that is equivalent to that of the 74LS86.

This is not all—the 74LS86 also gives four representations of the logical COINCIDENCE operator. Here are the symbols:

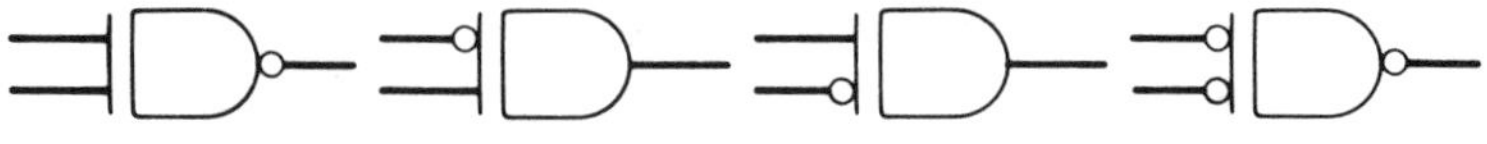

Notice the pattern: the EOR symbols have an even number of circles, whereas the COINCIDENCE symbols have an odd number of circles. This marvelous 74LS86 chip gives eight building blocks for our drafting kit. We may use any of the four EOR symbols, depending on the requirements for our particular circuit, and still perform EOR logic. For instance, Fig. 2–25 is a drawing of two syntheses of a logic equation involving EOR. With mixed logic, these designs arise naturally, with no mental effort wasted on logic–voltage interrelations.

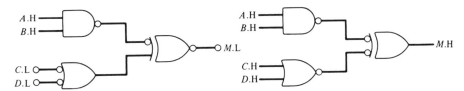

Figure 2–25. Mixed-logic circuits for $M = (A \cdot B) \oplus (C + D)$.

Such elegant and useful results send a thrill of joy through the mixed logician. These eight easily remembered symbols allow us to produce EOR and COINCIDENCE in a simple, efficient manner.

Open-Collector Gates

In typical TTL integrated circuits, we cannot connect two outputs directly together; such an act results in improper electronic circuit operation and may damage the gates. If one output is in its low-voltage state and the other is in its high-voltage state, an "output fight" ensues. High current passes from V_{CC} to ground through both chips. To merge logic signals, we normally use an OR or an AND gate; the choice depends on the function we wish to perform. However, the *open-collector* branch of TTL technology allows us to tie outputs together without using an explicit AND or OR gate, thereby creating a *wired OR* or *wired AND*. Open-collector gates are an interesting use of mixed-logic theory.

The technology of regular and open-collector TTL circuits is discussed in Chapter 12; for now it is sufficient to describe the effect of the open collector. In an open-collector gate, the output circuit is incomplete—the output has no internal path to V_{CC} as in regular TTL gates. A regular TTL gate supplies solid voltages for both high and low output states. The open-collector gate can supply a solid low-voltage level when this is desired, but in its high-output state the gate acts as if it is disconnected. Many open-collector outputs can be wired together without electronic conflict, even if the outputs are in different states; the lows dominate, and the highs act as if they are not there. Any low output voltage will cause the voltage of the entire output circuit to be low. To provide a valid high-voltage level, the designer of an open-collector circuit must supply a single connection to V_{CC}. This is done with a *pull-up resistor*. The function of the resistor is to establish a valid high-voltage level when none of the open-collector outputs is in the low-voltage state. With the addition of the pull-up resistor, the open-collector circuit produces a normal TTL-level output that can serve as an input to one or more TTL-compatible circuits.

With this voltage behavior, what logic is performed? We think of each open-collector output as an input to a logic box that actually contains wires connected together and a pull-up resistor. What is the output of this logic box?

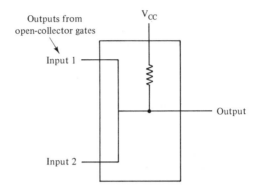

When two open-collector outputs serve as inputs to the box, the voltage table is:

$Input_1$	$Input_2$	Output
L	L	L
L	H	L
H	L	L
H	H	H

If we choose the T = H convention at all inputs and outputs, this voltage table transforms into the truth table for AND; when T = L at all inputs and outputs, we get the OR truth table. The results are the same when there are more than two inputs to the circuit. Here is the fundamental logical behavior of open collectors: open-collector circuits perform a wired AND when high voltage equals truth, and a wired OR when low voltage equals truth.

Let's consider a specific open-collector chip. The 7406 Open-Collector Inverter is a typical gate of this type. To get the wired-OR behavior, T must equal L at each 7406 output in the circuit. (Of course, this means that T = H at the inputs to the 7406, but this has no bearing on the open-collector behavior.) In Fig. 2–26 we show open-collector circuits for a wired OR and a wired AND, using the 7406 Inverter. The figure also shows the customary graphic notation for open-collector outputs—a vertical line within the gate symbol, near the output. In Fig. 2–26, the logic is emphasized with a dotted-line symbol; in actual circuit diagrams you may use such a notation or omit it.

Open-collector circuits are now used only in a few special applications. You will encounter several of these later in this book. Open-collector circuits are used only rarely to build combinational logic functions, but such uses provide such a satisfying display of the power of mixed logic that we include one example.

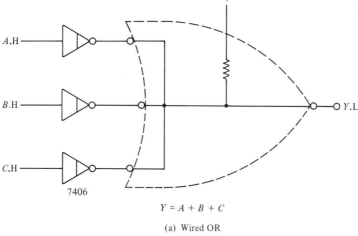

$$Y = A + B + C$$

(a) Wired OR

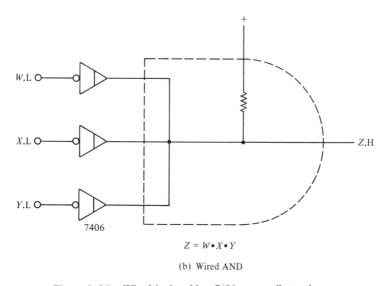

$$Z = W \cdot X \cdot Y$$

(b) Wired AND

Figure 2–26. Wired logic with a 7406 open-collector inverter.

Suppose we wish to design a circuit to satisfy the equation

$$OUT = A \cdot B \cdot C + D + E$$

using the 7406 Open-Collector Inverter. The mixed logician proceeds as follows. We require a three-input OR circuit, one of the inputs being a three-input AND. The wired AND and the wired OR will each require a pull-up resistor. At the output of the wired AND, $T = H$, but T must equal L at the contributors to the wired OR. Therefore, we know we must introduce a device to change voltage conventions, and this device must produce an open-collector output to feed into the wired OR. The 7406 is a perfectly good voltage inverter that produces an open-collector output, and so our schematic takes the form of Fig. 2–27. Since

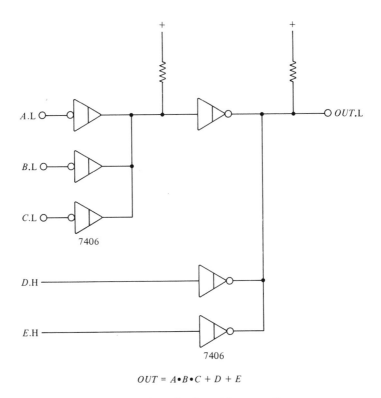

$$OUT = A \cdot B \cdot C + D + E$$

Figure 2–27. A logic realization with open-collector gates.

the 7406 contains six open-collector inverter gates, this design requires only one chip and two pull-up resistors.

The Less-Used Logic Functions

To close our discussion of the use of mixed-logic methods for simple logic functions we will examine how we might directly realize each of the Boolean functions of up to two variables. Hardware designers make heavy use of AND, OR, NOT, EOR, and COINCIDENCE, since each represents a construct that humans use in their everyday lives. There are other Boolean functions of two input variables, seldom used by designers but of interest to symbolic logicians. A truth table for a function with two inputs has four rows; $2^4 = 16$ different output functions are possible with two inputs.

Table 2–1 contains truth tables for these 16 functions, $Z0$ through $Z15$. You are already familiar with several of these: $Z1$ implements the AND function, $Z6$ is EOR, $Z7$ is OR, and $Z9$ is COINCIDENCE. We know that mixed logicians have several ways to handle these functions; one is to use the 74LS00 Nand Gate and the 74LS86 Exclusive Or Gate. Several other functions listed in Table 2–1 are available to mixed logicians "for free": $Z0$ and $Z15$ are the FALSE and TRUE logic functions, easily created by connecting a wire to an appropriate voltage source. $Z3$, $Z5$, $Z10$, and $Z12$ are really functions of a single variable,

TABLE 2–1 LOGIC FUNCTIONS OF TWO VARIABLES

A	B	$Z0$	$Z1$	$Z2$	$Z3$	$Z4$	$Z5$	$Z6$	$Z7$	$Z8$	$Z9$	$Z10$	$Z11$	$Z12$	$Z13$	$Z14$	$Z15$
F	F	F	F	F	F	F	F	F	F	T	T	T	T	T	T	T	T
F	T	F	F	F	F	T	T	T	T	F	F	F	F	T	T	T	T
T	F	F	F	T	T	F	F	T	T	F	F	T	T	F	F	T	T
T	T	F	T	F	T	F	T	F	T	F	T	F	T	F	T	F	T

representing the identity operation on A and B, and the inversion operation on B and A, respectively. These functions are available at no cost, with proper choices of voltage conventions.

The remaining six two-variable Boolean functions are less familiar, since they do not correspond closely to everyday concepts:

$$Z14 = A \text{ NAND } B$$
$$Z8 = A \text{ NOR } B$$
$$Z13 = A \text{ IMPLIES } B$$
$$Z11 = B \text{ IMPLIES } A$$
$$Z2 = \text{NOT}(A \text{ IMPLIES } B)$$
$$Z4 = \text{NOT}(B \text{ IMPLIES } A)$$

We found that the 74LS86 chip would generate two logic functions, EOR and COINCIDENCE, each in four different ways. All eight ways of choosing input and output voltage conventions produce realizations of these two functions. Now consider the 74LS00 Nand Gate. We have used only two of its eight possible choices of voltage representation—one to implement AND and another to implement OR. What logic functions will the other six ways implement?

In Fig. 2–28 we display all of the eight possibilities, using a logician's notation for the operators. Rearrange the rows of the truth tables into the more familiar form and compare the tables with those of Table 2–1. You can see that the 74LS00 Nand Gate will, if we desire, produce eight different logic functions! In fact, given the mixed-logician's freedom to select useful voltage representations of the logic variables, we now have designs requiring no more than a single gate for each of the 16 Boolean functions of two variables—an elegant result.

In Fig. 2–28, notice how the mixed logician produces the logical NAND function with the Nand gate: no voltage-polarity circles on any input or output. In the positive-logic convention, NAND logic generated by the 74LS00 appears with an inversion circle on the output.

You may use chips other than the 74LS00 for a variety of logic functions; ways are suggested in the exercises at the end of the chapter. In Chapter 3 you will see an elegant circuit that will produce all 16 functions.

Our discussion of the foundations and methodology of mixed logic is complete. You have a design tool of unsurpassed power for realizing logic expressions of arbitrary form with any gates you wish to use. In the next chapter we explore more complex combinational building blocks. Each of these has one or more mixed-logic symbols to add to your drafting tool kit.

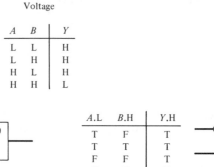

74LS00
Voltage

A	B	Y
L	L	H
L	H	H
H	L	H
H	H	L

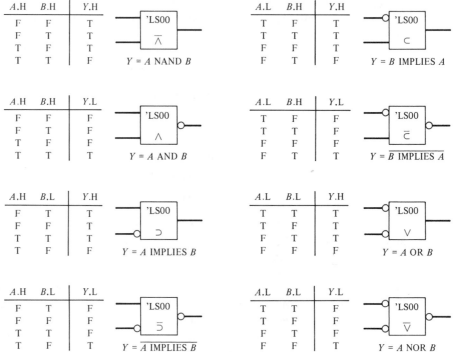

$A.H$	$B.H$	$Y.H$
F	F	T
F	T	T
T	F	T
T	T	F

'LS00 $\overline{\wedge}$
$Y = A$ NAND B

$A.L$	$B.H$	$Y.H$
T	F	T
T	T	T
F	F	T
F	T	F

'LS00 $\subset$
$Y = B$ IMPLIES A

$A.H$	$B.H$	$Y.L$
F	F	F
F	T	F
T	F	F
T	T	T

'LS00 $\wedge$
$Y = A$ AND B

$A.L$	$B.H$	$Y.L$
T	F	F
T	T	F
F	F	F
F	T	T

'LS00 $\overline{\subset}$
$Y = \overline{B \text{ IMPLIES } A}$

$A.H$	$B.L$	$Y.H$
F	T	T
F	F	T
T	T	T
T	F	F

'LS00 $\supset$
$Y = A$ IMPLIES B

$A.L$	$B.L$	$Y.H$
T	T	T
T	F	T
F	T	T
F	F	F

'LS00 $\vee$
$Y = A$ OR B

$A.H$	$B.L$	$Y.L$
F	T	F
F	F	F
T	T	F
T	F	T

'LS00 $\overline{\supset}$
$Y = \overline{A \text{ IMPLIES } B}$

$A.L$	$B.L$	$Y.L$
T	T	F
T	F	F
F	T	F
F	F	T

'LS00 $\overline{\vee}$
$Y = A$ NOR B

Figure 2–28. What logic does the 74LS00 Nand Gate perform?

READINGS AND SOURCES

DIETMEYER, DONALD L., *Logic Design of Digital Systems*, 2nd ed. Allyn & Bacon, Boston, 1978. Traditional positive-logic development.

FAST: Fairchild Advanced Schottky TTL. Fairchild Camera and Instrument Corporation, Digital Products Division, South Portland, Maine. Data book for the FAST series of TTL chips.

FLETCHER, WILLIAM I., *An Engineering Approach to Digital Design.* Prentice-Hall, Englewood Cliffs, N.J., 1980. Uses mixed logic.

HILL, FREDERICK J., and GERALD R. PETERSON, *Introduction to Switching Theory and Logical Design*, 3rd ed. John Wiley & Sons, New York, 1981. Traditional positive-logic development.

MALEY, G. A., and J. EARLE, *The Logic Design of Transistor Digital Computers*. Prentice-Hall, Englewood Cliffs, N.J., 1963. An influential work; look here (for example, at page 114) to see how difficult design was before the advent of mixed logic.

MANO, M. MORRIS, *Digital Design*. Prentice-Hall, Englewood Cliffs, N.J., 1984. Traditional positive-logic development.

PROSSER, FRANKLIN, and DAVID WINKEL, "Mixed logic leads to maximum clarity with minimum hardware," *Computer Design*, May 1977, page 111.

Systems Design Handbook, 2nd ed. Monolithic Memories, 2175 Mission College Blvd., Santa Clara, Calif. 95054, 1985. Manufacturer's data book espousing mixed logic (see pages 2–48).

The TTL Data Book. Texas Instruments, P.O. Box 225012, Dallas, Tex. 75265. A good source of information about TTL chips.

EXERCISES

2–1. Write the voltage tables for the following devices:

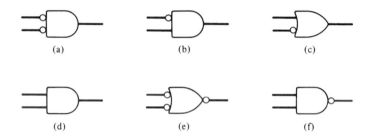

2–2. Using a TTL data book for reference, find physical devices (where possible) that correspond to the logic symbols of Exercise 2–1.

2–3. Many popular integrated circuits have 16 pins. What is the widest (most inputs) AND or OR gate that a 16-pin chip will accommodate? Base your answer solely on the number of pins. Look in a TTL data book to see if you can find any chips of this type.

2–4. What is positive logic? Negative logic? What is the positive-logic convention? The negative-logic convention? The mixed-logic convention?

2–5. Explain carefully the meaning of these mixed-logic notations:
LOADH
LOADH.H
LOADH.L

2–6. What is the difference in logic performed by the signals *XYZ*.H and *XYZ*.L?

2–7. True or false: The .H notation means that the voltage used to represent a given logic variable is high.

2–8. Without looking at a data book, give the probable complete chip name for an integrated circuit that contains elements represented by the following symbol:

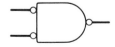

2–9. Why do we not wish to think of a 74LS00 Quad Two-Input Nand gate as generating the NAND logic function?

2–10. Give the dual of each of these mixed-logic symbols:

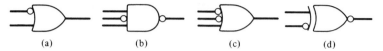

(a)   (b)   (c)   (d)

2–11. Why do we wish to maintain a strict separation of the concepts of logic and voltage?

2–12. How does a mixed logician denote logical NOT? What device is used?

2–13. Fill in all intermediate signal names on this mixed-logic circuit diagram:

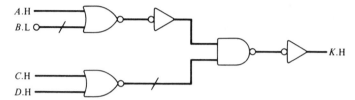

2–14. What logic does each of these circuit elements perform?

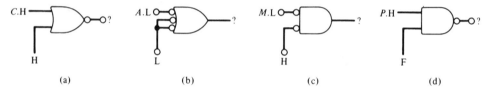

(a)   (b)   (c)   (d)

2–15. Here is a legal mixed-logic diagram:

What logic does this circuit implement? Why do we not prefer this notation? What is the preferred notation?

2–16. Explain the difference between synthesis and analysis.

2–17. State the prescription for analyzing mixed-logic circuits.

2–18. Analyze these circuits to produce the equations for the outputs:

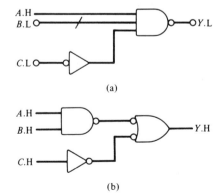

(a)

(b)

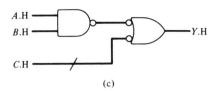

(c)

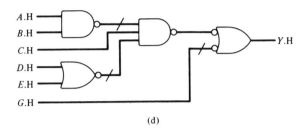

(d)

2–19. Synthesize the following equations using only the 74LS00 Quad Two-Input Nand Gate. The inputs are generated elsewhere as A.L, B.H, C.L, D.L, and E.H. The desired voltage representation for truth on the output is given with each equation.

    (a) $Y = A \cdot B + C \cdot D$ (Y.H)

    (b) $Y = (A + D) \cdot \overline{(B + E)}$ (Y.L)

    (c) $Y = A + C + B + E$ (Y.H)

    (d) $Y = \overline{A} \cdot B \cdot C + B \cdot E$ (Y.H)

    (e) $Y = A \cdot D$ (Y.L)

2–20. Repeat Exercise 2–19 with the same voltage conventions, but expand your catalog of available integrated circuits to include the 74LS02, 74LS04, and 74LS10 chips.

2–21. Synthesize the following equations, using any chip in the TTL data book. The available signals are A.H, B.H, B.L, C.H, D.H, and E.L. Let the output signal convention for each circuit be the natural one arising from your synthesis.

    (a) $Y = \overline{B} \cdot C \cdot D + E \cdot B \cdot A$

    (b) $Y = \overline{B} \cdot C \cdot D + \overline{E \cdot B}$

    (c) $Y = \overline{(A + B + C)} \cdot (D + E)$

    (d) $Y = (A + B + \overline{C}) \cdot D$

    (e) $Y = A + B + C + D + E$

    (f) $Y = (A \cdot B + \overline{C}) \cdot (D \oplus E)$

    (g) $Y = (A \cdot B + \overline{C}) \cdot (D \odot E)$

2–22. We may express any Boolean function using only the AND, OR, and NOT logic functions. Therefore, if we wish, we may build any Boolean function using only 74LS00 Nand gates. Why? Can we similarly implement any Boolean function using only 74LS08 And gates? Why or why not?

2–23. (a) Give expressions for $X \oplus Y$ and $X \odot Y$ in terms of AND, OR, and NOT operators.

    (b) By comparing truth and voltage tables, derive the four 74LS86 mixed-logic representations for $\oplus$.

    (c) In similar manner, derive the four 74LS86 representations for $\odot$.

2–24. Give efficient mixed-logic syntheses for the following logic equations, using only the specified gates and signal polarities. Do not perform any transformations on the logic equations.

(a) $OUT = (\overline{A} \cdot B + C) \odot (B \cdot \overline{D \cdot E}) + \overline{D}$, using two-input Nor gates (74LS02), two-input Exclusive Or gates (74LS86), and signals A.H, B.H, C.H, C.L, D.L, E.L, $\overline{OUT}$.H.

(b) $OUT = \overline{((X \odot (Y \cdot \overline{Z})) \cdot \overline{Z} + W)} \oplus P$, using two-input Or gates (74LS32), two-input Exclusive Or gates (74LS86), Inverters (74LS04), and signals P.H, P.L, W.H, X.H, Y.L, Z.H, and OUT.H.

(c) $X = A \oplus (\overline{B} + C \odot (\overline{D \cdot A}))$, using two-input Nor gates (74LS02), two-input Or gates (74LS32), two-input Exclusive Or gates (74LS86), Inverters (74LS04), and signals A.L, B.H, C.L, D.H, D.L, and X.L.

(d) $Y = (\overline{A \cdot B} + C) \cdot (D \oplus \overline{E} + B)$, using two-input Nand gates (74LS00), two-input Nor gates (74LS02), two-input Exclusive Or gates (74LS86), Inverters (74LS04), and signals A.L, B.H, C.L, D.H, D.L, E.L, Y.H, and Y.L.

2–25. Consider the function X of Exercise 1–33: X is true only when two 2-bit binary numbers A,B and C,D are different.

(a) Working directly from the definition of X, write a logic equation for X, making use of the EOR operator.

(b) Working directly from the definition of X, write a logic equation for $\overline{X}$, making use of the COINCIDENCE operator.

(c) Show how you may use De Morgan's law to transform the expression for X of part (a) into the expression for $\overline{X}$ of part (b).

(d) Show the equivalence of your results for parts (a) and (b) to the results of Exercise 1–33.

2–26. Derive a logic expression that is true if one 5-bit positive binary number A is less than another similar number B.

2–27. Many cars have an alarm buzzer that warns of unfastened seat belts, lights left on, and the key left in the ignition. The system might operate as follows:

> The alarm should sound if the driver's seat belt is not fastened when the motor is running, or if the passenger seat is occupied and the passenger's seat belt is not fastened when the motor is running, or if the lights are on when the key is not in the ignition switch, or if the key is in the ignition switch when the motor is not running and the driver's door is open.

Assign a meaningful name to each variable in the statement, and write a logic equation for the alarm buzzer's control signal. You may find it useful to break the statement of the buzzer's behavior into several statements and combine these statements to produce your final result.

2–28. You are to design a 5-input circuit, 4 inputs of which form the binary representation of a decimal digit (in other words, the BCD code D,C,B,A for a digit 0 through 9). The fifth line is a control signal CL. If the control signal is false, the single output of the circuit should be true only if the decimal input number is 4 or greater. If CL is true, the output should be the inverse of input bit C. You should be able to write an equation for the output without writing the large (32-row) truth table. Express the equation in a circuit, using chips of your choice.

2–29. In circuit diagrams, the small circle means dramatically different things to a mixed logician than to a positive logician. Explain.

2–30. To a positive logician, what logic does the inverter chip perform? To a mixed logician, what logic does the inverter chip perform?

2–31. Here is a circuit with the *positive-logic convention*.

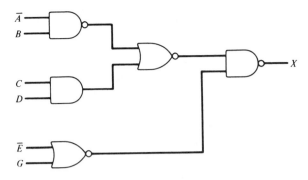

(a) Write a logic equation directly from the diagram.

(b) Convert the diagram to mixed-logic notation.

(c) Analyze the mixed-logic diagram, and show the equivalence of the result with that obtained in part (a).

2–32. Exercise 1–36 produced logic equations for the seven segments a through g of a decimal digit display. Design a circuit for each of these seven equations, making $T = H$ at all inputs and $T = L$ at each output. Use only 74LS00, 74LS02, and 74LS04 chips, and try to minimize the total *chip count* in the entire design. Don't overlook the possibility of sharing common terms among the seven equations.

2–33. Design a circuit embodying $OUT = A \cdot \overline{B} \cdot C + \overline{D} + E$, using only the 7406 Open Collector Inverter. You may choose the voltage representations of truth at the inputs and outputs.

2–34. Read the section in Chapter 12 on computing resistor values for open-collector circuits. Specify suitable values of any resistors in your solution to Exercise 2–33.

2–35. The Boolean implication function, denoted by $\supset$, is important in the study of symbolic logic, but not in digital design. From Fig. 2–28, you saw that it is possible to realize logical implication with a 74LS00 gate. Show a mixed-logic symbol that realizes $Y = A \supset B$, using a single 74LS02 Nor gate.

2–36. Suppose a 74LS32 Or Gate were used as indicated by the mixed-logic diagram

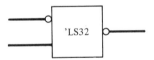

What Boolean logic function (of two inputs) does the output produce?

2–37. At one time, designers developed a complete logic family of inverters, AND and OR gates, and so on, based on fluid flow. These devices were proposed to immunize military devices from nuclear radiation. It was also suggested that fluid-flow devices might drive mechanical devices such as printers and card readers directly with fluid logic, bypassing the expensive conversion of electronic digital signals to mechanical control signals. The scheme is now only of historical interest, but it is interesting to contemplate how this class of device might fit in with our modern design methods. Suppose you are given a family of fluid-flow devices and are asked to build a digital system using them.

(a) What definition of fluid flow (either at rest or moving) would you choose to represent logic truth?

(b) Could you still use our standard drafting symbols to represent designs with fluid logic elements?

(c) Would there still be little circles? If so, what would they mean?

(d) On our standard logic diagrams, a line represents a wire or an electrical path. What would a line represent in a fluid-logic circuit?

2–38. Read the sections in Chapter 12 on integrated circuit data sheets and performance parameters. Now consider a 74LS04 Inverter chip. What can you say about the voltage on one of the inverter outputs if its input receives:

(a) A voltage of 0.6V?

(b) A voltage of 1.6V?

(c) A voltage of 2.6V?

2–39. Using the triangle notation of Fig. 2–17 for displaying mixed-logic voltage polarities, redraw Fig. 2–13.

3

Building Blocks for Digital Design

The construction of most digital systems is a large task. Disciplined designers in any field will subdivide the original task into manageable subunits—building blocks—and will use the standard subunits wherever possible. In digital hardware, the building blocks have such names as *adders, registers,* and *multiplexers.*

Logic theory shows that all digital operations may be reduced to elementary logic functions. We could regard a digital system as a huge collection of AND, OR, and NOT circuits, but the result would be unintelligible. We need to move up one level of abstraction from gates and consider some of the common operations that digital designers wish to perform. Some candidates are:

(a) Moving data from one part of the machine to another.
(b) Selecting data from one of several sources.
(c) Routing data from a source to one of several destinations.
(d) Transforming data from one representation to another.
(e) Comparing data arithmetically with other data.
(f) Manipulating data arithmetically or logically, for example, summing two binary numbers.

We can perform all these operations with suitable arrangements of AND, OR, and NOT gates, but always designing at this level would be onerous, lengthy, and error-prone. Such an approach would be comparable to programming every software problem in binary machine language. Instead, we need to develop building blocks to perform standard digital system operations. The building blocks will allow us to suppress much irrelevant detail and design at a higher level. The procedure is analogous to giving the architect components such as doors, walls, and stairs instead of insisting that he design only with boards, nails, and screws.

INTEGRATED CIRCUIT COMPLEXITY

In Chapter 2, you studied low-level building blocks—AND, OR, and NOT. These come in integrated circuit packages containing a few gates that are not interconnected and that can be used in synthesizing elementary circuits from Boolean algebraic equations. The industry's name for devices of this complexity, containing up to 10 gates, is *SSI (small-scale integration)*.

Digital designers find that operations such as those we listed above—(a) to (f)—occur in nearly any system design. As a result, manufacturers have provided integrated circuit packages of interconnected gates to perform these operations. Chips that contain from 10 to about 100 gates are called *MSI (medium-scale integration)*. Circuits with a nominal complexity of about 100 to 1,000 gates are classified as *LSI (large-scale integration)*, and circuits with more than the equivalent of 1,000 gates are called *VLSI (very-large-scale integration)*. Microprocessors and memory arrays are examples of VLSI integrated circuits, as are many special-purpose chips. The boundaries between the categories are only suggestive; in fact, the use of the term LSI as a measure of chip complexity is waning.

Each class of integrated circuit—SSI, MSI, LSI, and VLSI—is important at its appropriate level of design. The digital designer should try to design at the highest conceptual level suitable to the problem, just as the software specialist should seek to use prepackaged programs or a high-level language instead of assembly language when possible. In software programming, the accomplished problem solver has not only a knowledge of Fortran, but also of computer organization, system structure, assembly language, and machine processes. Similarly, to achieve excellence in solving problems with digital hardware, we need skill in using all our tools, from the elementary to the complex.

COMBINATIONAL BUILDING BLOCKS

Combinational and Sequential Circuits

In this chapter, we will develop a set of building blocks that have hardware implementations of the MSI or LSI level of complexity, and have no internal storage capacity, or "memory." Such circuits, with outputs that depend only on the present values of the inputs, are called *combinational*. The important class of circuits that depend also on the condition of past outputs is called *sequential*. We will present sequential circuits and sequential building blocks in Chapter 4. Table 12–1, at the end of Chapter 12, is a list of useful SSI and MSI integrated circuits.

The Multiplexer

A *multiplexer* is a device for selecting one of several possible input signals and presenting that signal to an output terminal. It is analogous to a mechanical switch, such as the selector switch of a stereo amplifier (Fig. 3–1). The amplifier switch is used for selecting the input that will drive the speaker. Except for the moving hand, the electronic analog is easily constructed. We use Boolean variables instead of mechanical motion to select a given input. Consider a two-position switch with inputs A and B and output Y, such as shown in Fig. 3–2. Introduce a variable S to describe the position of the switch and let $S = 0$ if the switch is up and $S = 1$ if the switch is down. A Boolean equation for the output Y is

$$Y = A \cdot \overline{S} + B \cdot S$$

Using this equation, we can build an electronic analog of the switch; Fig. 3–3 is one design. There, S is the *select* input.

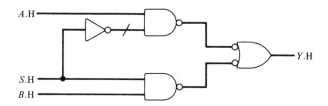

Figure 3–1. A mechanical selector switch.

Figure 3–2. A two-position mechanical switch.

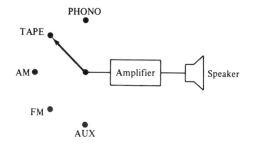

Figure 3–3. Implementation of the electronic switch $Y = A \cdot \overline{S} + B \cdot S$.

Commercial MSI devices correspond to the more complex switch shown in Fig. 3–4. The right-hand switch, called an *enable* (or *strobe*), acts as a Boolean AND function. If we call the closed position of the enable switch G and the open position $\overline{G}$, then

$$Y = G \cdot (A \cdot \overline{S} + B \cdot S) \tag{3-1}$$

Figure 3–4. An enabled selector switch.

Figure 3–5 is a schematic for a device based on this equation that behaves like common commercial devices. Such a circuit is called a 2-input multiplexer (*mux*); its mixed-logic symbol and the representation of Eq. (3–1) are shown in Fig. 3–6. Figure 3–7 is a diagram of a typical commercial device, the 74LS157 Quad Two-Input Multiplexer. This chip has four 2-input multiplexer units (designated 1 through 4) packaged in a single integrated circuit chip, with each mux sharing a common select input S and a common enable input En.

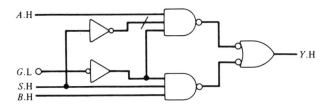

Figure 3–5. An enabled two-position electronic selector switch.

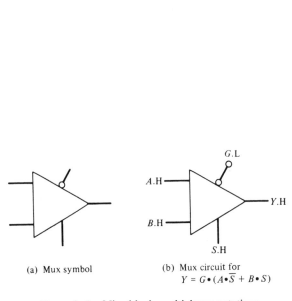

(a) Mux symbol

(b) Mux circuit for
$Y = G \cdot (A \cdot \overline{S} + B \cdot S)$

Figure 3–6. Mixed-logic multiplexer notations.

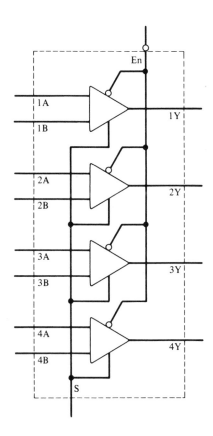

Figure 3–7. The 74LS157 Quad Two-Input Multiplexer.

If we desire to select an output from among more than two inputs, the multiplexer must have more than one select input. The select inputs to a mux form a binary code that identifies the selected data input. One select line has $2^1 = 2$ possible values; two select lines allow the specification of $2^2 = 4$ different values. For instance, if we have select lines $S1$ and $S0$, the pair $S1, S0$ represents a binary number that may identify one of four possible inputs:

$S1$	$S0$	Selected input position
0	0	0
0	1	1
1	0	2
1	1	3

Commercial integrated circuits provide 2-, 4-, 8-, and 16-input multiplexers, with a variety of inverted and noninverted outputs, separate and common enable inputs, and so on. You should consult the manufacturers' integrated circuit data books for detailed information about these useful devices.

The multiplexer select code represents an *address,* or *index,* into the ordered inputs. We may view the data inputs to the mux as a vector or table, and the select lines as an address. A multiplexer is thus a hardware analog of a 1-bit software "table look-up." Figure 3–8 illustrates the analogy. In systems design, table look-up is an important concept which hardware designers have not exploited to the same extent as programmers. In subsequent chapters, you will see many powerful uses of this concept. When you are faced with selecting, looking up, or addressing one of a small number of items, think MUX.

Table look-up, or input selection, is convenient for up to 16 inputs, using the appropriate MSI integrated circuit chip. For more than 16 inputs the solution is not as neat but is nevertheless systematic. Suppose that you need to do a look-up in a table of 32 entries. Divide the inputs into groups, say four groups of 8 entries each. Then we may view the 32-element look-up as consisting of two look-ups, one to select the proper group of 8, and the second to pick the correct member of that group. The selection index for the 32-element look-up is a 5-bit binary code of the form $S4,S3,S2,S1,S0$. Each group of 8 inputs requires 3 bits for specifying an input and 2 bits for picking the proper group of 8. Figure 3–9 shows a realization that uses four 8-input multiplexers and one 4-input mux.

The conventional symbol for a multiplexer shows the inputs as having T = H, but the output may be either high- or low-active, depending on the particular chip. As mixed logicians, we realize that we may present all the inputs in T = L form without affecting the circuit; then the output will be of opposite polarity to that in the conventional symbol for the device. In Fig. 3–10 we show the two equivalent mixed-logic forms for an element of a 74LS352 Dual Four-Input Multiplexer, a circuit whose output polarity is inverted. Changing the polarity of the inputs affects all the input lines and the output (all the *data* paths) but has no effect on the selection or enabling systems.

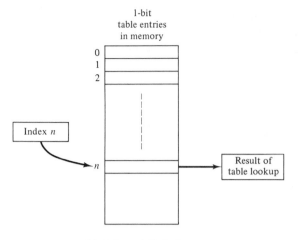

(a) Software table lookup

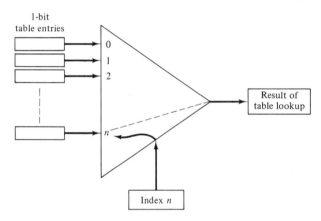

Figure 3–8. A hardware analog of a software table lookup. A single multiplexer provides a 1-bit lookup. Several multiplexers addressed by a common signal form a multibit lookup.

(b) Hardware table lookup using multiplexer

The Demultiplexer

A *demultiplexer* sends data from a single source to one of several destinations. Whereas the multiplexer is a data selector, the demultiplexer is a data distributor or data router. A mechanical analog is the switch used to route the power amplifier output of an automobile radio either to a front or a rear speaker, as illustrated in Fig. 3–11. This switch is the same type of two-position mechanical switch shown in Fig. 3–1. A mechanical switch can transmit a signal in either direction, whereas the electronic analog can transmit data in only one direction. Since we cannot use a multiplexer in the reverse direction, we are forced to provide a demultiplexer to handle this operation.

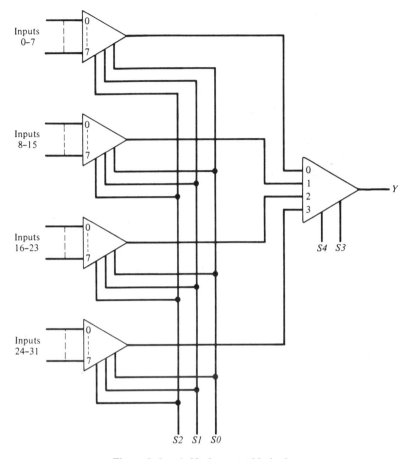

Figure 3–9. A 32-element table lookup.

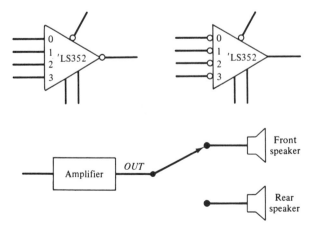

Figure 3–10. Mixed-logic symbols for the 74LS352 4-input multiplexer.

Figure 3–11. A mechanical distributor switch.

The Boolean equations for the switch in Fig. 3–11 are

$$\text{Front speaker} = OUT \cdot \overline{S}$$

$$\text{Rear speaker} = OUT \cdot S$$

S = T when the switch is down. The electronic gate equivalent of these equations is so simple that no commercially packaged version is available.

The smallest demultiplexer that is available in an integrated circuit package is a dual 4-output device. Figure 3–12 is the mixed-logic symbol for one of the two identical and independent demultiplexer elements in the 74LS139 Dual One-to-Four Demultiplexer chip. Inputs B and A are the routing controls for the data source. Just as the multiplexer has a binary code for the selection of an input, the demultiplexer has a similar code for selecting a particular output. All the outputs on the demultiplexer must have a value of FALSE except the one selected by the routing code. The selected output will be T or F, following the condition of the input G. The 74LS139 produces a high voltage level at all unselected outputs (T = L), so the mixed-logic symbol must have small circles on the outputs. Since this demultiplexer routes the input data to an output unchanged, there must also be a circle on the demultiplexer input.

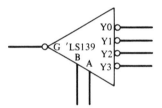

Figure 3–12. An element of the 74LS139 Dual One-to-Four Demultiplexer.

The truth table for a 4-output demultiplexer and the voltage table for the 74LS139 are

| | Demultiplexer logic | | | | | | | 74LS139 Voltage | | | | | | |
G	B	A	$Y0$	$Y1$	$Y2$	$Y3$	G.L	B.H	A.H	$Y0$.L	$Y1$.L	$Y2$.L	$Y3$.L
F	X	X	F	F	F	F	H	X	X	H	H	H	H
T	F	F	T	F	F	F	L	L	L	L	H	H	H
T	F	T	F	T	F	F	L	L	H	H	L	H	H
T	T	F	F	F	T	F	L	H	L	H	H	L	H
T	T	T	F	F	F	T	L	H	H	H	H	H	L

The logic equations for the outputs follow either from the truth table or from the description of the operation of the demultiplexer

$$Y0 = \overline{B} \cdot \overline{A} \cdot G$$

$$Y1 = \overline{B} \cdot A \cdot G$$

$$Y2 = B \cdot \overline{A} \cdot G$$

$$Y3 = B \cdot A \cdot G$$

For practice, you may wish to design a mixed-logic SSI circuit for this demultiplexer, using Nand gates and inverters. This circuit has three inputs (*B*, *A*, and *G*) and four outputs Y_i.

Eight-output demultiplexers are available as MSI chips. For example, the 74LS42 chip, usually described as a "decoder," functions as a demultiplexer. Figure 3–13 is its symbol, plus the logic equations for each output. There are actually two more outputs on this chip:

$$Y8 = \overline{C} \cdot \overline{B} \cdot \overline{A} \cdot \overline{D}$$
$$Y9 = \overline{C} \cdot \overline{B} \cdot A \cdot \overline{D}$$

These outputs do not correspond to any function of the demultiplexer and they do not appear on the mixed-logic symbol. We will encounter them in the next section, when we discuss decoders.

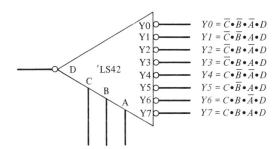

$Y0 = \overline{C} \cdot \overline{B} \cdot \overline{A} \cdot D$
$Y1 = \overline{C} \cdot \overline{B} \cdot A \cdot D$
$Y2 = \overline{C} \cdot B \cdot \overline{A} \cdot D$
$Y3 = \overline{C} \cdot B \cdot A \cdot D$
$Y4 = C \cdot \overline{B} \cdot \overline{A} \cdot D$
$Y5 = C \cdot \overline{B} \cdot A \cdot D$
$Y6 = C \cdot B \cdot \overline{A} \cdot D$
$Y7 = C \cdot B \cdot A \cdot D$

Figure 3–13. The 74LS42 demultiplexer and the equations for its outputs.

To summarize, the demultiplexer building block routes a single source to one of several destinations. A routing code is supplied to the control inputs to select the destination.

The Decoder

In digital design, we frequently need to convert an encoded representation of a set of items into an exploded form in which each item in the set has its own signal. The concept of "encoded information" pervades our lives. Encoding is a useful way of specifying a single member of a large set in a compact form. For instance, every decimal number is a code for a particular member of the set of natural numbers. In everyday affairs, we usually do not need to decode the code explicitly, but sometimes the decoding becomes necessary.

Suppose you walk into a store in a foreign country to buy a coffee cup. You choose a cup on the shelf, so you tell the clerk that you want the fourth cup from the left. You have used a code (4) to identify the desired cup, but the clerk does not know English and cannot pick out the correct cup. Since the clerk is unable to decode your "4" into a specific item, you point to the cup. Your pointed finger means "This one." You were forced to decode your code.

Whenever we use a number to designate a particular object, decoding must occur. Usually, we do this implicitly or intuitively, without thinking about it, but sometimes, as in the china shop, the decoding becomes very explicit.

In hardware, codes are frequently in the form of binary numbers and, in most cases, the decoding required to gain access to an item is buried within a building block. For example, an 8-input multiplexer has a 3-bit select code to specify the particular input. We purposely include within the mux the decoding of the select code—the mux building block contains the circuitry to translate "input 4" on the control lines to "*this* input."

In computer programming we specify a memory location by giving its address. In the hardware (the memory unit of the computer), this numeric address must be decoded to gain access to the particular memory cell.

Another common use of codes is in the operation code of a typical computer instruction. Most computers allow only one operation to be specified in each instruction, and the operation code describes the particular operation. In Part II of this book you will study the art of digital design and will participate in the design of a minicomputer modeled after the PDP-8. The PDP-8 instruction has a 3-bit operation code field that specifies one of eight possible operations. For now we will call the 3 bits of this field *C, B,* and *A*. The operation codes and their instruction mnemonics are

Operation code	Bits C B A	Instruction
0	0 0 0	AND
1	0 0 1	TAD
2	0 1 0	ISZ
3	0 1 1	DCA
4	1 0 0	JMS
5	1 0 1	JMP
6	1 1 0	IOT
7	1 1 1	OP

From the viewpoint of the computer programmer, the decoding of the operation code is buried inside the computer. But we are studying hardware design, and we must face the decoding problem squarely. To implement this instruction set, we require eight logic variables (*AND . . . OP*) to control the specific activities of each instruction. Only one of these eight variables will be true at any time. The translation from the operation code into the individual logic variables is a decoding. We could build the decoding circuits from gates, using the methods of the previous chapters. For instance, the logic equations for two of the variables are

$$TAD = \overline{C} \cdot \overline{B} \cdot A$$
$$JMS = C \cdot \overline{B} \cdot \overline{A}$$

Decoding is so common in digital design that our appropriate posture is to package the decoding circuitry into a logical building block. The *decoder* building block has the characteristic that only one output is true for a given encoded input and the remaining outputs are false. Integrated circuits for decoders are available in several forms, typically with 2, 3, or 4 inputs. The main limitation on the size of the input code is the number of pins required for the outputs, since the number of outputs grows exponentially with the size of the code. For instance, a 3-bit binary decoder has 8 outputs ($2^3 = 8$), whereas a full 4-bit binary decoder has 16 outputs ($2^4 = 16$), requiring a larger integrated circuit package. Decoding a 5-bit binary number would produce 32 outputs—too many for useful packaging as an MSI chip.

The 74LS42 Four-Line-to-Ten-Line Decoder is a typical MSI decoder. This chip will decode a decimal number 0 through 9, expressed as a 4-bit binary code, into one of 10 individual outputs. The binary representation for 9 is 1001; the 4-bit codes from 1010 through 1111 do not arise from the encoding of the decimal numbers. By eliminating 6 of the possible 16 output pins, the 74LS42 circuit can be made to fit conveniently into a 16-pin chip. Figure 3–14a is the mixed-logic symbol for the 74LS42. The outputs are all low-active (T = L)—a characteristic of most commercial MSI decoders.

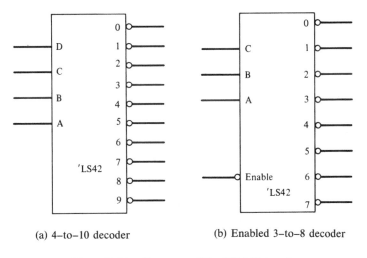

(a) 4–to–10 decoder   (b) Enabled 3–to–8 decoder

Figure 3–14. Two uses of the 74LS42 decoder.

The 74LS42 serves as a 3-bit decoder when the high-order bit (*D*) of the input code is false. The 8 outputs from the resulting 3-bit decoding are *Y0* through *Y7*.

Another important use of the 74LS42 is as an enabled 3-to-8 decoder. The enabling feature is similar to that found on multiplexers in that outputs are always false unless the circuit is enabled. Whenever the *D* input to the 74LS42 is H, outputs *Y0* through *Y7* are false, regardless of the condition of the inputs *C, B,* and *A*; thus the 8 outputs for the 3-bit code are false and the chip is disabled. When input *D* is L, the chip is enabled for 3-bit decoding, and exactly 1 of the 8 outputs *Y0* through *Y7* is true. When the normal mixed-logic representation for the decoder is used, the *D* code bit acts as a disabling or not-enabling signal (with T = H). To use the 74LS42 as an enabled decoder, we usually represent the D input as an enabling signal (with T = L) rather than as a part of the input code. The mixed-logic notation for this enabled 3-to-8 decoder building block is shown in Fig. 3–14b.

In addition to the 74LS42 chip's uses as a decoder, this same chip serves as a demultiplexer, as we saw in the previous section. In that application, we viewed the A, B, and C inputs as a select code, and the D input as a data signal to be routed to a selected destination. This duality of function is characteristic of decoders and demultiplexers. In practice, the decoding applications far outnumber those of demultiplexing.

The Encoder

The converse of the decoding operation is encoding—the process of forming an encoded representation of a set of inputs. This operation does not occur in digital design as frequently as decoding, yet it is of sufficient importance to be a candidate for one of our standard building blocks. In strict analogy with decoding, we should require that exactly one input to an encoder be true. Since there is no way that an encoder building block can enforce this restriction on input signal values, encoders always appear in the form of *priority encoders*. This variation, which is more useful than the regular encoder, allows any number of inputs to be simultaneously true, and produces a binary code for the highest-numbered (highest-priority) true input.

A well-designed priority encoder should provide some way to denote a situation in which *no* input is true. There are two approaches to this problem. Method 1 is to number the input lines beginning with 1 and reserve the output code 0 to indicate that no inputs are true. Method 2 is to number the inputs beginning with 0, but provide a separate output signal which is true only when no input is true. The first method requires fewer output lines but uses up a code pattern to indicate no active inputs. The second method requires an extra output but allows all the code values to represent true conditions at the input.

As a small illustration of priority encoding, consider circuits that produce a 2-bit code from a set of individual inputs. The first method will handle only 3 input lines, whereas the second method accommodates 4 inputs. Here are truth tables for the two styles of priority encoders (remember, X in the truth table means "both values" and – means "don't care"):

Method 1					Method 2						
D3	D2	D1	B	A	D3	D2	D1	D0	B	A	W
F	F	F	F	F	F	F	F	F	–	–	T
F	F	T	F	T	F	F	F	T	F	F	F
F	T	X	T	F	F	F	T	X	F	T	F
T	X	X	T	T	F	T	X	X	T	F	F
					T	X	X	X	T	T	F

Equations for the output variables can be derived by the methods described in Chapter 1. For instance, the logic equations for the outputs for method 2 are

$$B = \overline{D3} \cdot D2 + D3 \quad = D2 + D3$$
$$A = \overline{D3} \cdot \overline{D2} \cdot D1 + D3 = \overline{D2} \cdot D1 + D3$$
$$W = \overline{D3} \cdot \overline{D2} \cdot \overline{D1} \cdot \overline{D0}$$

Commercial priority encoders come in a variety of forms representing both of these methods, sometimes also having an enabling control input, and occasionally with extra inputs and outputs to permit several chips to be cascaded. Customarily, the inputs and code outputs are low-active (T = L). Figure 3–15 is the mixed-logic symbol for a typical chip, the 74LS147 Ten-Line-to-Four-Line Priority Encoder, which conforms closely to method 1 above and produces a 4-bit output code, *D, C, B, A*. The chip has only 9 input lines, not 10 as the name suggests. The tenth line, corresponding to the output code 0000, is inferred from the absence of any true input.

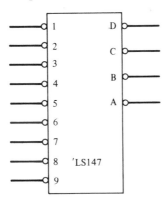

Figure 3–15. The 74LS147 Ten-Line-to-Four Line Priority Encoder.

Priority encoders are frequently used in managing input–output and interrupt signals. The encoder produces a code for the highest-priority true signal. This code may serve as an index for branching or for table lookup in a computer program.

The Comparator

Comparators help us to determine the arithmetic relationship between two binary numbers. Occasionally, we need to compare one set of *n* bits with another reference set of *n* bits to determine if the first set is identical to the reference set. The proper way to determine identity is with a logical COINCIDENCE operation. For instance, to find if a single bit *A* is identical to a reference bit *B*, we use

$$A.EQ.B = A \odot B \qquad (3\text{--}2)$$

For a pattern of *n* bits, we need the logical AND of each such term:

$$A.EQ.B = (A0 \odot B0) \cdot (A1 \odot B1) \cdot \ldots \cdot (An \odot Bn) \qquad (3\text{--}3)$$

We can make an important distinction based on whether the reference set of bits is an unvarying (constant) or a varying pattern. Expanding the single-bit Eq. (3–2) into its AND, OR, NOT form, we have

$$A.EQ.B = A \cdot B + \overline{A} \cdot \overline{B}$$

If B is constant, this equation can be simplified into one of two forms:

$$A.EQ.B = A \text{ if } B = \text{T}$$
$$A.EQ.B = \overline{A} \text{ if } B = \text{F}$$

Consider a comparison of an arbitrary 4-bit A with a fixed 4-bit B = T, F, F, T. Equation (3–3) can be reduced to

$$A.EQ.B = A0 \cdot \overline{A1} \cdot \overline{A2} \cdot A3$$

which can be realized with a 4-input AND element.

If the reference pattern is not fixed, we are stuck with Eq. (3–3). Figure 3–16 is a circuit for 4-bit inputs, using common SSI chips. This type of circuit is a candidate for a building block, and there are MSI chips that perform multibit arithmetic comparisons.

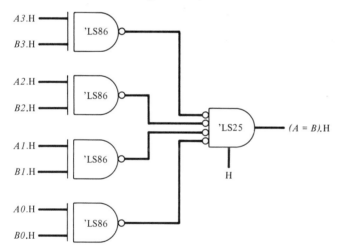

Figure 3–16. An SSI implementation of the equality comparison in Eq. (3–3).

We frequently treat arrays of bits as representations of positive binary numbers. Sometimes we need to determine if one such representation A is arithmetically greater than, equal to, or less than another reference pattern B. In addition to chips that perform only the equality comparison, there are also several arithmetic magnitude comparators, which have outputs capable of simultaneously showing the values of the conditions $A < B$, $A = B$, and $A > B$. A common MSI chip of this type is the 74LS85 Four-Bit Magnitude Comparator. This chip has three status inputs that permit several chips to be cascaded so as to yield comparisons of multiples of 4 bits. Figure 3–17 contains the arrangement

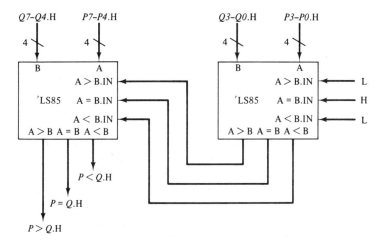

Figure 3–17. An 8-bit magnitude comparison using the 74LS85.

of 74LS85 comparators to accomplish a comparison of 8-bit quantities P and Q. The status outputs of a comparator stage connect to the corresponding status inputs of the next most significant comparator. The results of the final comparison are available as outputs of the most significant stage. It is necessary to pass to the least significant chip the information that the previous (nonexistent) comparison showed equality; this is done by asserting T (a high voltage) on the $A=B.IN$ pin, and F on the other two.

There are several diagrammatic conventions in Fig. 3–17. To avoid cluttering functional diagrams, we often show a group of similar signal lines as a single line with a *numbered* slash across it. The notation for inputs $Q7$.H, $Q6$.H, $Q5$.H, and $Q4$.H appears as $Q7–Q4$.H with a "4" mark on the wire. Figure 3–17 also contains three other similarly collected groups of inputs. This use of the numbered slash is a widely accepted convention; the *numbered* slash has nothing to do with inversion.

Figure 3–17 forced us to make a choice. We usually desire that signals move toward the right in a circuit diagram, with an output on the left feeding inputs to the right. We also usually wish for the least significant part of numerical information to be on the right and the more significant parts on the left. In this figure, variables $Q7–Q0$ and $P7–P0$ represent numbers. We cannot easily accommodate both these goals in the diagram without creating a nightmare of lines, so we usually choose to show the numbers in their customary order rather than adhere to the convention of rightward-moving signals. In practice you will encounter both conventions.

Since Fig. 3–17 is the first instance of this drafting custom in the book, we have placed arrows on all the signal lines to show the direction of the signals' travel. Although we frequently use arrows in high-level functional diagrams, in most instances we would not use such arrows on an actual circuit diagram; the chip's nomenclature shows which pins are inputs and which are outputs.

A Universal Logic Circuit

We have gates for implementing the specific logic operations AND, OR, EOR, and so on. These gates are useful when we know at the time we design what logic we must implement in a given circuit. But in many applications we must perform various logic operations on a set of inputs, based on command information that is not available when we are designing. The best example is the digital computer, which must be designed to meet the requirements of any of its set of instructions. Just as we may select an input with a multiplexer, so must we be able to select a logic operation with a suitable circuit.

Let the inputs to this circuit be A and B, and the output Z. To select the particular logic operation, we must have some control inputs S_i. Figure 3–18 shows this black box. We may require that the black box be able to perform any possible Boolean logic function of its two inputs. We routinely use several of these logic functions: AND, OR, NOT, EOR, and COINCIDENCE. As we mentioned in Chapter 2, there are 16 functions of two variables. They are enumerated in Table 3–1. Some are already familiar:

$$Z1 = A \cdot B \qquad Z7 = A + B \qquad Z10 = \overline{B}$$
$$Z6 = A \oplus B \qquad Z9 = A \odot B \qquad Z12 = \overline{A}$$

TABLE 3-1 LOGIC FUNCTIONS OF TWO VARIABLES

A	B	Z0	Z1	Z2	Z3	Z4	Z5	Z6	Z7	Z8	Z9	Z10	Z11	Z12	Z13	Z14	Z15
0	0	0	0	0	0	0	0	0	0	1	1	1	1	1	1	1	1
0	1	0	0	0	0	1	1	1	1	0	0	0	0	1	1	1	1
1	0	0	0	1	1	0	0	1	1	0	0	1	1	0	0	1	1
1	1	0	1	0	1	0	1	0	1	0	1	0	1	0	1	0	1

There are some others in the table that at first sight appear to be uninteresting but are in fact useful:

$$Z0 = F \qquad Z3 = A \qquad Z5 = B \qquad Z15 = T$$

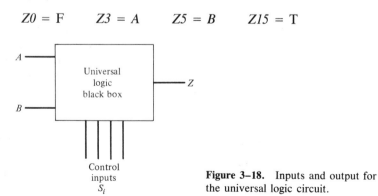

Figure 3–18. Inputs and output for the universal logic circuit.

If we can produce such a comprehensive black box, we will have a circuit that can:

(a) Ignore both inputs and produce a fixed FALSE or TRUE output ($Z0$, $Z15$).

(b) Pass input A or input B through the circuit unchanged ($Z3$, $Z5$).

(c) Perform our important logic functions ($Z1$, $Z6$, $Z7$, $Z9$, $Z10$, $Z12$).

(d) Perform the remaining functions of two variables ($Z2$, $Z4$, $Z8$, $Z11$, $Z13$, $Z14$). These last operations include the NAND and NOR logic functions and four variations of the logical implication function that play no important role in our study of digital design but which we list for completeness.

You will see later that such a general-purpose device is a "natural" at the heart of the digital computer. Computers usually operate on two numbers to produce a result. Not only could this device perform useful logic operations upon two inputs, but it could transmit either input, unaltered or inverted. In addition, it could be a source of T and F bit values.

Many designs for producing the 16 Boolean functions are known, but from our viewpoint the most elegant is a single 4-input multiplexer. To produce a function, our circuit must receive a 4-bit code specifying the particular function Zi. The obvious code values are 0 to 15, corresponding to $Z0$ through $Z15$. Call the code $Z.CODE$, with the code bits designated $Z.CODE_0$ through $Z.CODE_3$. Notice, in Table 3–1, how the definition of each function Zi is exactly the binary representation of the corresponding $Z.CODE$. In an unusual interpretation of the multiplexer in its table-lookup role, we may use the "data" variables A and B as the select inputs of the mux, and feed the function-specifying code $Z.CODE$ into the data inputs:

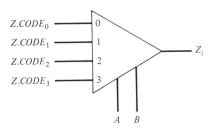

Thus we may produce all 16 Boolean functions of two variables with one-half of a 74LS153 Multiplexer chip. This is tight design!

The universal logic circuit is elegant, but it is capable of performing logic operations only. If it is to be used as the heart of a computer, it should also be able to perform arithmetic operations. Let us leave our universal logic circuit for a moment and discuss the structure of circuits that can perform arithmetic on binary numbers. Later we will consider circuits that can perform both logic and arithmetic.

Binary Addition

The full adder. We assume that you are familiar with the process of binary addition and the representation of numbers in the two's-complement notation. For each bit position, the truth tables defining the addition process are given as Table 3–2. A and B are the bits to be added, CIN is the carry bit generated by the previous bit position, SUM is the sum bit for the current bit position, and $COUT$ is the carry generated in the current bit position. A device for summing three bits in this manner is called a *full adder*. (A similar circuit without the CIN input is called a *half adder*.)

TABLE 3–2 TRUTH TABLE FOR BINARY ADDITION

CIN	A	B	SUM	COUT
0	0	0	0	0
0	0	1	1	0
0	1	0	1	0
0	1	1	0	1
1	0	0	1	0
1	0	1	0	1
1	1	0	0	1
1	1	1	1	1

The full-adder truth table yields Boolean equations for the sum and carry bits:

$$SUM = \overline{CIN} \cdot \overline{A} \cdot B + \overline{CIN} \cdot A \cdot \overline{B} + CIN \cdot \overline{A} \cdot \overline{B} + CIN \cdot A \cdot B$$
$$= (A \oplus B) \cdot \overline{CIN} + (A \odot B) \cdot CIN$$
$$= (A \oplus B) \cdot \overline{CIN} + (\overline{A \oplus B}) \cdot CIN$$
$$= A \oplus B \oplus CIN$$
$$COUT = \overline{CIN} \cdot A \cdot B + CIN \cdot \overline{A} \cdot B + CIN \cdot A \cdot \overline{B} + CIN \cdot A \cdot B$$
$$= A \cdot B + CIN \cdot (A + B)$$

You may derive the simplification of $COUT$ from the K-map:

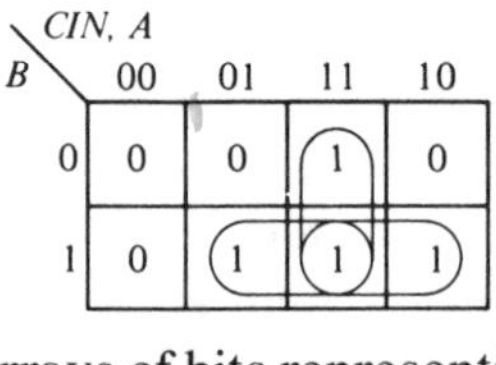

To perform addition on arrays of bits representing unsigned binary numbers, we may connect full adders together as in Fig. 3–19. As a concrete example, let's add two 3-bit binary numbers A and B, where $A = 101$ and $B = 110$. The result of the binary addition is 1011. The corresponding values that would be present on the wires of the hardware are shown in Fig. 3–20.

This method of connecting full adders is called the *ripple carry* configuration,

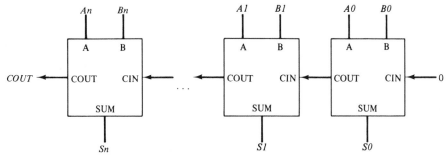

Figure 3-19. Addition with cascaded full adders.

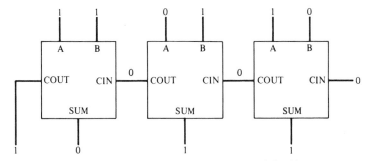

Figure 3-20. 101 + 110 = 1011, using full adders.

since stage zero must produce output before stage 1 can become stable. After stage one becomes stable, stage two will begin to develop its stable outputs. In other words, the carry does indeed ripple down the chain of adders. This is the simplest but slowest way to perform binary addition. Presently we will look at ways of speeding up the process.

Cascading single-bit full adders is not a particularly useful way to perform addition. In digital design we need to add numbers whose binary representations span several bits, and we wish to have building blocks suited to this task. The 74LS283 Four-Bit Full Adder is a useful MSI chip constructed with four full adders packaged as an integral, interconnected unit. The device accepts two 4-bit inputs and a single carry-in bit *CIN* into the low-order bit position; it produces a 4-bit sum and a carry-out bit *COUT* from the most significant bit position. By feeding the *COUT* of the chip into the *CIN* of the next-most-significant stage, it is easy to produce binary adders for any reasonable word length. In the typical application, the *CIN* to bit 0 would be forced to be false, representing numeric 0. In Fig. 3-21, we show the data paths for a 12-bit full adder composed of three 4-bit full adders.

Signed arithmetic. The multibit full adder circuit of Fig. 3-21 does binary addition on 12-bit positive numbers. If the inputs *A* and *B* represent signed integers in the two's-complement notation, the circuit of Fig. 3-21 can perform signed arithmetic. In the two's-complement notation, the leftmost bit represents the sign of the number, and so the circuit shown in Fig. 3-21 can handle 11-bit integers plus a sign.

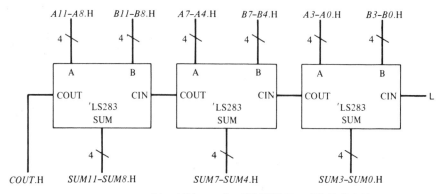

Figure 3–21. A 12-bit addition, using 74LS283 Four-Bit Full Adders.

When the circuit receives two integers, it produces the (signed) sum: A PLUS B. The circuit performs subtraction if the B input receives the two's-complement of the subtrahend: A MINUS B = A PLUS (MINUS B). Incrementing and decrementing are useful special cases: A PLUS 1, and A MINUS 1 = A PLUS (MINUS 1).

Similar results occur if the A and B inputs are represented in one's-complement form. The same circuit performs addition on positive integers and on signed integers represented in one's- or two's-complement notation. The hardware did not change—only the interpretation of the data.

The Arithmetic Logic Unit

This building block, as its name implies, combines the logic capability of the universal logic circuit with a general set of binary arithmetic operations built around cascaded full adders. With the multibit full adder, we may produce arithmetic operations such as subtraction and incrementing only by manipulating the input data; the adder circuit itself only adds.

If we are to have a general arithmetic capability as a building block, such special preparation of the input data is inappropriate; the arithmetic unit should allow us to select an operation. This is a similar situation to that which led us to the universal logic circuit for the performance of arbitrary logic operations on its inputs.

For operands A and B, each consisting of several bits, some useful types of binary arithmetic operations are

Addition:	A PLUS B
Subtraction:	A MINUS B
Incrementing:	A PLUS 1
Decrementing:	A MINUS 1
Negation:	MINUS A

To supplement operations of this type, we might wish to have a source of special constants, such as 0, plus 1, minus 1, and so on.

It would be nice to include multiplication and division in our list, but these operations prove to be quite complex. Although some LSI integrated circuit chips perform multiplication and division, we will exclude these operations from the present discussion.

We might wish to include operations such as *B* MINUS *A*, MINUS *B*, *A* PLUS *A*, and so on—close relatives of the basic operations given above. In any event, it appears that the number of basic arithmetic operations in our list will not exceed 16, so a 4-bit control input would suffice to select any desired operation. Aiming toward a 4-bit arithmetic unit, we would have two 4-bit data inputs, a 4-bit output for the result, an input for carry-in and an output for carry-out, and a 4-bit control input to select the operation.

Can we combine these arithmetic operations with the logic capability of the universal logic implementer to produce an *arithmetic logic unit*? Our universal logic circuit requires four control inputs to specify a code for any of its 16 logic functions. To include the arithmetic operations within a similar control structure will require one additional control bit, producing a 5-bit code. We may use this fifth bit to separate the 32 possible operations into two groups—the 16 logic operations and 16 operations that are arithmetic in nature.

An arithmetic logic unit (ALU) circuit would provide a nice building block for our bag of design tools, and there are a number of integrated circuit chips that approximate the structure developed above. The original chip of this type is the 74LS181 Four-Bit Arithmetic Logic Unit. It has five control inputs and provides the 16 logic functions of two variables, operating simultaneously on each pair of bits. It also provides 16 other operations that have an arithmetic character. The 74LS181 does not have all the arithmetic operations in our wish list, but it is capable of adding, subtracting, incrementing, and decrementing numbers in the two's-complement representation. Several of the "arithmetic" operations of the 74LS181 are somewhat bizarre, and are of little interest to us, but do no harm.

The 74LS181 allows the cascading of 4-bit chips to provide arithmetic capability for long words. The arrangement of the data paths is the same as in Fig. 3–21 for the full adder.

Most of the building blocks described in this chapter are tools for performing a single useful function, such as decoding or multiplexing. The ALU is a step toward more powerful LSI building blocks that perform a variety of functions on a small number of bits, under the control of a set of input signals that we may view as an operation code. Such building blocks are frequently called *bit slices*, alluding to their ability to be ganged together to process larger numbers of bits. Most bit-slice devices have internal registers and are designed to support the basic operations required in modern computer processors. You will encounter additional bit-slice components in Chapter 4 and later in this book.

Speeding Up the Addition

Bit-slice circuits such as the 74LS283 Four-Bit Adder and the 74LS181 Arithmetic Logic Unit are helpful digital building blocks, but if they depended on the simple ripple-carry scheme for binary addition they would be very slow. Binary addition is a combinational process. You know that, at least in theory, any combinational process can be expressed as a truth table and implemented as a two-level sum-of-products function. This approach has only limited practical value in binary arithmetic, since the truth tables for a multibit sum become too large to manage. For instance, the truth table for a 12-bit sum has 24 input variables (25 if we allow for separately specifying the initial carry-in to bit position 0).

Truth tables are a way of specifying in detail the outputs for each combination of input values. In nonarithmetic work we can usually find a simple repetitive pattern of one or two bits that serves as a model for the behavior of the entire circuit, and we can express the repeating function as a small truth table or as an equation. This works well in logic operations, since for each bit the result of an operation depends only on the data entered for that bit, and not on the data in adjacent or more distant bits. Unfortunately, arithmetic does not have this simple property, because of the complex way in which the carry bits affect the result. So two-level binary addition, although desirable because of its speed, is intractible when there are more than a few bits. Within a small bit-slice, however, it is sometimes feasible to produce two-level addition—for instance, four-bit data inputs yield five 9-input truth tables, for the 4 bits of the sum and the single carry-out bit. Each of these truth tables has $2^9 = 512$ rows—painful but not impossible to produce if the rewards are great enough. But you can see that this is hardly a promising general approach.

Ripple-carry is a serial method, slow and simple; two-level circuits are fully parallel, fast but difficult. We need an intermediate technique that provides *some* parallelism with a reasonable effort. A widely used approach is to cast the problem of addition into terms of *carry generate* and *carry propagate* functions. For the moment, consider a one-bit full adder, with inputs A_i, B_i, and C_i, and outputs S_i and C_{i+1}. We will focus on some properties of the data inputs A_i and B_i. We introduce a carry-generate function G_i that is true only when we can guarantee that the data inputs will generate a carry-out. We introduce a carry-propagate function P_i that is true only when a carry-in will be propagated as an identical carry-out. For a 1-bit sum, the truth tables for G_i and P_i are

A_i	B_i	G_i	P_i
0	0	0	0
0	1	0	1
1	0	0	1
1	1	1	0

Using these functions, we may express the carry-out and sum:

$$C_{i+1} = G_i + P_i \cdot C_i \qquad (3\text{--}4)$$

$$S_i = P_i \oplus C_i \qquad (3\text{--}5)$$

(To verify the equation for S_i, you may wish to refer to Table 3–2, our original definition of the full adder.) These equations express the sum and carry-out in terms of just the generate and propagate operators and the carry-in—an important property that we will use when we extend these concepts to bit-slice adders. With these equations, we may implement multibit full adders, but the ripple-carry effect is still present, since each bit's carry-in depends on the preceding bit's carry-out. However, we may expand the equation for C_i in terms of the equations for less-significant bits, to achieve a degree of *carry look-ahead*. For instance

$$C_1 = G_0 + P_0 \cdot C_0 \tag{3-6}$$

$$C_2 = G_1 + P_1 \cdot (G_0 + P_0 \cdot C_0) \tag{3-7}$$

We can derive similar equations (of increasing complexity) for C_3 and C_4. Again, each equation involves only the generate and propagate operators and the original carry-in.

When the generate and propagate operators apply to one-bit slices

$$G_i = A_i \cdot B_i \tag{3-8}$$

$$P_i = A_i \oplus B_i \tag{3-9}$$

These equations involve only the arithmetic data inputs for the ith one-bit unit. By substituting and simplifying, we may derive the following equations for the carry bits:

$$C_1 = A_0 \cdot B_0 + A_0 \cdot C_0 + B_0 \cdot C_0$$
$$C_2 = A_1 \cdot B_1 + A_1 \cdot A_0 \cdot B_0 + A_1 \cdot A_0 \cdot C_0 + A_1 \cdot B_0 \cdot C_0 + B_1 \cdot A_0 \cdot B_0$$
$$+ B_1 \cdot A_0 \cdot C_0 + B_1 \cdot B_0 \cdot C_0$$

These equations yield two-level results. The expansion of C_3 and C_4 are more lengthy, and will be left as homework problems. Within a bit-slice chip, for instance the 74LS181, these equations could be used for high-speed implementations of the sum outputs S_0 through S_3 and the carry-out bit C_4.

If we are building a large adder or ALU from smaller bit-slices, we are still faced with the ripple-carry problem across the boundaries of each chip, even though within each chip the carry-out is being computed rapidly. For instance, in Fig. 3–21, the most-significant carry-out cannot be computed until its corresponding carry-in (into bit 8) is stable, which in turn must await the stabilization of the carry-in to bit 4. We have carry look-ahead within the chips, but not among the chips.

The concept of carry generate and propagate, as typified by Eqs. (3–6) and (3–7), may be extended to larger bit slices. The 74LS181, in addition to producing the carry-out C_4, also produces two outputs G and P that are equivalent to our one-bit G_3 and P_3. The 74LS181's G output is true whenever the data inputs assure that its carry-out is true; the P output is true only when the data inputs are such that the block carry-out is the same as the block carry-in. These G and P functions are considerably more complex than our one-bit Eqs. (3–8) and (3–9), but still involve only the arithmetic data inputs for the chip.

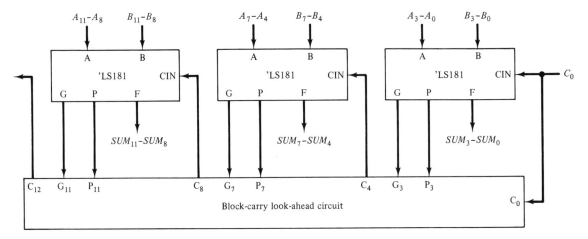

Figure 3–22. A 12-bit ALU containing a block-carry look-ahead circuit.

Figure 3–22 is the circuit for a 12-bit adder constructed with 74LS181 ALUs and a block-carry look-ahead circuit. Instead of relying on each 74LS181 to send its computed carry-out on to the next most significant 74LS181, we send the G and P outputs to the block-carry look-ahead box. The look-ahead box is a combinational circuit that accepts all the G's and P's and the initial carry-in, and simultaneously computes all the carry-outs that must be sent to the 74LS181s. The more significant 74LS181 chips do not have to wait for their carry-in signals to ripple in from lower stages.

We may build the look-ahead box from equations such as the following, which can be inferred from the intuitive meaning of the generate and propagate operators in Fig. 3–22.

$$C_4 = G_3 + P_3 \cdot C_0$$
$$C_8 = G_7 + P_7 \cdot G_3 + P_7 \cdot P_3 \cdot C_0$$

Integrated circuits that perform the block-carry look-ahead functions are available. The 74LS182 Look-Ahead Carry Generator supports the look-ahead process in up to four 74LS181 chips. Furthermore, the 74LS182 produces its own version of G and P, so that look-ahead circuits of more than 16 bits may be constructed by adding levels of 74LS182s. Two levels of 74LS182 will support 64-bit addition. Each new level of look-ahead circuitry increases the time required for the adder outputs to stabilize, but only by about 10 nanoseconds. Since the 74LS181 requires about 20 nanoseconds to perform a binary addition and produce its generate and propagate outputs, the 12-bit adder in Fig. 3–22 requires about 30 nanoseconds and a 64-bit adder would require only about 40 nanoseconds. On the other hand, the 74LS181 requires about 25 nanoseconds to compute its carry-out. Were the 74LS181s used in a ripple-carry configuration, as in Fig. 3–21, each stage would require 25 nanoseconds. The 12-bit addition would require 50 nanoseconds, with longer words requiring an additional 25 nanoseconds for each additional 4 bits.

Arithmetic is a vital function in most computer applications, and much effort has gone into producing fast and efficient arithmetic circuits. Multiplication and division present their own sets of difficulties; fast division is a particularly challenging problem. We will not cover these specialized areas; consult either the textbooks listed at the end of the chapter or the technical literature on specific computers.

DATA MOVEMENT

One of the most important operations in digital design is moving data between a source and a destination. This often occurs inside computers, and is a major activity of peripheral devices. Frequently, the data itself involves several bits— an n-bit byte or word—that must move through the system in parallel. Typical data paths in modern digital design are 8 to 64 bits wide.

If there is only a single source of data and a single destination, we have no problem. We simply run n wires from the source's output to the destination's input. With several sources and various destinations, the situation becomes more complex and requires that we allow data to be moved from any source to any destination. One alternative is separate data paths from each of S sources to each of D destinations. An item that is a source of data at one time may be a destination of data at other times. In Fig. 3–23, we show the data paths in two configurations: three sources $S1$–$S3$ with three different destinations $D1$–$D3$, and four common sources and destinations, A, B, C, D.

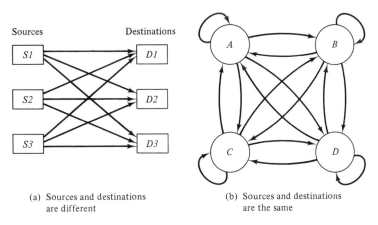

(a) Sources and destinations are different

(b) Sources and destinations are the same

Figure 3–23. Moving data by complete connection of sources and destinations. All paths are n bits wide.

There are obvious drawbacks to this scheme. The number of wires becomes very large as the number of sources and destinations increases, in general being $S \times D \times N$. Adding new nodes to the system involves massive rewiring, affecting each source and destination. A completely connected system allows several simultaneous data transactions to occur over the independent data paths, and this is the main advantage of the scheme. When high-speed parallel movements of data are vital, we would expect to pay the price of inflexibility and complex wiring, and would adopt some form of the completely connected system.

The Bus

In most applications, especially when the number of sources and destinations is not fixed at design time, we need a more flexible solution to the problem of moving data. By giving up parallelism, we may achieve this flexibility. Suppose we run all sources into a single node and take all destination paths from this node. We call this configuration a *data bus,* or just a *bus.* Figure 3–24 consists of two ways to represent a bus, Fig. 3–24b being the more common way. (Remember that all the data paths are actually *n* bits wide, although we usually only draw a single path representing the entire word or byte.) The bus structure permits only one data transfer to occur at a time, since all data paths funnel through the bus node. However, adding or deleting elements on the bus is simple, and this overwhelming advantage accounts for the widespread use of this configuration in digital computers.

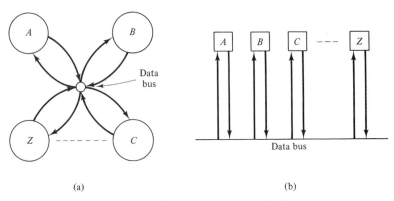

(a) (b)

Figure 3–24. Two ways to view a data bus.

Controlling the bus. We have a building block to move data—the bus—that takes the form of just *n* wires. How do we regulate the traffic over these wires? In any scheme with more than one source or destination, there is a need to *control* the movement of the data. This control takes two forms: who talks, and who listens. On the bus, there must be no more than one talker (source) at a time, but several destinations may listen.

The responsibility for listening on the bus (receiving data) is part of each destination device and is not directly a part of the bus operation. All destinations are physically capable of listening; whether they actually accept data is under their control. Maintaining control over the bus sources, to assure only one talker at a time, is very much a concern of the designer of the bus. We shall mention four control mechanisms, two of which you have already encountered.

Bus access with the multiplexer. Our job is to select one source from several candidates. The digital designer, when encountering the concept of selection, has a knee-jerk response—the multiplexer. For each bit of the bus's data path, attach a multiplexer output to the bus, making each source an input to the mux. We control this collection of n multiplexers with a common source-select code feeding into the multiplexers' select inputs. We show the idea in Fig. 3–25. In this approach, we collect the control for access to the bus in one spot, and assure that only one source is talking at a time—both important advantages. Further, it is easy to debug, since we maintain explicit centralized control over which source has access to the bus. On the other hand, the data mux method of bussing requires considerable hardware; we use an S-wide mux for each of the n data bits in the bus path. If n is large, we have a boardful of data multiplexers. Adding new sources is convenient as long as we do not exhaust the input capability of our muxes. If we exceed this capacity, we have a difficult hardware-modification job. For instance, with 8-input multiplexers, we may manage up to eight sources, but the ninth source causes great agony. Thus, the data multiplexer method of bussing suffers from a certain inflexibility and is not very conserving of hardware. Nevertheless, it is a good method of bussing a moderate number of sources and we will use it in the design of a minicomputer in Chapters 7 and 8.

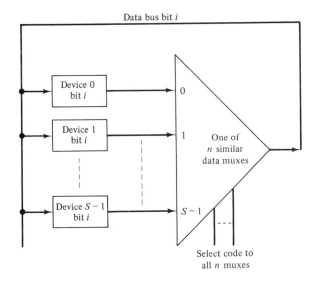

Figure 3–25. Bus control with multiplexers. The mux selects a single source, and the bus routes it to all receivers.

The remaining three methods lack the security of the multiplexer's encoded selection control.

Bus access with OR gates. A primitive form of bus control is to merge all sources into the bus data path, using OR gates. For S n-bit sources, we would have n OR gates, each accepting S inputs. This produces the merging required to give all sources access to the single bus path, but it does not provide the control needed to allow only one source onto the bus at a time. Each bit of each source is either T or F; we must arrange for all sources but one to have all their bits false, while the one designated source presents its T or F data through the OR gates onto the bus. This approach places the responsibility for access with each source, rather than directly with the bus as in the multiplexer method. Each source must have its own gating signal to open or close the gate on its data bits. Typically, the sources have some form of AND gate on each data bit: the data forms one input and the control signal forms the other. (It is this usage that gave rise to the "gate" terminology in digital circuits.) The method is shown in Fig. 3–26.

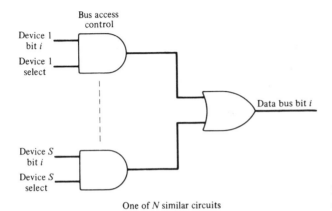

One of *N* similar circuits

Figure 3–26. Controlling access to the data bus with OR gates.

The OR-gate method has little to recommend it. Electronically, it performs the same functions as the mux method, with the mux circuits split into OR gates and AND gates. We might view this as a "poor man's mux," although its components will cost more than those of the actual mux method. It suffers from the same inflexibility of input-size as the mux method and lacks the certainty of control provided by the multiplexer's encoded selection process.

The remaining two methods introduce new concepts of digital building blocks.

Open-collector gates. Open-collector technology provides a way to implement the OR logic function, and thus can be used in bussing applications. We must adhere to the stipulation that open-collector gates produce wired-OR when truth is represented by a low voltage. In Fig. 3–27, we show the 7407 Open-Collector Buffer used as a bus driver.

Since their primary use was for bussing, where several destinations may listen in on the bus, open-collector chips usually can carry more current than their normal TTL counterparts. This provision of extra power is called *buffering,* and such chips are called *buffers* or *drivers*.

The advantage of open-collector circuits is the elimination of the wide OR gates. As long as only one source at a time is on the bus (at most one input is asserting truth), we may connect a large number of open-collector outputs together. Aside from this, achieving proper control of the bus with open-collector wired-OR logic involves the same concerns as the ordinary OR-gate method: we must still control each of the sources so that at most one is talking at a time.

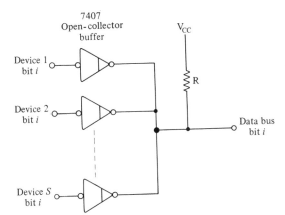

Device 1 bit i

Device 2 bit i

7407
Open-collector
buffer

V_{CC}

R

Data bus
bit i

Device S bit i

Figure 3–27. Controlling access to the data bus with open-collector buffers.

Three-state outputs. The *three-state output* has, as its name implies, three stable states instead of the customary two. In addition to the usual high and low voltage levels, the third state provides a *high-impedance* mode, usually called *Z*, in which the output appears as if it were disconnected from its destinations. The three-state output requires an enabling three-state control input. When the output is *enabled,* the circuit transmits the normal H or L signal presented at the input of the three-state circuit. If the output is *disabled,* the circuit output is for practical purposes not there at all. (Logicians should note that three-state outputs are not the same as ternary logic, which is a true base-3 system.)

Many SSI, MSI, and LSI chips incorporate three-state data outputs. The fundamental use is in bussing, so three-state outputs often provide power buffering like their open-collector cousins. We might select the 74LS244 Octal Three-State Buffer as our prototype three-state chip. This is one of several SSI building blocks that perform no logic but simply afford three-state control of their inputs. This particular chip has two three-state sections—a two-buffer unit controlled by another three-state-enable input.

We find three-state outputs in many useful building blocks. The multiplexers discussed earlier in this chapter have an enable input that holds their output false when the chip is disabled. In the three-state varieties, the output is "disconnected" when the chip is disabled. In Chapter 4, you will see more examples of three-state outputs in MSI building blocks.

The uses of three-state output control in data bussing are substantial. We may connect almost any number of three-state devices together and, with proper three-state enabling of only one source at a time, control access to the bus. Often the chips providing the bus's source data will have three-state output control built in; in other cases, we may need to add buffers such as the 74LS244 to achieve three-state control. Figure 3–28 is a typical three-state bussing configuration.

There are two drawbacks to three-state bus control. First, as in the last two bussing methods, the control of access is decentralized, residing with the sources rather than with the bus structure itself. This makes more difficult the task of assuring that only one source at a time is talking on the bus. Second,

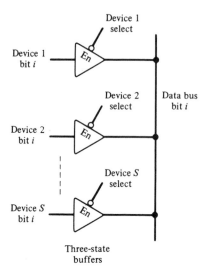

Device 1
select

Device 1
bit i

En

Device 2
select

Data bus
bit i

Device 2
bit i

En

Device S
select

Device S
bit i

En

Three-state
buffers

Figure 3–28. Controlling access to the
data bus with three-state buffers.

the three-state bus is more difficult to debug than the bus formed from multiplexers. The three-state bus itself is just a collection of n wires. A debugger who sees bad data on the bus cannot easily identify which source or sources are contributing. Failure of the three-state circuitry often requires tedious disconnecting of sources from the bus until the guilty party identifies itself by a change in the bus's behavior. In the multiplexer method, we may check inputs, controls, and output directly, and quickly determine which element is misbehaving.

These drawbacks to three-state bus control are insufficient to counteract the tremendous advantages that this technology offers, and three-state control is used in most modern applications of data bussing.

One caveat: Do not try to use three-state control at the output of a control signal. Control signals must be either true or false (asserted or negated) at all times, and we cannot afford to have them simply not there. Only with data whose use is *governed* by control signals do we have the opportunity to have certain data sources disconnected some of the time.

READINGS AND SOURCES

BLAKESLEE, THOMAS R., *Digital Design with Standard MSI and LSI,* 2nd ed. John Wiley & Sons, New York, 1979.

FAST: Fairchild Advanced Schottky TTL. Fairchild Camera and Instrument Corp., Digital Products Division, South Portland, Maine.

FLETCHER, WILLIAM I., *An Engineering Approach to Digital Design.* Prentice-Hall, Englewood Cliffs, N.J., 1980. Chapter 4 discusses MSI building blocks.

HILL, FREDERICK J., and GERALD R. PETERSON, *Digital Logic and Microprocessors.* John Wiley & Sons, New York, 1984.

MANO, M. MORRIS, *Digital Design.* Prentice-Hall, Englewood Cliffs, N.J., 1984.

The TTL Data Book. Texas Instruments, P.O. Box 225012, Dallas, Texas 75265.

WIATROWSKI, CLAUDE A., and CHARLES H. HOUSE, *Logic Circuits and Microcomputer Systems,* McGraw-Hill Book Co., New York, 1980.

EXERCISES

3–1. What are SSI, MSI, LSI, and VLSI?

3–2. What distinguishes a combinational circuit from a sequential one?

3–3. Explain the structure and the function of the multiplexer. What are the two major types of output enable found in MSI multiplexers?

3–4. Figure 3–5, which corresponds to the commercial equivalent of the enabled multiplexer, is not a direct counterpart of Fig. 3–4 or Eq. (3–1). Give a logic equation that best describes the device shown in Fig. 3–5.

3–5. By consulting a TTL data book, make a list of the available MSI multiplexers and their distinguishing characteristics. You must inspect the data sheets to discern the details; chip names alone are not sufficient. (Since the multiplexer is such an important digital building block, this effort is well worthwhile.)

3–6. Why not have one select input for each multiplexer data input rather than encoding the select information?

3–7. The 74LS251 Eight-Input Multiplexer has a three-state output-enable feature. Construct a 16-input multiplexer building block from two 74LS251 chips and an inverter.

3–8. Build the 4-input multiplexer in Fig. 3–10, using SSI gates.

3–9. Show how to construct a 64-input multiplexer building block using eight 74LS251 chips and a 74LS42 decoder.

3–10. The 4-input multiplexer symbol below looks like a mixed-logic notation. Why do we not find this symbol useful?

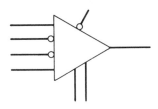

3–11. The 74LS42 serves as a demultiplexer and a decoder. Characterize the difference in these two views of it.

3–12. Build the 4-output demultiplexer in Fig. 3–12, using SSI gates. Will your design also serve as a decoder? If so, how?

3–13. What is the most important characteristic of the outputs of a decoder?

3–14. Explain how we may view the 74LS42's decoding capability as either a 4–to–10 or an enabled 3–to–8 decoder.

3–15. The 74LS42 is sometimes called a BCD (binary-coded decimal) decoder. Why is this an appropriate name?

3–16. Construct a building block that will decode a 4-bit binary code into one of 16 outputs, using 74LS42 decoders and any necessary gates.

3–17. What is the purpose of an encoder? Why are practical encoders *priority* encoders?

3–18. Explain the difference between the concepts of encoding and decoding.

3–19. Using SSI gates, design a priority encoder that accepts five inputs and produces a 3-bit output code. Use method 1 of the text.

3–20. *Parity* is an important concept, frequently used in error-detection circuits within digital systems. The parity of a group of bits is odd if there are an odd number of 1-bits in the group; even parity implies an even number of 1-bits. Although rapid parity-computing circuits are available, the EXCLUSIVE-OR function provides the basis for parity computation.
  (a) Show that the EXCLUSIVE OR of two bits computes odd parity.
  (b) Show that, in general, $A_1 \oplus A_2 \oplus A_3 \oplus \ldots \oplus A_n$ expresses an odd-parity function of n bits.

3–21. Multiplexers offer another interesting approach to parity computation. In the following, the output should be asserted if the parity is odd.
  (a) Show how a 4-input multiplexer (for instance, one-half of a 74LS153) can be used to compute the parity of a 3-bit group.
  (b) Write the logic equation (in terms of AND, OR, and NOT) for the actions of the multiplexer in part (a), and show that this equation is equivalent to the EXCLUSIVE-OR use suggested in the preceding exercise.
  (c) Show how to use four 4-input multiplexers to compute the parity of a 9-bit group. How many 74LS153 chips would be required for this design?

3–22. The 74LS280 Parity Generator/Checker accepts 9 bits of data and reports the parity. Use this chip and any necessary SSI gates to design a circuit that will assert an output when the parity of a 10-bit group is odd.

3–23. Derive logic equations for determining if one 4-bit positive binary number is greater in magnitude than another. Design a circuit for this logic, using SSI gates.

3–24. The 74LS85 Magnitude Comparator has outputs for designating $A < B$, $A = B$, and $A > B$. How may we determine if $A \leq B$? $A \geq B$? $A \neq B$?

3–25. Draw a circuit for 10-bit magnitude comparison, using 74LS85 chips.

3–26. Performing arithmetic comparisons on signed numbers is more complex than comparing magnitudes. Consider two 4-bit signed numbers A and B, recorded in signed-magnitude notation. (This notation denotes a negative number with a 1 in the leftmost bit position and a positive number with a 0 in that bit position; the other bits record the magnitude of the number.) Develop logic equations to determine if $A < B$, $A = B$, and $A > B$ in this notation. Explore whether the 74LS85 Magnitude Comparator is useful in realizing these equations. Produce a circuit (either with or without the 74LS85) for generating the three comparisons.

3–27. Design a 3-bit full adder equivalent to Fig. 3–19, using 1-bit full adders fabricated from SSI gates.

3–28. Modify Fig. 3–21 to perform the operation A (+) B (+) 1.

3–29. Using 74LS283 Four-Bit Full Adders and any necessary SSI gates, design a circuit that will accept a 12-bit signed number in the two's-complement representation and produce the negative of that number.

3–30. Devise a circuit that will accept a 12-bit signed number in the two's-complement representation and produce the absolute value of that number.

3–31. Verify Eq. (3–5) for the full-adder sum expressed in terms of the carry-generate and carry-propagate operators.

3–32. Derive equations for C_3 and C_4, similar to Eqs. (3–6) and (3–7), using the carry-generate and carry-propagate operators.

3–33. Derive the expressions for the 74LS181's G and P outputs. (Hint: consider the generate and propagate operators for the bits within the 4-bit slice; ask yourself what conditions must apply to obtain truth on the overall G and P.) Verify your results by consulting a 74LS181 data sheet.

3–34. Describe a data bus. Give the main advantages and disadvantages of this method of moving data. Why is the bus such a widely used concept?

3–35. Discuss the merits of controlling a bus with:
  (a) Multiplexers.
  (b) OR gates.
  (c) Open-collector buffers.
  (d) Three-state buffers.

3–36. Three-state control of outputs is common. Why do we not employ three-state control of inputs?

3–37. Design bussing systems similar to Fig. 3–25 for six 4-bit devices, using:
  (a) Open-collector bus drivers.
  (b) Three-state bus drivers.

3–38. The multiplexer bus control method shown in Fig. 3–25 has the desirable property that only one source can be talking on the bus at any time. Devise a three-state bus control system that also has this "guaranteed single-talker" feature.

3–39. A logic probe is a small laboratory instrument that, when touched to a point in a digital circuit, indicates the digital voltage level at that point. A typical logic probe shows a low voltage level as a green light and a high voltage level as a red light. If the voltage level is outside the acceptable ranges, the probe shows no light or indicates invalidity in some other way. Read the sections in Chapter 12 on integrated circuit data sheets and performance parameters. Using this information, specify the following voltages for a logic probe designed to operate with the 74LS family of integrated circuits:
  (a) The highest voltage that will light the green light.
  (b) The lowest voltage that will light the red light.

4

Building Blocks with Memory

In the preceding chapters, we tacitly assumed that electronic devices are infinitely fast and that they generate outputs that depend only on the present input values. In this chapter, we explore the interesting consequences of violating these assumptions.

First, we examine what can happen when there are finite propagation delays within gates. Output signals from assemblies of gates sometimes have spurious short pulses that are not predicted by standard Boolean algebra. These spurious pulses are seldom useful, but we must contend with them, usually by waiting until they have gone away.

Next, we explore gate circuits that include feedback. Some of these circuits exhibit *memory,* which is an essential tool for the system designer. We consider useful sequential (memory) building blocks: flip-flops, registers, counters, and so on. These are basic tools for developing the digital architectures in the coming chapters of Part II.

We then discuss large memory arrays—RAMs, ROMs, and allied solid-state memories. Then, after a treatment of programmable logic devices, we conclude with a description of some timing devices that are needed when designing with these large memories.

THE TIME ELEMENT

Hazards

The outputs of real gates cannot change instantaneously when an input is changed. Integrated circuits operate by movement of holes and electrons within some physical material, usually silicon. Not even very light particles such as electrons can move at infinite speeds, and their movement will always involve delays. The time between a change in an input signal and a corresponding change in an output is called the *propagation delay* of the circuit. When inputs change, an output may undergo a change from L to H or from H to L. The corresponding propagation delays are denoted t_{pLH} and t_{pHL}. Propagation delays depend on the input waveforms, temperature, output loadings, operating power, logic family, and a host of other parameters. Single-gate propagation delays are about 5 nanoseconds in TTL low-power Schottky devices.

Another source of delay is the wire carrying signals between gates. Electricity in a wire can travel only about 8 inches in a nanosecond, so when wires become long, the *interconnection delays* may become serious.

Our purpose here is to show how these delays can create spurious outputs called *hazards*. Consider the following simple circuit that changes the voltage polarity of a signal:

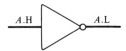

Assume that the voltage at the input *A* has been stable for a long time. The output will also be stable and of the opposite voltage level. If the voltage at the input changes, the output will change a short time later. When an input changes from L to H, the output will change from H to L after a propagation delay t_{pHL}; similarly, a H $\rightarrow$ L transition in the input will produce a L $\rightarrow$ H output transition after a time t_{pLH}. Figure 4–1 is a *timing diagram,* a graph of input and output values (either voltage or logic) as a function of time. Each variable's graph is called a *waveform*.

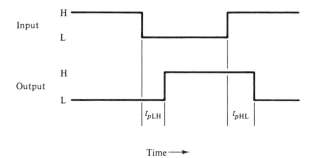

Figure 4–1. Timing diagram showing propagation delays in a logic circuit.

To see what can happen when we introduce time into Boolean algebra, consider the following circuit, whose output is $A + \overline{A}$:

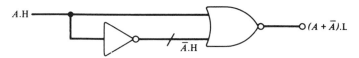

Of course, we know that $A + \overline{A}$ = T regardless of the logic value of A, and we predict, from Boolean algebra, that the output of the circuit will always be L. But assume that each circuit element has a propagation delay t_p for any transition. If A changes from T to F, the voltage pattern in Fig. 4–2 will prevail; there is a spurious high-voltage (F) output that lasts for one gate delay.

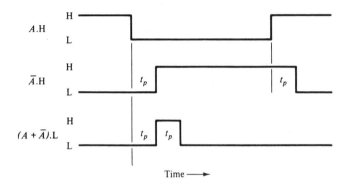

Figure 4–2. A hazard caused by propagation delay in an inverter.

These spurious outputs of combinational circuits, called *hazards* or *glitches,* are common in digital systems. Fortunately, given sufficient time they will die out and the outputs of gates will assume the values predicted by classical Boolean algebra.

Occasionally, it is necessary to generate gate outputs that are *clean*—that have no hazards. It can be shown that a function *may* have a hazard if the function's Karnaugh map has adjacent 1's not enclosed in the same circle. The preceding example, when plotted on a one-variable K-map, becomes

The two adjacent 1's do not share a common circle, and indeed the circuit has a hazard. If we circle both 1's in the K-map, we have the TRUE function, which is hazard-free.

The following function is a more complex example

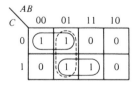

The theory is that a circuit based on the two solid loops may or may not contain a hazard; however, if we build a circuit that includes the dashed loop, we can be sure that the circuit will have no hazards. Using the dashed loop requires extra hardware (additional AND and OR gates), a necessary penalty when we cannot tolerate hazards.

This technique of eliminating hazards works in simple sum-of-products circuits derived from K-maps. In more general circuits, the elimination of hazards is quite complex, and therefore we must use finesse instead of brute force. Rather than use design techniques that require hazard-free signals, we will make our designs *insensitive* to the hazards that occur when combinational inputs are changing. A standard technique is to wait a fixed time after the inputs of the gates change, during which time the hazards will die out. We may then proceed to use the stable signals. This idea is the basis of synchronous (clocked) design, which we introduce in Chapter 5.

Circuits with Feedback

In the preceding section, we discussed purely combinational circuits. Except for momentary hazards, the behavior of the circuits is adequately described by the Boolean algebraic or truth-table methods used in the previous chapters. After a sufficient time to "settle," the circuit's outputs become a function only of the inputs. We now consider another class of circuits, in which the value of the outputs after the settling time depends not only on the external inputs but also on the original value of the outputs. Such circuits exhibit *feedback:* the output feeds back to contribute to the inputs of earlier elements in the circuit.

Feedback yields curious results in some circuits. The following circuit, which has no external inputs, consists of three inverters and feedback:

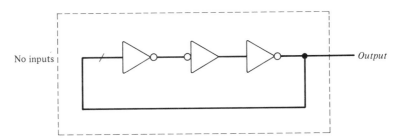

The voltage at the output is fed back into the input where, after a short time, it appears inverted on the output. The new voltage causes a similar inversion; the output voltage *oscillates* rapidly.

Remove one inverter from this circuit, producing the following circuit:

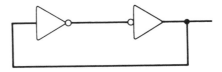

If you construct this circuit with real inverters and apply operating power, the output voltages of each inverter will go through a period of instability, during which one output will settle at a high level and the other at a low level. Although there is no way to predict which output will be high and which low, the circuit will remain stable after the settling time. You can verify the stability by tracing

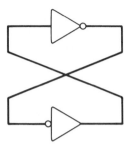

Figure 4–3. Memory displayed by a circuit with feedback.

voltages around the circuit. Redrawing the circuit, as in Fig. 4–3, helps to illustrate the stability. Since neither of the inverter feedback circuits shown above has external inputs, Boolean algebra is powerless to describe the circuit's behavior.

SEQUENTIAL CIRCUITS

The circuit in Fig. 4–3 exhibits a primitive form of memory: the circuit "remembers" the resolution of the initial voltage conflict. Without external inputs, this memory is useless. In contrast, certain feedback circuits with external inputs not only exhibit memory, but also allow the designer to control the value stored in the memory. Controllable memory is the digital designer's most powerful tool. Digital systems with memory are called *sequential circuits.*

Sequential devices may be synthesized from gates, but this procedure is not within the scope of this book, except in that it shows the typical structure of some simple memory elements. Manufacturers have packaged proven gate designs of various sequential circuits, and we can use these as building blocks once we know their behavior. Sequential building blocks have names such as *latch, flip-flop,* and *register.*

Unclocked Sequential Circuits

The latch. The latch is the simplest data storage element. Its logic diagram is in Fig. 4–4. To describe the action of the latch, we must introduce time as a parameter. This was not necessary in combinational logic, but it is always necessary in sequential logic. The timing diagram is frequently used to portray sequential circuit behavior. To analyze the latch circuit, consider the several cases shown in the timing diagram, Fig. 4–5.

Case A: *HOLD* = F. In this case, *Y = DATA.*

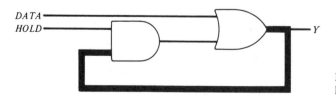

Figure 4–4. A latch circuit; the heavy line is the feedback path.

Tools for Digital Design   Part I

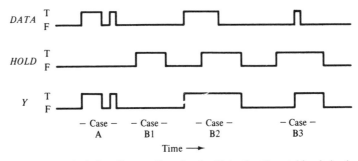

Figure 4–5. A timing diagram for a latch. Note the 1's catching behavior.

Case B: *HOLD* = T. Any occurrence of *DATA* = T will be captured, and the output will thereafter remain true until *HOLD* becomes false. We consider three subcases:

Case B1: *DATA* is false throughout the period when *HOLD* is true. Then *Y* is false.

Case B2: *DATA* is true when *HOLD* is true. When *HOLD* becomes true, the latch captures the (true) value of *DATA* and stores it as long as *HOLD* remains true. (After *HOLD* becomes false, case A applies.)

Case B3: *DATA* is false when *HOLD* becomes true. At the beginning, *Y* is false. The first occurrence of a true signal on the *DATA* line will cause *Y* to become true; the output will *remain true* until *HOLD* becomes false.

The latch has the property of passing true input data to its output immediately. This behavior is sometimes useful in digital design, but it can be quite dangerous. Suppose that while *HOLD* is true, a glitch or noise pulse on the *DATA* line causes *DATA* to become true momentarily. This momentary true, or 1, will cause output *Y* to become true and remain true as long as *HOLD* is true. This behavior is sometimes called *1's catching;* it is useful only rarely.

The latch circuit in Fig. 4–4 is not frequently used, and it is not generally available as an SSI integrated circuit. A true latch is a memory element that exhibits combinational behavior at some values of its inputs. There are other varieties of latch; unfortunately, designers use the term loosely to describe various signal-capturing events. We will soon develop more satisfactory memory devices.

Timing diagrams may be used to show gross voltage or logic behavior, or to show fine detail. The timing diagrams in Figs. 4–1 and 4–2 show the fine detail of gate delays. On the other hand, the timing diagram in Fig. 4–5 shows only the gross behavior of the latch circuit and is accurate only when the time scale is sufficiently large. On a fine time scale, the output *Y* in Fig. 4–5 would be shifted slightly to the right to account for the delays incurred while changes in *DATA* or *HOLD* are absorbed by the gates in the circuit.

The asynchronous RS flip-flop. The feedback circuit in Fig. 4–3 exhibits a peculiar form of memory: it remembers which inverter had a low output after "power-up." The circuit has two stable states, and is indeed a memory, albeit a useless one, since there is no way to change it from one state to the other.

By changing the inverters to two-input Nor gates, we obtain a useful device known as the *asynchronous RS flip-flop* (see Fig. 4–6). We will study voltage behavior in this circuit before we introduce the concept of logic truth.

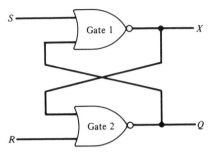

Figure 4–6. An asynchronous RS flip-flop constructed with Nor gates.

The RS flip-flop is a *bistable* device, which means that in the absence of any inputs it can assume either of two stable states. To see this, assume that $R = S =$ L, and assume that the output X of gate 1 is L. Gate 2 will then present a high voltage level to Q. When this H feeds back to the input of gate 1, it will produce an L at X, which is consistent with our original assumption about the polarity of X. We can describe this behavior by saying that the circuit is in a *stable state* when gate 1 outputs L and gate 2 outputs H. Once the circuit assumes this state, it will remain there as long as there are no changes in the R and S inputs.

There is another stable state during which gate 1 outputs H and gate 2 outputs L. We could predict this from the symmetry of the circuit, but you should verify it by tracing signals as we just did.

We have shown that the circuit of two cross-coupled Nor gates can exist in two stable states. We call one of the stable states the *set state* and the other the *reset state*. By convention, the set state corresponds to $Q =$ H, and the reset state to $Q =$ L.

The conventional representation of a flip-flop is a rectangle from which Q.H emerges at the upper right side. Most flip-flops produce two voltages of opposite polarity and the second output appears below the Q.H output. In data books, the second output is usually called $\overline{Q}$. Since this output behaves like Q with a voltage inversion, mixed logicians prefer to designate the signal as Q.L, the alternative voltage form of Q.H. Nevertheless, the nomenclature within the flip-flop symbol, like our other building blocks, must conform to normal data book usage so that there will be no confusion about the interpretation of the pins of the chip. The interior of the symbol serves to identify pin functions; the external notations for inputs and outputs represent specific signals in a logic design. Thus, if we have a flip-flop whose output is a logic variable *RUN,* our standard notation for the output is

Q ——— *RUN*.H

$\overline{Q}$ ——o *RUN*.L

Now we will consider the S and R inputs to the RS flip-flop. We know that as long as S and R are low, the flip-flop remains in its present state. We may use the S and R lines to force the flip-flop into either state. S is a control input that places the RS flip-flop into the set state ($Q = $ H) whenever $S = $ H. Analogously, $R = $ H resets the flip-flop by making $Q = $ L. The obvious association of truth and voltage is T = H at S, R, and Q, so that we set the flip-flop by making $S = $ T, and we reset by making $R = $ T. This leads us to our usual mixed-logic notation for an RS flip-flop constructed of Nor gates:

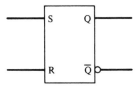

Figure 4–7 is a similar asynchronous RS flip-flop designed with Nand gates. This figure, a mixed-logic diagram of the cross-coupled gates, emphasizes that T = L at the inputs of this flip-flop.

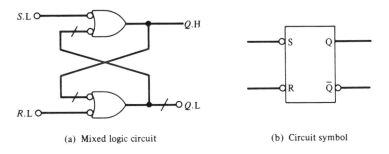

(a) Mixed logic circuit   (b) Circuit symbol

Figure 4–7. An asynchronous RS flip-flop constructed with Nand gates.

The term *asynchronous* associated with the RS flip-flop implies that there is no master clocking signal that governs the activity of the flip-flop; suitable changes of S or R cause the outputs to react immediately. Asynchronous means *unclocked*. Its counterpart is a *clocked*, or *synchronous*, circuit. (Some workers refer to all the unclocked storage elements as latches; we will not adopt this practice.) The asynchronous RS flip-flop is sensitive to noise, or glitches, at the S input when in the reset state, and at the R input when in the set state. This sensitivity is occasionally useful, but in general you should avoid using asynchronous devices, since glitches are undesirable byproducts of gate delays and noise is usually unpredictable in digital systems. Part of our goal is to develop design techniques that bypass these inevitable problems. Therefore, one of our dictums will be: don't use asynchronous RS flip-flops as a general design tool.

Switch debouncing. There is one standard use of the RS flip-flop—as a *switch debouncer*. It is an unfortunate fact that mechanical switches do not make or break contact cleanly. At closure there will be several separate contacts over a period of many microseconds. The same is true during switch opening. The switch bounces. Since we do not wish to use a bouncy or spiky signal in our digital designs, we need a way to clean up the switch output.

Whenever a mechanical switch changes its position, we wish the associated digital signal to undergo one smooth change of voltage level. The asynchronous RS flip-flop is well suited for this. Figure 4–8 contains two switch debouncing circuits. In Chapter 12 we discuss the electrical details of the input circuits; here we will be satisfied to state that the resistors keep the control inputs inactive unless the voltage from the switch forces one input to become active. When the switch is off, it is constantly resetting the flip-flop, producing a constant F output. As the switch moves toward the on position, there will be a period of oscillation or bounce on the R input, caused by the mechanical switch breaking and making its contact with its off terminal. The S input is false throughout all of this, and the repeated resetting does not affect the false output of the flip-flop. There follows a "long" period when the switch moves between its off and on positions, during which time both S and R are false. Then the switch begins its bouncy contact with the on terminal. The first contact causes S to become true, which sets the flip-flop to its true state, where it remains throughout the on-position bounce and until the switch is returned to off.

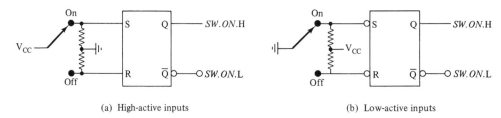

(a) High-active inputs   (b) Low-active inputs

Figure 4–8. Mechanical switch debouncing circuits using asynchronous RS flip-flops.

Ambiguous behavior in the RS flip-flop. Of the four voltage combinations of the S and R inputs, we have used three: to hold, set, and reset. What happens when S and R are simultaneously true? In the Nor-gate version, the voltages at both outputs of the flip-flop will be low—a disturbing situation. In the Nand-gate version, both will be high. Although this deviation from voltage complementarity is unwelcome, it nevertheless represents a well-defined and stable configuration of the flip-flop. But watch what happens when we try to retreat from this configuration of inputs. If we change only one of the inputs, the flip-flop enters either the set or reset state, without difficulty. But if we try to change both inputs simultaneously (in an attempt to move to the hold state), the flip-flop is in deep trouble. Consider the Nor-gate version of the RS flip-flop, Fig. 4–6. If the voltages at S and R are both high, then they are low at both X and Q. If the voltages at S and R both become low simultaneously, then after one gate delay both gates in the flip-flop will produce high outputs. These high outputs, feeding back to the inputs of the Nor gates, will result in low gate outputs after one more gate delay. And so on. The circuit oscillates rapidly, at least at the beginning, with both outputs producing either high or low voltage levels "in phase." The resulting changes occur so rapidly that the flip-flop is forced out of the digital mode of operation for which it was designed, and the output voltages quickly cease to conform to reliable digital voltage levels—an example of *metastable behavior* that is discussed in Chapter 12. Eventually, the slight differences in the physical properties of the two gates will allow the flip-flop to drop into the set state or the reset state. The time required for the voltages to settle and the final result are uncertain, so this behavior is of no use to designers. Therefore, it is considered improper design practice to allow R and S to be asserted at the same time.

Excitation tables. Timing diagrams are useful for displaying the time-dependent characteristics of sequential circuits, but for most purposes a tabular form is better. The *excitation table* is the sequential counterpart of the truth table or voltage table for combinational circuits. The excitation table looks much like a truth table, but it contains the element of time. In a sequential circuit, the new outputs depend on the present inputs and also on the present values of the outputs. We can display the behavior of the RS flip-flop of Fig. 4–6 in the following excitation table:

S	R	$Q_{(t)}$	$X_{(t)}$	$Q_{(t+\delta)}$	$X_{(t+\delta)}$	
L	L	q	x	q	x	Hold
L	H	q	x	L	H	Reset
H	L	q	x	H	L	Set
H	H	q	x			Disallowed

$Q_{(t)}$ is the value of output Q at time t; $Q_{(t+\delta)}$ is the value of Q at a small time δ after t, where δ is sufficiently long for the effects of the gate delays to settle down.

The excitation table is also useful for displaying the logical behavior of sequential circuits. For instance, the following excitation table describes the logical behavior of RS flip-flops, using a modification of the previous notation:

S	R	Q	Q'	
F	F	q	q	Hold
F	T	q	F	Reset
T	F	q	T	Set
T	T	q		Disallowed

In the literature, notations for excitation tables vary greatly and in this chapter we will use a variety of forms. You should be able to recognize these notational differences.

Clocked Sequential Circuits

Asynchronous flip-flops are 1's catchers. A more useful class of flip-flop is available for general digital design. In these flip-flops, outputs will not change unless another signal, called the *clock,* is asserted. Since the activity is synchronized with the clock signal, these flip-flops are called *synchronous.* Digital systems usually have a repetitive clock with a square waveform. The clock signal alternates between its H and L signal levels. Depending on the application, we may view either H or L as representing truth on the clock line, although in almost all our applications we shall use the T = H assignment for clock signals. In Chapter 12 we discuss ways of generating this important signal, and you will encounter clocked circuits throughout the remainder of this book.

Clocked RS flip-flop. We can derive a clocked flip-flop from an asynchronous RS flip-flop by gating the R and S input signals to restrict the time during which they are active, as in Fig. 4–9. The flip-flop outputs may change whenever the clock is true—a potentially risky situation similar to the 1's catching of the latch circuit. In digital systems, flip-flop outputs often contribute to combinational circuits that produce inputs to other flip-flops. Shortly after the rise of the clock, the system is in "shock" owing to the changing of flip-flops. During this period of shock, hazards may be present that can feed erroneous signals into flip-flop inputs while the clock is *still true,* resulting in false setting or resetting of the flip-flops.

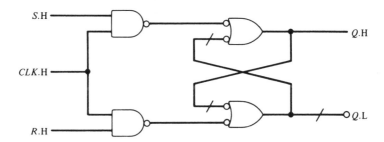

Figure 4–9. A clocked RS flip-flop circuit.

It is natural to try to avoid this problem by making the true portion of the clock signal as narrow as possible. Unfortunately, this is not a good solution, since the system's behavior is crucially dependent on the quality of the clock and narrow clock signals are difficult to generate and distribute.

The aim is to reduce the time during which the flip-flop outputs respond to the inputs. Since altering the clock waveform leads to difficulties, can we achieve the goal by further modification of the flip-flop circuit itself? Can we devise a flip-flop that will recognize R and S only at a single instant and ignore the inputs at other times? Such behavior would be desirable because all flip-flops would change at precisely the same time if they were clocked from the same source. This would mean that we could arrange for all the R and S inputs on all flip-flops to be stable at the time of clocking, and the flip-flops would not be influenced by the shock of the changes induced just after clocking.

Flip-flops that allow output changes to occur only at a single clocked instant are called *edge-driven* or *edge-triggered*. An edge is a voltage transition on the clock signal, and may be either a positive edge (L → H) or a negative edge (H → L). The clocked circuit in Fig. 4–9 is *level-driven*, since its outputs may change at any time during the true part of the clock cycle. In your designs of clocked sequential circuits, use only edge-driven devices.

Master-slave flip-flop. The master-slave flip-flop is a relic from the early days of integrated circuit technology, but is still widely used because of its pseudo-edge-driven characteristics. It is a relatively simple device that we can easily discuss at the gate level, so we will show how one is derived by extending the clocked RS flip-flop. Figure 4–10 is a master-slave flip-flop schematic. The master flip-flop will respond to inputs S and R as long as the clock signal is high. This period must be long enough to ensure that S and R are stable when the clock goes from high to low. This H → L transition, the negative clock edge, isolates the master flip-flop from the inputs S and R. The master flip-flop will now remain unchanged until the next positive clock edge.

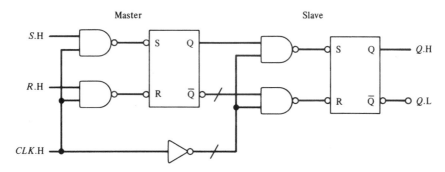

Figure 4–10. A master-slave clocked RS flip-flop.

Because of the voltage inverter, the slave flip-flop does not become sensitive to its input until one gate delay after the negative clock edge. At that time, it receives its S and R inputs from a stable master flip-flop. The net effect is that the outputs of the master-slave combination change only on the negative clock edge rather than during a clock level.

Pure edge-driven flip-flop. The master-slave flip-flop appears to be an attractive edge-driven device. Why are we not content with this design? Because the master flip-flop is still a 1's catcher during the positive half of the clock cycle. This means that R and S must stabilize during the negative half of the clock, since the master flip-flop will react to any T glitches during the positive clock phase. We could greatly simplify our digital circuit designs if we could eliminate the 1's-catching behavior. We need a flip-flop that samples its inputs only on a clock edge and changes its outputs only as a result of the clock edge. Such a device is called a *pure edge-driven flip-flop*. The F → T clock transition is called the *active edge*. It may be either the H → L or L → H transition, although in the most useful integrated circuits the L → H transition is the active edge.

The property of changing state and sensing inputs only at a given instant gives the designer a powerful tool for combatting glitches and noise. We can now choose the time to look at signals and can fix that time to allow adequate stabilization of the system. We will make constant use of pure edge-driven sequential circuits in our designs. The internal structure of these devices is rather complex, but for purposes of digital system design it is not necessary for us to examine their construction in detail. Hereafter, in all our discussions of clocked sequential circuits, we will assume the use of pure edge-driven devices.

Excitation tables for edge-driven flip-flops. Assume that the edge-driven flip-flop is subjected to a steady stream of active clock edges. Each clock edge will cause the flip-flop to enter either its set or its reset state, in accordance with the values of its inputs and the current value stored in the flip-flop. After the flip-flop has received n clock triggers, the value stored in the flip-flop is $Q_{(n)}$. If the flip-flop is in the set state after the nth clock edge, then $Q_{(n)} = $ T; if in the reset state, $Q_{(n)} = $ F. After the appearance of the next clock edge, the value of Q will be $Q_{(n+1)}$. The excitation table for edge-driven devices is a tabulation of $Q_{(n+1)}$ for all combinations of the exciting variables.

In the remainder of this chapter, we will use excitation tables to classify flip-flops. For the excitation table to be valid, we must ensure that the control inputs are stable for a short time before the active clock edge (the *setup time*), and perhaps for a short time after the active clock edge (the *hold time*). The input voltages may go through wild excursions prior to the onset of the setup time and after the hold time, as long as they remain stable during the setup and hold times. (See Chapter 12 for a discussion of setup and hold times.)

CLOCKED BUILDING BLOCKS

In this section, we present the common SSI building blocks for clocked digital design. Table 12–1, at the end of Chapter 12, contains a selected list of useful integrated circuits for these as well as more complex building blocks.

The JK Flip-Flop

Whereas the RS flip-flop displays ambiguous behavior if both R and S are true simultaneously, the JK flip-flop produces unambiguous results in all combinations of its inputs. A logical excitation table for the basic JK flip-flop is:

Clock	J	K	$Q_{(n)}$	$Q_{(n+1)}$	
F	X	X	q	q	
T	X	X	q	q	
↑	F	F	q	q	Hold
↑	F	T	q	F	Reset
↑	T	F	q	T	Set
↑	T	T	q	$\bar{q}$	Toggle (complement)

J is the counterpart of the S input of an RS flip-flop, and K is the counterpart of R. The first two lines of the excitation table demonstrate the edge-triggered behavior of the flip-flop: when the clock signal is a stable false or true, the output of the flip-flop is insensitive to the other inputs. Often these lines do not appear in the excitation table, since such behavior is expected of an edge-triggered device. The remaining four lines in the table describe the flip-flop behavior when the clock undergoes its active (F → T) transition. The first three of these lines are analogous to the RS flip-flop. The last line shows that, if both control inputs are true when the clock fires, the flip-flop will complement its output. This behavior is called *toggling*.

Commercial JK flip-flops come in various forms. The most interesting variations are:

(a) Active clock edge: positive or negative. On all clocked devices, we show the clock input as a small wedge ▷ inside the device symbol. A negative edge-triggered flip-flop has a small circle (representing T = L) on the clock input: ◁.

(b) Active voltage level for J and K. We find flip-flops with both J and K active-high (T = H), and also a flip-flop with J active-high and K active-low. In the latter form, the K input has a small circle on the circuit symbol.

(c) Availability of asynchronous R and S inputs. These are often called *direct clear* or *preclear* and *direct set* or *preset*. One, both, or neither may be present on the chip. Direct set usually appears at the top of the flip-flop symbol, and direct clear at the bottom. Truth is usually a low voltage level, in which case these inputs will bear small circles. As long as an asynchronous input is asserted, it will override the normal synchronous behavior of the flip-flop.

The 74LS109 Dual JK Flip-Flop is our most-used version. It is positive-edge-triggered, which is compatible with the standard MSI sequential building blocks, and has a high-active J input and a low-active K input. As usual, when designing with JK flip-flops we think in terms of logical operations rather than voltages. It is useful to describe the primary logical operations on the JK flip-flop as the "set" and the "clear," setting Q to T and clearing Q to F. The 74LS109 flip-flop has two useful mixed-logic representations, shown in Fig.

4–11 with appropriate input and output signals. "Pr" is the asynchronous preset input; "Clr" is preclear. The symbol in Fig. 4–11a is conventional and causes no difficulty. The circuit shown in Fig. 4–11b is less easy to derive, but it gives us a degree of flexibility that repays our efforts. The voltage excitation table for the 74LS109 is

Preset	Preclear	Clock	J	K	$Q_{(n+1)}$	Action if Q is active-high
L	H	X	X	X	H	Direct set
H	L	X	X	X	L	Direct clear
L	L	X	X	X	—	Disallowed configuration
H	H	↑	L	L	L	Clear (Reset)
H	H	↑	L	H	$Q_{(n)}$	Hold
H	H	↑	H	L	$\sim Q_{(n)}$	Toggle
H	H	↑	H	H	H	Set

Here, $\sim Q_{(n)}$ means the voltage inversion of $Q_{(n)}$; as usual, X means "don't care." This excitation table contains yet another variation of notation, in which the monotonous input column for the present value of the flip-flop's output is eliminated.

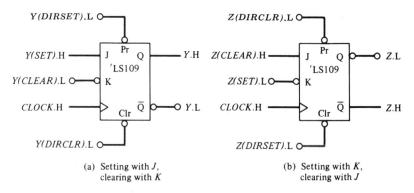

(a) Setting with J,
clearing with K

(b) Setting with K,
clearing with J

Figure 4–11. Two mixed-logic uses of the 74LS109 Dual JK Flip-Flop.

When T = H at Q, we derive the mixed-logic symbol in Fig. 4–11a, the usual form. If T = L at Q, the logical act of setting the flip-flop must result in an L output at $\overline{Q}$; logical clearing must yield Q = H. In order to match this behavior with the voltage excitation table, we are led to the conclusion that we must *set* the flip-flop with the K input and *clear* with the J input. In turn, this causes the preset input *Pr* to perform as a *logical direct clear,* and the preclear input *Clr* to perform as a *logical direct set.*

The advantage of the form used in Fig. 4–11b is its versatility, since we use a different voltage convention for setting and clearing than we do with the conventional symbol. This mixed-logic symbol for the 74LS109 is the most difficult of the common building blocks to derive, yet having once derived and mastered it, we may use either symbol for the 74LS109, as our use dictates, without further thought.

This exercise in mixed logic illustrates one aspect of good design: We try to define the general behavior of common circuit elements, and arrive at general solutions to common design problems. We move these recurring but perhaps difficult items up front, where we face them squarely, so that having dealt with their intricacies once, we may thereafter use the standard results in our design work. You will see this principle invoked many times in this book; it is the essence of top-down design.

The JK flip-flop is our most powerful SSI storage element, and you must master its use. There are several ways of using a single flip-flop, and later you will see many larger constructions based on this flexible element.

JK flip-flop as controlled storage. The most general use of the JK flip-flop, and the one that gives it such power and flexibility, is as a storage element under explicit control. In digital design, whenever we must set or clear or toggle a signal to form a specific value for later use, we usually think of a JK flip-flop. The penalty for this generality is the need to control two separate inputs.

JK flip-flop for storing data. The JK flip-flop is basically a controlled storage element. On occasion, we wish to adopt a different posture and view the JK flip-flop as a medium for entering and storing data. From the excitation table, we see that $Q_{(n+1)} = Q_{(n)}$ whenever $J = K = F$ at the clock edge. This is simply a data-storage mode. All that is necessary to continue holding data in the flip-flop is to ensure that $J = K = F$ during the setup time before each clock edge.

JK flip-flop for entering data. The J and K inputs are not data lines; they are control lines for the flip-flop storage. Nevertheless, we can view the JK flip-flop as a data-entry device. We can enter data in three ways:

(a) Clearing, followed by later setting if necessary.
(b) Setting, followed by later clearing if necessary.
(c) Forcing the data into the flip-flop in one clock cycle.

The rule for case (a) is:

> If you are sure that the flip-flop is cleared, you may enter data D into the flip-flop on a clock edge by having $J = D$, independent of the value of K.

Case (b) is analogous to case (a). The rule is:

> If you are sure that the flip-flop's output is true, you may enter data D into the flip-flop on a clock edge by having $K = \overline{D}$, regardless of the value of J.

You should verify the rules for cases (a) and (b).

As for case (c), the designer often cannot guarantee that a flip-flop will be in a given state. Proceeding as we did in cases (a) and (b) would waste one clock cycle for the initial clearing or setting operation. It would be nice to have a mode that would force data to enter the flip-flop at a clock edge, regardless of the present condition at the output. Such a data-entry mode is called a *jam transfer,* since the data is "jammed" into the flip-flop independent of prior conditions. Examination of the excitation table for the JK flip-flop shows that such a mode is indeed available. We enter data D as follows: If D = F, J must equal F and K must equal T. If D = T, J must equal T and K must equal F. Combining these conditions, we see that $Q_{(n+1)}$ will equal D whenever $J = D$ and $K = \overline{D}$. Now you see the utility of having opposite voltage conventions for truth on the J and K inputs of the 74LS109 flip-flop. With this device, we can connect the J and K inputs together to make $K = \overline{J}$ as required by the analysis above. Then, by connecting the input data D to the joined inputs of the flip-flop, we will enter D into Q at each clock edge.

The D Flip-Flop

The D (Delay) flip-flop has a simpler excitation table than the JK, and is used in applications that do not require the full power of the JK flip-flop. The symbol and excitation table for the D flip-flop are:

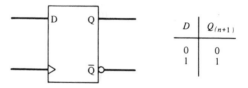

As an SSI device, the D flip-flop appears in these common varieties:

(a) The active clock edge can be either positive (L → H) or negative (H → L), which is shown by the absence or presence of a small circle on the clock terminal.

(b) Direct (asynchronous) set and clear inputs appear in these combinations: both, neither, or clear only. Almost always, these inputs, when present, are low-active, and appear in the diagram with the small circle. These asynchronous inputs are 1's catchers, and you should use them only with great caution.

(c) Some D flip-flops have only the Q output; others provide both polarities. Although it appears to be ideal for data storage, there are, in fact, just a few common uses of the D flip-flop in good design.

D flip-flop as a delay. As its name implies, the D flip-flop serves to delay the value of the signal at its input by one clock time. You will see such a use in Chapter 6 when we discuss the single-pulser circuit for manual switch processing.

D flip-flop as a synchronizer.   One natural application of the D flip-flop is as a synchronizer of an input signal. Clocked logic must sometimes deal with input signals that have no fixed temporal relation to the master clock. An example is a manual pushbutton such as a stop switch on a computer console. The operator may close this switch at any time, perhaps so near the next edge of the system clock that the effect of the changing signal cannot be fully propagated through the circuit before the clock edge arrives. If the inputs to clocked elements are not stable during their setup times, their behavior is not predictable after the clock edge: some outputs may change, others may not. We need some way to process this manual switch signal so that it changes only when the active clock edges appear. This is called *synchronization*. Since the output of a clocked element changes only in step with the system clock, we may use the D flip-flop as a synchronizer by feeding the unsynchronized signal to the flip-flop input. We deal with this matter more fully in later chapters.

D flip-flop for data storage.   The D flip-flop appears to be well suited to data entry and storage. Unfortunately, designers use it far too often for this purpose. The problem is that every clock pulse will "load" new data and this is seldom wanted. We usually need a device that allows us to control when the flip-flop accepts new data, just as we could with the JK flip-flop. With the D flip-flop, it seems natural to *gate the clock* by ANDing it with a control signal in order to produce a clock edge at the flip-flop only when we wish to load data. This is a dangerous practice, as you will see in later chapters. Clocked circuit design relies on a clean clock signal that arrives at all clock inputs simultaneously. We have the best chance of meeting these conditions if we use unmodified clock signals. This means that the devices will be clocked every cycle, so we must seek other ways of effecting the necessary control over the flip-flop activities.

The enabled D flip-flop.   To alleviate the problems caused by gating the clock input to a D flip-flop, we will construct a new type of device called the *enabled D flip-flop*. Figure 4–12 shows the principle. The circuit consists of a D flip-flop with a multiplexer on its input. A new control signal *LOAD* appears, in addition to the customary data input.

The system clock goes directly to the clock input, thereby avoiding the problems of a gated clock. As long as *LOAD* is false, the data selector selects the current value of the flip-flop output as input to the flip-flop. The net effect is that Q recirculates unchanged: the flip-flop stores data. When *LOAD* = T, the multiplexer routes the external signal *DATA* into the D input, where it will

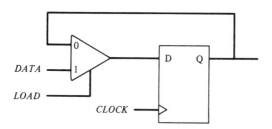

Figure 4–12.   An enabled D flip-flop.

be loaded into the flip-flop on the next clock edge. The loading process is a jam transfer.

The enabled D flip-flop is the element of choice for simple data storage applications. Although we can accomplish the same effect with the JK flip-flop, the enabled D device provides a more natural way of handling data. Curiously, integrated circuit manufacturers were slow to produce SSI elements containing several independent enabled D flip-flops. However, the concept is widely used in arrays of storage elements, and we have available a good selection of tools for registers.

REGISTER BUILDING BLOCKS

A *register* is an ordered set of flip-flops. It is normally used for temporary storage of a related set of bits for some operation. This is a common activity in digital design, especially when the system must process byte- or word-organized data. You are familiar with the use of the word *register* in the context of digital computers, but the notion is more general than just accumulators and instruction registers. Multiple-bit storage is such a desirable architectural element that it is a natural candidate for building blocks. Integrated-circuit manufacturers have provided an assortment of useful devices at the MSI level of complexity.

Data Storage

Enabled D register. The most elegant data storage element for registers contains the enabled D flip-flop. Chips are available with four, six, or eight identical elements per package with common clock and enable inputs. The 74LS378 Hex D Register with Enable is typical of this building block. This 16-pin chip provides six flip-flops, each with a data input and a single Q output. The 74LS379 Quad D Register with Enable has four flip-flops, each with both output signal polarities.

As you have seen, we favor the enabled D configuration because we may hook the system clock directly to the device's clock input. The apparently small point of not gating the clock line is really of great importance to the reliability of the system, and you should adopt the practice routinely.

Pure D register. There are a few occasions when a register of pure D flip-flops is the element of choice. We can always achieve this behavior with the enabled variety by setting the enabling input to the true condition. Pure D registers are also available, usually with a common asynchronous (direct) clear input. The only reason to choose such an element is if you want the direct clear feature; you know to be wary of its 1's-catching properties.

Counters

Modulus counting. Counting is a necessary operation in digital design. Since all binary counters are modulus counters, we will explore the concept of modulus counting before we examine the hardware for it.

Counting the positive integers is an infinite process. We have a mathematical rule for writing down the integer $n + 1$ if we are given the integer n. This may cause the creation of a new column of digits; for example, if n is the three-digit decimal number 999, then $n + 1$ is the four-digit number 1,000. In an abstract mathematical sense, the creation of the fourth digit is trivial. Not so in hardware. Hardware counters are limited to a given number of columns of digits, and thus there is a maximum number that a counter can represent. A three-digit decimal counter can represent exactly 10^3 different numbers, from 000 through 999. We define such a counter as a *modulus* (mod) *1000* counter. (A *number M modulo some modulus N*, written M modulo N, is defined as the remainder after dividing M by N.) Another way of viewing this is that the counter will count normally from 000 through 999, and one more count will cause it to cycle back to 000. An automobile's odometer behaves much the same way.

Counting with the JK flip-flop. The JK flip-flop, operating in its toggle mode, goes through the following sequence

Clock pulse number: 0 1 2 3 4 5 6 . . .
Flip-flop output Q: 0 1 0 1 0 1 0 . . .

We see that the flip-flop behaves as a modulo-2 binary counter. Counters of higher moduli can be formed by concatenating other binary counters. For instance, a modulo-4 counter made from two modulo-2 counters must behave as follows

Clock pulse number: 0 1 2 3 4 5 6 7 8 . . .
Counter outputs Q_1, Q_0: 00 01 10 11 00 01 10 11 00 . . .

Can we devise a logic configuration that will cause two JK flip-flops to count in this fashion? One answer is in Fig. 4–13. Here, for drafting convenience, we draw the least significant bit Q_0 on the left, whereas Q_0 appears on the right in the usual mathematical representation of the number Q_1, Q_0. Q_0 alternates in value (toggles) at each clock. At alternate clock edges, Q_1 is clocked when $Q_0 = T$; at these times the value Q_1 toggles.

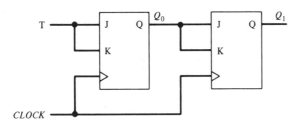

Figure 4–13. A two-bit binary counter. The least significant bit is on the left.

Figure 4–14 contains another solution that appears to give equivalent results. Again, Q_0 will toggle at each clock pulse, since $J = K = T$ on that flip-flop. This is necessary for a binary counting sequence. Every time Q_0 generates a T → F transition (H → L in this circuit), Q_1 will toggle since $J = K = T$ on that flip-flop also, and Q_0 provides the Q_1 *clock*. Figure 4–15 is a timing diagram for this circuit.

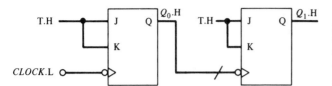

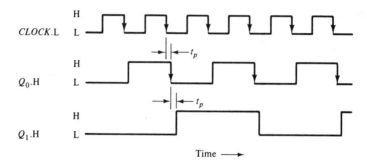

Figure 4–14. A binary ripple counter. The least significant bit is on the left. This circuit displays both logic and voltage, whereas the related Fig. 4–13 displays only logic.

The timing diagram for Fig. 4–13 is almost identical to Fig. 4–15; the difference is due to propagation delays. In Fig. 4–13, if we assume that t_p is the flip-flop propagation delay, both Q_1 and Q_0 will change t_p nanoseconds after the clock edge, since J and K were stable during the setup time of both flip-flops. We define such counters as *synchronous*.

Figure 4–15. A timing diagram for a 2-bit ripple counter. Each stage suffers a cumulative propagation delay. In synchronous counters there is only one delay.

By contrast, Q_1 in Fig. 4–14 cannot change until t_p nanoseconds after Q_0 has changed. Counters that change their outputs in this staggered fashion are called *asynchronous,* or *ripple,* counters, since a change in output must ripple through all the lower-order bits before it can serve as a clock for a high-order bit.

Ripple counters are easily extensible to any number of bits. Thus a modulo-16 ripple counter would be as in Fig. 4–16. This simple configuration is useful if you are not interested in the temporal relation of Q_3 to any lower-order bits. A common example is the digital watch, in which a 32,768-(2^{15})-Hz quartz crystal oscillator is the primary timing source. The watch display is driven at a rate of 1 Hz, using the output of a 15-stage ripple counter.

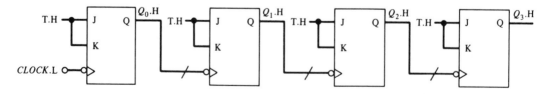

Figure 4–16. A 4-bit (modulo-16) ripple counter.

To discover the problems that can arise with ripple counters, consider a modulo-8 counter that is changing its count from 3 to 4. The ripple sequence from the initial clock edge would be

Time	Q_2	Q_1	Q_0
0	0	1	1
t_p	0	1	0
$2t_p$	0	0	0
$3t_p$	1	0	0

In a typical application, the count code is fed into a decoder to produce individual signal lines for each count. In this case, we would have momentary true hazards at decoder outputs 2 and 0, each lasting for a time t_p. These glitches are seldom useful and may be quite harmful. We can eliminate them by using a synchronous counter.

Figure 4–13 represents a 2-bit special case of synchronous counters. The rule for changing the nth bit of a binary counter is that all lower bits must be 1. Using this rule, we can construct a modulo-16 sychronous counter from JK flip-flops, as in Fig. 4–17. At the cost of the extra AND gates, we have manipulated the inputs to each flip-flop to cause the flip-flops to toggle at the proper time. Since a common clock signal runs to each flip-flop, the output changes will occur simultaneously, without ripple.

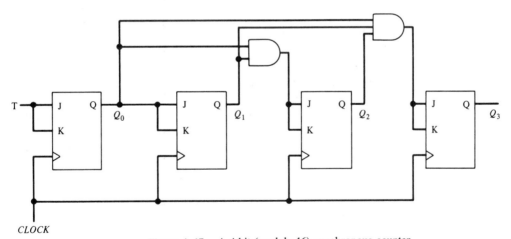

Figure 4–17. A 4-bit (modulo-16) synchronous counter.

MSI counters. Synchronous counters are so useful that manufacturers have prepared a wide variety as MSI integrated circuits, typically modulo-10 (decade) and modulo-16 (4-bit binary) devices with provisions for cascading to form higher-modulus counters. Some synchronous counters have an asynchronous-clear input. You must be careful to supply a noise-free signal to this terminal

Chap. 4   Building Blocks with Memory

to ensure reliable operation. Remember, it is a 1's catcher! Novices (and some experienced designers) tend to use the asynchronous clear feature too often. About the only time it can be safely used is during a power-up or master reset sequence to drive crucial flip-flops to a known state.

The ideal counter would be cascadable and have a synchronous clear input terminal, such as in the 74LS163 Four-Bit Programmable Binary Counter. With this device, a cascaded 12-bit synchronous counter would appear as in Fig. 4–18. Each 74LS163 chip is a synchronous counter. *CLOCK*.H is the system clock signal, and since it goes to each 4-bit chip, all output changes will be synchronous.

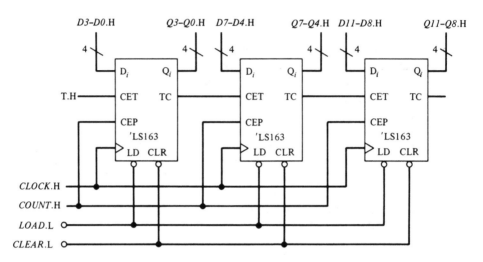

Figure 4–18. A 12-bit binary counter constructed with 74LS163 chips.

TC (Terminal Count) and *CET* (Count Enable Trickle) are individual device controls that permit proper counting in a cascaded configuration. The counting rule is that a given bit must toggle if all lower bits are equal to 1. This rule will yield the normal binary counting sequence. We may reinterpret the binary sequence as a hexadecimal sequence by grouping the binary bits into 4-bit units and giving each 4-bit unit a range of 0_{16} to F_{16}. The rule for counting a hexadecimal digit is:

all lower-order hexadecimal digits must equal F_{16}.

TC and *CET* implement this rule. The defining equation for *TC* is

$$TC = (COUNT = F_{16}) \cdot CET$$

The signal to the *CET* input comes from the *TC* output of the previous counting stage. When *CET* is true, it serves as a signal to the chip that all previous stages are at the terminal F_{16} count. *TC* is thus an output that notifies the next stage if all lower bits in the counter cascade are 1. You can see the proper cascading connections in Fig. 4–18.

CEP (Count Enable Parallel) is a master count enable signal which goes to all chips. It allows the designer to specify when the circuit should engage in counting activity.

The 74LS163 has two more controls: the synchronous clear input *CLR*, which permits clearing of the chip to zero at the next clock pulse, and the synchronous load input *LD*, which allows loading of the counter with the 4-bit pattern appearing at the data inputs. The existence of the load feature accounts for the "programmable" in the chip's name. The priority of operations is such that asserting *CLR* will override *LD*, which will override *CEP*.

It will be a useful exercise for you to derive the *J* and *K* inputs to each flip-flop in the 74LS163 counter. For an input data bit D_0, the (unsimplified) result for bit 0 is

$$J_0 = CLR \cdot F + \overline{CLR} \cdot LD \cdot D_0 + \overline{CLR} \cdot \overline{LD} \cdot CEP \cdot CET \cdot T$$
$$K_0 = CLR \cdot T + \overline{CLR} \cdot LD \cdot \overline{D}_0 + \overline{CLR} \cdot \overline{LD} \cdot CEP \cdot CET \cdot T$$

Many other synchronous counters are available as MSI chips. We have covered the 74LS163 in detail, since it is an example of a useful and well-engineered MSI building block. When dealing with circuits of MSI or LSI complexity, you must be cautious about adopting new designs. Frequently, a chip that appears to be exciting will have subtle features that make it less desirable or useless in good design. An example is the 74LS193 Four-Bit Binary Up-Down Counter. This chip has two serious flaws that may escape the attention of the "chip-happy" designer. First, its data-loading feature is asynchronous, which presents us with the necessity of keeping the data-load input signal clean at all times, to avoid its habit of reacting to any momentary true glitch. Second, the chip has two clocks, one for counting up and another for counting down. Using this chip requires very careful and arduous planning of the type that we choose to avoid entirely in our designs in Part II. Another up-down counter, the 74LS669, does not suffer from the 74LS193's deficiencies. Our point is that you must be alert to detect such deviations from good design and should choose your building blocks carefully and conservatively.

Shift Registers

A *shift register* performs an orderly lateral movement of data from one bit position to an adjacent position. We may construct a simple shift register from D flip-flops, as shown in Fig. 4–19. This circuit accepts a single bit of data *DATA* and shifts it down the chain of flip-flops, one shift per clock pulse. Data enter the circuit serially, one bit at a time, but the entire 4-bit shifted result is available in parallel. Bits shifted off the right-hand end are lost. Such a circuit is a primitive serial-in, parallel-out shift register.

In practice, we have need for four shift register configurations: serial-in, parallel-out; parallel-in, serial-out; parallel-in, parallel-out; and serial-in, serial-out. The parallel-in, parallel-out variety is the most general, subsuming the other forms. Let's design one.

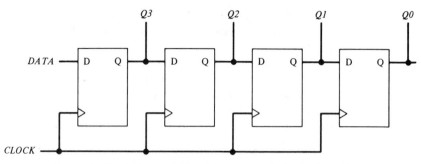

Figure 4–19. A simple shift register constructed with D flip-flops.

Assume that we are building a 4-bit general shift register. What features do we require?

(a) We must be able to load initial data into the register, in the form of a 4-bit parallel load operation.

(b) We must be able to shift the assembly of bits right or left one bit position, accepting a new bit at one end and discarding a bit from the other end.

(c) When we are not shifting or loading, we must retain the present data unchanged.

(d) We must be able to examine all 4 bits of the output.

Suppose we start with an assembly of four identical and independent D flip-flops, clocked by a common clock signal. Let the flip-flop inputs be D_3–D_0 and the outputs be Q_3–Q_0, from left to right. Let the external data inputs be $DATA_3$–$DATA_0$. We have four shift register operations: load, shift left, shift right, and hold. These will require at least 2 bits of control input to the circuit; let $S1$ and $S0$ be the names of two such control bits. Our task is to derive the proper input to each D flip-flop, based on the value of the control inputs $S1$ and $S0$. In our design of an enabled D flip-flop, we encountered a related problem, actually a subset of the present problem. There we had two operations, hold and load, that we implemented with one control input, using a multiplexer. We may employ the same technique here, using a four-input multiplexer to provide input to each flip-flop. We may then define codes $S1,S0$ for our four operations. Using $S1$ and $S0$ as mux selector signals, we may infer the proper inputs to the multiplexers. Here are the inputs for a typical bit i of the shift register:

Clock	$S1$	$S0$	Result desired	Selected mux position	Required mux input
↑	0	0	Hold present data	0	Q_i
↑	0	1	Shift right	1	Q_{i+1}
↑	1	0	Shift left	2	Q_{i-1}
↑	1	1	Load new data	3	$DATA_i$

In Fig. 4–20, the logic for the ith bit is displayed.

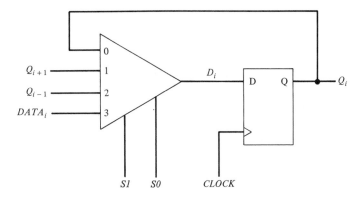

Figure 4–20. A typical bit Q_i of a general shift register.

This circuit makes a useful parallel-in, parallel-out shift register. These functions are incorporated into the 74LS194 Four-Bit Bidirectional Universal Shift Register in this way.

Providing both parallel inputs and parallel outputs requires many pins on an integrated circuit chip, so chip manufacturers make a variety of good shift registers for all four combinations of serial and parallel input and output, with up to 8 bits per package.

Three-State Outputs

In Chapter 3, you learned the advantage of three-state control of the inputs to a data bus. We may provide such control with three-state buffers, such as the 74LS244. However, some register chips provide built-in three-state control of their outputs. In bussing applications, this can be very convenient. When you are using such a chip but do not need the high-impedance state, you may permanently enable the outputs by wiring the three-state output enable line to a true value.

Bit Slices

In Chapter 3 we discussed combinational circuits that incorporated a complex set of functions for a several-bit array of inputs. The arithmetic logic unit was the central example of this purely combinational bit-slice circuitry. In Chapter 7 you will see in detail how to assemble registers, ALUs, multiplexers, shifters, and other components into a central processing unit for a computer. Experience has shown that the architecture of the processing units of a large class of register-based computers and device controllers is quite similar, even when the width of the computer words varies over a wide range. Figure 4–21 is an abstraction of this common architecture. A collection of registers, usually called a *register file,* provides inputs to an ALU. The output of the ALU passes through a shift unit before being routed back into the register file. The register file is a three-port memory that accepts three addresses and simultaneously reads from two addresses and writes into the third. Within this common architecture, the width

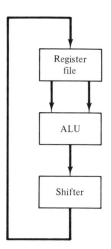

Figure 4–21. The architecture of a typical processor.

of the data paths, the internal structure of the register file, the operations performed by the ALU, and the complexity of the shifter vary. Several manufacturers have abstracted a *processor bit slice* from this architecture, and have provided integrated circuits that capture 4 or 8 bits of the architecture. These chips are frequently general enough to span a range of likely designs of processors, and powerful enough to eliminate much of the MSI-level "glue" usually found in such designs. By stacking the bit slices side-by-side, it is often possible to design a high-speed, simple, and powerful processor.

One of the first chips to be designed as a bit-slice component was Advanced Micro Devices' Am2901 Four-Bit Processor, which incorporated a 16-word register file, 16 arithmetic and logical operations, and rudimentary provisions for supporting the extra storage required for multiplication and division. Bit-slice architecture has evolved into quite powerful ALU slices, supported by large families of auxiliary chips for processor sequencing, input-output support, and so forth. The Am2903 and the TI888 are representative of the central elements in such bit-slice families. The TI888 has an auxiliary multiplier-quotient register to support multiplication and division, and also supports a wide range of logical and arithmetic operations, including binary multiplication and division and binary-coded decimal arithmetic. It has provisions for attaching an external register file, to enlarge the capacity of the holding store.

To build structures based on conventional processor architecture, you should investigate the use of processor bit-slice chips. But be aware that these are highly complex circuits whose data sheets will require considerable study before you understand the details of the architecture, set of instructions, and timing.

LARGE MEMORY ARRAYS

If a few bits of register storage are good, would 1,024 bits be better? How about 4K, 64K, or 1M? For data storage alone, the more bits per package the better. But with such a large number of bits stored, we would have to give up the ability to gain access to all the bits simultaneously, and we would need some way of specifying which bit or group of bits we wish to look at. Several forms of large-scale solid-state memories are available. Such devices have revolutionized computer technology by making inexpensive, fast storage readily available to the computer architect. Integrated circuit manufacturers are doubling the number of bits per package roughly every three years, whereas the price per bit is steadily declining. This trend will continue, and memories will be in the forefront of new electronic technology, because manufacturers can spread engineering and development costs over millions of identical units.

Random Access Memory

A large memory that requires the same time to access each data bit is called a *random access memory* (RAM). All RAMs share some common features. The unit of storage is the bit, which is built into the surface of a thin silicon wafer. The area of silicon devoted to a bit is a *cell*. Several cell structures are in use: some closely resemble the D flip-flop and others store a bit by the presence or absence of an electric charge on a microscopic capacitor embedded in the silicon surface.

A typical RAM has so many cells that it would be impossible to connect each cell to its own integrated circuit pin. To conserve pins, RAMs contain a demultiplexer to distribute an incoming data bit along an internal bus to the correct cell. Similarly, there is a multiplexer to select one cell's output and route it to the output pin. Both the demultiplexer and the multiplexer receive their control from an address supplied to the chip on a set of address lines. The address is encoded: eight pins devoted to an address can select one of 256 cells; 12 pins will handle 4,096 cells.

RAMs are characterized by their total number of storage cells, for example 4K (4,096), and by the size of the words they contain. A $4K \times 1$ RAM contains 4K 1-bit words, whereas a $1K \times 4$ RAM still contains 4K cells, but the cells are organized as 1,024 four-bit words. The 1-bit-per-word organization saves integrated circuit package pins as compared to the 4-bits-per-word structure. (You should figure out why this is so, and how many pins are saved.) Thus in large memories, 1-bit-per-word RAMs are common, whereas designers of smaller memories often use the 4-bits-per-word type to keep down the number of chips devoted to memory.

Large RAMs require many address bits to specify the cells. A $64K \times 1$-bit RAM requires 16 address bits; a $1M \times 1$-bit RAM requires 20 address bits. In each case there are too many address bits to allocate each bit to a separate pin on the chip. To alleviate this problem, the address bits in large RAMs are split into two parts, a *row address* and a *column address*. At the proper time in the memory cycle, each part is fed to the RAM over the same address inputs—that is, the row address and column address are *time-multiplexed*. (The row-and-column nomenclature derives from the internal structure of the RAM, which can be viewed as a large, two-dimensional array with elements addressed by row and column indices.) A 256K RAM has 9 input pins devoted to the address;

the row address and the column address each have 9 bits. The multiplexing of address inputs saves valuable pins on the chip, but at the cost of considerable additional complexity in the timing of the memory. The multiplexing of addresses permits physically smaller integrated circuit packages, which saves valuable space on printed circuit boards having many RAM chips. We discuss aspects of RAM memory timing later in this chapter.

Memory system organization. Just as RAM chips have an internal bus structure, so is each chip designed to be a component of a bus-organized memory system in which only one unit of memory is available at any time. Suppose we are given the task of building a $3M \times 2$ memory (3 million words, 2 bits per word), using $1M \times 1$ RAM chips. We would create a pair of system busses, data in and data out, and hang the memories onto the busses as shown in Fig. 4–22. As in all bus-organized systems, we must have some way of selecting a given element while ensuring that all other devices stay off the bus. RAMs provide a *chip enable* (*CE*) for this purpose. The system in Fig. 4–22 enables RAMs in pairs, since the original specification called for 2 bits per word. The input and output busses may be either open-collector or three-state, depending on the type of chip. For most applications, we prefer the three-state systems because they are faster and eliminate the open-collector's pull-up resistors.

Having developed an architecture for our memory system, we must now

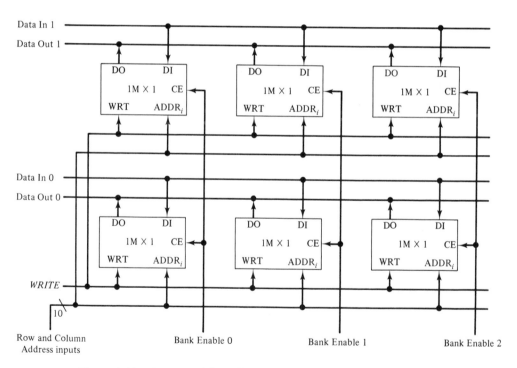

Figure 4–22. A 3M-word by 2-bit memory constructed with 1M × 1 RAM chips.

determine how to address a particular word in the memory. To derive this address, we shall back away from the chips, look at the memory system as a whole, and ask how we specify a location within the system. Our memory device will need a 22-bit address, since $2^{21} = 2M$, and $2^{22} = 4M$. Each 1-megabit RAM chip requires 20 bits of address; the remaining 2 bits will specify which of the three *banks* of 1M-bit RAM chips is being selected. Although the partition of the address is arbitrary, it is common to select the bits for the bank field from the most significant positions in the address. Since 1M-bit RAM chips require multiplexed address inputs, the 20 bits of RAM chip address must be presented as a 10-bit row address followed by a 10-bit column address. The division of the 20 bits of chip address into row and column addresses is arbitrary, and is often decided by the ease of layout on the printed circuit board. Here is one straightforward choice for the system memory address:

$$\textit{System Memory Address}(A_{21} \cdots A_0) = (A_{21}A_{20})(A_{19} \cdots A_{10})(A_9 \cdots A_0)$$

Bank	Column	Row
Address	Address	Address

We can now route the 10 RAM address lines to all RAMs in parallel, and use the bank address $A_{21}A_{20}$ as a code for the proper bank. A decoder can produce individual enable signals for each bank from the 2-bit bank address code:

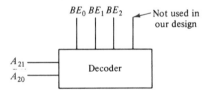

BE_0 enables the chips in bank 0, BE_1 enables bank 1, and so on. A bank contains two RAM chips, since there are 2 bits per word. (The row-select and column-select control lines are not shown in the diagram.)

The last system-wide signal is the read-write selector *WRITE*. RAMs can both read and write, so we must supply a logic signal for the operation to be performed when the chip is enabled. By convention, *WRITE* = F implies reading; *WRITE* = T means writing.

RAM timing requirements. RAMs are unclocked (asynchronous) devices that require detailed specification of the timing of their control and data signals. Timing is a function of the internal technology of the chip, and varies widely with the type of memory. To achieve reliable RAM operation, you must adhere rigidly to the manufacturer's requirements. In this section, we will present a few of the major timing parameters. The crucial parameters are:

(a) The period of stability of the address lines.
(b) The period when the chip enable *CE* is active.

(c) The value of the *WRITE* signal.

(d) During RAM writes, the period of stability of the write-data input.

(e) During RAM reads, the period of stability of the read-data output.

There are three parameters that characterize all RAMs. The *read cycle time* is the minimum time that must elapse after the start of a read operation until another operation may begin. This usually determines the time during which the address must be stable. Read-cycle times may range from a few nanoseconds to several hundred nanoseconds, depending on the type of chip. The *write cycle time* is the corresponding parameter for the write operation. It is usually similar in magnitude to the read-cycle time. *Read access time* is a measure of the time that must elapse after the start of a read operation before read data is available for use. The access time for read is equal to or less than the cycle time for read.

For RAM write operations, the data sheet specifies when the *WRITE* signal may become true relative to the address, chip enable, and data signals, and for how long *WRITE* must remain stable to complete the write operation.

Besides these fundamental measures of RAM performance, the data sheets usually contain a welter of other timing figures, specifying times between various possible signal changes. You can simplify all this by making such reasonable design assumptions as:

(a) Chip enable (*CE*) becomes true at the same time as the address becomes stable.

(b) During a RAM write operation, the data to be written stabilizes at the same time as the address lines.

With these assumptions, provided you choose the most conservative values for the timing parameters, you can usually reduce the number of relevant timing figures to a handful.

RAM timing parameters fall loosely into two groups: setup times and hold times. *Setup* implies that a signal must be stable prior to some event; it is similar to the setup time for inputs to clocked circuits. *Hold* time means that an input signal must remain stable for a time after some event.

Read access time is an example of a setup time; it describes the required period of stability of the address prior to the appearance of stable data at the output. The data sheets also contain setup times for chip enable. The intervals during which address, chip enable, and data must remain stable after *WRITE* becomes false are examples of hold times. Conversely, in RAM reading, the data is often stable for a period after the address or *CE* changes; such a period is also called a hold time.

In using RAMs in digital designs, you must remember that there will also be delays in the host system. These will arise from combinational propagation delays, bus driving delays, and so on. The RAM timing computations begin with the arrival of stable signals *at the RAM,* so you must take into account any delays in the host when you determine how fast your system will really be.

Static RAM. The *static RAM* contains bit cells that are similar to a type D flip-flop, similar enough that you may use the D flip-flop as a model for predicting static RAM behavior. As long as the address is constant and the *WRITE* line is false, the selected cell will continue to put its read data onto the output data bus. A transition from false to true on the *WRITE* line serves as a clock signal to the RAM to cause input data to be written into the specified RAM cell. As long as the power is on, the static RAM will retain its contents without the need for intervention. The address lines of most static RAMs are not multiplexed—the entire address is presented to the chips, considerably simplifying the RAM control circuitry. These properties make static RAMs simple to incorporate into designs, once you have mastered the basic timing requirements. We will use a static RAM for the memory in our minicomputer design in Chapters 7 and 8.

Dynamic RAM. This device stores data on a tiny capacitor within each bit cell. The dynamic storage cell is much smaller than a cell in a static RAM, typically allowing four times as many bits per unit area of silicon. This factor of 4 becomes of overwhelming importance in large memory systems and nearly all large systems use dynamic RAM chips. The penalty is a significantly increased complexity of the control of the memory; nevertheless, you should consider dynamic RAMs for systems that would require more than, say, 32 static RAM chips.

Most dynamic RAMs have such a large storage capacity that their address inputs are multiplexed to save pins. The multiplexing complicates the control of the RAM—each half of the address must be presented separately to the RAM. The RAM has additional *row address strobe* and *column address strobe* control pins, which the circuit that controls the memory must assert at the proper time in the memory cycle to announce that the row address or column address is stable on the address input pins.

The storage element in a dynamic RAM is a capacitor which, like all capacitors, is an imperfect holder of charge. After a period of time—a function of the temperature and geometry of the device and of circuit technology—the charge will leak away. To preserve the data, the system using a dynamic RAM must periodically read and rewrite the stored contents. This periodic restoration is called *refreshing:* the entire memory must be refreshed at intervals of several milliseconds. Dynamic RAMs allow an entire column of bits to be refreshed at once. A typical RAM controller will cycle through the column addresses, inserting a refresh cycle at regular intervals so that each column is refreshed in a timely fashion. The control circuitry of dynamic RAMs may be contained within each RAM chip or in a separate controller chip.

Read-Only Memory

Read-only memory (ROM) is actually a write-once, read-thereafter memory. Since the writing takes place during the manufacture of the chip, large numbers of identical ROMs can be fabricated at low cost. Changing the write pattern is very expensive, and ROMs are therefore only appropriate when we can amortize the initial cost by purchasing several thousand identical chips.

The bussing and timing requirements tend to be similar to those of the static RAM, with the omission of the now-superfluous read-write line.

The ROM has important uses in digital design as a permanent memory, a code converter and, surprisingly, a logic function generator.

Firmware memory. Consider the ubiquitous pocket calculator with transcendental functions such as square root and trigonometrics. A debugged square-root routine need never change, and could therefore be committed to ROM and made a part of the calculator's address space. Such an application is called *firmware*. ROM firmware remains in the memory when the power is off, and is ready to use as soon as power is restored.

Code conversion. As you learned in Chapters 2 and 3, we may use conventional Boolean algebraic techniques to synthesize outputs from inputs. In a sense, we are using gates to *compute* the outputs. Consider Exercise 1–36, in which a 4-bit BCD code generates the outputs to drive a seven-segment lamp. To aid your recollection, we will sketch the Boolean algebraic solution:

(a) List the BCD bit patterns for digits 0 through 9, and draw the corresponding segments to be lit for each digit:

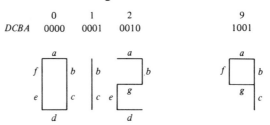

(b) Plot each segment *a* through *g* on a 4-bit Karnaugh map with the unused codes 10 through 15 plotted as "don't cares." Derive equations for each of the seven segments. For example, the result for segment *a* is

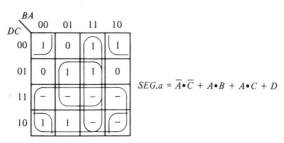

$$SEG.a = \overline{A} \cdot \overline{C} + A \cdot B + A \cdot C + D$$

(c) Assemble gates to compute the functions *a* through *g*. In other words, synthesize logic circuits corresponding to the equations for segments *a* through *g*.

Viewing the problem in this light, we have a large combinational logic circuit that accepts four inputs and computes seven outputs. We may package this as a building block if we wish; such a seven-segment lamp driver is available as an MSI chip.

From another viewpoint, the seven-segment lamp driver is a problem in code conversion, in which we must convert a 4-bit BCD code into a 7-bit lamp driver code. A seven-function truth table with four inputs and seven outputs is a convenient way to display the code conversions; refer to Table 4–1. We may consider the truth table as representing the contents of an addressable memory: Each row of the truth table is an address, and the corresponding set of output values is the contents of the addressed memory element. Table 4–1 then represents a 16-word memory, with 7 bits in each word. The outputs (the contents of the memory) do not change, so we can realize the truth table as a 16-word by 7-bit ROM; such a ROM requires four address inputs: the BCD code.

TABLE 4–1. TRUTH TABLE FOR A BCD-TO-SEVEN-SEGMENT CODE CONVERTER

| Row | Inputs | | | | Outputs | | | | | | |
	D	C	B	A	a	b	c	d	e	f	g
0	0	0	0	0	1	1	1	1	1	1	0
1	0	0	0	1	0	1	1	0	0	0	0
2	0	0	1	0	1	1	0	1	1	0	1
3	0	0	1	1	1	1	1	1	0	0	1
4	0	1	0	0	0	1	1	0	0	1	1
5	0	1	0	1	1	0	1	1	0	1	1
6	0	1	1	0	0	0	1	1	1	1	1
7	0	1	1	1	1	1	1	0	0	0	0
8	1	0	0	0	1	1	1	1	1	1	1
9	1	0	0	1	1	1	1	0	0	1	1
10	1	0	1	0	–	–	–	–	–	–	–
11	1	0	1	1	–	–	–	–	–	–	–
12	1	1	0	0	–	–	–	–	–	–	–
13	1	1	0	1	–	–	–	–	–	–	–
14	1	1	1	0	–	–	–	–	–	–	–
15	1	1	1	1	–	–	–	–	–	–	–

The conversion of a 7-bit ASCII code to the corresponding 12-bit Hollerith card code can be accomplished with a ROM. The ASCII code contains 128 characters; each character has a Hollerith representation as a set of holes (1's) and blanks (0's) in the 12 rows of a card column. The ROM will contain $2^7 = 128$ words, each of 12 bits. Every ASCII code represents the address of one ROM word; the content is the corresponding Hollerith code. Each word of ROM contains a unique code.

ROM may also be used to convert a valid 12-bit Hollerith code into 7-bit ASCII. This ROM has 12 address inputs and consists of $2^{12} = 4,096$ words of 7 bits. Of the 4,096 rows in the Hollerith-to-ASCII truth table, only 128 are relevant; the rest correspond to don't-care outputs, present in the ROM but of no interest in this code conversion.

Generating logic functions. The foregoing descriptions show that a ROM provides a general way to generate an arbitrary logic function. Every logic function has a truth-table representation. A ROM gives the function value, T or F, for every row in some canonical truth table; the ROM explicitly contains each bit of information in the full truth table. In a synthesis of logic functions using gates, we explicitly use only a portion of the truth-table's information— the product terms leading to true values of functions (or, equivalently, those leading to false values). Furthermore, we often can simplify the logic expressions to reduce the number of terms.

Such simplifications are not useful, and in fact are disadvantageous, in ROM synthesis of logic functions. Although the ROM approach is completely general, it suffers severely because the size of the ROM is doubled by each additional input variable. In logic synthesis, ROMs are best used in situations similar to code conversion, where highly encoded information must be transformed into a large number of output functions. Other, more appropriate, devices are available to support the uniform synthesis of logic functions. We will discuss these devices later in this chapter.

Field-Programmable Read-Only Memories

In the preceding section you learned some of the uses of the ROM in digital design. A ROM must be programmed by the manufacturer and is therefore not a suitable tool in the developmental phases of a design, nor in systems in which the read-only material must occasionally be altered. When only a few copies of a system are required, or occasional changes in the memory are required, we need permanent or semipermanent memories that can be programmed by the designer in the laboratory. Several types of *programmable memories* are available.

PROM. The *programmable read-only memory,* or PROM, is a ROM in which the one-time writing process has been deferred to the end user. During manufacture, the bits of the PROM have microscopic metallic or polysilicon fuses that set all the bits to 1. The user can blow these fuses by selecting a given bit and then applying a pulse to a special programming pin, thereby creating a 0. The process is inexpensive and relatively easy, and PROM programming devices are readily available. In use, the access to a PROM is rapid, equivalent in speed to ROM. PROMs are also physically similar to ROMs, and the layout of the circuit board is simplified in systems that will eventually contain ROMs. Once the PROM is programmed, it cannot be reprogrammed.

EPROM. The *erasable programmable read-only memory* (EPROM) is an even more flexible design based on the ROM. As its name implies, the EPROM allows the designer to erase the contents and start over. Bits are stored by electric charges that are trapped in the silicon of the chip. When erased, all the bits of the EPROM become 1's. To write a zero value into a bit, the bit receives a high voltage that temporarily makes the silicon a conductor, allowing charge to accumulate at the site of the bit. After the high voltage is removed, the trapped charge remains, essentially forever. EPROMs are programmed with PROM programmers, using techniques similar to, but more complex than, those used to program PROMs.

A transparent quartz lid covers the silicon of the EPROM chip. Erasure occurs when high-intensity ultraviolet light makes the silicon a weak conductor, allowing the trapped charge to bleed slowly away from the bits in the memory. To be erased, the chip must be placed in an *EPROM eraser* and exposed to ultraviolet light for about a half-hour. The EPROM will withstand many erasures before it becomes unreliable.

The EPROM is convenient in system development because the designer may write the content of the memory, test the design and, if necessary, erase the old pattern and install a new one. The housing of microcomputer firmware is a common and important use of EPROMs, but we can use them with equal facility to generate digital logic functions. The device is slower than the PROM, but its capacity for erasure makes it a potent design tool. EPROMs are physically larger than ROMs or PROMs, and have longer access times. Sizes range from 16K ($2K \times 8$) to 256K ($32K \times 8$).

EEPROM. The *electrically-erasable programmable read-only memory* (EEPROM) may be altered without removing it from the circuit in which it is used. Sequences of voltages of 5V or greater are applied to special programming pins to permit the rewriting of selected bytes of the memory. This device and its cousins are important in designs in which in-place "writing with difficulty" is necessary—for instance in a unit housed in a remote area that must be manipulated at a distance. The structure and the pin assignment of EEPROMs are not compatible with other read-only devices, and only a meager selection of chips is available. A typical size is 16K, as exemplified by the $2K \times 8$-bit 2816 EEPROM.

As you have seen, the devices for "read-only memory," whether permanent or reprogrammable, are used in data and program storage, in code conversion, and even to generate random logic functions, but the first two uses of ROMs are by far the most important. We shall now describe programmable devices suitable for general logic applications in digital design.

PROGRAMMABLE LOGIC

Programmable logic provides a systematic way to generate complex logic functions. Programmable logic devices allow the designer to create arbitrary sum-of-product functions of many variables, in a highly structured and regular manner. The term *programmable* means that the functions may be specified after the chip has been manufactured.

Usually the programming is accomplished by blowing (breaking) fuses along the chip's internal data lines, as is done in the PROM. The fuse is a narrow ribbon of metal or other conductor deposited during the manufacture of the chip, and it serves a role similar to the three-state gate. When the fuse is intact, it allows an input signal to proceed unimpeded; when the fuse is blown, no signal can pass and the input is disconnected from the remainder of the circuit. Unlike a three-state circuit, once the fuse is blown, its input is forever disconnected.

As you saw in Chapter 1, it is always possible to express an arbitrary combinational logic function in sum-of-products form. This is the starting point of the application of programmable logic. To construct an arbitrary sum-of-products function, we need to be able to form the required AND (product) terms, and then form the OR of the product terms to create the appropriate sum.

The PLA

The most general programmable logic structure is the *programmable logic array* (PLA), which allows the programming of arbitrary products and arbitrary sums. Within a VLSI design, the PLA is an important tool for creating complex circuits on a single chip. The PLA may also be packaged as an LSI-level integrated circuit, to be used at the level of design that we are now studying. But because of its extreme generality, requiring many inputs, many outputs, and much internal circuitry, the PLA is not widely available as a discrete integrated circuit.

Consider the programmable logic implementation of a single function of three variables

$$TEST = \overline{A} \cdot \overline{B} + \overline{A} \cdot C + A \cdot B \cdot \overline{C}$$

In this example, the programmable device must have at least three inputs and one output, and must have the capability of specifying at least three product terms within its structure. Within the device, each input passes through a non-inverting and an inverting buffer, thereby providing the true and complemented forms of each input variable. Figure 4–23 is a schematic of the portion of the PLA required for this example. Every vertical line represents a product term, and each horizontal crossline is an input to the AND gate that forms the product. The horizontal line at the bottom represents the single sum of the product terms; each product term is an input to the OR gate that forms the sum. Before programming, each crossing has a fuse that determines if the input will contribute to its product or sum. In Fig. 4–23, each vertical line initially generates the trivial case of

$$A \cdot \overline{A} \cdot B \cdot \overline{B} \cdot C \cdot \overline{C}$$

which, of course, is identically false. The act of programming the PLA destroys the unwanted fuses. In Fig. 4–23, the large dots represent the remaining fuses; all the other fuses have been blown.

Usually, the designer wishes to create several functions using the same input variables. Figure 4–24 is a PLA pattern for producing four functions of five input variables. This PLA is capable of receiving up to six input variables, generating up to twelve different product terms, and producing up to five different

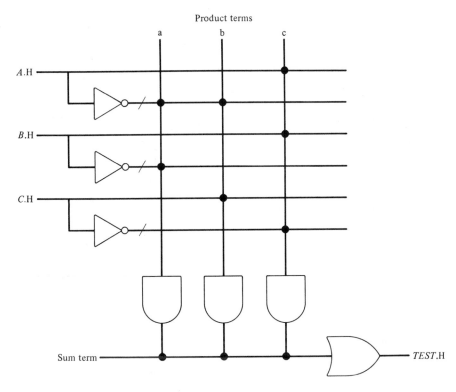

Product terms

Figure 4–23. A PLA realization of $TEST = \overline{A} \cdot \overline{B} + \overline{A} \cdot C + A \cdot B \cdot \overline{C}$.

sums of these products. One input, three product terms, and one sum output are unused. The equations implemented by this PLA are

$$Y_1 = I_3 + \overline{I}_1 \cdot I_2 + I_1 \cdot \overline{I}_2 \tag{4-1}$$

$$Y_2 = \overline{I}_3 \cdot I_4 \cdot \overline{I}_5 + I_1 \cdot \overline{I}_3 \cdot I_4 \tag{4-2}$$

$$Y_3 = \overline{I}_3 \cdot I_4 \cdot \overline{I}_5 + I_2 \cdot I_3 \cdot I_4 \cdot I_5 + \overline{I}_2 \cdot \overline{I}_4 \tag{4-3}$$

$$Y_4 = I_1 \cdot \overline{I}_3 \cdot I_4 + \overline{I}_2 \cdot \overline{I}_4 + \overline{I}_3 \cdot I_4 \cdot \overline{I}_5 + I_3 \cdot \overline{I}_4 \cdot I_5 \tag{4-4}$$

A typical PLA that is available as an integrated circuit is the National Semiconductor DM7575, which accepts 14 inputs, supports 96 product terms, and generates 8 logic functions as outputs.

The PROM as a Programmable Logic Device

We have already mentioned that the programmable memory devices, exemplified by the PROM, can be used to generate logic functions. We usually think of a PROM as a memory device whose n address inputs select one of 2^n words of memory. To perform this selection, the PROM must contain a complete decoding of its address inputs. Think of a truth table with the address lines as the inputs.

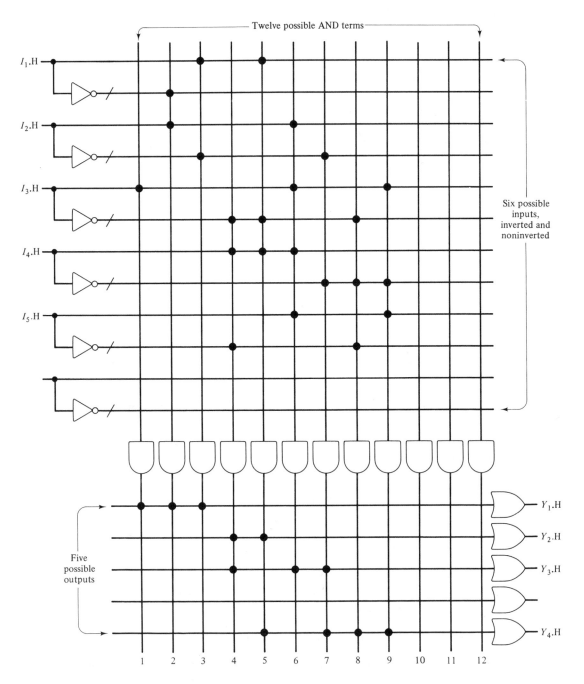

Figure 4–24. A PLA realization of Eqs. (4–1) to (4–4). The sixth input, the fourth output, and AND gates 10, 11, and 12 are not used. This PLA is smaller than commercial chips.

For a PROM, the truth table is canonical—it explicitly contains all 2^n rows. Thus, the PROM provides *all possible product terms* of its inputs, whether we like it or not. Each bit of the PROM's output can represent a logic function—a column in the output section of the truth table. To generate a logic function, we must specify whether each canonical product term is to be a contributor (the value in the memory is 1) or not (the value in the memory is 0). In such applications as code conversion, this is quite useful, since we are concerned with converting all possible patterns of input bits. For random logic, the PROM is of limited attractiveness.

Monolithic Memories, a leader in the development of programmable logic devices, has coined the term *PLE, programmable logic element,* for such applications. Manufacturers produce a limited variety of PLEs, with a fixed array of ANDs (fixed product terms) but with less than the full canonical set found in PROMs.

The PAL

The PLA generates a general sum of general products; within the limits of the design, the designer may specify the detailed structure of each product term and each sum of products. However, the great generality of the PLA makes for difficulties in its manufacture and use. The PROM (or PLE) provides a predetermined set of product terms, and the designer may specify the nature of each sum of products. This structure is useful as a memory and for logic operations requiring the full decoding of inputs, but the PROM has limited application as a tool for programmable logic.

The PAL (*programmable array logic*) allows the designer to specify the nature of the product terms, but the ways in which the products may be formed into sums is fixed in the chip. In practice, the PAL is by far the most useful of the trio for generating random logic functions. The programmable product array, coupled with a limited summing capability, fulfills the requirements of digital designs well. The term PAL is a registered trademark of Monolithic Memories, which originated this application. PALs are available from many manufacturers in a wide variety of useful configurations. The PAL is indeed a pal of the designer of digital circuits.

The PAL18L4, shown schematically in Fig. 4-25, is typical of the PALs that generate combinational logic functions. As suggested by its number, the PAL18L4 has 18 input pins and 4 output pins. The "L" signifies that the outputs are produced in low-active (T = L) form. Two of the 4 outputs, pins 18 and 19, each provide the sum of four product terms. The other two outputs, pins 17 and 20, each provide the sum of six product terms. Within the chip, each of the 18 inputs is split into its true and complemented form. Each of the twenty product terms available in this chip can contain any combination of the true and complemented forms of the 18 inputs. Fuses appear at each intersection on the product-term lines. The programming task consists of blowing the unwanted fuses.

18L4

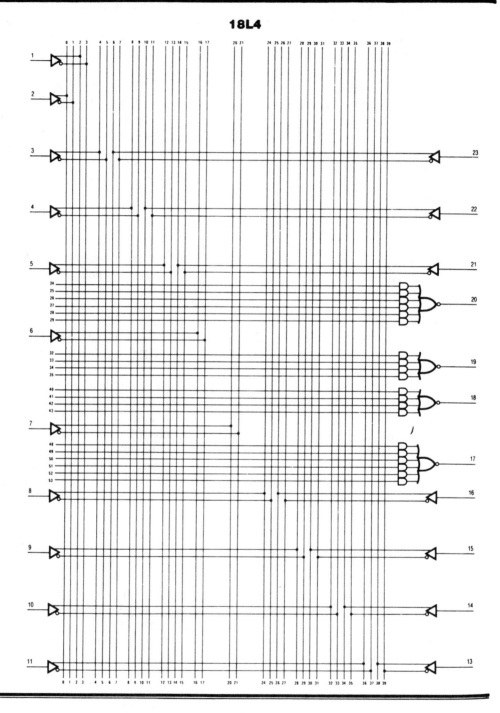

Figure 4–25. A PAL18L4. (Courtesy of Monolithic Memories, Inc.)

Figure 4–26 shows an implementation of *Y1, Y2,* and *Y3* in Eqs. (4–1) through (4–3), using the PAL18L4. Unused portions of the PAL do not appear in the figure. *Y1*.L, *Y2*.L, and *Y3*.L are formed as sums of products, drawn directly from the equations (the high-active *Y2*.H is discussed later). Each bold dot denotes a contribution to a product term, and represents a fuse that is to remain unblown. The *unused* product terms in a sum have been blacked in. Inputs of either voltage polarity are easily handled, using standard mixed-logic notations. Within the chip, the PAL generates sums of products with T = H. The manufacturer's diagram shows the input buffers with the inverted voltage form of the input emerging below the noninverted form in each buffer. If T = H at the input, the logic inversion occurs on the bottom of the two buffer outputs; if T = L at the input, we redraw the input buffer to show that the logic-inverted form emerges from the upper buffer. Figure 4–26 contains illustrations of the notational changes to accommodate low-active inputs.

The PAL18L4 produces low-active sum-of-products outputs. If we desire a high-active output, we mixed logicians have several choices. We may add a voltage inverter to the output signal to change its polarity—a reasonable but inelegant solution. If the PAL has an unused output and an unused input, we may feed the low-active form of our signal back into the PAL and produce the high-active form on the unused output. We may use a different PAL, such as a PAL18H4, that produces high-active outputs. Another approach, useful when one is striving to use minimal hardware, is to generate a sum-of-products form

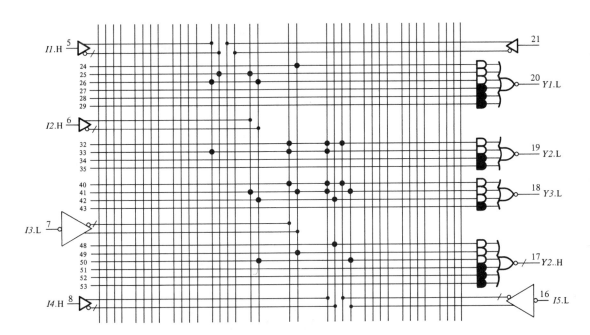

Figure 4–26. An implementation of Eqs. (4–1) through (4–3), using the PAL18L4. Note the mixed-logic changes to the diagram.

of the *inverse* of the desired function. The mixed logician immediately recognizes that the output represents a high-active version of the desired function

$$FUNCTION.\text{H} = \overline{FUNCTION}.\text{L}$$

For instance, suppose we wish to produce the Y_2 of Eq. (4–2) in high-active form, using the PAL18L4. To generate the inverse of Y_2, we may plot a Karnaugh map of the function, circle the 0's, and write an equation for $\overline{Y}_2$:

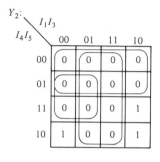

$$\overline{Y}_2 = \overline{I}_4 + I_3 + \overline{I}_1 \bullet I_5$$

Figure 4–26 contains the resulting implementation of $Y_2.\text{H}$.

Many PALs are available with clocked flip-flops and three-state control at their outputs. An example is the PAL16R8, shown in Fig. 4–27. The R8 in the chip number signifies a PAL with eight "registered" outputs. Each of the eight flip-flops in the register receives a clock signal from a single input pin, and each flip-flop's output is buffered by a three-state inverter controlled by another input pin. The 16 inputs implied by the chip number arise from 8 external input pins and 8 signals fed back from the flip-flop outputs.

PAL, PLA, and PLE Programming

Programmable logic devices are a powerful tool for the designer, providing the equivalent of complex, tedious logic and architecture within compact and relatively inexpensive integrated circuit chips. These devices must be programmed and, with the exception of reprogrammable devices such as EPROMs and certain PALs, the programming is a one-time, unalterable act. The programming requires special equipment for the presentation to the chip of a complex sequence of properly timed voltages. Modern "programmers" are costly, but they provide many necessary capabilities conveniently and reliably. Most such programming devices are capable of accepting programming data from an external source— valuable when one is transmitting computer-generated programming files. The fuse patterns become the raw data for the programmer, but developing fuse patterns from diagrams such as Fig. 4–26 is a tedious and error-prone task. Many programmers for PALs permit the designer to enter logic equations directly; these equations, expressed in a suitable notation, provide the input to a software translator that produces fuse patterns.

16R8

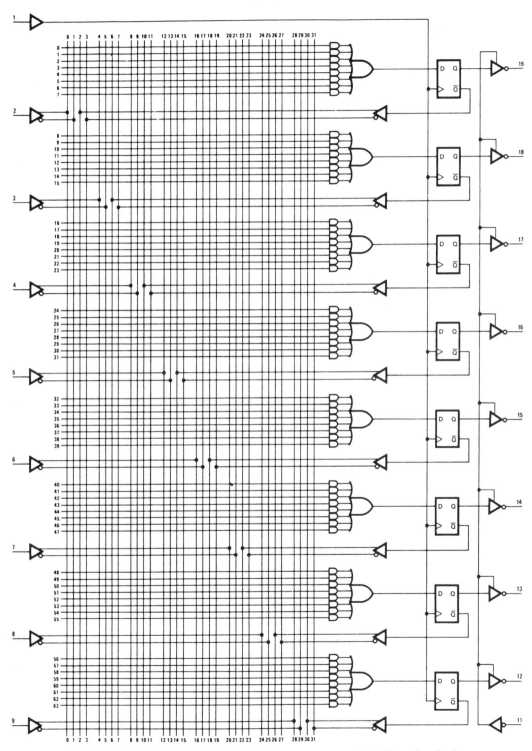

Figure 4–27. A PAL16R8. (Courtesy of Monolithic Memories, Inc.)

Programmable logic devices are powerful design tools that provide some of the desirable attributes of VLSI for the chip-level designer, without the extraordinary commitment necessary to develop a custom-made VLSI chip. The size and complexity of programmable logic devices are growing rapidly—PALs with 64 inputs and 30 registered outputs, and containing more than 32,000 fuses, are available. You will read about several applications of PALs and PROMs in later sections of this book.

TIMING DEVICES

Such systems as RAMs and ROMs have rigid timing requirements that a designer must obey. In this section, we discuss two nondigital devices, the single shot and delay line, that help the designer to develop appropriate signals for timing.

We might use a single shot or a delay line when we must derive a timing event of arbitrary duration and with respect to some arbitrary starting point. For instance, in RAM reading, we must produce a timing signal for the host system that starts when the RAM read operation begins, and ends only after the read access time has elapsed. This time is independent of the system clock; the RAM's timing requirements remain the same whether the system clock is slow or fast.

The idea is simple, but the simplicity of the concept does not mean that the corresponding hardware is simple to use. Another way of describing the arbitrary starting time and duration of the operation is to say that the timing is *asynchronous*. Synchronous devices derive their timing from some external source, usually the system clock. Asynchronous devices are *internally* timed. This is not inherently bad, but the result is that independent timings are spread throughout the system. We have relinquished central control, and for this reason alone we should avoid asynchronous devices unless absolutely necessary.

Centralized control is nice, but we cannot always have it. Memory devices, for example, require internal timing that is independent of the host system. We cannot do without devices for memory so we are forced to generate their arbitrary timing locally. The alternative is distinctly worse: we could have central control if we rigidly fixed our system clock to match the local asynchronous timing requirements. This would eliminate our most powerful hardware debugging tool, the ability to slow the central clock to zero speed, thereby freezing our system in a given condition.

The Single-Shot

The *single-shot,* or monostable multivibrator, is based on the electrical properties of capacitors. A capacitor stores a charge (electrons) as a function of the voltage impressed across the capacitor. The amount of stored charge q is proportional to the impressed voltage V; the proportionality constant is the capacitance C. Thus

$$q = CV$$

Capacitors require a period of time for the charge to build up or decay, and it is this behavior that allows the single shot to function as a timed delay element.

A single-shot behaves like the electrical circuit below

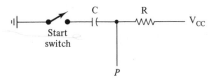

Normally, the single-shot switch is open and no current is flowing. The voltage at point P is V_{CC}, which causes the single-shot output Q to be false. When we trigger the single-shot by (digitally) closing the switch, charge begins to rush into the capacitor C. Because of this current flow through R, the voltage at point P becomes low, which in turn causes the single-shot output Q to assume a true value. As the flow of current charges the capacitor, the voltage at P slowly rises back toward V_{CC}. When the voltage at P reaches a certain level, the single-shot turns off, bringing output Q to false. The time during which Q is asserted is the single-shot delay time.

By choosing capacitor C and resistor R according to tables or formulas in the single-shot data sheet, we may select any desired single-shot delay time within a wide range of values. Typical delays, using the popular 96L02 Dual Single-Shot, range from about 50 nanoseconds to 10 seconds.

In circuit diagrams, we draw a single-shot as a rectangle similar to a flip-flop and add the external timing capacitor and resistor (see Fig. 4–28).

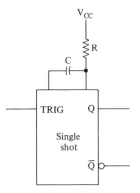

Figure 4–28. A single-shot circuit symbol.

Single-shots respond to a F → T transition on the trigger input, so they will start up if any glitch occurs at the input. Therefore, you must be careful to drive the single shot with a clean trigger to avoid spurious results. The value of the single-shot delay will drift somewhat, and 5 percent stability is about all you should expect.

The Delay Line

Delay lines, on the other hand, generate highly stable delays. Their stability derives from the fact that they are passive devices—they have no transistor amplifiers built into them, as do all gates, flip-flops, and so on. This explains their lack of power-supply pins, since with no active amplifiers they do not need operating power.

Delay lines are available in standard integrated circuit packages. Internally, they are constructed from a series of stable inductors and capacitors. Whatever waveform is put into the delay line will appear, after a given delay, on the output. There is no concept of a trigger; the waveform at the input is simply reproduced after the delay.

In their most convenient form, delay lines have a series of taps (pins), often 10, that yield fractions of the nominal delay. For instance, in a 10-tap delay line with delay rated at t_d, the nth tap will produce a delay of $t_d \times n/10$. Such taps are useful in generating the timing in dynamic RAMs, which require a sequence of accurately timed pulses of varying duration. We may obtain the pulses by sending an edge down the delay line and using an Exclusive Or gate to detect the period during which the edge has arrived at the input but not yet reached the output. For example, suppose that we need a pulse that starts 200 nanoseconds after *START* becomes true and lasts for 150 nanoseconds. Figure 4–29 is a diagram of the circuit, using the standard delay-line symbol.

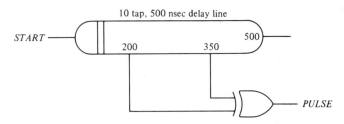

Figure 4–29. A circuit for a delay line to produce a 150-nsec pulse delayed by 200 nsec.

Delay lines are moderately expensive but can generate delays from about 2 nanoseconds to greater than 1 microsecond. Since a delay line faithfully transmits the input waveform, you must be sure the signal at the input is clean. Often the operating specifications of the delay line will require that the input signal be derived from a line driver, and that the output be properly terminated (see Chapter 12).

THE METASTABILITY PROBLEM

We began this chapter with a discussion of hazards, a nuisance created by the characteristics of physical devices used to implement logical concepts. In Chapter 5 you will encounter other design pitfalls rooted in physical behavior—pitfalls that arise through the interactions of several components of a design. In this chapter, there remains to discuss the most alarming physical problem of all—metastability. We will alert you to the problem and give some advice, but you should look to Chapter 12 for a more extensive treatment of this topic.

Digital devices are fundamentally analog devices that behave digitally only when stringent rules of operation are obeyed. Sequential devices contain amplifiers (gates) and feedback loops to achieve their storage properties. In addition to establishing proper voltage levels at the inputs, to assure proper operation of a sequential device you must adhere to the setup times, hold times, and other timing specified in the data sheets. When the operational requirements are met, the device's outputs will be proper digital voltage levels, and changes in the level of the output will occur quickly and cleanly. Except during the rapid period of transition, the circuit remains in one of its stable states. You have seen that there are difficulties associated with the RS flip-flop when one tries to move from the $R = S = $ T input configuration to the hold configuration, in which $R = S = $ F. The difficulties arose from the attempt to change both inputs simultaneously. As long as no more than one input is changing at a time, the sequential circuit performs well, but if the voltage level of more than one input is allowed to change at nearly the same time, the circuit is being required to perform outside the framework of design for digital operation and the result may be unpleasant. For the proper operation of clocked circuits, the setup and hold times require that certain inputs must not change too near the time that the clock signal is changing.

Violation of the timing requirements of a sequential circuit may throw the circuit into a *metastable state,* during which the outputs may hold improper or nondigital values for an unspecified duration. In one form of metastability, the output voltage lingers for an indefinite period in the transition region between digital voltage levels, before it eventually resolves into a stable value. In another form of metastability, the output appears to be a proper digital value, but after an unpredictable interval switches to another value. Metastability can be disastrous. In synchronous design, we sidestep the problem by never changing the inputs in the vicinity of the clock. As you will see, this allows vast simplification of the design of complex circuits. But every circuit is at some point exposed to external reality—other circuits with different clocks, unclocked or nondigital devices, and human operators, for instance. Signals from such sources are not tied to our clock and may change at any time during our clock cycle. Therefore, although we can simplify our design by using good practices, no amount of digital or analog wizardry will eliminate the problem of metastability. However, by proper design or choice of components, we may lower the probability of finding the circuit in a metastable state to a satisfactory level. In Chapter 12, we discuss metastability in more detail and offer guidelines for dealing with the problem.

CONCLUSION

You have completed Part I of this book, in which we have explored the fundamental tools underlying digital design. From basic combinational circuits we have developed a set of building blocks that range from simple logic gates to complex ALUs, from flip-flops to large memories. Now you are ready to begin the exciting activity of digital design. Part II introduces you to this process.

READINGS AND SOURCES

BLAKESLEE, THOMAS R., *Digital Design with Standard MSI and LSI,* 2nd ed. John Wiley & Sons, New York, 1979. Sound design practices.

DIETMEYER, DONALD L., *Logic Design of Digital Systems,* 2nd ed. Allyn & Bacon, Boston, 1978. Chapter 12: hazards. Chapter 13: traditional asynchronous design.

ERCEGOVIC, MILOŠ D., and TOMÁS LANG, *Digital Systems and Hardware/Firmware Algorithms.* John Wiley & Sons, New York, 1985. Good treatment of sequential systems.

FLETCHER, WILLIAM I., *An Engineering Approach to Digital Design.* Prentice-Hall, Englewood Cliffs, N.J., 1980. Chapter 5 contains a good discussion of flip-flops.

HILL, FREDERICK J., and GERALD R. PETERSON, *Digital Logic and Microprocessors.* John Wiley & Sons, New York, 1984.

HILL, FREDERICK J., and GERALD R. PETERSON, *Introduction to Switching Theory and Logical Design,* 3rd ed. John Wiley & Sons, New York, 1981. Good standard treatment of sequential circuits.

HWANG, KAI, *Computer Arithmetic—Principles, Architecture, and Design.* John Wiley & Sons, New York, 1979.

KLINGMAN, EDWIN E., *Microprocessor System Design.* Vol. 2, *Microcoding, Array Logic, and Architectural Design.* Prentice-Hall, Englewood Cliffs, N.J., 1982. Bit slices and programmable logic.

MANO, M. MORRIS, *Digital Design.* Prentice-Hall, Englewood Cliffs, N.J., 1984.

MICK, JOHN, and JAMES BRICK, *Bit-Slice Microprocessor Design.* McGraw-Hill Book Co., New York, 1980. A collection of design notes for the Advanced Micro Devices 2900 bit-slice family. This book is useful far beyond the Am2900 chips.

MYERS, GLENFORD J., *Digital System Design with LSI Bit-Slice Logic.* John Wiley & Sons, New York, 1980.

WIATROWSKI, CLAUDE A., and CHARLES H. HOUSE, *Logic Circuits and Microcomputer Systems,* McGraw-Hill Book Co., New York, 1980.

Data Books

Am29300 Family Handbook. Advanced Micro Devices, 901 Thompson Place, P.O. Box 3453, Sunnyvale, Calif. 94088. High-performance 32-bit building blocks.

Bipolar Microprocessor Logic and Interface. Advanced Micro Devices, 901 Thompson Place, P.O. Box 3453, Sunnyvale, Calif. 94088. Data book for the AM2900 family and its support devices.

Bipolar/MOS Memories. Advanced Micro Devices, 901 Thompson Place, P.O. Box 3453, Sunnyvale, Calif. 94088. Data book.

FAST: Fairchild Advanced Schottky TTL. Fairchild Camera and Instrument Corporation, Digital Products Division, South Portland, Maine. Data book.

LSI Databook. Monolithic Memories, 2175 Mission College Blvd., Santa Clara, Calif. 95954. PALs, memory products, arithmetic units, system building blocks.

Memory Components Handbook. Intel Corp., Literature Department, 3065 Bowers Avenue, Santa Clara, Calif. 95051.

PAL Programmable Array Logic Handbook. Monolithic Memories, 2175 Mission College Blvd., Santa Clara, Calif. 95054. Authoritative practical treatise on PALs and their uses.

Systems Design Handbook, 2nd ed. Monolithic Memories, 2175 Mission College Blvd., Santa Clara, Calif. 95054, 1985. Good discussion of DRAMs and their control in Section 10. Good discussion of multiplication algorithms in Section 8.

The TTL Data Book. Texas Instruments, P.O. Box 225012, Dallas, Tex. 75265.

EXERCISES

4-1. (a) Show that the following combinational circuit contains a hazard.

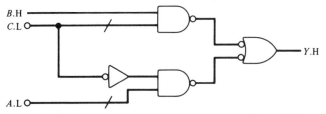

   (b) Write the logic equation corresponding to the circuit, and draw a K-map with circles corresponding to the circuit.
   (c) Most of the time our design techniques will nullify the bad effects of hazards; nevertheless, suppose that you must eliminate the above hazard from the circuit. Starting with the K-map you drew for part (b), produce a hazard-free map by making certain that adjacent 1's share at least one circle. Write the logic equation and draw the hazard-free circuit.
   (d) Prove, by using a timing diagram, that your new circuit is free of hazards.

4-2. Assume that each combinational circuit element has a propagation delay of t_p. What is the total (worst-case) propagation delay in the following circuit?

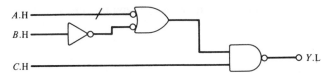

4-3. In Fig. 3-5, the circuit for the enabled multiplexer imposes the enabling operation on each of the initial AND gates, forcing them to have three inputs. Suggest why, in Fig. 3-5, the enabling operation was not designed as a single final AND gate with only two inputs.

4-4. A circuit consisting of a closed loop of an odd number of inverters (greater than one) can function as an oscillator. Assume that the propagation delay through an inverter is 10 nanoseconds.
   (a) With a timing diagram, show the oscillatory behavior of a loop of three inverters.
   (b) The oscillator consisting of a loop with just a single inverter is not stable. Speculate about why this circuit is unsatisfactory.

4-5. What is feedback in digital design? Draw a gate circuit that exhibits feedback with memory.

4-6. Why are combinational methods inadequate to deal with sequential circuits?

4-7. Explain "1's catching." Why is this behavior usually a disadvantage in digital design?

4–8. Explain the terms *asynchronous* and *synchronous*.

4–9. Show that the asynchronous RS flip-flop has two stable states.

4–10. Why do we usually avoid asynchronous flip-flops in digital design?

4–11. What is switch debouncing? Why can we usually not use a mechanical switch signal directly in a digital design? Draw a switch-debouncing circuit.

4–12. Using a timing diagram, analyze the behavior of the switch debouncer shown in Fig. 4–8a or 4–8b.

4–13. Assume that two (noisy) mechanical switches generate the *DATA* and *HOLD* signals for the latch in Fig. 4–4. Is there any sequence of switch closings and openings that would yield a clean output signal at *Y*?

4–14. The RS flip-flop exhibits anomalous output behavior if both *R* and *S* are true.
   (a) What is the anomaly?
   (b) Does the anomaly occur in outputs *X* and *Q* of Fig. 4–6?
   (c) In Fig. 4–6, assume that $R = S = T$. What is the value of *Q* if both signals become false, but *R* becomes false slightly before *S*?
   (d) Under similar conditions, what value does *Q* assume after precisely simultaneous $T \rightarrow F$ transitions of *R* and *S*?

4–15. What is an edge-driven flip-flop? Why is it desirable? What is the defect in the master-slave flip-flop? What is a pure edge-driven flip-flop? What kind of flip-flops do we use in digital design?

4–16. Consider an edge-driven JK flip-flop such as the 74LS109 with the direct set input and the *K* input asserted (true), and the direct clear input and the *J* input negated (false). What will be the flip-flop's output shortly after the next active clock edge arrives?

4–17. Suppose your design requires a 74LS109 JK flip-flop with output *FLG*. You wish to set the flip-flop with a logic variable *SETFLG* and clear the flip-flop with a variable *CLRFLG*. In the design, you find that the control inputs are available as *SETFLG*.L and *CLRFLG*.H. Draw the desired circuit with a 74LS109 flip-flop, using no inverters. Draw the mixed-logic diagram.

4–18. Repeat Exercise 4–17 under the condition that both *SETFLG* and *CLRFLG* are available with T = L, and inverters are available.

4–19. The text describes three cases in which the JK flip-flop may be used to store a bit. Two of these cases are (a) clearing, followed by later setting if the data bit is true; (b) setting, followed by later clearing if the data bit is false. Verify the text's rules for implementing these two cases.

4–20. Figure 4–30 shows the structure of a commercial 74LS74 D flip-flop. The notation is mixed logic; the asynchronous set and clear inputs have been deleted to simplify the diagram. In this exercise, you will demonstrate that the circuit does indeed exhibit pure edge-triggered behavior—it copies and stores its *D* input only when the clock fires, and it changes its output only as a result of the clock transition.

Assume (reasonably) that the propagation delay t_{pr} is the same through each gate in the circuit. To demonstrate the flip-flop's behavior, you will start from a known configuration of gate inputs and outputs and will manually simulate the effects of a given change in one input, in time units of t_{pr}, until no further changes occur in the circuit. Using the mixed-logic diagram, you may choose to deal with either logical variables or voltage signals. In the explanation below, we use logic rather than voltage.

Start, for instance, with both *Q* outputs, the *D* input, and the clock input all

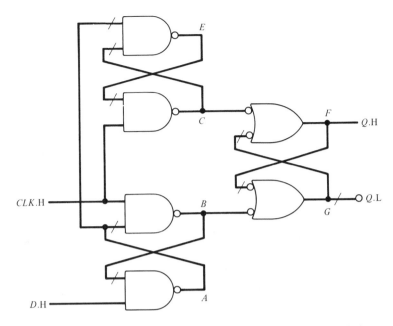

Figure 4-30. A 74LS74 D flip-flop without an asynchronous set and clear.

false, and establish the initial conditions of gate outputs A through G. Tabulate the values of all signals so that you will have a detailed record. Make the D input true and follow any changes in gate outputs until the circuit stabilizes. (You should, of course, observe that the outputs are unaffected by this activity.) Then, from this new configuration, bring the D input back to false and verify that the outputs do not change as the circuit stabilizes. Once again, bring the D input true to set the stage for some clocked activity. Make the clock input true and watch the circuit accept and store the value on the D input. Continue in this manner with other sequences of changes at the input.

4-21. Do you want to observe metastability in action? In the preceding simulation of the actions of the D flip-flop, you changed only one input at a time and observed the effect on the internal structure of the circuit. Changing both the input D and clock input at the same time is a violation of the flip-flop's setup specifications. Start at the same point as in the previous exercise, with the input D and clock input false, and with both Q outputs false. Set both inputs true simultaneously and simulate the behavior of the circuit. What behavior do you observe on the Q outputs?

4-22. Figure 4-31 is a mixed-logic diagram of the internal structure of a commercial 74LS109 JK flip-flop, with the asynchronous set and clear inputs deleted to simplify the diagram. Perform a hand simulation similar to that of Exercise 4-20 to show the edge-triggered operation of the JK flip-flop.

4-23. What is the difference between the names used for inputs and outputs *inside* a mixed-logic circuit symbol and the names appearing *outside* the symbol?

4-24. Draw an efficient mixed-logic circuit in which you use a 74LS109 JK flip-flop to synchronize a signal *TIME*∗.L. On your diagram, call the synchronized logic variable *TIME.SYNC*.

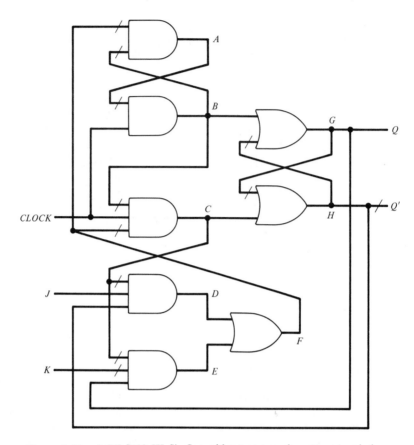

Figure 4–31. A 74LS109 JK flip-flop without an asynchronous set and clear.

4–25. There are four possible transitions $Q_{(n)}$ to $Q_{(n+1)}$ for a clocked flip-flop output: $0 \to 0$, $0 \to 1$, $1 \to 0$, and $1 \to 1$. These transitions are given the names t_0, t_α, t_β, and t_1, respectively. Consider the ways in which we can make a D flip-flop and a JK flip-flop execute each of these transitions. Fill in the missing elements in the following table:

| Transition | D flip-flop | | JK flip-flop | | |
	$Q_{(n)}$	$D_{(n)}$	$Q_{(n)}$	$J_{(n)}$	$K_{(n)}$
t_0	0	0	0	0	X
t_α					
t_β					
t_1					

[In each case there will be two ways that the JK flip-flop can execute the transition. For instance, the $0 \to 0$ (t_0) transition occurs by *clearing* the flip-flop to 0 (having

$J = 0$, $K = 1$), or by *holding* the previous 0 (having $J = 0$, $K = 0$). These cases give rise to the X (don't-care) entry in the table.]

4-26. Compare the asynchronous RS flip-flop and the synchronous JK, D, and enabled D flip-flops as to their best uses in digital design.

4-27. Two types of clocked flip-flop behavior that are occasionally useful are the T (toggle) and the SOC (set overrides clear) flip-flop modes. A toggle flip-flop changes its output Q only when its input *TOG* is true at the time of the clock edge. A SOC flip-flop behaves like a clocked RS flip-flop except that it ignores the value of input R whenever input S is true. Write excitation tables defining each type.

4-28. By means of external gates, convert the 74LS109 flip-flop into a type T (toggle) and a type SOC (set overrides clear) flip-flop.

4-29. By analogy with Fig. 4–12, construct a type T (toggle) flip-flop from a D flip-flop.

4-30. What is a register? How does it differ from a flip-flop?

4-31. Construct synchronous modulo-2, modulo-4, and modulo-8 counters using:
  (a) D flip-flops.
  (b) JK flip-flops.
  (c) T (toggle) flip-flops.

4-32. Repeat Exercise 4–31 with ripple counters instead of synchronous counters.

4-33. For a 4-bit ripple counter, demonstrate how the output ripple can produce hazards in circuits that receive the outputs.

4-34. Using a TTL data book, compare the characteristics of the 74LS160, 74LS161, 74LS162, and 74LS163 counters.

4-35. Use two 74LS163 counters to build a divide-by-24 circuit. The output of your circuit should be true during 1 of every 24 clock periods. This and similar circuits are *frequency dividers*.

4-36. Using two 74LS163 counters and any required gates, design a circuit that functions as an 8-bit counter or as an 8-bit multiply-by-2 circuit. The circuit should be controlled by two control signals, according to the following table:

CTL1	CTL0	
0	0	Do nothing
0	1	Count
1	0	Multiply by 2
1	1	Clear

4-37. There are many special counting sequences that are of some interest in digital design. The *binary counter* produces the sequence of binary integers. The *gray code counter* produces a sequence in which exactly one bit changes in moving from one element of the sequence to the next. For a 2-bit counter, the gray code is 00, 01, 11, 10. (Where have you seen this sequence in this book?) Build a series of 2-bit gray code counters using the following approaches:
  (a) Use logic gates to compute the inputs to D flip-flops.
  (b) Use multiplexers to look up the inputs to D flip-flops.
  (c) Use logic gates to compute the inputs to JK flip-flops.
  (d) Use multiplexers to look up the inputs to JK flip-flops.

4–38. The *moebius counter* produces another special sequence. The algorithm for N bits numbered $C_N \ldots C_1$ is

$$C_k \leftarrow C_{k+1} \qquad \text{when } k = N - 1, \ldots, 1$$
$$C_N \leftarrow \overline{C_1}$$

(a) Design a 4-bit moebius counter, using JK flip-flops as the storage elements.

(b) Design a 4-bit moebius counter using a shift register as the basic storage element.

(c) How many elements are in an N-bit moebius sequence that begins with 0? Determine the answer empirically.

4–39. Consider a 74LS163 4-bit Programmable Binary Counter used in a mixed-logic circuit with data outputs and inputs expressed as T = L.

(a) What are the logical effects of performing the 74LS163 count, load, and clear operations with such a circuit?

(b) With the T = L interpretation of the data signals, can we still cascade 74LS163 chips to provide larger counters?

(c) Why are we not free to alter the voltage representations of the *LD, CLR, CEP, CET,* or *TC* signals?

4–40. The programmable counters in the 74LS160 through 74LS163 series may act as enabled D registers. In such an application, what, if anything, should you do with the clear and count control inputs?

4–41. From a TTL data book, find an example of:

(a) A serial-in, serial-out shift register.

(b) A serial-in, parallel-out shift register.

(c) A parallel-in, serial-out shift register.

4–42. Use three 74LS194 shift registers to implement 12-bit shift registers with the following capabilities:

(a) Left shift with serial input; parallel output.

(b) Right shift with serial input; parallel load; serial output.

(c) Full left and right shifts with serial inputs; parallel load; parallel and serial outputs.

4–43. Use gates or multiplexers to modify the 74LS194 to include a fifth mode of operation. This new mode will preserve the most significant (leftmost) bit during a right shift; in other words, after the shift, the two leftmost bits will be the same. This is called an *arithmetic right shift*—useful in computing with signed two's-complement numbers.

4–44. Describe the principal characteristics of the RAM, ROM, PROM, and EPROM.

4–45. Describe the principal characteristics and applications of the PLA, PLE, and PAL.

4–46. Consider two static RAMs, each with a capacity of 16K (2^{14}) bits, one RAM organized in a 16K × 1 configuration, the other in a 4K × 4 format. Assume that each type of RAM has two power supply pins, one read-write-select pin, one chip-select pin, and one three-state-output-enable pin. Also assume that each data bit has a single bus-oriented pin for input and output. Determine the smallest number of pins in each type of RAM organization.

4–47. How do static and dynamic RAMs differ? What advantages do dynamic RAMs offer? What disadvantages?

4–48. Design PROMs that realize the following sets of logic functions:

(a)

$$X = A \cdot B \cdot C + A \cdot \overline{B} \cdot C + \overline{A} \cdot B \cdot C + \overline{A} \cdot \overline{B} \cdot \overline{C}$$

$$Y = A \cdot B \cdot \overline{C} + A \cdot \overline{B} \cdot C + \overline{A} \cdot \overline{B} \cdot C$$

$$Z = A \cdot B \cdot \overline{C} + \overline{A} \cdot B \cdot \overline{C} + \overline{A} \cdot \overline{B} \cdot \overline{C}$$

(b)

$$X = A \cdot B + A \cdot \overline{C} + \overline{A} \cdot C$$

$$Y = \overline{A} \cdot B + B \cdot C + \overline{A} \cdot B \cdot \overline{C}$$

$$Z = B \cdot C + \overline{B} \cdot \overline{C} + \overline{A}$$

4–49. Design PLAs that realize the sets of logic functions in Exercise 4–48.

4–50. Synthesize the logic equations in Exercise 4–48, using a PAL18L4. For each set of equations, use X and Y with T = L, and Z with T = H.

4–51. The PAL16R8, whose functions are shown in Fig. 4–27, produces T = L outputs in a natural way. A proper mixed-logic diagram for an output stage of this PAL is

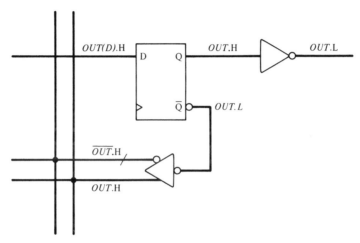

For emphasis, the diagram shows each signal and its polarity. If a T = H output is required, one useful technique is to produce the *inverse* of the required flip-flop input. Develop a standard mixed-logic diagram for an output stage of this PAL that produces an output having T = H.

4–52. The following prescription will convert an n-bit binary number into an n-bit gray code (n is the most significant bit):

$$\text{gray}_n = \text{binary}_n$$

$$\text{gray}_k = \text{binary}_k \oplus \text{binary}_{k+1} \qquad (k = n - 1, \ldots, 2, 1)$$

(a) Tabulate the 5-bit binary and 5-bit gray codes.

(b) Design a PROM that converts 5-bit binary numbers into 5-bit gray codes.

4–53. When $k = 1, 2, \ldots, n$, bit k of an n-bit binary number is equal to the EOR of the corresponding gray code bits from k through n (n is the most significant bit). That is

$$\text{binary}_k = \text{gray}_k \oplus \text{gray}_{k+1} \oplus \ldots \oplus \text{gray}_n$$

(a) Tabulate the 4-bit gray code and the 4-bit binary code.

(b) Design a PLA that converts a 4-bit gray code into a binary number.

4-54. Under what circumstances do we use a single-shot in digital design? How is the delay period for the single-shot established?

4-55. Refer to a data sheet for the 96L02 Dual Single-Shot. Determine suitable values for the resistor and capacitor to provide delays of:

(a) 100 nsec

(b) 1 msec

(c) 1 sec

4-56. Why does a delay line not require a power supply?

4-57. Using the delay line in Fig. 4-29, construct circuits that will do the following:

(a) Delay a signal *HURRY* by 400 nsec.

(b) Assert a signal *BINGO* for 100 nsec, beginning when the signal *GOMANGO* becomes true.

Design Methods

5

In Part I we presented basic design tools and introduced components used to build digital systems at the MSI level of complexity. The fascinating part of digital design lies before us. We now consider how we may assemble a complete system from building blocks. It is in this area that the designer can create elegance and beauty, or chaos and headaches. In Part II we present methods of designing systems of moderate complexity, using the fundamental integrated circuit building blocks. In Part III we will introduce programmed control of complex digital systems, using microprogramming and conventional microcomputers. The methods of Part II serve to support independent designs and to provide the "glue" required to design systems with programmed control.

To some extent, digital design is an art form. Most designers have had to develop a style of digital design by trial and error, and their efforts often have not converged to an efficient and aesthetic style. On the other hand, there are underlying principles that can be immensely useful to designers. Our goal in this book is to start you down the right path, and to take you far enough so that you can develop your own designs with a solid sense of good style.

Design style is a curiously neglected subject, perhaps because a traditional study of logic design emphasizes the microscopic transistor and gate aspects of the subject. We emphasize a macroscopic view of digital systems by starting from the original problem. The result is a top-down approach to design.

ELEMENTS OF DESIGN STYLE

Here are the guidelines for good design style:

(a) Design from the top down.

(b) Maintain a clear distinction between the controller and the controlled hardware (the architecture).

(c) Develop a clearly defined architecture and control algorithm before making detailed decisions about hardware.

Top-Down Design

A design starts with a careful study of the overall problem. At this stage, we deliberately ignore details and ask such questions as:

(a) Is the problem clearly stated?

(b) Could we restate the problem more clearly or more simply?

(c) If we are working with a subsystem of a larger system, what is its relationship to its host? Would a different partitioning of the entire system yield a simpler structure?

At this stage our concerns are global, and we must stay at that level until we have hammered out a sensible statement of the problem and have digested the problem to the point where we understand what we must solve. This is essential, since any difficulties at this level are serious. No amount of wizardry with components can remedy errors in the understanding of the problem.

After we have clearly specified the problem at the global level, we seek a rational way to partition the problem into smaller pieces with clearly defined interrelationships. Our goal is to choose "natural" pieces in such a way that we can comprehend each piece as a unit and understand the interaction of the units. This partitioning process proceeds to lower levels until finally we choose the actual integrated cicuits.

Unfortunately, many designers reverse this process by rushing to integrated circuit data books to find the chip that will "solve" their problem. Often they find a circuit that solves a slightly different problem. Thus enters the infamous "patch" to force the problem, which is itself not well defined in the designer's mind, to the integrated circuit. This process proceeds from the bottom up, often in a divergent manner. Many commercial designs bear unmistakable traces of this method of design.

Separation of Controller and Architecture

One of the first steps in a top-down design is to partition the design into (a) a *control algorithm* and (b) an *architecture* that will be controlled by this algorithm. The top-down analysis will suggest a rough preliminary version of the system's architecture, involving abstract building blocks such as registers, memories, and data paths. Since the architecture is specific to the particular design, there is no general prescription for writing down this preliminary architecture. The main guidelines are to make the architecture natural to the problem and to design with high-level units rather than with chips and voltages. The examples of design in Chapter 6 will illustrate the art of specifying the rough architecture.

Next, we work out the details of the control algorithm at an *abstract level*. The control algorithm is often surprisingly independent of hardware. For example, if you were designing a computer, what operations would you expect your control algorithm to accomplish?

(a) Get the next instruction.

(b) Test to see if operands are needed and get the operands if required, making any necessary indirect memory references to indirectly addressed operands.

(c) Execute the individual instruction.

As you will see in Chapter 7, we may specify a complete flowchart (algorithm) for operations like this with almost no knowledge of the specific hardware.

You should explore the construction of the control algorithm until you have a clear understanding of your approach to the solution. The exploration may go through several iterations, but eventually you will complete the process, at which time you should turn your attention to the hardware for the architecture that is suggested by the control algorithm. The algorithm will guide you to the hardware. Note how powerful this concept is. We have a tool that allows us to solve a problem in a rational way instead of randomly looking at integrated circuits and wondering if they will fit into the design.

We can formalize the controller-architecture separation with the diagram in Fig. 5–1. The controller issues properly sequenced commands to the controlled device. These commands make the architecture perform the actions dictated by the control algorithm. Usually, the controller will need status information from the architecture that serves as decision variables for the control algorithm. As the design matures, the controller's command outputs and status inputs go from abstract concepts of control to Boolean variables, and finally to voltage representations of the Boolean variables.

Consider the following example. A problem requires that a word be written into a memory. The preliminary architecture for the problem is just a black box for the memory, the details of its inner construction being deferred until later. The memory will require four items of input: a memory address *MA* to tell where to write the data, a word of *DATA* for input, a line *R/W* to tell whether to read or write, and a *GO* signal to start the read or write operation. The only status returned by the memory will be memory cycle complete *CC*.

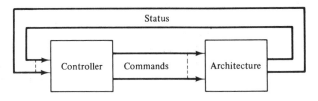

Figure 5–1. The structure of a state machine.

Figure 5–2 is the functional diagram corresponding to this analysis. The command lines are:

MA (*n* lines)

DATA (*m* lines)

R/W (1 line)

GO (1 line)

The parameters *n* and *m* are determined by the characteristics of the memory needed by the problem. For instance, a 1,024-word by 8-bit memory would have $n = 10$ (1,024 $= 2^{10}$) and $m = 8$ (each word of memory has eight bits). The status line is *CC*.

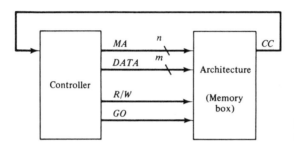

Figure 5–2. A diagram of a memory-write machine.

We now have a good idea of what signals the controller and architecture must generate and accept, even before we know what hardware we will use to build the controller or what the memory box is like inside. We choose the memory command lines by realizing that we must have data to write into memory and a location where it must be written. The structure in Fig. 5–2 is not affected by the actual type of memory.

We can say something about the nature of the control algorithm that initiates a memory write operation, without knowing exactly how we will translate that algorithm into hardware. The algorithm must look something like Fig. 5–3. The purpose of the first step **STW** is to issue a *GO* signal to the memory, along with the necessary data and commands to initiate a writing operation. The next step **WAIT**s until the memory has finished the writing.

We can accomplish an amazing amount of the design by a general consideration of the problem. Carry the top-down analysis as far as you can, because decisions at this level are more easily altered than they will be once the hardware has intruded.

Refining the Architecture and Control Algorithm

We are now ready to move down one level. We have sketched out the algorithm by ignoring the hardware as much as possible; instead we were trying to reduce our problem to high-level, abstract statements of control and architecture. The only consideration about hardware introduced thus far were general ones that

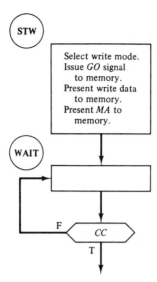

Figure 5-3. Algorithm for writing to a memory.

we could state from a knowledge of the variables required for the operation of abstract building blocks such as memories, registers, and arithmetic units. We now begin to refine the control algorithm by introducing more detail.

As we refine algorithm, we need a more detailed knowledge of the architecture of the system. We therefore begin to elaborate the architecture as a set of building blocks, moving carefully through a set of high-level building blocks toward a selection of the major hardware. We choose the architectural elements by asking what specific building blocks the developing control algorithm requires; we do not select elements by looking in a data book and saying, "Hey, this is a neat chip—I must fit it into my next design."

A good architecture will be simple, clear, and easy to control. If it is not, there is no way to rescue the resulting mess with exotic integrated circuits or Boolean algebra. If the architecture is clear and simple, the rest of the design will be relatively straightforward. This step—the first introduction of hardware—is an important point in the design.

After we choose the major building blocks, we know what control signals they will require. At this point, we tabulate these signals and then quit worrying about the hardware. In fact, we suppress consideration of the hardware lest it capture our thought processes and cause us to lose sight of the algorithm we are trying to develop. Let the algorithm drive the design process as much as possible. Now we can continue with an elaboration of the algorithm, whose function is to provide a properly sequenced set of commands to the architecture. Spend a large fraction of the total time for the project on the detailed algorithm-architecture phase.

After we have completely specified the algorithm, we can reconsider the architecture and our choice of building blocks. The detailed construction of the algorithm will usually reveal areas of the design that we can simplify or speed up by using slightly different architectural components. We incorporate these changes into the architecture and make the corresponding changes in the algorithm. At this point the process should have converged to a final solution. We should have:

(a) *The architecture.* This should include a detailed set of components and data paths for the controlled device—usually a specification of the actual integrated circuits for the major components such as registers, ALUs, and memories, and a statement of the command signals that these components require and the status signals that they produce. Our specification of the architecture does not include any logic required to generate these commands, since the generation of commands is assigned to the control algorithm.

(b) *The algorithm.* This will produce a properly sequenced set of command signals to make the architecture perform the original problem. It does not include the hardware to implement the algorithm. We can derive the hardware from the algorithm in a straightforward and mechanical way, as you will see later in this chapter.

If by this time the process has not converged to a stable solution, you probably had trouble at an earlier stage of the design process. In such circumstances, proceeding further is fruitless; you should go back to the beginning and start over. Do not "kludge" your solution. You have not yet wrapped any wires or drawn any hardware circuit diagrams, so beginning anew is relatively painless.

And now, a secret: whether in hardware or software, no one designs a system strictly from the top down. A knowledge of low-level components and techniques always influences the design, even at the highest levels. The best top-down hardware designers have an intimate knowledge of hardware, and this knowledge tempers and guides the high-level design decisions. As their understanding of the design expands, good designers use their knowledge of lower-level technology to avoid unproductive approaches. The designer dips repeatedly into lower and more detailed levels for short excursions, but invariably returns to the present top level. The top-down approach has the great virtue of providing the discipline that keeps one thinking at the highest useful level. We like to imagine that a complex design proceeds linearly from the top to the bottom, but that is rarely so. But whenever you dip down, your top-down training will pull you back up as soon as possible.

ALGORITHMIC STATE MACHINES

The control algorithm plays a major role in a digital design, so we need a good notation for expressing hardware algorithms. The notation should assist the designer in expressing the abstract algorithm and should support the conversion of the algorithm into hardware. There are several ways of describing the control. For synchronous circuits, the ASM chart technique is the superior notation.

ASM stands for *algorithmic state machine*.* The name is appropriate, since all controllers are state machines, and we are trying to translate algorithms into controllers. The ASM chart is a flowchart whose notations superficially bear a strong resemblance to the conventional software flowchart. The ASM chart expresses the concept of a sequence of *time intervals* in a precise way, whereas the software flowchart describes only the sequence of events and not their duration.

States and Clocks

An algorithmic state machine moves through a sequence of *states*, based on the position in the control algorithm (the state) and the values of relevant status variables. The concept of a state implies sufficient knowledge of present and past conditions to determine future behavior. It is the task of the *present state* of the system to produce any required output signals and to use appropriate input information to move at the proper time to the *next state*. In most of this book we are dealing with synchronous systems whose state times are determined solely by a master clock. The most convenient form of clock is a periodic square-wave voltage:

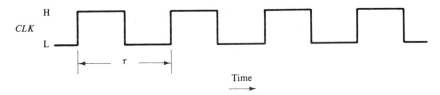

The clock event that triggers the transition from one state to another and other actions of the system is called the *active edge,* and in synchronous systems is usually the rising (L → H) edge of the clock. In a synchronous system, the clock will thus have T = H. Commonly, the clock period τ ranges from 20 nanoseconds to several microseconds. The frequency f of the clock is its number of oscillations per second. Frequency and period are related by the expression

$$f = \frac{1}{\tau}$$

The unit of frequency is the hertz (Hz), which is defined as one oscillation per second. Convenient units for fast waveforms are kilohertz (KHz = 10^3 Hz) and megahertz (MHz = 10^6 Hz).

The clock's frequency may vary, and the clock may even stop—desirable during the debugging phases of a design. Within reasonable limits, the duration of the high portion of the clock waveform in relation to the total clock period

---

* T. E. Osborne developed the ASM chart notation and C. R. Clare describes the method in his book, *Designing Logic Systems Using State Machines* (New York: McGraw-Hill Book Co., 1973).

(the *duty cycle*) is unimportant. The crispness and reliability of the active clock edge *is* of extreme importance, and designers of systems pay close attention to the production and distribution of an excellent clock signal.

ASM Chart Notations

States. Each active transition of the clock causes a change of state from the present state to the next state. The ASM chart describes the control algorithm in such a way that, given the present state, the next state is determined unambiguously for any values of the input variables. The symbol for a state is a rectangle with its symbolic name enclosed in a small circle or oval at the upper left corner:

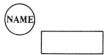

We would represent a purely sequential algorithm as an ASM chart of a sequence of states, as in Fig. 5–4. The corresponding division of the time axis would be

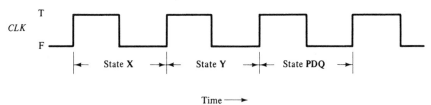

A sequence is an inherent property of an ASM chart; state **Y** follows state **X**, and so on. It is cumbersome to show time relations with a timing diagram such as the above; therefore, you should learn to think of time as rigorously implied in the ASM chart notation.

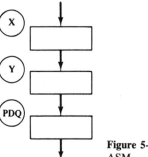

Figure 5–4. A purely sequential ASM.

Outputs. The function of a controller is to send properly sequenced outputs (voltage command signals) to the controlled device according to some algorithm. To indicate an output, we place the command description within the appropriate state rectangle. In this book we use several notations for outputs. Depending on our depth of understanding of our design and on the level of detail we wish to convey, we may use informal expressions of actions or detailed statements of particular output operations. Figure 5–5 contains some examples. In state **PRINT.LINE**, the expression "Start print cycle" represents a set of actions, as yet not fully elaborated, that initiates a printer cycle. The arrows in the next two lines imply actions that are to be consummated at the *end of this state;* at that time *LINE* is to be loaded into *PRINTBUF,* and the *AC* register is to be cleared. The fourth line, *MOVING,* calls for the assertion of the signal *MOVING* (making *MOVING* true) during this state. The last line means that the output variable *STATUS* is to have the value of the variable *ERRFLAG* (T or F) during this state. Other output notations may be useful; improvising notations is fine as long as you define your terminology.

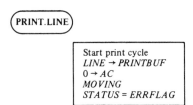

Figure 5–5. ASM output notations.

Branches. Purely sequential ASMs are of little interest because they are usually not powerful enough to describe useful algorithms. We need some way to express *conditional branches* so that the next state is determined not only by the present state but also by the present value of one or more test (status) inputs. Our symbol is the same as in conventional flowcharts for software: the diamond or diamond-sided rectangle

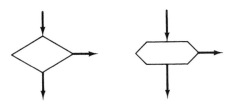

We incorporate this symbol into the ASM chart by appending it to a state rectangle, placing the description of the test input inside the diamond, as in Fig. 5–6. In this case a portrayal of the time line would be

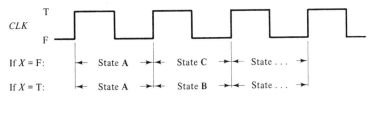

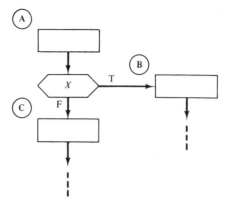

Figure 5–6. An ASM with a conditional branch.

The decision to jump to one of the two states **B** or **C** is made during state **A**, and the jump occurs at the end of state **A**. In hardware implementations the voltage representing input X must be stable for some time before the decision. Ideally, X should be stable for the entire clock period of state **A**, for then the hardware that decides to jump to **B** or **C** has the maximum time to settle. It is important to realize that a test does not require a separate clock period—it is done "in parallel" with the actions of the parent state rectangle and thus is part of the parent state.

We are not limited to two-way branches from a state. We may draw sequences of test diamonds or we may have more than two paths coming from the same diamond. Two ways of representing a three-way branch are shown in Fig. 5–7. The test structure in Fig. 5–7a is a diagrammatic representation of a truth table in which neither P nor Q appears to dominate. Figure 5–7b conveys the feeling that the test of variable P is of higher priority than the test of Q. Which form is preferable depends on the designer's thought process. Use the ASM notations that best describe your design.

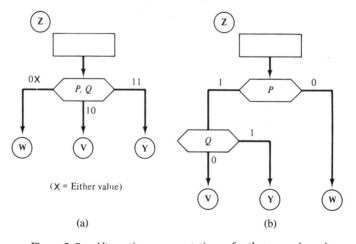

(a)   (b)

Figure 5–7. Alternative representations of a three-way branch.

Conditional outputs. A command written within a state rectangle indicates that the controller is to produce the output whenever the algorithm is in that state. Sometimes we want a command to occur only when some other condition also exists. We call such a command a *conditional output* and specify it within an oval, as in Fig. 5–8. Command *CMD1* will appear for one state time whenever the ASM is in state **P**. The command *CMD2* will occur during one state time whenever the ASM is in state **Q**, but when the ASM is in state **P**, *CMD2* will occur *only if test input Z is false*. In this example, *CMD2* is an *unconditional output* in state **Q** and a *conditional output* in state **P**.*

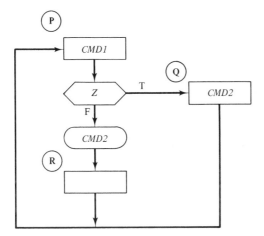

Figure 5–8. An ASM with a conditional output.

Test inputs may serve two functions in ASM charts: they may help specify the next state and they may control the issuing of conditional outputs. Ovals for conditional outputs and diamonds for tests of inputs belong to the parent state, since the activities will occur during the same state time. A state thus consists of its rectangle, which is always present, and any test diamonds and conditional output ovals associated with that state. Unconditional outputs are a function only of the parent state; conditional outputs depend on both the state and the path within that state. It would be appropriate to draw a dashed line around the entire structure for each state, but we usually do not do this because the chart defines each state without this aid.

This is the entire ASM chart notation. Our goal is to use it to help us build digital circuits. In the following chapters we emphasize the design phase, the difficult part of our work. Once we have an architecture and a control algorithm, we must then implement them. In this chapter we next present some standard and systematic methods for realizing any ASM chart control algorithm.

---

* An ASM with only unconditional outputs is equivalent to the Moore machine of traditional sequential circuit theory; the traditional Mealy machine has conditional outputs. The ASM formulation subsumes both traditional cases.

REALIZING ALGORITHMIC STATE MACHINES

Once we have expressed the control algorithm as an ASM chart, it is a simple job to express the flow of control as hardware. We describe four methods—a traditional technique and three style-driven methods.

Our task is to construct a *state generator* for a given ASM. In any state machine, the concepts of present state and next state are vital. The state generator's task is to record the present state and generate the next state. State machines are sequential circuits, and to keep track of the present state we need a memory. In this part of the book, we use flip-flops as the state memory. There are two ways to express the present state in a flip-flop memory. We may assign a binary number to each state and express the present state as an encoding, using its binary number. In this scheme, n flip-flops will encode up to 2^n states. We may describe a state by its name or by its number in binary or in decimal, as we find convenient. Alternatively, we may avoid the encoding by assigning one flip-flop to each state. We will use each of these approaches.

Traditional Synthesis from an ASM Chart

The traditional technique for state generation is to use an encoded representation of the present state and *compute* the code for the next state. The bits of the code are the *state variables: n* state variables describe up to 2^n states. The term state variable used in this way is unfortunate, since there is a more important use for this term—to specify the name of a logic variable for each state. Nevertheless, in this section, we use the term in the traditional way. We make an arbitrary *state assignment* of binary state variable values to states. On the ASM chart, we show the binary assignment for each state above its state rectangle on the right-hand side. We might choose the state assignment shown in Fig. 5–9. We need not label the test diamond or the conditional output oval since they are part of state 00. The state assignment is arbitrary. There may be more hardware associated with one state assignment than another, but this is not an

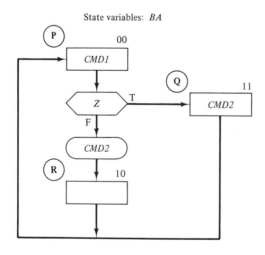

Figure 5–9. A simple ASM with a state assignment.

important factor. The two state variables B and A in Fig. 5–9 specify an address that points to the present state. If we could compute the next address and put it into the state flip-flops, we would then be pointing at a new state. As you might guess, we can use gates to build a combinational circuit to compute the next address. Figure 5–10 is the model of this process. This figure displays only the control portion of the digital system, not the architecture.

We can use either JK or D flip-flops for the state variables. JK flip-flops usually result in less combinational logic than D flip-flops, but they also require twice as many input lines. The JK form is more compact but yields more obscure results. In this example, we use D flip-flops to provide a more direct comparison with the methods to follow.

For the ASM chart in Fig. 5–9, the state generator model of Fig. 5–10 has one status input Z, two command outputs $CMD1$ and $CMD2$, and two state flip-flops B and A. The combinational logic must compute the value of the next-state address:

Present			Next	
B	A	Z	$B(D)$	$A(D)$
0	0	0	1	0
0	0	1	1	1
0	1	X	0	0
1	0	X	0	0
1	1	X	0	0

The condensation of rows on variable Z arises because the move from states 10 or 11 does not depend on Z.

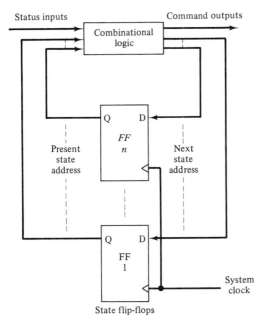

Figure 5–10. A model of an encoded ASM state generator.

The state assignment 01 is a possible pattern of the flip-flop's outputs but does not label any state in the algorithm of Fig. 5–9. Hardware is perverse and may get into this state, for example during power-up, when flip-flops may settle into random values. In our example, pathological behavior would result if the next-state logic computes 01 whenever the present state is 01. We would be locked into state 01 and could not get out unless we did something drastic, such as shut off system power. Clearly, if we ever get into state 01, we must get back to the main algorithm loop. Thus, we have arbitrarily chosen to go to state 00. We must always take into account all unused state assignments in encoded designs of state generators.

We may write the equations for state flip-flop inputs $B(D)$ and $A(D)$ by inspecting the logic table above. They are

$$B(D) = \overline{B} \cdot \overline{A} \cdot \overline{Z} + \overline{B} \cdot \overline{A} \cdot Z = \overline{B} \cdot \overline{A}$$
$$A(D) = \overline{B} \cdot \overline{A} \cdot Z$$

When we are designing more complex state machines, K-maps may help to simplify the expressions for the state flip-flop inputs.

This completes our discussion of the design of the state generator. But the purpose of the algorithm was to produce properly sequenced outputs. An examination of Fig. 5–9 yields these equations for the outputs:

$$CMD1 = STATE.P = \overline{B} \cdot \overline{A} \tag{5–1}$$
$$CMD2 = STATE.P \cdot \overline{Z} + STATE.Q = \overline{B} \cdot \overline{A} \cdot \overline{Z} + B \cdot A \tag{5–2}$$

The hardware for the ASM is shown in Fig. 5–11.

Comments. The approach we used in this section—to *compute* the code for the next state—is a traditional method. JK flip-flops used to store the state variables may lead to somewhat more compact next-state logic than D flip-flops, since the JK flip-flop is the more flexible device.

Unfortunately, the traditional method results in no obvious correspondence between the hardware for the state generator logic and the algorithm it represents. This is true of D flip-flops, and even more true of JKs. Every change in the algorithm, no matter how minor, requires a fresh design of the next-state combinational logic. The traditional approach violates our goal of clarity in design. Next-state generation is the standard implementation process associated with ASM charts, and we would like both the synthesis and the analysis of our state generators to be as straightforward and mechanical as possible. The next technique, while using the same structure for expressing states with encoded state variables, differs dramatically in the method by which the next-state combinational logic is generated.

The Multiplexer Controller Method for ASM Synthesis

Here is a simple method of synthesizing the combinational logic for any controller. The method has several desirable attributes:

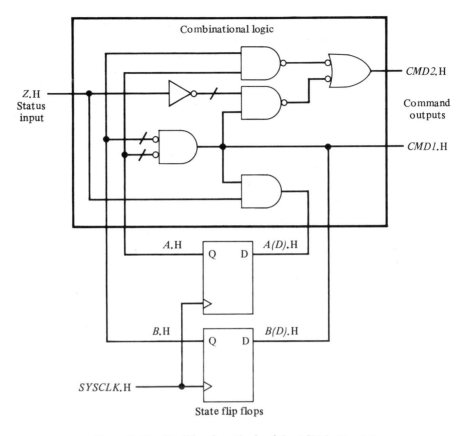

Figure 5-11. Traditional synthesis of the ASM in Fig. 5-9.

(a) It produces a design having a direct correspondence with the algorithm that generated it.

(b) It is a standard method that can be applied to any ASM chart.

(c) It forces the designer to complete the ASM chart before the hardware is constructed.

(d) It is easy to learn.

The design progresses in the traditional way through the state assignment: we select an appropriate number of D flip-flops for the state variables and we make an arbitrary assignment of binary values to the states in the ASM chart. Now consider the model of state generation in Fig. 5-10. Given the code for the present-state address and the status variable inputs, we must produce the code for the next state.

The traditional way is to use gates to *compute* the next-state code. An alternative way is to *look up* the next-state code in a table. Rather than compute each bit of the next-state code, in this simplified method we will look up the value of each bit. In Chapter 3, we described the multiplexer as our table-lookup building block.

The multiplexer controller method uses a mux for the input to each state flip-flop. Each mux produces the new input to its state flip-flop. The assembly of multiplexers, each providing the input to its respective flip-flop, yields the code for the next state. We choose a multiplexer wide enough to have an input for each state of the ASM. The present-state address code, which is the ordered output of the state flip-flops, feeds into the select inputs of each multiplexer, where it selects the mux input appropriate for the present state of the system.

Our design task is to see that for each present state the mux inputs provide the 1 or 0 necessary to produce the next-state code. In effect, we have, for each present state, a table entry that produces the code for the next state.

Let's use the multiplexer controller method for building the state generator for the simple ASM in Fig. 5–9. A four-input multiplexer produces the input to each state flip-flop B and A. We prepare Table 5–1, showing the conditions for reaching any possible next state from each present state. For present state 0 (state **P**), look at the column for the next value of state variable A. If the ASM is in state 0, the A-mux must produce truth (1) only when Z is true. Thus the equation for this multiplexer input is

$$MUXA(0) = Z$$

For present states 1, 2, or 3, the table shows that the A-mux inputs are always false; we always produce a 0 for the A bit of the next-state code from these states.

TABLE 5–1. STATE TRANSITION DATA FOR THE ASM IN FIG. 5–9

| Present state: | | Next state: | | | Condition for |
Number	Name	Name	B	A	transition
0	P	R	1	0	$\overline{Z}$
		Q	1	1	Z
2	R	P	0	0	T
3	Q	P	0	0	T
1	—	P	0	0	T

Now consider the B-multiplexer. In present state 0, the required expression for the input is

$$MUXB(0) = \overline{Z} + Z = T$$

In the remaining states 1, 2, and 3, the B-mux inputs are again all 0, as they were in the A-mux. Figure 5–12 shows the resulting state generator. Equations for the outputs $CMD1$ and $CMD2$ are the same as in the traditional method: Eqs. (5–1) and (5–2).

This was a fairly simple illustration. With some familiarity with the multiplexer controller method, you could have read off the mux inputs directly from the ASM chart without using the table. Let's do a more complex example—the ASM in Fig. 5–13. This four-state machine will require two state flip-flops B

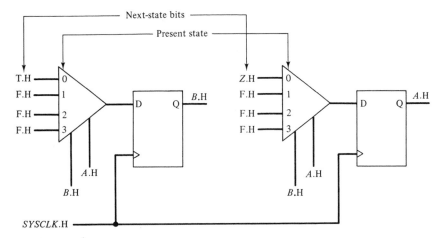

Figure 5–12. A multiplexer controller for the ASM in Fig. 5–9.

State variables: *BA*

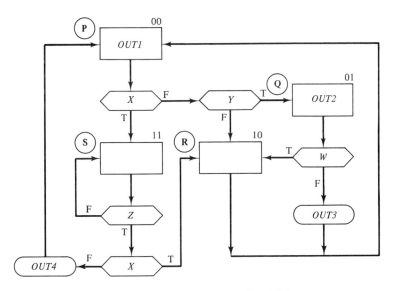

Figure 5–13. A more complex ASM.

and *A*. Again, two four-input multiplexers will provide the proper table-lookup environment for the inputs of the state flip-flop.

Table 5–2 contains the ASM chart information in a convenient form for the preparation of the multiplexer inputs. From the table, we derive these equations

$$MUXA(0) = \overline{X} \cdot Y + X = X + Y$$
$$MUXA(1) = \text{F}$$

$$MUXA(2) = F$$
$$MUXA(3) = \overline{Z}$$
$$MUXB(0) = \overline{X} \cdot \overline{Y} + X = X + \overline{Y}$$
$$MUXB(1) = W$$
$$MUXB(2) = F$$
$$MUXB(3) = Z \cdot X + \overline{Z} = X + \overline{Z}$$

TABLE 5-2. STATE TRANSITION DATA FOR THE ASM IN FIG. 5-13

| Present state: | | Next state: | | | Condition for |
Number	Name	Name	B	A	transition
0	P	Q	0	1	$\overline{X} \cdot Y$
		R	1	0	$\overline{X} \cdot \overline{Y}$
		S	1	1	X
1	Q	P	0	0	$\overline{W}$
		R	1	0	W
2	R	P	0	0	T
3	S	P	0	0	$Z \cdot \overline{X}$
		R	1	0	$Z \cdot X$
		S	1	1	$\overline{Z}$

Figure 5-14 shows the completed state generator for this ASM.

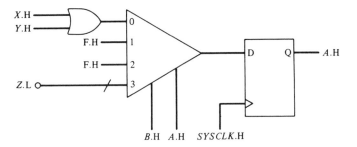

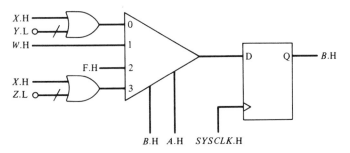

Figure 5-14. A multiplexer control for the ASM in Fig. 5-13.

This example has four outputs—two unconditional and two conditional. We may write equations directly from the ASM chart:

$$OUT1 = P \qquad (5\text{--}3)$$

$$OUT2 = Q \qquad (5\text{--}4)$$

$$OUT3 = Q \cdot \overline{W} \qquad (5\text{--}5)$$

$$OUT4 = S \cdot Z \cdot \overline{X} \qquad (5\text{--}6)$$

To make use of these equations, we need implementations of the logic variables P, Q, and S, which correspond to states **P**, **Q**, and **S**, respectively. We may expand these logic variables in terms of the state variables B and A, as we did in the earlier example. A more systematic, and therefore usually better, technique is to use a decoder with inputs B and A to produce the individual logic variables for each state. The outputs of the decoder will be, in order, P, Q, R, and S. The generation of Eqs. (5–3) through (5–6) for the ASM outputs proceeds easily, using standard mixed-logic techniques. You will see examples of this use of a decoder at several points in this book.

Comments. The multiplexer controller method provides a systematic, easily documented, and easily altered way of implementing ASM controllers. We prefer this method to the traditional method in virtually every case. How big a system will this method handle? The number of multiplexers, being the same as the number of state flip-flops, increases only as the logarithm of the number of states, whereas the size of each mux increases linearly. For example, systems with five to eight states require three 8-input multiplexers, nine to sixteen states require four 16-input multiplexers, and 17 to 32 states must have five 32-input muxes. Commercial muxes with up to 16 inputs are available; we may fabricate larger multiplexers from smaller components, using the methods in Chapter 3. In our experience, for systems of 17 to 32 states, the method is practical but unwieldy. For systems of up to 16 states, the mux method provides a comfortable approach.

The traditional method, using either D or JK flip-flops, becomes a pile of spaghetti beyond eight states and gives none too clear a result in smaller systems.

Most of the systems we wish to handle with the methods of Part II have fewer than 17 states. What do we do about the occasional system with more states? Our last method of state generation provides a way of extending the useful range of hardware state generators to more states than the mux method will handle easily.

The One-Hot Method of ASM Synthesis

In this method of generating states, we use one D flip-flop for each state. There is no encoding of the states, so there is no need to specify a state assignment as we did in the previous methods. Since we must always be in only one state at a time, we must arrange for only one of the state flip-flops to be true during each state time. Therefore, we must compute with combinational logic the value

T or F of each flip-flop's input to provide the one true input required to produce the next state of the system. This property of exactly one flip-flop being true at a time gives the method its name, "one-hot."

We may make exactly one flip-flop true with the aid of a tabular presentation slightly different from the one used in the multiplexer method. Consider the one-hot implementation of the ASM in Fig. 5–13. For this four-state system, we need four D flip-flops labeled with the states' names. Table 5–3 contains the information needed to produce the inputs to the one-hot flip-flops.

TABLE 5–3. STATE TRANSITION DATA FOR A ONE-HOT IMPLEMENTATION OF THE ASM IN FIG. 5–13

Next state	Present state	Condition for transition
P	Q	$\overline{W}$
	R	T
	S	$Z \cdot \overline{X}$
Q	P	$\overline{X} \cdot Y$
R	P	$\overline{X} \cdot \overline{Y}$
	Q	W
	S	$X \cdot Z$
S	P	X
	S	$\overline{Z}$

The equations for the flip-flop inputs follow from the table:

$$NEXT.STATE.P = P(D) = Q \cdot \overline{W} + R + S \cdot Z \cdot \overline{X}$$
$$NEXT.STATE.Q = Q(D) = P \cdot \overline{X} \cdot Y$$
$$NEXT.STATE.R = R(D) = P \cdot \overline{X} \cdot \overline{Y} + Q \cdot W + S \cdot X \cdot Z$$
$$NEXT.STATE.S = S(D) = P \cdot X + S \cdot \overline{Z}$$

Figure 5–15 is an implementation of this controller, assuming that the status inputs W, X, Y, and Z are available in both voltage forms. For the ASM outputs, we may generate Eqs. (5–3) through (5–6) directly, since the logic variables P, Q, and S are available from the one-hot flip-flops.

Initializing the one-hot controller. The first two methods of generating states require an encoded representation of the present state; the code inherently specifies only one state at a time. In the one-hot method, we must take care to initialize the system in such a way that exactly one flip-flop—representing the starting state—is true and the rest false. Once started, the one-hot controller will propagate from state to state in the proper manner. We can envision any one-hot circuit with a master reset signal built into its design to provide for initializing the state generator. (You will see that virtually all designs should

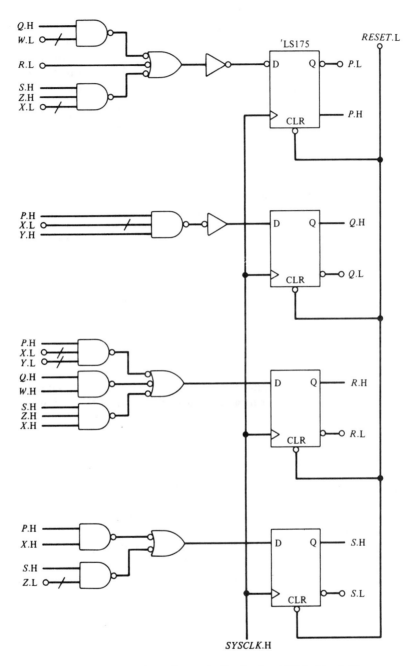

Figure 5–15. A one-hot controller for the ASM in Fig. 5–13.

have such a reset facility; resetting is the only routine use of an asynchronous input to our systems.)

The implementation of the state flip-flops will usually contain a set of D register chips having a common asynchronous clear input. In our example, we use a 74LS175 Quad D Register to provide the four flip-flops needed for state storage. Asserting the asynchronous "clear" input on this chip will clear the Q outputs to low voltages. We may use our mixed-logic notation to good advantage here, to provide for initializing exactly one flip-flop to true. In Fig. 5–15, we have assumed that state **P** is the desired starting state. Note how a master reset signal "clears" the P flip-flop to true and the Q, R, and S flip-flops to false. This is a routine application of mixed logic; the diagram shows exactly what is happening.

Comments. The one-hot method has the advantages of ease of design and clarity of circuit. The inputs to the state flip-flops directly describe the conditions under which each state is the next state. As you know, the mixed-logic notation provides for ease of analysis of circuits, so we may read off the next-state conditions from the circuit diagram. The size of the state generator circuit does not grow rapidly with the number of states. At first you may be horrified at the idea of using a flip-flop for each state instead of the more compact encoded scheme used in the other methods. However, the packaging of MSI integrated circuits gives us four, six, or eight D flip-flops on a chip, so the package count stays low. In medium-size to large designs, the one-hot method compares favorably in total package count with the other methods and usually wins the count.

The main disadvantage of the one-hot method is the need to take great care to initialize the system to a starting state. Also, if the hardware (or the design, heaven forbid!) is faulty, more than one state flip-flop may be true at once. This situation is difficult to debug because the system follows several ASM paths at the same time.

Our preference is to use the multiplexer controller method in systems of up to 16 states and the one-hot method for larger hardwired systems. You will see in Part III that as the systems become complex, involving many states, we shift gears and adopt the method of microprogramming for designing systems rather than using hardwired controllers.

Occasionally, an ASM of a special form will suggest a special type of state generator. For example, a cyclic ASM offers the opportunity to use a binary counter as its state generator. This special form appears in several of the examples in Chapter 6.

The ROM-Based Method of ASM Synthesis

Our fourth method for synthesizing ASMs is based on table-lookup from a permanent memory such as a ROM, PROM, or EPROM. This method has the advantage that it is highly regular in its approach, but it suffers from a serious explosion in the size of the ROM as the ASMs become more complex. In this method, we use the now-familiar encoded state generator shown in Fig. 5–10.

The combinational logic box, which produces the next-state address and the ASM outputs, consists of a ROM of suitable size. The inputs to the combinational logic box form the address of the ROM, and the outputs are the outputs of the ROM. To realize an ASM, we must choose a suitable ROM and specify its contents.

TABLE 5–4. STATE GENERATOR AND OUTPUTS FOR THE ASM OF FIG. 5–9, USING A ROM

Address			Output			
B	A	Z	B(D)	A(D)	CMD1	CMD2
0	0	0	1	0	1	1
0	0	1	1	1	1	0
0	1	0	0	0	0	0
0	1	1	0	0	0	0
1	0	0	0	0	0	0
1	0	1	0	0	0	0
1	1	0	0	0	0	1
1	1	1	0	0	0	1

We will illustrate the method by implementing the ASM in Fig. 5–9. The ROM will have three address inputs—B, A, and Z—so we will need an 8-word ROM. We require 4 bits of output—$B(D)$, $A(D)$, $CMD1$, and $CMD2$—so an 8-word × 4-bit ROM is appropriate. The contents of the ROM are listed in Table 5–4, and Fig. 5–16 is the implementation.

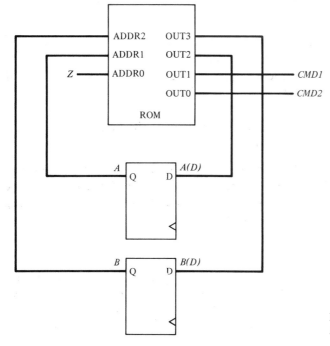

Figure 5–16. Implementation of the ASM in Fig. 5–9, using a ROM.

The synthesis of an ASM using a ROM appears straightforward, systematic, and elegant. There is a problem, however. As the complexity of the ASM grows, the size (number of words) of the ROM increases quickly. The state variables do not present a problem, since their number increases only logarithmically with the number of ASM states, and the ROM's size therefore increases only linearly with the number of ASM states. But the ROM address must contain a bit for each test input in the design, and thus the ROM's size increases exponentially with the number of test inputs. The ASM in Fig. 5–12, with four states and four distinct test inputs, would require a ROM with 8 address inputs, having 64 words. Complex ASMs with lots of test inputs can require huge ROMs with this method.

Another mildly unpleasant characteristic of the ROM-based method for synthesizing an ASM is the enormous redundancy of information. Since the ROM address is fully decoded, the ROM contains a word for each possible combination of address bits. Each test input tends to be used only in a few states, yet the ROM must contain a word for every combination of all test inputs and state variables. For small systems, this method can be effective; large systems will usually require computer software support to generate the ROM's contents. To sum up: the ROM-based method is attractive because of its regularity, but its use is limited by the exponential growth of the ROM's size as the number of test inputs to the ASM increases.

DESIGN PITFALLS

In our study of design, we have made several important assumptions about our systems.

(a) We have assumed that our ASMs are synchronous, with changes in state and other actions governed by a master clock.

(b) We have assumed that at the time of a state change, all inputs to the ASM are stable; in other words, inputs change synchronously with the system clock.

(c) We have assumed that the system clock edge reaches each element in the circuit simultaneously.

Let's investigate the effect of violating these conditions. Conditions (a) and (b) are design decisions of great importance, violation of which will lead to serious problems. Condition (c) is a subtle matter that we enforce by good construction practices. Let's consider (c) first.

Clock Skew

In a synchronous system, it is important that every clocked element in the system receive its clock edge at precisely the same time. To see why this is so, consider the general model of a controller with just two state flip-flops, shown in Fig. 5–17. Proper synchronous operation results when the *CLKA* and *CLKB* active

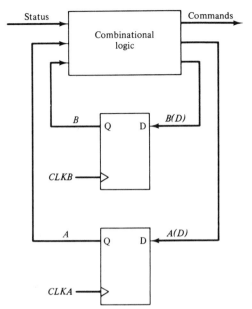

Status → Combinational logic → Commands

B Q D $B(D)$

$CLKB$ ──▷

A Q D $A(D)$

$CLKA$ ──▷

Figure 5–17. A model of a digital system for studying clock skew.

edges occur at the same instant. In that case, the combinational logic network will compute new values for $A(D)$ and $B(D)$ based on the current values of A and B and the status signals and, a few nanoseconds after the clock edge arrives, A and B will assume new values equal to the old values of $A(D)$ and $B(D)$. The changing flip-flop outputs will throw the combinational logic network into shock as it adjusts to the new inputs and computes new values of $A(D)$ and $B(D)$. The combinational logic outputs will experience hazards and delays caused by the finite propagation time of the gates. $A(D)$ and $B(D)$ may therefore have momentary wrong values but will eventually settle to levels predicted by Boolean algebra and will then wait for the next clock edge to come along.

Now suppose that $CLKB$ is delayed with respect to $CLKA$. Signal A changes when $CLKA$ fires; this will throw the gates into shock as before, and both $A(D)$ and $B(D)$ may have momentary wrong values for a few nanoseconds. Suppose that the "late" $CLKB$ edge comes during this time of instability; then B can record a false value. Even more galling: suppose that before $CLKB$ fires the combinational logic stabilizes to "new" values of $A(D)$ and $B(D)$, based on the new value of A and the old (unchanged) value of B. At this time, both $A(D)$ and $B(D)$ can be incorrect. Then, when $CLKB$ fires, an incorrect B is stored, and this change ripples through the logic. This phenomenon, *clock skew,* occurs when the clock edges do not appear simultaneously at all clock inputs. Clock skew can arise from gates in the clock path or from different wire lengths between the clock source and the clock inputs.

Don't gate your clock. Gating a clock is bad practice because it introduces skew (and may introduce hazards on the clock line). Suppose that flip-flop A responds to a positive clock edge but that we use a different type of flip-flop

Chap. 5   Design Methods   **189** for *B* that acts on a negative clock edge. We may be tempted to create the *CLKB* signal by running an inverter from *CLKA*, but this is just a case of gate-created clock skew. To avoid this type of skew, it is best to drive all flip-flops with the same active clock edge, thereby eliminating the need for inverters in part of the clock system. In all our clocked building blocks for synchronous design, we use positive-edge clocks.

Beware different length of clock paths. It is also desirable to have the clock distribution lines spread radially from the clock source to the separate elements of the system rather than linking them together in one long chain. When designing large systems you may have to buffer the clock lines to build up sufficient power, as shown in the radial clock distribution system in Fig. 5–18. This "gating" of the clock lines is acceptable if the three buffers are all of the same kind and, preferably, in the same integrated circuit package, so that they all have precisely the same propagation delays. The distribution wires should also be of the same length, within a few inches, since 8 inches of wire represents about 1 nanosecond of signal propagation time.

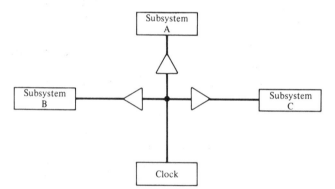

Figure 5–18. A buffered radial distribution system for the master clock.

Asynchronous Inputs and Races

Design assumption (b) is that all the inputs to an ASM change synchronously with the master clock. In practice, inputs often arise from sources *outside* our digital circuit, and the timing of changes in these inputs is beyond our direct control. These inputs are *asynchronous,* and we usually append an asterisk ∗ to their variable name to indicate their asynchronous nature. To see why asynchronous behavior is troublesome, consider the three-state ASM fragment in Fig. 5–19. We assume that we have made the (encoded) state assignment shown in the figure and that the sole test input *IN*∗ is asynchronous. For the moment, ignore the conditional output.

Transition races. Each state flip-flop requires that its input be stable for a few nanoseconds prior to the clock edge. This allows the input values to circulate through the internal circuitry of the flip-flop and stabilize to await the clock edge. If the flip-flop inputs change during this "setup time," the value of

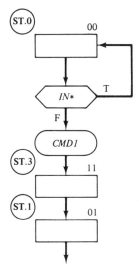

Figure 5–19. An ASM fragment that illustrates races.

the flip-flop's output after the clock edge will be unpredictable. After settling down, the output will assume either a T or F value, but which value is uncertain.

Now assume that the ASM in Fig. 5–19 is in state 00, and $IN* = T$. Then the inputs to both state flip-flops are 0, and the system is preparing to move next to state 00, the same state as before. If $IN*$ changes to F, the inputs to flip-flops A and B will change to 1, in preparation for the move to state 11. If the change in $IN*$ occurs during the flip-flop setup time, we cannot predict the changes in the flip-flops. Thus the next-state code may be 00, 01, 10, or 11, depending on the outcome of the *race* at the flip-flop inputs. Although we might argue that either state 00 or state 11 is an acceptable next state, clearly to reach states 01 or 10 is a calamity. This situation, in which the next state depends on the exact timing of the flip-flop input changes, is called a *transition race*. The situation is obviously intolerable and you must be certain that such races do not appear in your designs. Before considering the solution to the problem, let's investigate another type of race.

Output races. Now see what happens to the conditional output *CMD1* in the ASM of Fig. 5–19 when $IN*$ changes at an awkward time. For the moment, ignore the possibility of transition races. In state 00, when $IN*$ is true, *CMD1* is false, whereas when $IN*$ is false, *CMD1* is true. A change in $IN*$ will cause a corresponding change in *CMD1*. If $IN*$ changes from T to F late in the state 00 time, *CMD1* will be true for only a short part of a clock time, before the system moves into state 11. The possibility of a "runt pulse" for *CMD1* is in itself a serious matter, since the output *CMD1* may be used in situations that cannot tolerate such a short pulse. This problem is called an *output race*; it is a direct result of an ASM output being conditional on an asynchronous input. For example, suppose that the purpose of *CMD1* is to set a flip-flop that lights a light announcing that we have left state 00. If $IN*$ changes so late in the clock cycle that the flip-flop is not set to true, we will be in state 11 and the light will still be off.

The combination of the transition race and the output race in this ASM may lead to numerous ludicrous results, depending on the exact reactions of the flip-flops to the changing $IN*$ input. We could end up in state 00 with the light on, or in state 01 with the light off, and so on.

Avoiding races. Asynchronous inputs are at the root of the problem of races. Asynchronous inputs are fatal, dangerous, or at best difficult to use safely. There is no way to avoid output races except to avoid conditional outputs that depend on asynchronous inputs. The engineering literature is full of elaborate methods for skirting around the transition race problem by tinkering with the state assignments. The proper approach for good design style is to eliminate the cause of the problem—the asynchronous input. We may do this by *synchronizing* the input using a D flip-flop clocked by the system clock. The ASM will test the output of this flip-flop, and since this output only changes synchronously with the clock, there will never be a race in the ASM caused by that input. So we have a golden rule for synchronous design:

> Don't allow dangerous asynchronous inputs into your ASM chart.

You must be alert to identify asynchronous inputs—they have a habit of sneaking into the design. In our example, the input was easy to detect, since we had added an * to the signal name, but in practice, adding the asterisk is *your* responsibility, whether or not the original name was so equipped. Many useful synchronous integrated circuit chips have asynchronous control inputs for clearing, setting, or loading. The only routine use that we make of such inputs is as a *master clear* signal to be asserted when power is first applied or when the system "hangs up." In the one-hot controller method you saw an illustration of this usage; most controllers will require one such master clear signal. In other circumstances, avoid using asynchronous control signals.

Asynchronous ASMs

Our ASMs have all been synchronous, with a master clock to define the times for state transitions. It is possible to build asynchronous state machines, which depend not on a clock but on changes in the inputs themselves to create transitions between states. Troubles abound in this form of design, primarily because of these factors:

(a) All inputs must be clean, with no glitches, since any instability or noise on the input signals may induce spurious state transitions.

(b) The theory of asynchronous circuits is complex and diverse, involving numerous special cases and usually invoking unacceptably restrictive design conditions.

(c) The debugging of asynchronous systems is difficult.

We will not instruct you in the theory of asynchronous circuits, beyond the microscopic view we took in Chapter 4. Rather, we wish you to avoid this mode of design wherever possible. In Chapter 9 you will encounter a form of asynchronous ASM used to connect a peripheral device to a minicomputer, but this special case is as far as we will take the subject. If at some later stage in your design career you wish to investigate asynchronous circuits, you will find a rich literature. However, we wager that, after your investigation, you will still refrain from designing circuits in this mode.

Sidestepping the Pitfalls

Getting into trouble in digital design is easy. Asynchronous methods constantly expose the control algorithm to every signal change, intentional or accidental. Designing each circuit to be secure against such an unceasing attack requires enormous effort. Asynchronous control, despite its tempting generality, is too tedious to use as a major tool in design.

Synchronous methods ease the pressure on the designer by isolating the sensitive periods into small, regular intervals preceding clock edges. Expressing problems synchronously effectively moves the asynchronous difficulties away from the algorithm into the clock. This is a great simplification. We must work hard to make the synchronous clock system reliable, but the procedure is the same for all designs. In return, we gain breathing room in the algorithm. Synchronous design causes specialization of the hardware, where extreme versatility is rarely needed, yet it decreases the number of special cases in the process of algorithm design, giving designers valuable systematic methods.

As you saw in Chapter 4, the problem of metastability in the outputs of sequential circuits is inherent in every design. Wherever two or more inputs may change at the same time, metastability is possible. In asynchronous design, virtually every input may present this problem. In synchronous design, the asynchronous external inputs, which we routinely run through synchronizing circuits, are trouble spots. By synchronizing these inputs we have greatly simplified the internal structure of our control algorithm, but we have not eliminated the possibility of metastability. In Chapter 12 we offer recommendations for dealing with this irritating issue.

Debugging Synchronous Systems

Not only do synchronous methods ease the designer's worries, but they also support a powerful debugging technique. The system clock controls the speed of a synchronous design. Consider the benefits of a design that will behave properly not only at high clock speeds but also at slow speeds, even at zero speed. We are particularly interested in the zero-speed case, because with this feature we may freeze the system in any state by stopping the system clock. We may then debug the logic at leisure. Compare this technique with a system that requires a closely controlled clock frequency. Error conditions may be observable for only one clock cycle. If maintenance engineers cannot slow or stop the machine at will, they must troubleshoot it in real time. Debugging systems at high speed is much more complicated than freezing the machine in the erroneous state.

Synchronous designs that work at a variety of clock speeds, including zero, are called *static*. The benefits of static systems are so great that we should strive to use this technique whenever possible. After you have debugged a few dynamic (nonstatic) systems, you will better appreciate the beauty of static designs. In Chapter 6, we develop a system clock module for use in static designs.

CONCLUSION

This ends our exposition of basic design methods. We have covered the basic building blocks and there were remarkably few types. Next, for expressing algorithms, we described a language that contained only three constructs: the state box, the conditional output oval, and the conditional branch diamond. With these simple tools we can create digital systems limited only by our imagination.

The basic elements of style emerge from a desire to achieve understandable designs. We have discovered through bitter experience that opaque designs are enormously expensive in the long run. A good designer will use a design approach that always promotes clarity. We mention three important aspects of such an approach:

(a) *Good documentation.* It is tempting to avoid the drudgery of documentation. After all, the real fun is in the design and debugging. It is hard enough to document a simple design; complex designs are seldom documented well enough for anyone but the designer to understand them. Often, after a few months, even the designer cannot fathom the design. A good designer will adopt techniques that encourage or require good documentation during the design process. Mixed logic, functional building blocks, and ASM charts are powerful aids to documentation, built into the design discipline.

(b) *Modular designs.* Nearly every design will require small changes during its useful life. Monolithic designs are hard to understand and modify. Our goal is to build more accessible designs, so that we may change part of the complete system without the change rippling through the rest of the design.

Too often, digital designers overlook the cost of servicing digital equipment. Since servicing usually falls to other people, poor designers are not forced to live with their abominations. Hardware *will* need repair. Digital devices should be simple and modular so that other people can perform the maintenance. The use of functional building blocks and the separation of architecture from control both encourage modular design.

(c) *Absence of tricks.* Digital design affords unbounded opportunity for clever tricks. Such trickery should not be, but often is, confused with good design. We can benefit from the experience of computer programmers who, after years of maneuvering bits in clever ways, have come to realize that systematic, clear methods yield far more dividends than cute but obscure tricks.

Perhaps we can sum up good design philosophy in a single phrase: common courtesy. Consider the users and maintainers of your system, and ask yourself what they will need in order to deal efficiently with your creation. Let courtesy be your guide.

Summary of Design Guidelines

Here we bring together the three forms of design guidance presented in this chapter.

Basic Approach to Solving a Digital Problem
   (a) Design from the *top down.*
   (b) Separate the *architecture* from the *control.*
   (c) *Refine the design,* letting the control algorithm and the architecture influence each other as you converge on the solution.

Technical Design Considerations
   (a) Use *synchronous* (clocked) design techniques.
   (b) *Avoid asynchronous inputs* in the algorithm.
   (c) Make your designs *static*—independent of clock speed.

A Courteous Philosophy
   (a) Develop good *documentation* during the design.
   (b) Keep designs *modular* and *simple.*
   (c) *Avoid* obscure *tricks.*

In Chapter 6, we will work out several examples of designs. In the process, you will see the design tools in action and study in their proper context a number of common design situations and their handling. In Chapter 7, we will embark on the construction of a complete minicomputer, using top-down techniques at the MSI level of complexity. Finally, in Part III, we will introduce the powerful techniques of microprogramming and software control for managing complex design problems.

So now begins the actual design!

READINGS AND SOURCES

CLARE, CHRISTOPHER R., *Designing Logic Systems Using State Machines.* McGraw-Hill Book Co., New York, 1973. The original exposition of the ASM approach.

DIETMEYER, DONALD L., *Logic Design of Digital Systems,* 2nd ed. Allyn and Bacon, Boston, 1978. Chapter 13: traditional asynchronous design.

ERCEGOVIC, MILOŠ D., and TOMÁS LANG, *Digital Systems and Hardware/Firmware Algorithms.* John Wiley & Sons, New York, 1985. Good treatment of sequential systems.

FLETCHER, WILLIAM I., *An Engineering Approach to Digital Design.* Prentice-Hall, Englewood Cliffs, N.J., 1980. Uses ASMs.

HILL, FREDERICK J., and GERALD R. PETERSON, *Introduction to Switching Theory and Logical Design,* 3rd ed. John Wiley & Sons, New York, 1981. Good traditional treatment of sequential circuits.

MALEY, G. A., and J. EARLE, *The Logic Design of Transistor Digital Computers.* Prentice-Hall, Englewood Cliffs, N.J., 1963. An influential early work; you can see how difficult design can be without systematic methods.

MANO, M. MORRIS, *Digital Design.* Prentice-Hall, Englewood Cliffs, N.J., 1984. Presents ASM charts as an alternative to traditional techniques.

MEALY, G. H., "A method for synthesizing sequential circuits," *Bell System Technical Journal*, Vol. 34, September 1955, page 1045. The Mealy state machine.

MILLER, RAYMOND E., *Switching Theory.* Vol. 2, *Sequential Circuits.* John Wiley & Sons, New York, 1966. An important early work.

MOORE, E. F., "Gedanken experiments on sequential machines," in *Automata Studies*, edited by C. E. Shannon and and J. McCarthy. Princeton University Press, Princeton, N.J., 1956. The Moore state machine.

WIATROWSKI, CLAUDE A., and CHARLES H. HOUSE, *Logic Circuits and Microcomputer Systems,* McGraw-Hill Book Co., New York, 1980. Uses ASMs. Sensible treatment of asynchronous ASMs.

EXERCISES

5–1. Sketch a general method for the top-down solution of a digital problem.

5–2. How does the ASM chart differ from a software flowchart? Using Fig. 5–8 as an illustration, explain the fundamental differences in viewpoint.

5–3. What is meant by "active clock edge"?

5–4. What is a state time? In a synchronous system, what determines the duration of the state time?

5–5. Draw diagrams to illustrate the following:
  (a) A four-state cyclic (sequential) ASM.
  (b) A three-state ASM with a fixed sequence of states containing a conditional output.
  (c) A two-state ASM with a two-way branch in one state and no conditional outputs.
  (d) A four-state ASM that can produce this sequence of states: **S1, S3, S2, S1, S4, S2, S1, S1, S1,** . . . .

5–6. Explain the difference between an ASM input and an ASM output.

5–7. What is the difference between an ASM conditional branch and an ASM conditional output? Does one imply the other?

5–8. In a synchronous ASM, an unconditional output is stable for virtually the entire duration of the state. For what period is a conditional output stable?

5–9. Produce ASM charts that perform each of the following software operations:
  (a) If $X = N$, then . . . .
  (b) If $X \neq N$, then . . . ; else . . . .
  (c) For X from A to B step C, do . . . .
  (d) While $X = Y$, do . . . .

5–10. Here is a two-state ASM:

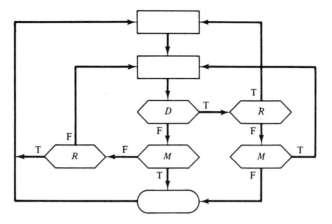

(a) Convert the ASM into a form that has a single decision box of the form below, with eight branches:

(b) Implement the original ASM and your modification. From the viewpoint of the implementer, which form is best?

5–11. Using timing diagrams, show the difference between these two ASMs:

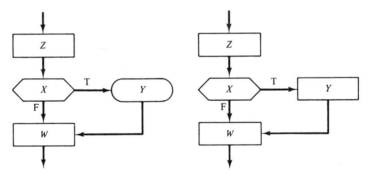

5–12. In many instances, we may remove a conditional output from an ASM by creating a new state dedicated to generating the old conditional output (see the diagrams in Exercise 5–11). Under what circumstances will this translation produce difficulties?

5–13. A conditional output is a function of both state and path. In a logic equation for a conditional output, what logic operator connects the state term with the path term? In other words, what logic operator corresponds to the box in the equation

$$\text{Conditional.output} = \text{State } \boxed{?} \text{ Path}$$

5–14. The ASM notation

can be used to indicate that "*A* assumes the value of *B* at this time." Show that this notation may be viewed as a shorthand for an ASM state that contains a test and a conditional output.

5–15. Consider the following fragments of an ASM chart. Carefully state what, if anything, is wrong with or objectionable about these notations.

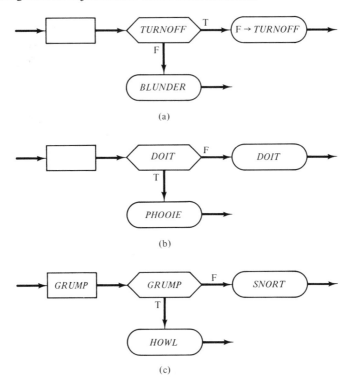

(a)

(b)

(c)

5–16. What is a state generator? With an encoded state assignment, how many states can four state variables specify?

5–17. Some older design methods lump the controller and architecture of Fig. 5–1 into the combinational logic of Fig. 5–10. Why is this poor practice?

5–18. Design a one-hot controller for Fig. 5–8.

5–19. Produce a realization of the ASM in Fig. 5–9 that is equivalent to Fig. 5–11 but with the state assignment **P** = 00, **Q** = 01, **R** = 10. Is there any difference in hardware complexity?

5–20. Can you make a state assignment in Fig. 5–9 that will simplify the hardware for generating *CMD1*?

5–21. Design a traditional state generator for the ASM in Fig. 5–20, using D flip-flops.

5–22. Design a multiplexer controller for the ASM in Fig. 5–20.

5–23. Design a one-hot controller for the ASM in Fig. 5–20. What special precaution must you take when using the one-hot method?

5–24. Design a ROM-based controller for the ASM in Fig. 5–20. Tabulate the contents of the ROM.

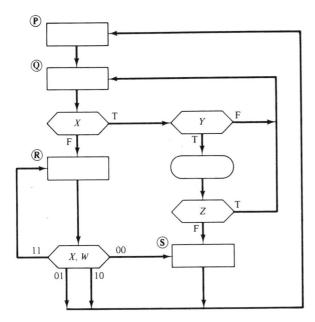

Figure 5–20.

5–25. Devise implementations, including circuit diagrams, for the synchronous ASM in Fig. 5–21, using each of the state-generation techniques given below.

    (a) Multiplexer controller, using SSI- and MSI-level components.

    (b) Multiplexer controller, using PALs.

    (c) ROM-based controller. Show the contents of the ROM.

    (d) One-hot controller, using SSI- and MSI-level components.

    (e) One-hot controller, using PALs.

5–26. Perform Exercise 5–25 for the ASM in Fig. 5–22. Generate W with a JK flip-flop and N with an enabled D flip-flop. The notation "$Z : Y$" implies "Z assumes the value of Y at this time" (see Exercise 5–14).

5–27. The logic equation for an ASM output has terms involving distinct logic variables for each ASM state in which the output appears.

    (a) For a state generator with encoded state assignments, show a standard and systematic method of transforming the state code into logic variables for each state, using a decoder.

    (b) Instead of using a decoder, we can generate ASM outputs with AND gates to decode the required logic variables for states from the state code. Demonstrate this method. When would you use this method in preference to the method of part (a)?

5–28. In the one-hot state generator method, show how logic equations are generated for conditional and unconditional outputs. Is output signal generation simpler with the one-hot method than with the multiplexer method?

5–29. Use two 8-input priority encoders to detect when more than 1 bit is true in an 8-bit quantity. (*Hint:* Connect the input signals to each of the encoders, but in opposite order.) Can this design be extended to more than 8 bits? Show a use of this circuit in the design of one-hot controllers.

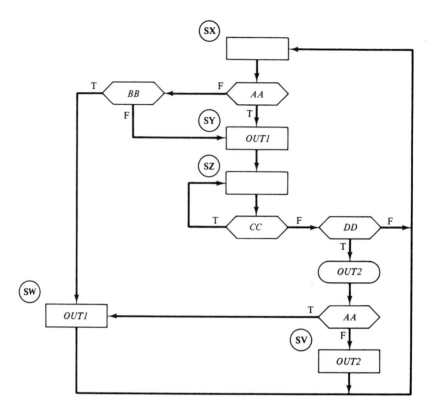

Figure 5–21.

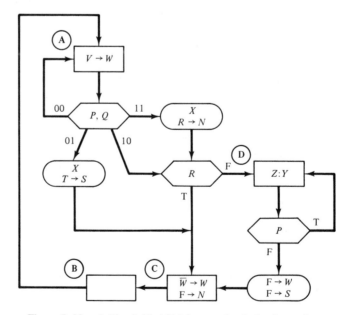

Figure 5–22. A "loaded" ASM for practice in implementing.

5-30. Consider the following eight-state cyclic ASM:

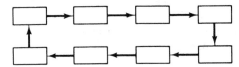

Design state generators for this ASM as a multiplexer controller, a one-hot, and a *binary counter*. Which method is simplest in this special case?

5-31. Consult a TTL data book, and design a state generator equivalent to Fig. 5–14, using 74LS00, 74LS02, 74LS04, 74LS175, and 74LS352 integrated circuits. Draw a complete circuit diagram, including pin numbers. How many chips are required?

5-32. Why is the clock such an important element in a synchronous design?

5-33. What is meant by "gating the clock"? Why is this practice dangerous?

5-34. For the ASM in Fig. 5–19, show with a timing diagram how the asynchronous input *IN*∗ can cause a transition race.

5-35. Using Fig. 5–19, demonstrate an output race.

5-36. How may you avoid races in your designs?

5-37. Suppose that, contrary to our basic design approach, you must deal with a synchronous ASM that tests an asynchronous input *T2*∗, as shown below. The logic-level timing diagram shows the condition of three signals over a period of six state times. Complete the logic-level timing diagram for the signals *OUT1*, *OUT2*, and *OUT3*.

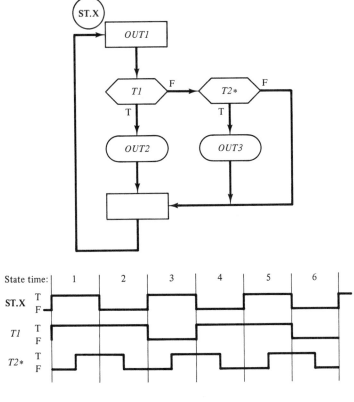

5-38. In Fig. 5–11, assume that the propagation delay of gates and flip-flops is t_p.

    (a) How much clock skew is tolerable between flip-flops A and B during the transition from state 11 to state 00?

    (b) Repeat part (a) for the transition $10 \rightarrow 00$.

    (c) Repeat part (a) for the transition $01 \rightarrow 00$.

Practicing Design

And now, let's do some design. In this chapter we present several detailed examples of digital design, from the small, yet important, to the substantial project. The goals are two: to illustrate good design methodology and to introduce important design problems and their treatment. There is repetition of some aspects of design, particularly the use of ASM charts, yet each example has fresh material. We will indulge in some excursions into interesting topics as they occur in the context of design. We hope that in this way the concepts will be more meaningful to you than they would be as separate, isolated subjects.

Examples 1 and 2 illustrate some basic design issues related to human interaction with a digital device. In examples 3 and 4 we develop circuits to support the conversion of information from a serial bit stream to a sequence of bytes and vice versa, common operations in data communications (in Chapter 11 we use these circuits as a part of a larger microcomputer-based project). The traffic-light controller and combination lock examples, 5 and 6, provide more practice in design. In example 7, a Black Jack Dealer, we bring together several design problems and techniques (this example reappears in a different design mode in Chapter 10).

DESIGN EXAMPLE 1: A SINGLE PULSER

Our first illustration of design is the development of a circuit for handling a common situation involving a human operator of a machine. Most real systems involve humans, usually at switches, pushbuttons, and lights. Digital systems generally (but not always) run at speeds many thousands of times faster than human reactions. When a machine operator presses a button to initiate some action, the digital device must detect this signal and perform the appropriate steps. In the typical case, the machine completes the actions in a flash, and is back interrogating the pushbutton signal again long before the operator can release the button. We must develop a scheme so that the machine processes a particular button depression only once. A circuit for this is called a *single pulser*; it delivers a pulse only a single clock cycle long when a button is pressed. If we have such a circuit, our digital machine may test the single-pulser output instead of dealing directly with the pushbutton signal and may thereby detect only one event as long as the button is down.

We will study this problem and its solution in several ways. First, we develop the most fundamental solution.

Algorithmic Solution of the Single Pulser

Statement of the problem.   We have a debounced pushbutton, on (true) in the down position, off (false) in the up position. Devise a circuit to sense the depression of the button and assert an output signal for one clock pulse. The system should not allow additional assertions of the output until after the operator has released the button.

Approach.   We will develop an ASM solution to the problem. Since the problem is stated in terms of clock pulses, we know we are dealing with a synchronous system (for which we are grateful). Let us name the important inputs and outputs. Clearly, the position of the pushbutton is of great importance to the algorithm. The pushbutton signal can change at any time, independently of the state of the system clock, so the signal is asynchronous, and our name for it should have a terminal ∗: for instance, *PB∗*. Call the output of the single-pulser circuit *PB.PULSE*.

We know to be cautious about allowing asynchronous test inputs to creep into our ASM charts. To avoid testing *PB∗*, we should synchronize the signal using a clocked D flip-flop. This flip-flop, with input *PB∗* and output *PB.SYNC*, becomes part of the architecture of our solution.

Control algorithm.   Now we may write an ASM chart to describe the algorithm. The algorithm will test *PB.SYNC* and produce an output *PB.PULSE*. The algorithm has two states: one for detecting the first moment that *PB.SYNC* becomes true (button goes down), and the other to wait until *PB.SYNC* becomes false (button goes up). Figure 6–1 is the ASM.

Implementing the design.   The equation for the output *PB.PULSE* is

$$PB.PULSE = FIND \cdot PB.SYNC$$

Once we have a way of producing the variable *FIND*, our problem is solved. To obtain the value of *FIND*, we must implement a state generator for our ASM. We will use the multiplexer method of state generation. The two-state ASM

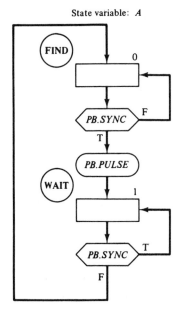

State variable: A

FIND

0

PB.SYNC — F

T

PB.PULSE

WAIT

1

PB.SYNC — T

F

Figure 6-1 A single-pulser ASM.

will require one flip-flop to record the state, and one two-input multiplexer to develop the next-state input to the flip-flop. In systems with more states, the state flip-flops hold an encoding of the current state and, in order to obtain logic variables for each individual state, we must decode the state code. However, in our present design, the output of one flip-flop has two distinct states, and this is sufficient to produce signals for the two states **WAIT** and **FIND** directly, without decoding.

Look at Fig. 6–1. Formally, the flip-flop output is a state variable A, $A = 0$ representing the **FIND** state and $A = 1$ the **WAIT** state. Thus

$$WAIT = A$$
$$FIND = \overline{A}$$

So we may just call the flip-flop output $WAIT$, and then we have

$$FIND = \overline{WAIT}$$

What are the two inputs to the next-state multiplexer? If the system is in state 0 (**FIND**), the condition for a true input to the flip-flop is the same as for moving next into state 1, namely, $PB.SYNC$. If the present state is 1 (**WAIT**), a true flip-flop input for the next state again requires that the next state be 1. The condition from the ASM chart is again $PB.SYNC$. Here is a multiplexer whose inputs are all the same! We can eliminate the mux from the state generator and run the signal $PB.SYNC$ directly into the flip-flop input. Now, our complete circuit consists of the synchronizing flip-flop $PB.SYNC$, the state flip-flop $WAIT$, and the logic to form $PB.PULSE$. Figure 6–2 is the single-pulser circuit.

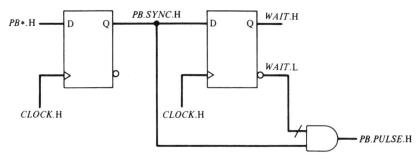

Figure 6–2 A single-pulser circuit.

A Combined Architecture-Algorithm Solution

Although the foregoing derivation is our most fundamental solution of the single-pulser problem, there is another approach that illustrates an important connection between the architecture and the algorithm in digital design. Suppose we reason as follows. Our architecture consists of a synchronizing flip-flop for *PB**, with output *PB.SYNC*, and another D flip-flop whose function is to delay the *PB.SYNC* signal by one clock time. The second flip-flop has input *PB.SYNC* and output *PB.DELAYED*. Then our single-pulser output signal *PB.PULSE* should be true only when *PB.SYNC* is true and *PB.DELAYED* is still false. As soon as *PB.DELAYED* becomes true, we must stop asserting *PB.PULSE*. Can we formalize these thoughts with an ASM? Of course; Fig. 6–3 is a one-state solution. This ASM requires no flip-flops for state generation, since there is only one state. The equation for *PB.PULSE* is

$$PB.PULSE = PB.SYNC \cdot \overline{PB.DELAYED}$$

The implementation looks identical to our original circuit in Fig. 6–2 except for a trivial change in the names in the interior of the circuit.

Implications. This is an interesting phenomenon. We first had a two-state system with minimal architecture (one flip-flop), and now we have a one-

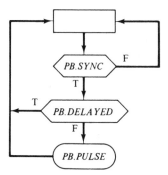

Figure 6–3 A single-pulser ASM in one state.

state system with more inputs and more architecture (two flip-flops). Both solutions yield the same hardware.

This reflects an important concept in digital design. The distinction between architecture and algorithm is arbitrary. In general, by enlarging the architecture, we may convert any ASM into one with fewer but more complex states. At the limit, it is always possible to describe any algorithm in a single state. For complex systems, this yields such a messy ASM that it is not useful; nevertheless, it is technically correct. The vital point is that architectures and ASMs are tools to assist us to understand our problem and to produce a clear, correct implementation. The tools are to serve us, not control us!

Whereas we prefer the first solution to our example as more fundamental, there is merit in the second solution also. Both result in equivalent (in this case, identical) hardware.

A Single-Pulser Building Block

Having developed a circuit for producing a single pulse from a long input signal, we may package the circuit in a black box and treat it as one of our design building blocks. The single-pulser black box has an input $PB*$ from an asynchronous source, and produces a one-clock-cycle true output $PB.PULSE$ when the input becomes true. Whenever we need this type of behavior, we may mentally plug in our black box, and when we build our circuit, we use hardware similar to Fig. 6–2 in the box. It would be nice if the single-pulser box were available in a single chip, but it is not. However, we still treat the single-pulser operation as a building block in our work.

Generalizing the Single Pulser

Think about the human-machine interaction implied by our single-pulser problem. When the operator presses the button, the single pulse promptly appears (and disappears). The machine must have been ready to act on the pulse signal. How did the operator know when it was okay to press the button? Somehow, the operator must infer the correct time from the condition of the machine. Usually this would mean a light or some combination of lights on the control panel. The circuit controls these lights, indicating its readiness to process a button depression. The single pulser works well in this common situation.

Now suppose we treat the operator-machine interaction differently. Let the operator press a button at will but require the operator to hold the button down until there is some indication that the machine has received the signal. Then the machine may be in any state when the button is depressed, and only when the circuit is ready to respond to the button will the light come on.

The single pulser will not work in this case, but we can handle the situation with an ASM structure that looks for the pushbutton depression, then lights the light until the operator releases the button, and then performs the desired operation. Figure 6–4 is a typical form of this ASM. (In Design Example 7 in this chapter, we discuss an additional variation of the single-pulser algorithm.)

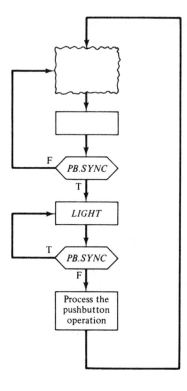

F PB.SYNC

T

LIGHT

T PB.SYNC

F

Process the
pushbutton
operation

Figure 6–4 Diagram of an ASM if the operator may press a pushbutton at any time. In this algorithm, the machine waits until the operator releases the button before it responds to the pushbutton.

DESIGN EXAMPLE 2: A SYSTEM CLOCK

The clock is the master pacer in digital systems, and we must design it carefully. A good clock is perhaps the most potent debugging tool a designer can put into a system. At the least, the clock must have an automatic mode whose frequency may be fixed at some high value and a manual mode that generates one hazard-free active clock edge when a pushbutton is depressed. To do this, there must be a mode switch to select automatic or manual status. Throwing the mode switch must not shorten a clock cycle in the automatic mode. If the mode switch were flipped 10 percent of the way into a clock cycle, thereby truncating the cycle at that point, our circuit could suffer from timing problems. We must let the last cycle run its normal course before the clock system enters the manual mode.

Our approach to design requires that we move difficult, universal concepts up front so that we can solve them once and then apply them to all projects. The system clock is an important concept, so let's design a circuit for it.

Statement of the Problem

(a) Design a hazard-free system clock that runs in two modes, automatic and manual. The automatic mode is a fixed-frequency mode derived from a continuously running clock or oscillator. (Chapter 12 shows ways of fabricating clock signals from oscillators.) The manual mode should produce a true clock output when a pushbutton is depressed, and a false output otherwise.

(b) Activating the mode switch must never cause truncation of a clock cycle.

(c) In the automatic mode, the clock circuit should ignore the manual pushbutton.

Digesting the Problem

There are two inputs, from a debounced mode switch and from a debounced pushbutton. We may formalize the variables in the problem. Let the mode switch output be *MAN*, and let *MAN* = T in the manual mode and *MAN* = F in the automatic mode. Let the manual pushbutton's output be *PB*, and let *PB* = T when the button is down and *PB* = F when the button is up. Let the clock output be *CLK*. Since we are dealing with a clock for synchronous edge-triggered systems, it is natural to let *CLK* = T be the high voltage level and *CLK* = F be the low voltage level, although this choice is not essential to the design.

In digesting the problem, we uncover a crucial point: the clock output must be free of hazards. In Chapter 4, you learned about the catastrophies that will occur when clocks deliver spurious edges. The most general way to avoid hazards is to avoid gates on the lines that must be hazard-free. How can we build something without gates? Commercial flip-flops are designed to have hazard-free outputs, and using a flip-flop output is the most convenient way to avoid gates. We will produce the clock output *CLK* directly from a flip-flop.

Now we are ready to derive an ASM chart for our system clock circuit.

Algorithm for the System Clock

Our system is a synchronous circuit clocked by a continuously running oscillator that will drive the system clock ASM. The output *CLK* has two levels, T and F, so we might use two ASM states, one for *CLK* = T and the other for *CLK* = F. In the automatic mode, we would expect to flip back and forth between the states constantly. Our design problem is to fit the manual-mode operations into this framework. Figure 6–5 is the ASM chart.

This ASM ignores the pushbutton unless the mode switch is in the manual mode. In the automatic mode, the active state alternates between **LO** and **HI**, producing a *CLK* output of half the ASM clock frequency, as shown in the following timing diagram:

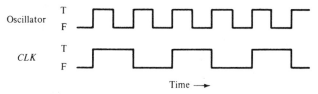

In the manual mode, the ASM moves to whichever state reflects the *PB* position.

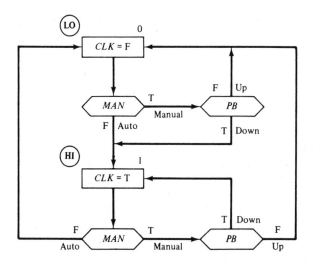

Figure 6-5 An ASM of a system clock.

The ASM formulation clearly demonstrates that the automatic mode overrides the pushbutton and that, whenever we switch from the automatic to the manual mode, the last automatic clock phase exists for its full duration, without any shortening.

But the inputs *MAN* and *PB* in the ASM are asynchronous to the ASM clock. In Chapter 5, you were told to avoid such signals in the ASM, since they introduce many dangers, and even when they are not dangerous, it is usually difficult to establish that their use is safe. Must we synchronize these signals to protect our ASM from harm? In this case, we can quickly demonstrate that we do not need to. Figure 6-5 is a two-state ASM with no conditional outputs. The machine is always either in one state or in the other. When the mode switch is changed from automatic to manual, the ASM either detects the change during the present ASM state, or it doesn't. If it does, at the next state transition it moves into the manual mode; if the ASM misses the change, it remains in the automatic mode for one more clock cycle before entering manual mode. In either event, in this simple ASM there is nothing to go wrong as a result of the asynchronous nature of the *MAN* signal. A similar argument applies to *PB*.

Implementing the Circuit

Let's use the multiplexer controller method for implementing the state generator. One flip-flop will encode the two states in our ASM. The purpose of the design is to produce the system clock output *CLK*. We have already decided to produce the system clock signal *CLK* as the output of a flip-flop. According to the ASM, *CLK* is true in state **HI** and false in state **LO**. In Fig. 6-5, we chose to represent **LO** by 0 and **HI** by 1. This is the same behavior as the *CLK* output: false in **LO** and true in **HI**. In this case, we get our desired *CLK* output from the same flip-flop used in the state generator. Very convenient.

For the state generator, we set up a table of next-state conditions:

Present state	Next state		Conditions
LO	LO	0	$MAN \cdot \overline{PB}$
	HI	1	$\overline{MAN} + MAN \cdot PB = \overline{MAN} + PB$
HI	LO	0	$\overline{MAN} + MAN \cdot \overline{PB} = \overline{MAN} + \overline{PB}$
	HI	1	$MAN \cdot PB$

Both voltage signal forms will be available for the debounced switch and pushbutton variables. In Fig. 6–6, we show a multiplexer controller for our clock machine. This completes our solution to the design problem.

Critique

All three individual gates in Fig. 6–6 come from a single 74LS00 Quad Two-Input Nand Gate chip, since a spare Nand gate will implement the voltage inversion conveniently. Our solution costs one chip for gates, one chip for the two-input multiplexer, and one chip for the flip-flop. To be sure, there are unused components in these packages, but for a fundamental module such as a system clock you probably don't want to share the chips with other parts of the system. In the unshared mode, the cost is three chips; sharing with other circuits, the cost is $\frac{3}{4} + \frac{1}{4} + \frac{1}{2} = 1\frac{1}{2}$ chips.

Advantages of using the mux controller are its simplicity of construction (synthesis) and its ability to display the original ASM structure in the circuit (analysis). The mux method seldom takes more chips than alternative solutions. In this simple case, using a JK flip-flop for the state generator can yield a solution requiring one flip-flop and two OR gates—a cost of two unshared chips or one shared chip. We have not taught you this JK method because it results in obscure circuits that bear little detectable relation to the original design and because any

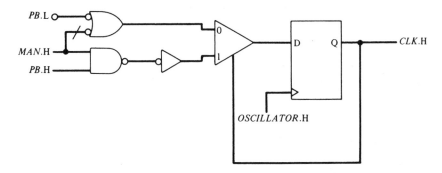

Figure 6–6 A circuit for a system clock.

change in the design requires a complete recalculation of the entire state generator. Also, in most complex designs, the JK method and the multiplexer method require comparable numbers of components.

The statement of the problem was the key step in its solution. Once we phrased the problem properly, the ASM chart followed naturally. In turn, the hardware followed easily from the ASM chart. The smoothness of the process is gratifying and is a sign that we are doing things right.

The original goal seemed deceptively simple. Our natural urge, and probably yours also, was to rush to the lab bench and try a few circuits. Almost surely we would have wound up with a circuit that appeared to work but contained undetected hazards. In earlier days, we tried various bottom-up approaches. This top-down circuit looks very different from our bottom-up designs, and it is unlikely that we would have stumbled onto it by random experimentation. Now we stay out of the lab until we have mastered the problem!

DESIGN EXAMPLE 3: A SERIAL BIT CLOCK

For our third example of digital design, let's look at a different type of clock. In the transmission of serial data, bits arrive at the receiving station one at a time on a single signal line, at some nominal rate. Typical rates are 1200, 9600, and 19,200 bits per second. The data transmitter uses a clock of the appropriate frequency to regulate the serial transmission of the data bits. The receiver, a completely separate device, has no knowledge of the transmitter clock other than the agreed-upon nominal bit rate.

The receiver picks off the incoming data bits by sampling the serial data line at points about midway into the interval for each data bit. Sampling at the midpoint avoids the problem of sampling while the data stream is undergoing a transition. To accomplish this midpoint sampling, the receiver needs a clock that delivers its active edge in the middle of each data-bit interval. Figure 6–7 shows a typical bit stream and the required waveform of the receiver's bit clock.

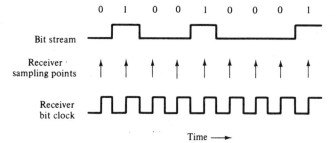

Figure 6–7 Deriving a receiver bit clock from the data stream.

Approaching the Problem

Our design problem is to produce a clock signal for use by a serial data receiver.

How can we create a clock for such a bit stream? What information do we have?

(a) We know the nominal bit rate.

(b) We can sense the incoming bits as voltage levels on the input line.

Item (a) tells us the basic frequency of the bit clock but not exactly when the clock edges should occur. Item (b) tells us where bit boundaries are by the voltage transitions on the input.

We will develop an ASM that will sample the incoming stream of bits often enough so that we will not miss any action. It will be simplest if the ASM samples the bits at a rate that is some multiple of the nominal bit rate. We choose a speed for our ASM that is 16 times the stated bit rate; call this ASM clock signal *X16CLK*. The sole output of our circuit will be a receiver bit clock signal *BIT.CLOCK*. Every time we detect a change in voltage level on the serial input, we will reset the bit clock to its false (inactive) level. After eight ASM clock cycles, we will change *BIT.CLOCK* to its true (active) level. At that point we have created a F → T edge on *BIT.CLOCK* that a data receiver can use to sample a data bit.

If, in the absence of a transition in the input data stream, we execute 16 ASM clock cycles after resetting the bit clock, we must again reset *BIT.CLOCK* to false, completing a bit-clock cycle. We use each transition of the input bit stream to create an inactive clock edge, and then after half a bit time, we create the active bit clock edge. If there is not another bit transition, then after a whole bit time elapses, we reset the bit clock and begin again.

There is one hitch: Whenever a string of 0's or a string of 1's arrives, the data stream will contain no transitions. In this case, there is no information in the input data stream to keep our bit clock aligned with the transmitted bit boundaries. Our method (and any method) will work only if the nominal bit clock frequencies of the transmitter and receiver are similar, so that the active edges of the generated receiver bit clock do not wander far from the center of each incoming bit. Designers of data-transmission systems are aware of this problem and usually avoid it by making sure that their message protocol produces some change in the values of the data at frequent intervals. Therefore, we will not worry further about this problem.

The Initial Architecture

In line with our general practices, we do not test the incoming stream of bits directly, since it changes asynchronously to our ASM clock. We will place a synchronizing D flip-flop into our architecture, with input *DATA∗* and output *DATA.SYNC*. The output of the bit clock circuit will be used as a clock by other systems, so we must be sure that the output is free of hazards. Therefore, we use a flip-flop to produce the output *BIT.CLOCK*. Since we must exert control over the value of *BIT.CLOCK* at several points in our algorithm, we will use a controlled (JK) flip-flop.

The Control Algorithm

At this point, the designer has a choice. We can create a 16-state ASM chart, or we can develop a one-state ASM that manipulates a 16-count counter in the architecture. We choose the latter, because we must always be on the lookout for a transition in the incoming data stream and putting such a test into each state of a 16-state ASM would require rather tiresome drafting. So we add a 4-bit binary counter to our architecture.

We know we are looking for transitions in the input data. How do we detect a transition? This requires a knowledge of the present value of *DATA.SYNC* and of its value during the ASM's previous state. To save the old value, we will need a D flip-flop with the input *DATA.SYNC* and the output *OLD.DATA*. Now, how do we tell if *DATA.SYNC* and *OLD.DATA* are different? The EX-CLUSIVE OR logical operator is ideal for this. The function EOR is true if and only if its two inputs are different. Thus the condition required for a transition on the input data line is

$$DATA.SYNC \oplus OLD.DATA$$

Now we may write our ASM to embody the analysis in the preceding paragraphs. Figure 6–8 is the one-state ASM for the receiver bit clock. Before implementing this ASM, however, we should ask a question. We should be suspicious of a clock circuit that has two different ways of creating one of the clock edges. Can the *BIT.CLOCK* output have any undesirable short clock phases? In the ASM, a data transition will preempt all other events. What if a data transition occurs when the ASM count is at 14 or 15, at 0 or 1?

Transitions at 14 are okay, since *BIT.CLOCK* has been true since the 7-count. A transition at 14 makes the T → F edge come slightly early, but that is necessary to keep *BIT.CLOCK* synchronized with the incoming bit stream. It is for exactly this purpose that we selected an ASM clock with a considerably

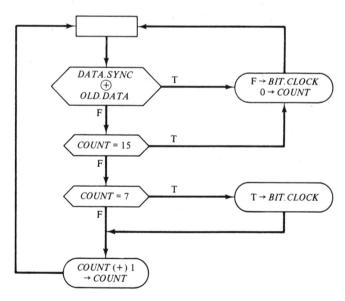

Figure 6–8 An ASM for a serial bit clock.

faster frequency than the bit frequency, so that early or late bit transitions will still result in rather even clock intervals. A transition at 15 is fine and is the most probable case since it means that the bit transitions and the ASM count are synchronized. Transitions at 0 or 1 will cause *BIT.CLOCK* to be set to false, whereas *BIT.CLOCK* underwent a T → F move just one or two ASM cycles earlier. Again, this is okay, because clearing an already-cleared flip-flop causes no change in the output. The inactive phase of *BIT.CLOCK* will last a little longer than usual but, again, this is proper to maintain synchronization.

If transitions occur at intermediate counts far from the normal time, then there is trouble in the input data stream and all we can do is try to follow the input data, which the ASM will surely do. Noise on the line—"glitches"—will cause problems. Our assumption about this circuit must therefore be that the data stream is free of noise. There are more sophisticated techniques involving phase-locked loops that help alleviate the problem of noisy lines, but these methods are beyond the scope of this book. The technique proposed in this example is quite acceptable for data transmission at moderate speeds over clean lines.

Implementing the ASM

Since we have a one-state machine, we need no state generator. Most of the detail is in the architecture. The 4-bit binary counter performs several operations: clearing to zero, counting, and testing for counts of 7 and 15. A fine choice for our counter building block is the 74LS163 Four-Bit Programmable Binary Counter, which has synchronous clear, count, and load inputs, and a "terminal count" status output to report a count of 15. With this knowledge, we may write the equations for the ASM outputs. Define an auxiliary variable *DATA.TRANS* as follows:

$$DATA.TRANS = DATA.SYNC \oplus OLD.DATA$$

Then

$$COUNT(CLR) = DATA.TRANS + \overline{DATA.TRANS} \cdot (COUNT = 15)$$
$$= DATA.TRANS + (COUNT = 15)$$
$$COUNT(CNT) = \overline{DATA.TRANS} \cdot \overline{(COUNT = 15)} \cdot (\overline{(COUNT = 7)}$$
$$+ (COUNT = 7))$$
$$= \overline{DATA.TRANS} \cdot \overline{(COUNT = 15)}$$
$$= \overline{COUNT(CLR)}$$
$$BIT.CLK(SET) = \overline{DATA.TRANS} \cdot \overline{(COUNT = 15)} \cdot (COUNT = 7)$$
$$BIT.CLK(CLR) = COUNT(CLR)$$

To complete the specification of the logic equations, we need expressions for *(COUNT=15)* and *(COUNT=7)*. The terminal count output *COUNT(TC)* of the 74LS163 gives us the former term directly; we will build the *(COUNT=7)* term with gates. Calling the data outputs of the counter $C3$ through $C0$, with $C3$ being the most significant bit, we have

$$(COUNT=15) = COUNT(TC)$$
$$(COUNT=7) \ = \overline{C3} \cdot C2 \cdot C1 \cdot C0$$

Look at the equation $BIT.CLK(SET)$. Our intuition tells us that if the count is 7, then it certainly is not 15, and so we should be able to toss out the term $\overline{(COUNT=15)}$ from the equation. We may confirm our intuition by expanding the meaning of the two terms

$$\overline{(COUNT=15)} \cdot (COUNT=7) = \overline{(C3 \cdot C2 \cdot C1 \cdot C0)} \cdot (\overline{C3} \cdot C2 \cdot C1 \cdot C0)$$
$$= (\overline{C3} + \overline{C2} + \overline{C1} + \overline{C0}) \cdot (\overline{C3} \cdot C2 \cdot C1 \cdot C0)$$
$$= \overline{C3} \cdot C2 \cdot C1 \cdot C0$$
$$= (COUNT=7)$$

In the second step we use De Morgan's Law [Eq. (1–6)], and in the third Eq. (1–5), $A \cdot \overline{A} = F$.

Now we may draft circuit diagrams. Figures 6–9 and 6–10 show the architecture and the control signals. We have assumed that on the data line $T = H$, although the choice does not affect the result.

Now we have a serial bit clock. In the next example, we will use this clock in a circuit for a data receiver that converts a serial stream of bits into a sequence of bytes.

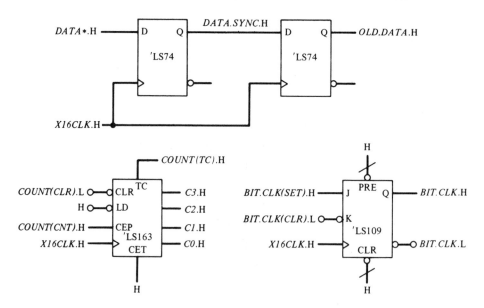

Figure 6–9 The architecture of a serial bit clock.

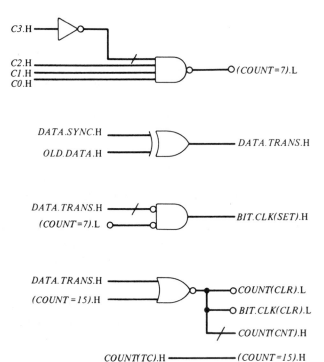

Figure 6–10 Control signals for a serial bit clock.

DESIGN EXAMPLE 4: SERIAL-PARALLEL DATA CONVERSIONS

Serial data transmission is a practical and economical way to move information between distant points. Each bit moves in serial fashion over a single signal path, according to some agreed-upon protocol. This method is conservative of signal wires but slower than if more bits were transmitted in parallel. Within a computer system, serial bit transfers are inefficient, since the basic unit of manipulation is the byte or word; within computers, data is transferred in parallel. To accommodate these different modes of moving data, we need the ability to convert serial data into parallel form and parallel data into serial form.

In this example, we design circuits for serial-parallel conversions. (We will use the designs in Chapter 11 to support a more complex microcomputer-based data multiplexing scheme.) Our example has two independent circuits: a serial-to-parallel converter and a parallel-to-serial converter. Both converters deal with 9-bit parallel bytes, an unusual byte size dictated by the structure of the hardware at the other end of our serial data lines. (LSI chips are available to assist the serial-parallel conversions of the more common byte sizes, such as 8 bits. You will see an example in Chapter 9. For this 9-bit application, no LSI chip is available.)

Specifying the Problem

Parallel-to-serial conversion. The parallel-to-serial (P → S) converter accepts 9-bit bytes from some source (in Chapter 11, this source will be a microcomputer) and transmits the bits serially on a serial-out line (*SO*) at a specified bit rate. As long as the system is running, the serial bit rate is constant, and the P → S process never stops. In our example, we may use a rate of 9600 bits per second.

The need to maintain a valid serial stream of data at this fixed rate places a severe constraint on the device that supplies the 9-bit bytes: whenever the P → S converter needs a new byte, the new byte must be present. But the byte supplier is running at its own speed, engaged in its own duties, only one of which is to supply bytes to the P → S converter. A simple way to handle such a situation is to provide a *one-byte buffer* register in the converter to hold the incoming byte whenever the sending device supplies it. Then our converter can move the data to another spot when it is ready to process the byte, thus freeing the buffer to hold another byte.

(The term *buffer* has two meanings in digital design. Previously in this book, the word has meant a source of power for a logic signal. Here, we use the term in the software sense of a temporary storage area to accommodate differences in the operating characteristics of a source and a destination.)

We need some way to notify the supplier of bytes when it is time to fill the converter's buffer with a new byte. The byte supplier must not provide a new byte too soon, lest the new byte destroy the previous byte in the buffer before the P → S converter has processed it. On the other hand, a basic presumption of this system is that the byte supplier *must not fail* to supply a byte on time; otherwise, we cannot supply the continuous stream of bits required by the serial-out line. We may use a simple one-way signal to tell the byte supplier to fill up the buffer. No response from the byte supplier (other than filling the buffer!) is necessary, since we cannot tolerate any slippage and our converter could do nothing about a failure if one occurred. Let's use a variable *FILLIT* as a signal to the byte supplier; whenever the converter accepts a new byte from its buffer, the converter will toggle (complement) *FILLIT*. At a time safely before the converter needs the next byte, the byte supplier must sense this *change* in *FILLIT*, and provide a new byte into the buffer. *FILLIT* behaves like a modulo-2 counter.

Serial-to-parallel conversion. The serial-to-parallel (S → P) converter captures the bits arriving serially on the serial-in line (*SI*), accumulates 9 bits into a byte, and places the 9-bit byte into a buffer register. An external byte receiver (again, in Chapter 11, the microcomputer) will accept the bytes for further processing. We insist that the byte receiver keep up with the incoming stream of bytes. The receiver must accept the buffer data before the converter puts the next byte into the buffer. In analogy with the P → S case, the S → P converter can toggle a variable *READIT* to announce to the byte receiver that a new byte is available. Since the byte receiver is committed to keep up, it need not reply to the *READIT* signal, except to empty the buffer.

There are two complications in serial-to-parallel conversion:

(a) How do we find the serial bits in the incoming voltage waveform on the serial-in line?
(b) How do we locate byte boundaries among the stream of bits?

The devices at the two ends of the serial line will be set to an agreed-on nominal rate of incoming serial bits; in our example it is 9600 bits per second. With this information, we may use the serial bit clock produced in Design Example 3 to signal the S $\rightarrow$ P converter when to accept a bit.

Locating byte boundaries requires that the incoming serial bit stream have some special characteristic that our system can recognize. In our protocol, we assume that a special 9-bit pattern, *SYNC*, will periodically arrive in the serial input stream. Our S $\rightarrow$ P converter must detect this *SYNC* pattern and use it to establish the byte boundaries. We will defer to the byte receiver the decision as to when to seek out this *SYNC* pattern. With this division of responsibilities, we can add an input signal *RESET* to the S $\rightarrow$ P converter. Whenever the byte receiver asserts this signal, the S $\rightarrow$ P converter will suspend normal byte transfers until it locates a *SYNC* pattern among the incoming bits, at which time it will resume placing bytes into the byte buffer and toggling *READIT*.

Building the P $\rightarrow$ S Converter

Design. The architecture will have a 9-bit buffer register to hold a byte from the byte supplier, and a 9-bit shift register to hold a byte while it is being disassembled and shipped out bit by bit over the serial-out line. Also, there will be a controlled flip-flop *FILLIT* to tell the byte supplier when to deliver another byte.

The basic timing element for parallel-to-serial conversion is the serial bit time, so a clock operating at the P $\rightarrow$ S bit rate (9600 Hz) is a natural system clock for our synchronous design. (To assist in debugging, we would probably use the system clock of Design Example 2, which has both automatic and manual modes; the automatic clock frequency would be fixed at 9600 Hz.)

There are two approaches to the design of the control algorithm. We could draw a nine-state ASM chart that produces a serial bit for *SO* in each state. This ASM could load a byte from the buffer into the shift register in one state and shift the byte one position to the right in each of the other eight states.

Alternatively, we can view the ASM as having a single state, with the architecture containing a binary counter capable of counting from 0 through 8, to distinguish the nine activities per byte. Let's adopt this latter view and add a counter to our architecture. The ASM in Fig. 6–11 is then self-explaining; Fig. 6–12 shows the supporting architecture.

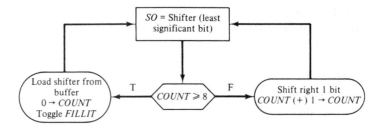

Figure 6-11   An ASM for parallel-to-serial conversion.

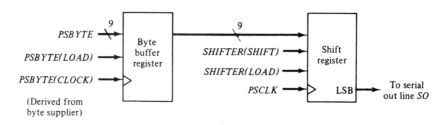

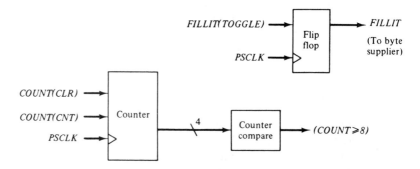

Figure 6-12   The logical structure of a parallel-to-serial converter.

Implementation.   The ASM yields the following equations for the P → S control outputs:

$$COUNT(CLR) \quad = (COUNT{\geqslant}8)$$
$$COUNT(CNT) \quad = \overline{(COUNT{\geqslant}8)}$$
$$SHIFTER(LOAD) = (COUNT{\geqslant}8)$$
$$SHIFTER(SHIFT) = \overline{(COUNT{\geqslant}8)}$$
$$FILLIT(TOGGLE) = (COUNT{\geqslant}8)$$

Realizing the P → S converter is straightforward. Here are some hints for selecting the chips. Nine-bit registers are not available in standard MSI, so we must form them from other units. We may make the byte buffer in several ways, perhaps with two 74LS378 Hex Enabled D-Register chips; the buffer register will be clocked and loaded by the *byte supplier*. There are also several ways to fabricate the 9-bit shift register. For this parallel-to-serial conversion, we need a parallel-in serial-out register. Although not quite the most economical in chip count, two 74LS165 Parallel-Load Eight-Bit Shift Register chips form an easily understood arrangement. In addition to the loading and shifting controls and the 8 data inputs, these chips have a serial input into the most significant data bit and a serial output from the least significant bit. This allows for the cascading of units to form longer registers. (As usual, you should consult a TTL data book for the exact specifications.) The output from bit 0 (the least significant bit) is the serial data signal used to produce *SO* from the P → S converter.

FILLIT can be a 74LS109 Dual JK Flip-Flop. Our faithful 74LS163 Four-Bit Programmable Binary Counter can be the counter required by the ASM. With the 74LS163 counter, we note that the *(COUNT≥8)* variable is just the most significant bit of the counter output; the "counter compare" box in Fig. 6–12 is therefore trivial.

Building the S → P Converter

Design.   The architecture requires a 9-bit shift register to accumulate a byte from the serial-in line, and a 9-bit buffer register to hold a byte for the byte receiver. The signal *READIT* will require a flip-flop with its output available to the byte receiver. At this stage, it is not clear if the incoming *RESET* signal line from the byte receiver also requires a flip-flop.

The S → P architecture has a "*SYNC* byte identifier" black box that constantly compares the 9-bit pattern currently in the shift register with the fixed *SYNC* pattern and produces a true output only when the patterns match.

Timing of the serial-to-parallel conversion is provided by the serial bit clock in Design Example 3, which produces a clock signal that is synchronized with the incoming bit stream.

Our ASM will have two states: **RUN**, for the normal serial-to-parallel byte assembly analogous to the P → S ASM; and **START**, for locating the *SYNC* pattern in response to a *RESET* signal from the byte receiver. The *RESET* signal arrives as an asynchronous signal from the byte receiver; it signals a precipitous retreat from normal operations. As the S → P converter ASM in Fig. 6–13 shows, we enter the **START** state as soon as we receive a *RESET* signal; we remain (synchronously) in **START** until *SYNC* appears. Once the algorithm locates the *SYNC* pattern, we move to **RUN** for regular accumulations and transfers of bytes.

To help protect the asynchronous *RESET*∗ signal from noise, we will capture the byte receiver's *RESET* signal in a flip-flop, and use the flip-flop output as the S → P converter's *RESET*∗ signal. The *RESET*∗ signal will cause an immediate transition to the **START** state, where the S → P system will remain until after the receiver has transmitted $\overline{RESET}$ to clear the flip-flop. The architecture of the S → P converter is shown in Fig. 6–14.

Implementation.   Once in state **RUN**, the ASM remains there until a reset signal causes a precipitous transfer to state **START**. The state generator for this

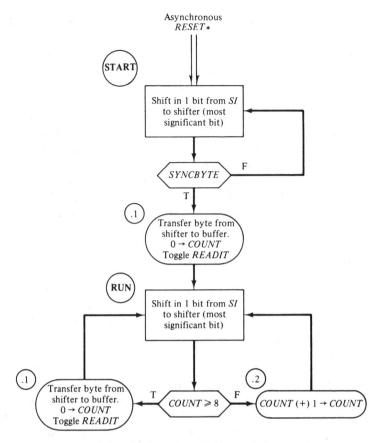

Figure 6–13 An ASM for serial-to-parallel conversion.

two-state machine is a JK flip-flop with output *RUN*. The reset signal must perform a direct (asynchronous) clear function on the flip-flop, and is thus labeled *RESET**.

In Fig. 6–13, we have labeled each conditional output oval. These labels are not state names, but just represent positions within the parent state. For the conditional output term ①) in state **START**, we will create a logic variable *START.1*; similarly, the notations ①) and ②) in state **RUN** will result in variables *RUN.1* and *RUN.2*. Conditional output labels are excellent aids in the systematic implementation of complex ASM charts.

From the S → P ASM chart in Fig. 6–13, we derive the following equations:

$$START.1 \qquad = START \cdot SYNCBYTE$$

$$RUN.1 \qquad = RUN \cdot (COUNT \geqslant 8)$$

$$RUN.2 \qquad = RUN \cdot \overline{(COUNT \geqslant 8)}$$

$$START \qquad = \overline{RUN}$$

$$RUN(SET) \qquad = START.1$$

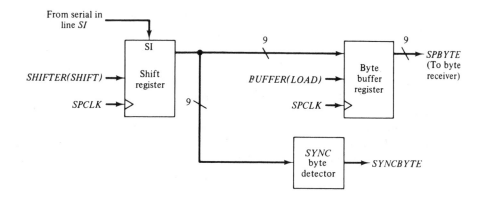

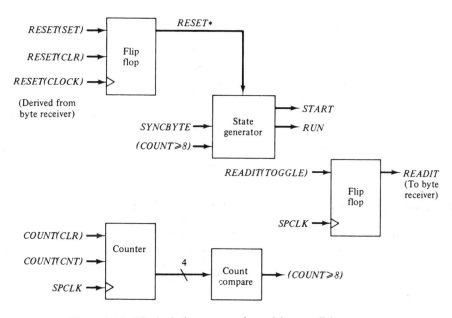

Figure 6–14 The logical structure of a serial-to-parallel converter.

$$RUN(DIRECT.CLEAR) = RESET*$$

$$SHIFTER(SHIFT) \quad = T$$

$$COUNT(CLR) \quad = START.1 + RUN.1$$

$$COUNT(CNT) \quad = RUN.2$$

$$BUFFER(LOAD) \quad = START.1 + RUN.1$$

$$READIT(TOGGLE) \quad = START.1 + RUN.1$$

Here are some suggestions for chips to implement the S $\rightarrow$ P converter. The 9-bit buffer register may be two 74LS378 Hex Enabled D-Register chips.

The shift register is a serial-in, parallel-out variety; two 74LS164 Eight-Bit Parallel-Out Serial Shift Register chips would serve. The counter may be the 74LS163 Four-Bit Programmable Binary Counter, and once again the determination of $(COUNT \geqslant 8)$ for this device is trivial, just as it was in the similar $P \rightarrow S$ circuit. The flip-flop for *READIT* should be a 74LS109 JK Flip-Flop.

The *SYNC* byte detector must produce truth whenever the current 9-bit data pattern matches the fixed *SYNC* pattern. In Chapter 3, we showed how to compare a fixed quantity with a variable one. For this circuit, we need a 9-bit AND function; a 74LS133 Thirteen-Input Nand Gate would do the job. Note that although the *SYNC* byte detector constantly checks for the *SYNC* pattern, the result is used only when the ASM is in the **START** state.

Critique

This completes our design of serial-to-parallel and parallel-to-serial converters. We have left unspecified the detailed nature of the devices on the serial and the parallel ends of our converters. (In Chapter 11, we will incorporate the converter circuits into their larger environment.)

Our protocol is a rather strange one, with its 9-bit byte. For the more ordinary 8-bit conversions, available complex LSI chips perform many of the same activities as our circuits. When appropriate, we would use these single-chip solutions in preference to building a multiple-chip MSI circuit. In this case, the available LSI chips were not suitable, and the exercise gives you an understanding of serial-parallel conversion.

DESIGN EXAMPLE 5: A TRAFFIC-LIGHT CONTROLLER

This example was inspired by a similar problem in Carver Mead and Lynn Conway's pioneering book, *Introduction to VLSI Systems*. We will solve the problem with our structured design techniques.

Statement of the Problem

A busy highway is intersected by a little-used farm road, as shown in Fig. 6–15. The farm road contains sensors that cause the signal *CARS* to go true when one or more cars are on the farm road at the positions labeled 'C.' We wish to control the traffic signals at the intersection so that, in the absence of cars waiting on the farm road, the highway light will be green. If a car activates the sensor at either position C, we wish the highway light to cycle through yellow to red and the farm-road light then to turn green. The farm-road light is to remain green only while the sensors indicate the presence of one or more cars, but never longer than some fraction of a minute, after which it is to cycle through yellow to red and the highway light is to turn green. The highway signal is not to be interrupted again for farm-road traffic until some fraction of a minute has elapsed.

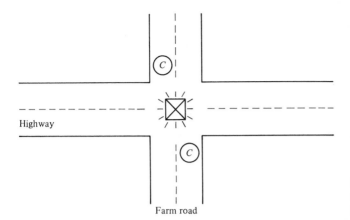

Figure 6–15 The location of the traffic signal and the sensors in Design Example 5.

Highway

Farm road

Preliminary Considerations

The highway traffic is given priority, but not to the extent that the farm-road traffic can be stalled indefinitely. Since the default condition is a green light on the highway, we do not need any sensors in the highway lanes. To keep the example uncluttered, we will assume that the outputs of the sensors are combined external to our design to produce the single signal *CARS*, and that this signal satisfactorily indicates the presence of cars desiring to enter or cross the highway. We might ask what signals the traffic lights must receive to activate their three colors, but we defer such inquiries because we would like our solution to be independent of any particular brand of traffic signal until we are ready to specify one.

The control of the traffic signals involves four intervals: the minimum time the highway light will be green, the maximum time the farm-road light will be green, the duration of the highway's yellow signal, and the duration of the farm-road's yellow signal. For simplicity, we assume that the first two intervals are the same and that both yellow-light intervals are the same. To generate signals representing these two intervals, we plan to have a timer, driven by a (high-speed) master clock and initiated by a starting signal *START.TIMER*. Whenever the timer is started, two output signals, *THOLD* and *TYEL*, are negated. If the timer is not interrupted, the two signals will be asserted after their respective intervals have elapsed; the signals will remain asserted until the timer is restarted with *START.TIMER*. The same timing unit and the same enabling signal will suffice for both timings, since the two intervals do not overlap.

Our preliminary architecture is shown in Fig. 6–16.

The Control Algorithm

The control of the traffic signals breaks naturally into four events, which will result in four ASM states:

State **HG**: Highway light green (and farm-road light red).

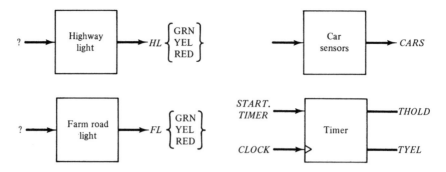

Figure 6-16 The preliminary architecture of the traffic-light controller.

State **HY**: Highway light yellow (and farm-road light red).

State **FG**: Farm-road light green (and highway light red).

State **FY**: Farm-road light yellow (and highway light red).

When the highway light is green (state **HG**), the controller must be alert for farm-road traffic, and, if cars are on the farm road and sufficient time has elapsed, must cycle the lights through state **HY** to state **FG**. In state **FG**, when the farm-road light is green, the controller must be prepared to cycle through state **FY** to state **HG** whenever no cars remain on the farm road or if the stipulated time has elapsed. These observations lead quickly to the ASM in Fig. 6-17. Each state tests one of the two intervals *THOLD* or *TYEL*, and so, as we enter each state, we must start the timer unit.

In state **HG**, the ASM describes the mutual requirement of farm-road cars and a sufficiently long interval, using a single test of the product of *THOLD* and *CARS*. In state **FG**, the ASM uses separate tests of *THOLD* and *CARS* to describe the logic resulting in the escape to the next state. These constructions seem natural to us, so we used them. Other ways of representing the test conditions will lead to exactly the same output equations and state generator; for instance, follow the synthesis below, and then repeat the synthesis with these tests:

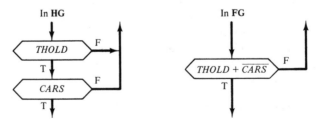

Realizing the ASM

From the ASM chart, we derive equations for the outputs and we tabulate information that will lead to the construction of a state generator. The natural

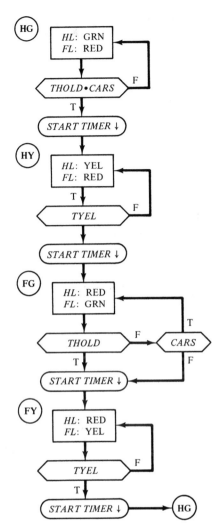

Figure 6–17 The ASM for the traffic-light controller.

parameters to describe the condition of a traffic signal are, of course, the colors of the signal. We derive

$$HL.GRN = HG \qquad FL.GRN = FG$$
$$HL.YEL = HY \qquad FL.YEL = FY$$
$$HL.RED = FG + FY \qquad FL.RED = HG + HY$$

The signal for starting the timer unit is

$$START.TIMER =$$
$$HG{\cdot}THOLD{\cdot}CARS + HY{\cdot}TYEL + FG{\cdot}(THOLD + \overline{CARS}) + FY{\cdot}TYEL$$

Let's use the multiplexer method to implement the state generator. In this method, we assume an encoded state generator, which will require two D flip-flops. The encoding is arbitrary: we use B and A as the state variables, and the following assignment:

State	B	A
HG	0	0
HY	0	1
FY	1	0
FG	1	1

For such a simple ASM, we could write down the state generator by inspection. Nevertheless, in almost every design we find it useful to tabulate the conditions for changes in state. Table 6–1 shows the next-state conditions for our traffic light controller. Figure 6–18 shows the state generator.

TABLE 6–1 CONDITIONS FOR STATE TRANSITIONS IN THE TRAFFIC-LIGHT CONTROLLER

Present state	Code	Next state	BA	Condition
HG	0	**HG**	00	$\overline{THOLD \cdot CARS}$
		HY	01	$THOLD \cdot CARS$
HY	1	**HY**	01	$\overline{TYEL}$
		FG	11	$TYEL$
FY	2	**FY**	10	$\overline{TYEL}$
		HG	00	$TYEL$
FG	3	**FG**	11	$\overline{THOLD \cdot CARS}$
		FY	10	$THOLD + \overline{CARS}$

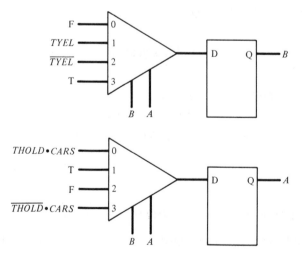

Figure 6–18 A multiplexer state generator for the traffic-light controller.

Choosing a Particular Traffic Signal

The ASM's outputs that are to control the traffic signals are expressed in terms of the active color of the lights. Now suppose we select a particular traffic signal and, reading the instructions, find that each signal is controlled by a 2-bit code that specifies which of the three colors is active. In Fig. 6–19 we show this particular choice of a traffic light, adopting T = H for the code. Now we may derive logic equations that express the bits of the manufacturer's code in terms of our traffic-light variables:

$$HL1 = HL.YEL$$
$$HL0 = HL.RED$$
$$FL1 = FL.YEL$$
$$FL0 = HL.RED$$

Now that the behavior of the particular traffic signal has been expressed in our nomenclature, there remains only to plug in the particular expressions for each light's colors to complete the implementation. Using the ASM output equations, we get

$$HL1 = HY$$
$$HL0 = FG + FY$$
$$FL1 = FY$$
$$FL0 = HG + HY$$

Our refusal to commit ourselves to a particular traffic signal left us with an incomplete early statement of the architecture in Fig. 6–16 but allowed us to form a solution independent of the brand. When we finally settled on a brand, in Fig. 6–19, we completed the solution *without altering our original work.* This is an important technique, and represents good top-down design.

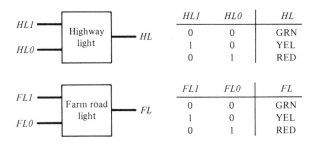

HL1	HL0	HL
0	0	GRN
1	0	YEL
0	1	RED

FL1	FL0	FL
0	0	GRN
1	0	YEL
0	1	RED

Figure 6–19 A particular traffic light for Design Example 5.

DESIGN EXAMPLE 6: A SIMPLE COMBINATION LOCK

Now let's use our design tools to build an electronic combination lock. Combination locks come in two varieties, parallel and serial. An example of a parallel lock is a bicycle-chain lock with four disks that we must rotate to the correct combination.

We can change each disk independently without having to start over if we have a wrong number. A serial lock is the common dial type: we rotate a single dial to three or more preset numbers in sequence. Any wrong number requires us to start over.

There are electronic equivalents of most components of locks. Rotary 10-position switches replace the disks and dial. These switches are available as side-by-side stackable units called *thumbwheel switches*, which look like the individual disks of a parallel bicycle lock. As we rotate a thumbwheel switch, numbers from 0 through 9 appear in a small window and appropriate contacts close on four lines to represent the decimal number in binary form. We can use a pushbutton switch to replace the manual pull test of a correct lock combination. We can then use a lamp driver to light a lamp if the try was successful. (In Chapter 12, we show how to take a low-power logic signal such as the lamp input and amplify it to do useful work, for example to drive the lamp or to provide power to a solenoid to pull the latch on a door.)

Let's build a simple lock to explore the design principles. Since we are digital designers, we will enter the lock's combination in binary form, one digit at a time. In this way we can build up the combination a bit at a time, starting from either end. A wrong digit should abort the entire procedure and send us to an error state, where the system waits for a reset signal to send it back to the start.

Stating the Problem

Build a 3-bit serial combination lock that lights a light when the correct combination is entered.

Digesting the Problem

We should have several questions. For instance:

  (a) How is the data entered?
  (b) What is the combination?
  (c) How do we reset the lock to start a new sequence?
  (d) How do we enter the combination—left to right or right to left?
  (e) How do we know when we have made an error?

We might answer these questions in the following way:

  (a) Binary data (0 or 1) is set on a toggle switch. The operator will signal that the switch data is ready to enter by pressing a pushbutton switch Read.
  (b) Assume that the combination is 011_2.
  (c) We will need a Reset switch of some sort.
  (d) Right to left.
  (e) We can do clever things with the error condition. The worst thing would be to light a Wrong light the first time the machine detected a bad digit.

This would allow a thief to build up the correct combination a digit at a time by writing down the preceding correct string of bits and experimenting with the current bit. The best procedure is to wait until all 3 bits are entered before signalling that an error was made.

We do not have to tell the thief how many bits to enter, and someone trying to use our lock might therefore try 5, 6, or more digits. We can add this feature with a Try pushbutton to test the combination already entered. If Try is pressed any time except when the machine has processed 3 correct bits, we should terminate the operation in an error state.

Developing the Algorithm

Figure 6–20 is a possible ASM for our lock problem. The essence of this chart is to test each fixed bit of the combination in a separate state and end up in either a correct state or an error state. Some states loop back to themselves while waiting on a human action. During the looping, these states must detect any erroneous actions and also any attempt to reset the system. For example, while the loop in state **BIT1.OK** waits for the *READ* signal, it must provide loop exits for an erroneous *TRY* signal and for a *RESET* signal. (Why is this important?) State **BIT3.OK,** which paves the way for a success, loops on *TRY,* and must recognize a *READ* signal as an error.

This is a fairly boring ASM chart. The only interesting part is the state assignment. Since so many paths lead to an error state, we give state **BAD** a state assignment of 000. The multiplexer controller uses gates to evaluate those paths that lead to state assignments containing 1's. By making the largest number of paths lead to a state with all 0's, we can sometimes simplify the multiplexer inputs.

This ASM chart has two objectionable features:

(a) The combination is built into the chart by the bit tests. Altering the bit pattern of the combination requires a tedious reexamination of the ASM chart and modification of the gate structure of the control mux system. Changing the number of bits in the combination is a major undertaking— distinctly poor design.

(b) Tests of the reset signal appear in every state and clutter up the ASM chart.

Let's rework the algorithm to try a more elegant approach.

An Elegant Combination Lock ASM

The reset tests in Fig. 6–20 are correct but unpleasing. A reset is often a cataclysmic action that is intended to throw the system into a known state, regardless of the present state and independent of the ASM's clocking. Although our intended reset action is not quite this dramatic, the result is much the same. Any time the operator hits Reset, the machine should go to the initial state and

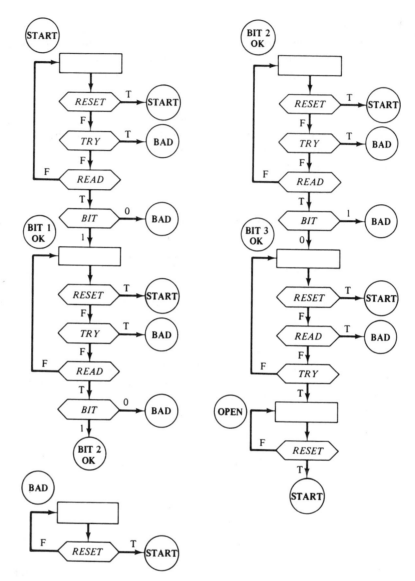

Figure 6–20 An ASM for a primitive combination lock.

start the lock algorithm anew. Nothing needs to be saved in the ASM. In this problem, we may make one of our few standard uses of asynchronous signals: treat the reset as an asynchronous event that will blast the ASM into the initial state. By doing this, we can remove the synchronous tests of *RESET* and replace them with a master asynchronous entry to the initial state by means of the signal *RESET∗*.

If we intend to reset the state flip-flops asynchronously, we must choose a state assignment that lets the initial state conform to the results of the reset. Usually, we assign 000 to the destination of the asynchronous reset.

Meeting the other objection, the fixed nature of the combination, requires a more subtle and more significant answer. First, we will get rid of the wired-in combination. Instead of branches such as

we could compare the current value of the toggle switch with a reference value supplied by the architecture and not by the ASM chart itself. If we call this reference value for a particular bit *REF,* then the ASM chart would contain

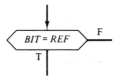

Now, by altering the reference bits we may change the lock's combination easily.

Next, we will put the bit test in a loop. The number of bits in the combination will determine when the loop terminates. Clearly, a 3-bit lock is not very safe, since the probability of opening it in one try is $2^{-3} = \frac{1}{8}$. It would be much nicer to have a 10-bit lock, which would reduce the probability to $1/1024$. Our modification will easily support such a change.

The new ASM, incorporating the proposed changes, is shown in Fig. 6–21. This is a much better algorithm. By coincidence, it has the same number of states as before, yet it is able to handle a combination of *M* bits as easily as three. Furthermore, we have regularized the branch out of the bit test—branching occurs only when the input *BIT* is not equal to the reference bit. This systematic feature allows us to embed the test inside a loop.

We have to pay something for the increased power of this algorithm:

(a) We now need a loop counter.
(b) We now need a table of reference bits—in other words, the combination.

Can we find hardware for this algorithm? The counter is no problem, since it is one of our standard building blocks. For example, the 74LS163 Four-Bit Programmable Binary Counter is able to support clearing, loading, and counting, and can handle lock combinations with up to 16 serial bits. The combination lookup table could be in a ROM (read only memory), a PLA (programmable logic array), or a multiplexer. (Remember the discussions of building blocks in Chapters 3 and 4.) The multiplexer is the cheapest solution for short tables. We use *CNT* as the table index to find the current bit. For a combination of 10011110, the table lookup suggests an eight-input multiplexer:

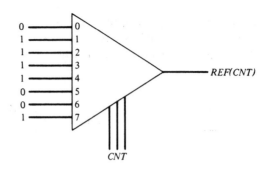

CNT

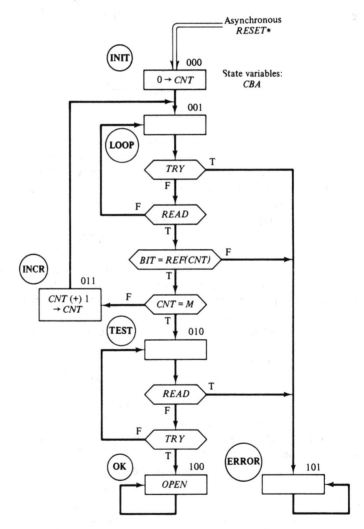

Figure 6–21   An ASM for an elegant combination lock.

Notice the great similarity of our hardware algorithm in Fig. 6–21 to a flowchart loop in software programming. All the ingredients are there; initialization of the loop counter, the loop body, the test of the loop counter, and the modification of the counter. Although we must refrain from drawing too close an analogy with software, it is gratifying to find, in good hardware design, many of the same concepts as software.

It is worthwhile considering the effect of placing the loop-counter incrementing step in different locations in the ASM. What is wrong with putting the counter modification operation into state **LOOP**? (*Hint:* The loop has an embedded loop.) Why not put the **INCR** state at the head of the loop instead of at the end? (*Hint:* Consider the table-lookup's indexing.)

Architecture

Normally, the operator of a digital system will follow the progress of the work by observing status responses such as lights or motor noises; the operator can thus determine when to perform various actions. In our combination lock, we wish the machine to be quiet, yielding no information except lighting an Open light following a successful Try. If the machine's clock is faster than human reactions, we may avoid "handshakes" between the operator and the machine, since the machine will always be ready to process the operator's input.

The clock speed of our lock machine may be as slow as 0.2 sec or 5 Hz. If we are driving our machine with batteries, for instance in a car, we would use an oscillator of perhaps 100 Hz for our clock. (We discuss this technology in Chapter 12.)

Our ASM chart tests the *TRY* and *READ* signals, which originate from debounced pushbutton signals *TRY∗* and *READ∗*. We will pass these signals through a single pulser to achieve synchronization and to assure that only a single ASM action results from the push of a button. The single-pulser outputs are *TRY* and *READ*.

The bit-entry switch for the serial combination may be a plain toggle switch *BIT*. (Why does it not need to be debounced or synchronized or single-pulsed?) The specifications call for a Reset pushbutton with the output *RESET∗*. The sole output of the machine is a light, Open, so we include a black box in the architecture to drive the lamp.

Buried within our hardware, inaccessible to the operator, will be a binary counter with output *CNT* to control the loop count. A 74LS163 Four-Bit Programmable Binary Counter will do nicely; *CNT* will then consist of 4 bits, *CNT3-CNT0*. The architecture will also contain a multiplexer, with an output *REF(CNT)*, of sufficient size to provide an indexed table lookup for the bits of the combination. For our illustration, let's use an 8-input mux such as the 74LS151. We do not really need to commit ourselves to particular chips this early. All we need at this point is a knowledge of what operations will be easily possible with each building block.

Last, we may include a black box with output *(CNT=M)* and inputs *CNT* and *M* to determine when the loop count *CNT* has reached the terminal value

M. Since the loop starts with a count of zero, *M* must be one less than the number of bits in the combination. We must build this black box. If we think it unlikely that the number of bits in the combination will change, we could implement the output *(CNT = M)* with a simple gate. For example, in an 8-bit combination, *M* would be 7, and we would have

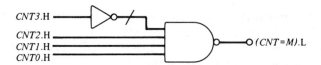

However, a more elegant approach is to allow the black box's input *M* to vary. Then we would use one of our standard building blocks, an arithmetic magnitude comparator such as the 74LS85 Four-Bit Magnitude Comparator. With 4-bit inputs *CNT* and *M*, this circuit can provide a signal *(CNT = M)* when the input values are identical. We could then complete the architecture by including assemblies of tiny toggle switches for *M* (4 bits) and for the combination (8 bits). We would hide these switches inside the combination lock machine, away from prying eyes.

Figure 6–22 is the architecture of our elegant combination lock.

Implementing the Control

There are few outputs in our ASM, and the equations are simple:

$$CNT(CLR) \quad = INIT$$
$$CNT(COUNT) = INCR$$
$$OPEN \qquad = OK$$

All the ASM test inputs are available in the architecture except for *(BIT = REF(CNT))*. How do we determine if one arbitrary logic signal is equal to another? In Chapter 3, we introduced the logic function COINCIDENCE, which produces truth if and only if its two inputs are the same. We may express the bit test as

$$(BIT = REF(CNT)) = BIT \odot REF(CNT)$$

Now for the state generator. We use the multiplexer controller method and the state assignment given in Fig. 6–21. The destination of the ASM asynchronous reset has state assignment 000 to allow easy use of an asynchronous clear input on the state generator flip-flop assembly. Table 6–2 contains the state transitions required by our ASM. The six states require a 3-bit code. A 4-bit D register with asynchronous clear, such as the 74LS175, and three 8-input multiplexers, for example the 74LS151, will support the structure.

To provide the three state signals *INIT*, *INCR*, and *OK* required by the ASM output equations, we may use a 74LS42 Four-Line-to-Ten-Line Decoder. This one-chip solution is more convenient than constructing the three signals from gates.

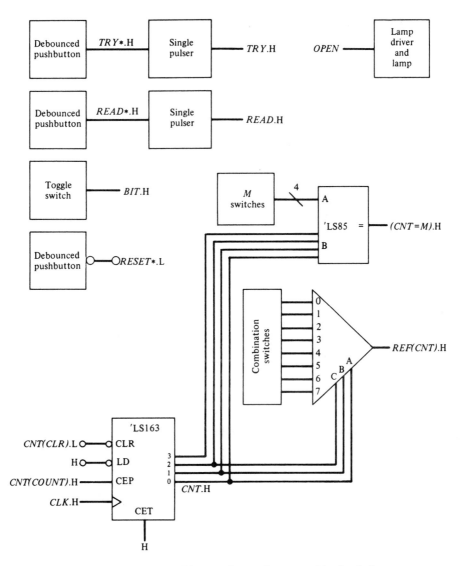

Figure 6-22 The architecture for an elegant combination lock.

Figure 6-23 is the state generator's structure. Figure 6-24 shows implementations of a few of the mux inputs. You should draft the rest.

DESIGN EXAMPLE 7: THE BLACK JACK DEALER

As our last example in this chapter, we will design a machine to simulate the dealer's actions in a black jack game. Black jack is a familiar card game involving a dealer and one or more players. The players can exercise their judgment but the rules specify what the dealer will do with each new card he receives.

TABLE 6–2 ELEGANT COMBINATION LOCK ASM STATE TRANSITIONS

	Present state	Next state	CBA	Condition from ASM
0	**INIT**	**LOOP**	001	T
1	**LOOP**	**LOOP**	001	$\overline{TRY} \cdot \overline{READ}$
		TEST	010	$\overline{TRY} \cdot READ \cdot (BIT = REF(CNT)) \cdot (CNT = M)$
		INCR	011	$\overline{TRY} \cdot READ \cdot (BIT = REF(CNT)) \cdot \overline{(CNT = M)}$
		ERROR	101	$TRY + \overline{TRY} \cdot READ \cdot \overline{(BIT = REF(CNT))}$
				$= TRY + READ \cdot \overline{(BIT = REF(CNT))}$
2	**TEST**	**TEST**	010	$\overline{READ} \cdot \overline{TRY}$
		OK	100	$\overline{READ} \cdot TRY$
		ERROR	101	$READ$
3	**INCR**	**LOOP**	001	T
4	**OK**	**OK**	100	T
5	**ERROR**	**ERROR**	101	T
	(any)	**INIT**	000	$RESET*$ (asynchronous)

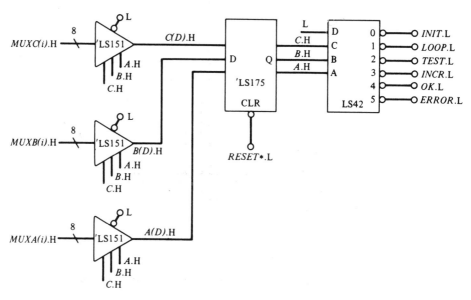

Figure 6–23 A state generator for an elegant combination lock.

The Rules of Play for the Dealer

The cards have values of 1 (ace) to 10 (10 and face cards). An ace may have the value of 1 or 11 during the play of the hand, whichever is advantageous. The dealer deals himself cards one at a time, counting ace as 11, until his score is greater than 16. If the dealer's score does not exceed 21, he "stands," and his play of the hand is finished. If the dealer's score is greater than 21, he is "broke" and loses the hand. The dealer must revalue an ace from 11 to 1 to

The Art of Digital Design   Part II

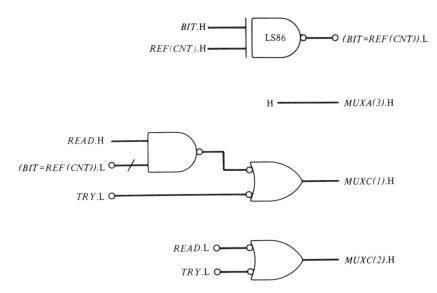

Figure 6–24 Some control mux inputs for an elegant combination lock.

avoid going broke but must then continue accepting cards ("hits") until the count exceeds 16.

Stating the Problem

These rules understood, we may state our hardware design problem. With a human operator to present cards to the Black Jack Dealer machine, play the dealer's hand to produce Stand or Broke.

Digesting the Problem

We will call the hardware device Dealer. The basic questions we might ask in our first stab at an architecture are: How will the operator present cards to Dealer? How will the operator know what the result of each card is? We can see that the operator interacts closely with the machine:

  (a) Dealer must signal the operator when to deal a new card.
  (b) The operator must signal Dealer when the new card's value is ready for processing.
  (c) Dealer must tell the operator if Dealer stands, went broke, or if the hand is still in progress.
  (d) When Dealer stands, the operator must be able to see the point value of Dealer's hand.

These thoughts suggest that an important aspect of Dealer's design is the interaction between the machine and the human. Assuming that the operator knows the binary number system, we choose a set of four toggle switches to hold a card value of from 1 to 10 in binary. These toggle switches will act as a register for the input data. A set of five lights is a simple way to display Dealer's score, which cannot exceed 21. To control the interaction, we can give the machine a set of status lights: Hit, to tell the operator when to enter a new card for Dealer; Stand and Broke, to inform the operator of the final results of the hand. We could provide a New.Game status light to show when to begin a new hand, but this is unnecessary, since either Stand or Broke must be signalled when the previous hand is complete, and thus the operator knows when a new game may begin without a special New.Game light.

We must not forget to give the operator a way to tell Dealer when to process a card; therefore, we will specify a pushbutton switch Card.Ready which the operator can press.

At this point we have a fairly good idea of the interface between Dealer and the human operator. Figure 6–25 shows the general plan. This is an important step in our design, for at this point we may go to our potential operators and describe how they will use the Dealer machine. Presumably, the operators don't much care what Dealer is like inside, but they will be interested in the operating instructions—the user's manual. Talking to users at this stage in the design can help avoid agonizing redesign later, in case we have misunderstood the problem.

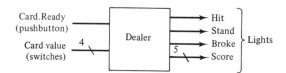

Figure 6–25 The Black Jack Dealer's interface with the operator.

Most of the architecture is not yet specified—only the interface signals. Now we can begin work on the control algorithm. To gain a clear understanding of the problem, it is useful to write a "software" version of the algorithm. This suppresses most of the machine details, including the detailed state timings that we must eventually specify. If we don't understand the problem at an operational level, we surely cannot build a correct machine! Here is a high-level statement of the dealer's algorithm. The variables have obvious meanings, except for *ace11flag*, to remember if the algorithm has valued an ace as 11 points.

Operational Algorithm for the Black Jack Dealer:

A1: (*prepare to start a new game*) *score* := 0; *ace11flag* := false; *stand* := false; *broke* := false;

A2: (*make a hit*) accept a card; *score* := *score* + *card value*; (*check for ace*) **if** *card* = *ace* and *ace11flag* = *false* **then** {*score* := *score* + 10; *ace11flag* := true}

A3: (*check for hit*) **if** *score* ≤ 16 **then goto** *A2* **else** (*check for stand*) **if** *score* ≤ 21 **then** {*stand* := true; display *score*; **goto** *A1*} **else** (*check for broke*) **if** *ace11flag* = true **then** {*score* := *score* − 10; *ace11flag* := false; **goto** *A3*} **else** (*indicate a broke*) {*broke* := true; **goto** *A1*}.

Initial ASM for the Black Jack Dealer

With the operational algorithm as a guide, we may propose a hardware algorithm, using the ASM notation. As usual, our design will be synchronous, running from its own internal clock at a speed independent of the human operator's actions. As we develop the ASM, we will gain insight into the internal architecture required by Dealer. Figure 6–26 is a first attempt to describe an ASM for Dealer. The ASM assumes the following architectural elements:

(a) Memory flip-flops for the *HIT, STAND,* and *BROKE* signals.
(b) A register to hold the current card value.
(c) A register to hold *SCORE.*
(d) A flip-flop for *ACE11FLAG.*
(e) A black box to add, subtract, and clear the *SCORE.*
(f) A black box to report if *CARD = ACE,* if *SCORE >* 16, and if *SCORE > 21.*

Reducing the Number of States

Our ASM has quite a few states, and you may wonder if this is desirable. States are the fundamental element of a hardware control algorithm, but they may have two undesirable side effects:

(a) Each state requires a clock cycle to execute.
(b) In a hardwired design such as this, excess states can enlarge the state generator's circuitry.

In the Black Jack Dealer machine, neither of these objections to states is apt to be serious. It is likely that the human operator is much slower than the clock, so superfluous clock cycles will not cause difficulty. Also, we have straightforward methods of developing state generators from ASM charts of any reasonable size.

Nevertheless, as practice in dealing with more complex and demanding designs, we will attempt to reduce the 10 states to a smaller number. Our basic technique is to introduce conditional outputs into an existing state, to replace the unconditional outputs of a separate state. States are candidates for collapse into the previous state if the state's activities (particularly its outputs) do not need to follow the activities of its predecessor state sequentially. Such state-saving moves do not necessarily save hardware. Although the number of flip-flop memory elements for states may decrease, the command output logic usually becomes more complex. Experience shows that a moderate effort to save states is worthwhile, as long as we maintain clarity in our design.

Another technique to reduce the number of states is to create new test inputs to direct a single state down several new paths. We encountered this technique in the single-pulser example at the beginning of this chapter. An extreme example would be to place the entire ASM into a single state: take any ASM and its set of state variables and create a new single-state ASM, using the

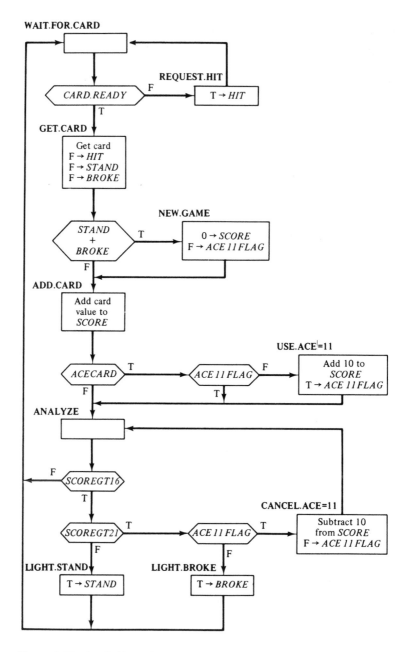

WAIT.FOR.CARD

REQUEST.HIT

CARD.READY F

T → *HIT*

T

GET.CARD

Get card
F → *HIT*
F → *STAND*
F → *BROKE*

NEW.GAME

STAND
+
BROKE T

0 → *SCORE*
F → *ACE 11 FLAG*

F

ADD.CARD

Add card
value to
SCORE

USE.ACE = 11

ACECARD T *ACE 11 FLAG* F

Add 10 to
SCORE
T → *ACE 11 FLAG*

F T

ANALYZE

F *SCOREGT16*

T

CANCEL.ACE = 11

SCOREGT21 T *ACE 11 FLAG* T

Subtract 10
from *SCORE*
F → *ACE 11 FLAG*

F F

LIGHT.STAND LIGHT.BROKE

T → *STAND* T → *BROKE*

Figure 6–26 A primitive ASM for a Black Jack Dealer, with too many states and an error.

The Art of Digital Design   Part II

State variables: BA

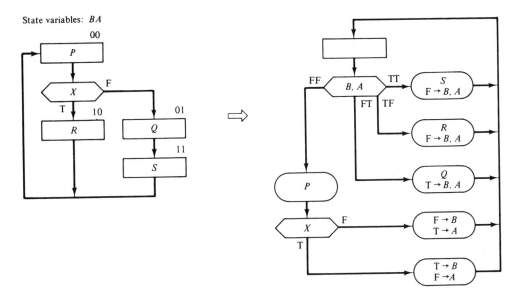

Figure 6–27 Converting a four-state ASM into a one-state ASM.

old state variables as test inputs. Figure 6–27 is an example. Obviously, this recasting of the algorithm, although technically equivalent to the original four-state version, is neither as clear nor as meaningful. The test inputs B and A are artificial and have no real meaning in the algorithm. Since we want maximum clarity, we reject this particular single-state ASM. There are occasions when saving states through the introduction of new test inputs is helpful. An example is a purely sequential ASM—a cycle of states with no tests, as in Fig. 6–28a. A simple state generator uses a binary counter, which counts up to the maximum state value and then resets to 0. On occasion, we may increase the clarity by

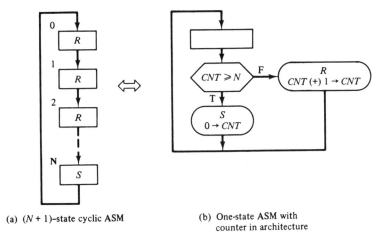

(a) $(N + 1)$–state cyclic ASM

(b) One-state ASM with counter in architecture

Figure 6–28 Equivalent formulations of a cyclic process of $N + 1$ elements.

making the counter a part of the architecture, as in Fig. 6–28b. In both designs, the hardware is the same. Choose the algorithm that seems clearer for your problem. Design Examples 3 and 4 in this chapter also illustrate this point.

In general, use moderation in eliminating states, having increased clarity as your goal. In the Black Jack Dealer ASM, we may easily incorporate the outputs of states **LIGHT.STAND**, **LIGHT.BROKE**, and **CANCEL.ACE = 11** into conditional outputs within state **ANALYZE**. We may also eliminate the **NEW.GAME** state.

Errors in the Algorithm

The ASM in Fig. 6–26 contains two errors. Can you find them? They are both in the interface with the operator, and you have seen both earlier in this chapter. Consider the Card.Ready button and its use. First, we assume by now that you will have debounced the *CARD.READY* signal. (We always debounce mechanical switch signals that are used to provide test inputs in our ASM.) But *CARD.READY* changes with the operator's actions and is not synchronized with the ASM's clock. In our ASM, we should therefore label this signal *CARD.READY*∗ (∗ for asynchronous). We hope you noticed this error, since asynchronous inputs are a common problem and you must be alert to them. Our treatment of this asynchronous test input is immediate and ruthless. Without pausing to investigate whether this asynchronous signal will cause problems, we eliminate it. We synchronize *CARD.READY*∗ with a D flip-flop operating synchronously with our ASM. In so doing, we add an element to our architecture. Call the output of the flip-flop *CARD.RDY.SYNC*. It is this new synchronized signal that the ASM tests in the **WAIT.FOR.CARD** state.

Races, Again

Let's again explore the problems introduced when an ASM tests an asynchronous input. Assume that we have allowed the *CARD.READY*∗ signal to remain in the ASM and that we have made an encoded state assignment. Figure 6–29 is the relevant part of the ASM. In state 0000, when *CARD.READY*∗ = F, the state generator logic will be preparing to change the state variable *C* from 0 to

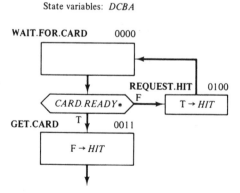

Figure 6–29 A segment of an ASM with a transition race.

1. If *CARD.READY** becomes true in state 0000, the state generator logic will switch *C* back to 0, and *B* and *A* to 1. If *CARD.READY** changes to true very close to the transition point for leaving state 0000, the inputs to the state flip-flops will be changing when the clock edge occurs, and the resulting outputs of flip-flops *C*, *B*, and *A* are unpredictable. The next state might be any of eight possibilities: 0000, 0001, 0010, 0011, 0100, 0101, 0110, or 0111! In this particular example, states 0000, 0011, and 0100 would be tolerable, but the rest are clearly erroneous. The situation is a transition race, and you must eliminate it.

By modifying our ASM chart in a straightforward way (but retaining the asynchronous *CARD.READY** signal), we can illustrate another type of race, the output race. If we manipulate *HIT* in conditional outputs in state **WAIT.FOR.CARD** instead of in separate states, we have the partial ASM in Fig. 6–30. Should *CARD.READY** change from F to T near the clock edge, not only is the next state in doubt, but since the input to the *HIT* flip-flop is changing from T to F, the *HIT* output is also uncertain. We may reach the **GET.CARD** state successfully, but the Hit light may still be on! The output race is characteristic of outputs that are conditional on asynchronous test inputs. You must eliminate these races.

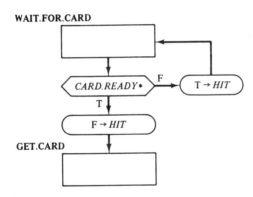

Figure 6–30 A segment of an ASM with an output race.

There is a welter of traditional and complex techniques for dealing with the problem of races by manipulating state assignments. The straightforward way, as you know, is to eliminate the cause of the problem! If all test inputs are synchronous with the ASM clock, there can be no races. It is tempting to study your particular ASM to see if a particular asynchronous input can cause trouble. Usually, this is wasteful, mind-cluttering activity. Synchronize the input and move ahead.

This sweeping simplification works because the ASM's action depends on only one (synchronized) input at a time. If the control of the ASM requires the simultaneous synchronization of more than one asynchronous input, no method will produce reliable results. This is bad design of the external interface. Modern practice requires that all relevant inputs be stable (unchanging) at the time the single status signal announces that an event is to occur. For our Black Jack Dealer, we require that the card switches be set prior to the single *CARD.READY** announcement.

Process Synchronization

Our problems are not quite over. Suppose the operator presses the Card.Ready button. What happens if, as is likely, the ASM completes its actions in response to the *CARD.RDY.SYNC* signal and returns to the **WAIT.FOR.CARD** state before the operator has released the pushbutton? Dealer will process the same card again, since *CARD.RDY.SYNC* is still true. This is the problem studied in this chapter's first example, the single-pulser. Because of its importance and the subtlety of some of the implications, we will discuss the subject again, from a different perspective.

Handshakes. The problem is a failure to complete a *full handshake* between the operator and the dealer. A *full handshake* is an important mechanism for controlling the activities of two independent but cooperating processes. Frequently one device (A) must issue a request for action to another device (B). Since the speed and state of device B are unknown to device A (and vice versa), we need a general method by which device A can request action of device B and can be certain that device B has recognized the request. The sequence of events in the full handshake is:

(1) Device A senses that device B is not still acknowledging a previous request and requests an action (device A extends its hand).
(2) Device B senses the request and acknowledges receipt of the request (device B extends its hand).
(3) Device A senses device B's acknowledgment and drops its request (device A drops its hand).
(4) Device B senses that device A has recognized device B's acknowledgment and drops its acknowledgment (device B drops its hand).

The success of the handshake depends heavily on the *sequence* of events but does not depend at all on the duration of any step. In our Black Jack Dealer, the operator and Dealer must shake hands in the process of requesting and entering a new card. *HIT* requests a new card; *CARD.READY*∗ is the operator's acknowledgment of the request. The desired sequence is:

(1) Dealer senses that the (synchronized) *CARD.READY* signal is false, and asserts *HIT*, keeping *HIT* true at least until *CARD.READY* goes true.
(2) Operator sees the Hit light on, and (after preparing a new card) presses the Card.Ready button, keeping it on at least until the Hit light goes off.
(3) Dealer detects that the (synchronized) *CARD.READY* signal is true, and drops *HIT* (and proceeds to process the new card), keeping *HIT* false at least until *CARD.READY* goes false.
(4) Operator sees the Hit light off, and releases the Card.Ready button, keeping it released (and hands off the card switches) at least until the Hit light goes on.

The Single-Pulser Revisited

What is wrong with our preliminary design? The operator's role appears correct (we would, of course, fully describe the rules in the user's operating instructions). The Dealer ASM performs step (1) properly, and most of step (3), but fails to keep *HIT* false until it detects that the operator's button is released. Dealer is failing to observe the dropping of the operator's hand. We can recast this requirement by saying that Dealer must respond once and only once to each action of the operator. Earlier in this chapter you studied several ways of describing and handling this common phenomenon of "once and only once." Here, let us express the solution yet another way. We incorporate the one-state single-pulser algorithm (Fig. 6–3) directly into the ASM, so that we may add our own specialized conditional outputs to the test branches. You have already seen that we will have a *CARD.RDY.SYNC* flip-flop in the architecture; to accommodate the single pulser we will include another flip-flop with output *CARD.RDY.DELAYED*. Then in the control algorithm, we explicitly test the values of these two flip-flops in order to isolate one and only one recognition of the operator's button push. The handshake signal *HIT* arises from a conditional output whenever the pushbutton is up. The relevant part of this ASM is in Fig. 6–31.

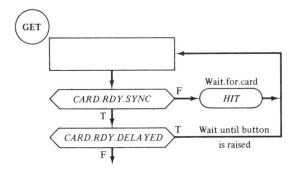

Figure 6–31 A specialized single-pulser for the Black Jack Dealer.

This algorithm solves the difficulty of accomplishing the full handshake: *HIT* will only be asserted in the Wait.for.Card branch of state **GET**. If the button is up, Dealer is happy to request a new card in state **GET**. Whenever **GET** detects that the button is pressed, *HIT* will go false. The first time **GET** sees the button pressed, the ASM will process a new card. If control returns to **GET** while the button is still down, the algorithm simply waits (with *HIT* still false) until the operator releases the button.

The Final ASM for the Black Jack Dealer

The algorithm for the Black Jack Dealer now seems to be developing nicely. We have found that we may eliminate some of the states in our original proposal without jeopardizing clarity. We have explored in detail the synchronization requirements of certain inputs and the larger handshaking requirements between the operator and Dealer. Figure 6–32 is the improved ASM for Dealer. In this figure, as in Fig. 6–13, the labels on the conditional output boxes describe conditional output terms—logic terms useful in developing a systematic implementation of a complex ASM. The conditional output terms are not state names; they merely represent positions within their parent state. For instance, the circled label Ⓓ on the conditional output in state **GET** is a shorthand for the label

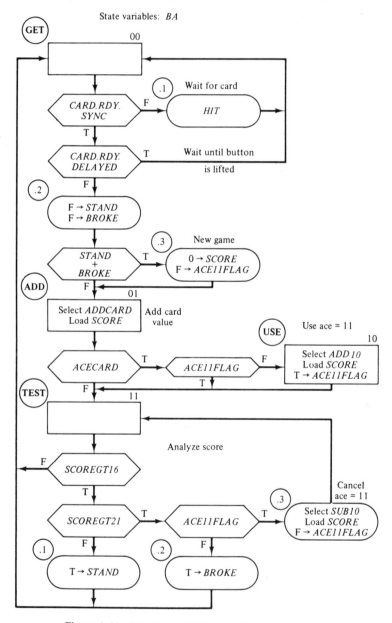

Figure 6–32 The final ASM for the Black Jack Dealer.

GET.1. During the implementation of the control algorithm, we will derive equations for the logic variable *GET.1* as well as each of the other terms.

The Final Architecture of the Black Jack Dealer

Figure 6–33 is the functional architecture of Dealer. Along the way, as we developed a more detailed understanding of the problem, we made certain modifications to the original architecture; these are reflected in the figure. The significant modifications or elaborations include the following:

(a) Since the operating procedures stipulate that the card switches are stable throughout the time that Dealer processes the card (in other words, as long

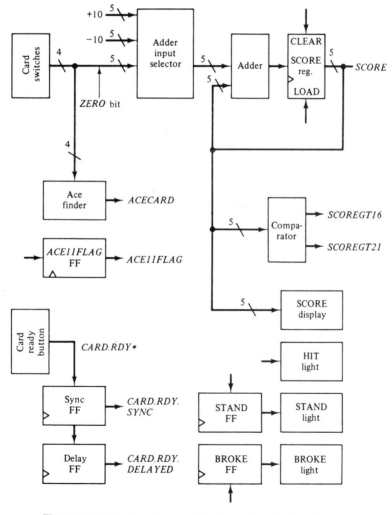

Figure 6–33   The functional architecture of the Black Jack Dealer.

as *HIT* is false), we do not need a separate register to hold the current card value; the card switches themselves will suffice.

(b) The ASM manages the *HIT* signal directly, without a flip-flop to preserve its value across states.

(c) Our treatment of *CARD.READY*∗ introduces two new flip-flops, the components of the single pulser.

(d) We have elaborated on the black box for preparing the input to the *SCORE* register. There are four operations that modify *SCORE*: clearing *SCORE* to 0, adding *CARD* to *SCORE*, adding +10 to *SCORE*, and subtracting +10 (adding −10) from *SCORE*. A 5-bit adder can add the appropriate value to *SCORE* if we can select the proper value. You know how to manage selection, so you will probably guess that we will use multiplexers. The important point to realize at this stage is that we can select one input from several, without worrying about the exact chips. With *SCORE* as one input to the 5-bit adder and the other input selected by a selector black box, the circuit is nearly specified. A judicious choice of the chips for *SCORE* should allow us to clear this register separately from the register-load operation. On this basis, the original nebulous black box has separated into three components: an adder building block, a selector building block, and a control input for clearing the *SCORE* register building block.

Surely now we will sit down and define the exact chips to support the dealer architecture. This would be a reasonable step to take at this time, but as a further illustration of the power of the methods you are learning, let's see how far we can carry the implementation of the control algorithm without specifying the exact chips in the architecture.

Implementing the Control Algorithm

It is the task of the ASM to generate the necessary commands (outputs) to control the architecture and to provide the outputs to the external world. In Fig. 6–34 we show how the ASM controls the architecture. The inputs and outputs of the ASM are still specified at a somewhat abstract level. We will develop the logic equations for each signal, prior to selecting chips. Such a development depends on our understanding of how to convert building blocks into chips, just as implementing a software flowchart requires a knowledge of the programming language. As we have stressed repeatedly, the goal is to think of the problem, not of the chips, for as long as possible.

Our design now involves two tasks: Implementing the flow of the ASM (the state generator), and implementing the outputs (commands). First, we will define the conditional output terms, which will be useful parameters for many of the remaining equations. Reading directly from the ASM, we have the following logic equations

$$GET.1 = GET \cdot \overline{CARD.RDY.SYNC}$$
$$GET.2 = GET \cdot CARD.RDY \cdot SYNC \cdot \overline{CARD.RDY.DELAYED}$$

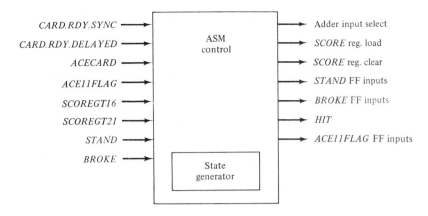

Figure 6–34 The functional control of the Black Jack Dealer.

$$GET.3 = GET.2 \cdot (STAND + BROKE)$$
$$TEST.1 = TEST \cdot SCOREGT16 \cdot \overline{SCOREGT21}$$
$$TEST.2 = TEST \cdot SCOREGT16 \cdot SCOREGT21 \cdot \overline{ACE11FLAG}$$
$$= TEST \cdot SCOREGT21 \cdot \overline{ACE11FLAG} \qquad \text{(by observation)}$$
$$TEST.3 = TEST \cdot SCOREGT16 \cdot SCOREGT21 \cdot ACE11FLAG$$
$$= TEST \cdot SCOREGT21 \cdot ACE11FLAG \qquad \text{(by observation)}$$

Next, we implement the state generator. We again choose a multiplexer controller, leaving the one-hot method as an exercise. Figure 6–32 shows a suitable (arbitrary) state assignment. Four states require two flip-flops for the encoding of the state variables. A four-input multiplexer feeds each flip-flop. Table 6–3 is a summary of the ASM state transitions.

Notice the handy use of the conditional output terms, which results in a simplification of the control multiplexer inputs and achieves a savings of hardware while increasing clarity. For example, the full condition for moving from state **GET** to state **ADD** is

$$FROM.GET.TO.ADD = CARD.RDY.SYNC \cdot \overline{CARD.RDY.DELAYED}$$

TABLE 6–3 STATE TRANSITIONS IN THE BLACK JACK DEALER ASM

	Present state	Next state		Condition from ASM
0	**GET**	**GET**	00	$\overline{GET.2}$
		ADD	01	$GET.2$
1	**ADD**	**USE**	10	$ACECARD \cdot \overline{ACE11FLAG}$
		TEST	11	$ACECARD \cdot ACE11FLAG$
2	**USE**	**TEST**	11	T
3	**TEST**	**GET**	00	$\overline{TEST.3}$
		TEST	11	$TEST.3$

But conditional output term *GET.2* is just this expression ANDed with *GET*. Producing the conditional output in the oval at position GET.2 will require that we implement the conditional output term *GET.2*, so we know that this term is available in the final circuit. Appending *GET* to the expression for *FROM.GET.TO.ADD* has no logical effect, since the expression already implies that the ASM is in state **GET**. It is therefore permissible, and convenient, to use the conditional output terms as required to develop the state generator logic.

As usual, it is convenient to decode the state flip-flop outputs, producing in this case the four state terms *GET*, *ADD*, *USE*, and *TEST*. These terms complete the requirements of the logic equations developed thus far.

As a last step in the ASM synthesis, we derive the output signals. Reading almost directly from the ASM, we have

$$
\begin{aligned}
HIT &= GET.1 \\
STAND(SET) &= TEST.1 \\
STAND(CLR) &= GET.2 \\
BROKE(SET) &= TEST.2 \\
BROKE(CLR) &= GET.2 \\
ACE11FLAG(SET) &= USE \\
ACE11FLAG(CLR) &= GET.3 + TEST.3 \\
SCORE(LOAD) &= ADD + USE + TEST.3 \\
SCORE(CLEAR) &= GET.3
\end{aligned}
$$

Deriving the controls for the adder selector requires one further set of parameters. We need 3 data inputs to the selector: *CARD*, +10, and −10. Each data path is 5 bits wide, so five 4-input multiplexers will serve nicely as the basis for this selector, each mux being controlled by the same select signals, *S1* and *S0*. We shall assign +10 as the input to the 0-position of the multiplexers, −10 to the 1-position, and *CARD* to the 2-position. Our task is to derive logic equations for the selector controls *S1* and *S0*. With these specifications and the ASM chart, we develop Table 6–4. From this tabulation we easily derive the select input equations:

$$
S1 = ADD
$$
$$
S0 = TEST.3
$$

TABLE 6–4 ADDER INPUT SELECTION FOR THE BLACK JACK DEALER

Mux position	Select inputs S1	S0	Data inputs	Notation in ASM	ASM condition
0	0	0	+10	ADD10	USE
1	0	1	−10	SUB10	TEST.3
2	1	0	CARD	ADDCARD	ADD
3	1	1	—	Don't care	—

We now have equations for all the control signals, still without a commitment to specific chips. The last step, hopefully straightforward and bug-free, is to select chips and draft the final circuit diagrams for the architecture, state generator, and output signals. Here is the first point at which voltage enters our design! Using mixed-logic drafting conventions, you can implement the logic equations with whatever gates are handy. JK flip-flops are an obvious choice for the storage elements for those signals requiring controlled setting and clearing: *STAND*, *BROKE*, and *ACE11FLAG*. As you learned in Chapter 4, you may use either the *J* or the *K* input to represent the set condition.

Summing Up

This completes the Black Jack Dealer problem. The methodology is important and well worth your close study. The documentation of this design should include most of the figures and tables developed in this exercise that relate to the final design. Think what a help this high-level documentation would be to you if you were presented with a real, nonfunctioning circuit and told to get it running or to modify it in some way. Once again, note that although our thinking was conditioned by our knowledge of the available types of integrated circuits and their characteristics, it was only at the end of the design that we introduced actual voltages and chips.

READINGS AND SOURCES

ERCEGOVIC, MILŎS D., and TOMAS LANG, *Digital Systems and Hardware/Firmware Algorithms.* John Wiley & Sons, New York, 1985. Good treatment of sequential systems.

FLETCHER, WILLIAM I., *An Engineering Approach to Digital Design.* Prentice-Hall, Englewood Cliffs, N.J., 1980. Contains numerous system-level examples of design.

GLASSER, LANCE A., and DANIEL W. DOBBERPUHL, *The Design and Analysis of VLSI Circuits.* Addison-Wesley Publishing Co., Reading, Mass., 1985.

HILL, FREDERICK J., and GERALD R. PETERSON, *Digital Logic and Microprocessors.* John Wiley & Sons, New York, 1984.

HILL, FREDERICK J., and GERALD R. PETERSON, *Introduction to Switching Theory and Logical Design,* 3rd ed. John Wiley & Sons, New York, 1981.

HWANG, KAI, *Computer Arithmetic—Principles, Architecture, and Design.* John Wiley & Sons, New York, 1979.

LALA, PARAG K., *Fault Tolerant and Fault Testable Hardware Design.* Prentice-Hall International, London, 1985.

LSI Databook. Monolithic Memories, 2175 Mission College Blvd., Santa Clara, Calif. 95954. PALs, memory products, arithmetic units, and system building blocks.

MEAD, CARVER, and LYNN CONWAY, *Introduction to VLSI Systems.* Addison-Wesley Publishing Co., Reading, Mass., 1980. The first VLSI textbook.

Memory Components Handbook. Intel Corporation, Literature Department, 3065 Bowers Avenue, Santa Clara, Calif. 95051.

MYERS, GLENFORD J., *Digital System Design with LSI Bit-Slice Logic.* John Wiley & Sons, New York, 1980.

WAKERLY, JOHN F., *Logic Design Projects Using Standard Integrated Circuits*. John Wiley & Sons, New York, 1976. A good source of laboratory projects.

WESTE, NEIL, and KAMRAN ESHRAGHIAN, *Principles of CMOS VLSI Design: A Systems Perspective*. Addison-Wesley Publishing Co., Reading, Mass., 1985.

EXERCISES

6–1. Show that any ASM may be expressed as an ASM with only one state. Why do we not do away with state generators by always designing with single-state ASMs? Discuss the advantages and disadvantages of this approach.

6–2. Using a 60-Hz periodic logic signal, produce a signal that can serve as a 1-Hz clock.

6–3. Although we find that we must debounce manual control switches, we usually do not need to debounce manual data entry switches. Why?

6–4. Build a four-state ASM that emits one of two BCD number sequences, depending on the value of a control variable *NINESCOMP*. Each sequence has a cycle of four digits:

When *NINESCOMP* = F, the sequence is 0, 1, 2, 3, 0, 1, ...

When *NINESCOMP* = T, the sequence is 9, 8, 7, 6, 9, 8, ...

6–5. Exercises 1–36 and 2–32 deal with the seven-segment numeric display. Assume that the display integrated circuit requires discrete signals for each segment *a* through *g*. Build a four-state ASM to repeatedly display the first four prime numbers in proper sequence: 2, 3, 5, 7, 2, 3, 5, ...

6–6. Using seven-segment decimal display chips, design the two-digit "second" display of a digital clock. The clock must cycle continuously from 00 through 59, except when a signal from a pushbutton forces the display to 00. Assume that a 1-Hz synchronous clock signal is available.

6–7. Extend the previous exercise to produce a full 24-hour clock display. How will you set the correct time? (*Hint:* Commercial LSI digital clock units often provide a speed-up mode, in which the displayed time goes through a complete cycle in less than 1 minute.)

6–8. Add an alarm feature to the digital clock.

6–9. Design an implementation of Fig. 6–1 using the formal one-hot method with no simplifications. Show how this implementation may be rigorously transformed into the circuit of Fig. 6–2.

6–10. Implement the system clock of Design Example 2 using a JK flip-flop instead of the D flip-flop.

6–11. Consider Figs. 6–9 and 6–10 of Design Example 3. Show that if you have DATA* with T = L, the mixed-logic circuit diagram notations change but the hardware remains the same.

6–12. Implement the traffic-light controller of Design Example 5 with PALs.

6–13. Add a blinking-light feature to the traffic-light controller in Design Example 5. Assume that a new *BLINK* signal is available and that, when *BLINK* is asserted, the highway lights blink yellow and the farm-road lights blink red.

6-14. Finish the detailed design of the combination lock in Design Example 6. Produce circuit diagrams suitable for actual construction of the lock's electronics.

6-15. In the elegant combination lock of Design Example 6, we used a 74LS163 counter and a comparator to determine when the proper number of combination digits have been entered (see Fig. 6–22). Eliminate the comparator by using the two's complement of the count and counting down instead of counting up.

6-16. Complete the detailed design of the Black Jack Dealer of Design Example 7. Produce circuit diagrams suitable for construction of the electronics.

6-17. For the Black Jack Dealer, the initial ASM in Fig. 6–26 required a flip-flop for the output *HIT*. Show how the further development of the control algorithm led to the elimination of the *HIT* flip-flop from the architecture. This is a typical example of the algorithm modifying the architecture.

6-18. In Fig. 6–29, state variables *C*, *B*, and *A* are all involved in transition races. Why is state variable *D* not similarly involved?

6-19. Binary patterns that differ in exactly 1 bit are said to be a *unit distance* apart. Consider an ASM state (the "predecessor") that has branch paths to several "successor" states. (Note that one possible successor is the predecessor state itself.)

(a) If all successor states have encoded state assignments at most a unit distance from the predecessor, show that no transition races will arise, even if asynchronous test inputs are present in the predecessor state.

(b) Show that the condition in part (a) is not sufficient to preclude output races.

6-20. In the Black Jack Dealer, the architecture contains black boxes that produce such signals as *ACECARD*, *SCOREGT16*, and *SCOREGT21*. For example, see Fig. 6–33. These black boxes involve only simple combinational circuits. Why do we choose to use the black boxes during the design process, instead of showing the circuits directly?

6-21. The ASM for the Black Jack Dealer (Fig. 6–32) tests several signals (*ACECARD*, *SCOREGT16*, *SCOREGT21*) that are generated by combinational logic within architectural black boxes. Since these black boxes are not clocked, how do we know that their outputs are synchronous and therefore suitable for testing in the ASM?

6-22. Convert the algorithm for the Black Jack Dealer (Fig. 6–32) into an ASM with only one state. Since such a move simplifies the state generator circuitry, why do we not choose a one-state ASM for the Black Jack Dealer?

6-23. Define carefully the use of a full handshake to synchronize two independent processes. Show how the need for process synchronization arises whenever a human operator interacts with a machine. Illustrate these concepts with the Black Jack Dealer.

6-24. To synchronize events in two cooperating but independent processes, designers have used a variety of techniques, many of which we may describe as "incomplete handshakes." For instance, consider the following incomplete handshake:

(1) Device A requests an action (device A extends its hand).

(2) Device B senses the request and acknowledges receipt of the request (device B extends its hand).

(3) Device A senses device B's acknowledgment and drops its request (device A drops its hand).

(4) Device B drops its acknowledgment at any time after asserting its acknowledgment (device B drops its hand).

This looks like a "complete" handshake, but there are circumstances in which the handshake may be incomplete. Draw timing diagrams for the possible behaviors of the request and acknowledge signals. Discuss the effectiveness of the above protocol for the following conditions:

(a) Device A is a machine; device B is a human being.

(b) Device A is a human being; device B is a machine.

(c) Both devices are machines.

(d) Both devices are human beings.

6–25. Implement a one-hot state generator for the Black Jack Dealer.

6–26. In the Black Jack Dealer example, the signals *SCOREGT16* and *SCOREGT21* arise from a comparator architectural element. Write logic equations for these two variables. Show by Boolean algebraic manipulation that the term *SCOREGT16*•*SCOREGT21* reduces to *SCOREGT21*, thus rigorously demonstrating the validity of the simplifications of *TEST.2* and *TEST.3* performed "by observation" in the text.

6–27. Design a synchronous digital circuit with the following properties:

Inputs:

(a) Two 4-bit binary numbers, *A* and *B*, in signed magnitude notation (1 sign bit, 3 magnitude bits).

(b) A *GO* signal from a manual pushbutton.

Outputs:

(a) A 4-bit binary number *C* in signed magnitude notation.

(b) A signal *EVEN* for a display lamp.

Task:

(a) Wait for the *GO* signal to be asserted.

(b) When *GO* appears, clear the signal *EVEN* to false, and load the values on the input lines *A* and *B* into two 4-bit registers *RA* and *RB*. [Hereafter, (RA) means "contents of *RA*," etc.]

(c) Then, produce an output *C* in register *RC*, as follows:

If $(RA) > (RB)$, then transfer the quotient of $(RA)/2$ to *RC*.

If $(RA) \leq (RB)$, then transfer (RB) to *RC*.

(d) If $(RA) > (RB)$ and the remainder of $(RA)/2$ is 0, then assert *EVEN*; otherwise, *EVEN* remains false.

(e) Return to step (a) to await another *GO* signal.

6–28. Construct an ASM that will turn on a light as the first person enters a room, and turn off the light as the last person leaves. Assume that there is a single door fitted with two photocells that generate TTL-compatible outputs. One photocell is on the inner side of the door and the other is on the outer side. Light beams shine on each photocell, producing a false output from the cell; a true output from a photocell arises when the light beam is interrupted. Assume that once a person starts through the door, the process is completed, and that only one person enters or leaves at a time.

6–29. Design a versatile timer circuit. The circuit has two input codes:

(1) A 3-bit code describing the unit of counting: code 0 = 100 nsec, code 1 = 1 μsec, code 2 = 10 μsec, . . . , code 7 = 1 sec.

(2) An 8-bit code describing the number of counts.

Asserting an input signal *START* will cause the timer to begin counting the specified number of counts, each count being of the specified duration. The only output from the timer is a signal *TIMESUP*, which becomes false when timing begins and becomes true when the specified interval has elapsed. The timer can time an interval from 100 nsec to 255 sec. Use a 100-nsec clock to drive the timer. Consider using the 74LS162 decade counter as the basic counting element. Use good synchronous design techniques in determining how to clock and advance the decade counters. Timers similar to this are available as LSI integrated circuits.

6–30. A popular microcomputer requires a 1-μsec square-wave clock for normal operation. Advanced modes of operation for high-speed input-output and for refreshing dynamic RAM require that the inactive (false) portion of the clock be stretched as follows:

Normal operations:	Inactive clock time = 0.5 μsec
Refresh request (*RFRQ*):	Inactive clock time = 1.0 μsec
Input-output request (IORQ):	Inactive clock time = 1.0 μsec
Both requests simultaneously:	Inactive clock time = 2.0 μsec

The active (true) portion of the clock signal is always 0.5 μsec. Design a microcomputer clock that responds to *RFRQ* and *IORQ* as shown above. The clock output must be free of hazards.

6–31. In high-speed input-output operations on a microcomputer, the input-output device may assert a signal that will put the microprocessor "to sleep," allowing the input-output device to have uninterrupted access to the microcomputer's memory until the transfer of data is complete. If the device is faulty, it is possible to put the processor to sleep for an indefinite period. To protect the processor from such faulty behavior, we may introduce a timer that will issue a "wakeup" signal to the processor after a period of memory inactivity. The timer should conform to these requirements:

(a) The timer should be enabled only when the processor is asleep.

(b) Timing should begin whenever the processor goes to sleep, and upon the successful transfer of each data word between the memory and the input-output device.

(c) While the processor is asleep, if no memory read or write occurs for 10 msec, the timer should assert its *WAKEUP* signal.

(d) The *WAKEUP* signal should remain asserted until the processor asserts an acknowledge signal *ACK*.

(e) *WAKEUP* should be synchronous with the microcomputer system clock, and should be free of hazards.

Design the timer. Use a single-shot for the 10-msec delay. Is the poor stability of the single-shot (about 5 percent) a disadvantage in this application?

6–32. Repeat Exercise 6–28, using decade counters to generate the 10-msec delay. Assume a 1-μsec microcomputer system clock.

6–33. In this exercise, you will design a circuit to resolve conflicts in memory accesses. Two microprocessors driven by the same system clock share access to a single dynamic RAM. The RAM can complete a memory operation in one clock cycle. Consider three types of memory request: for dynamic RAM refresh (*RFRQ*), for microprocessor 1 (*CPU1RQ*), and for microprocessor 2 (*CPU2RQ*). Memory request conflicts are resolved as follows:

(a) RAM refresh has highest priority.

(b) If both microprocessors simultaneously request memory access, the microprocessor that more recently received memory service has the lower priority.

Each device contending for memory access receives a status signal (*RFGNT*, *CPU1GNT*, or *CPU2GNT*) that specifies when memory access is granted to the device.

Design a device that accepts requests and responds with status signals according to the preceding rules.

6–34. The *stack* is a software data structure often implemented in hardware. A stack is an ordered set of elements analogous to a stack of plates in a cafeteria. Only the top element (plate) is accessible. Removal of the top element exposes the next-to-top element, which then becomes the top. Addition of an element to the stack causes the former top element to become next-to-top; the added element becomes the top. The operation of adding an element to a stack is called *push*. The removal operation is called *pop*. These are the only allowed stack operations. A stack is sometimes called a *LIFO* (last in, first out) memory.

In hardware, we may implement a stack in RAM or with an array of discrete registers.

(a) Using RAM, design a stack that accepts push and pop operations and properly adds or removes an element from the top of the stack.

(b) Design a small five-element stack using MSI registers.

6–35. Repeat Exercise 6–34 for a stack that also provides two status signals:

EMPTY: Asserted when the stack contains no elements. Your stack should ignore a command to pop an empty stack.

FULL: Asserted when the stack contains a predetermined maximum number of elements. Your stack should ignore an attempt to push a full stack.

6–36. The *queue* is a software data structure that is sometimes implemented in hardware. The queue has a front and a rear, like a line for tickets at a theater. A *write* operation adds an element to the rear of the queue; a *read* operation removes the element at the front of the queue. No other operations are allowed. The queue is also called a *FIFO* (first in, first out) memory. A queue has two status indicators:

EMPTY: Asserted when the queue contains no elements. The circuit should ignore an attempt to read an empty queue.

FULL: Asserted when the last available memory location is occupied with a queue element. The circuit should ignore an attempt to write into a full queue.

(a) One approach to implementing a queue in RAM is to maintain two pointers *FRONT* and *REAR* as RAM addresses to the extremities of the queue. *WRITE* increments *REAR* and adds an element to the rear of the queue; *READ* extracts the front queue element and increments *FRONT*. *FULL* becomes true when *REAR* points to the highest memory location in the RAM. Design such a queue. You may alter the foregoing suggestions as long as you still implement a queue.

Why is this project more difficult than designing the stack of Exercises 6–34 and 6–35? In this implementation, when *FULL* is true, is all of the memory filled with queue elements? If *EMPTY* is true, what should be the values of *FRONT* and *REAR*?

(b) Another approach to a RAM implementation of a queue is similar to that in part (a) but allows the queue to go "around the corner," so that *REAR* and *FRONT* may advance from the highest memory address to address zero. Design such a queue. When does *FULL* become true?

6–37. Design a controller for an elevator in a six-story building. Your controller must respond to call switches on each floor and floor-select switches within the car.

6–38. Design a four-way traffic-light controller that will keep traffic moving efficiently along two busy streets that intersect. In this exercise, consider only straight-through traffic.

6–39. Extend Exercise 6–38 to include left-turn signals at each approach to the intersection.

6–40. Extend Exercise 6–38 to include pedestrian crosswalk signals.

7

Designing
a Minicomputer

You are ready to tackle a really substantial project to round out your study of hardwired design. Nothing will sharpen your design skills more than wading through the design of a complex project from start to finish. Thus far, you have studied pieces of the design process; in the next three chapters we will help you forge your knowledge into an integrated and workable design tool. What project should we choose? Such an undertaking should be detailed yet elegant, large yet not too large. Let's design a computer!

Our aim is to design an entire operational computer system, taking no shortcuts, leaving nothing out. Most computers, even the smallest microcomputers, are highly complex structures—too complex to be a suitable teaching illustration at the MSI and LSI level of design. Instead, we choose the first minicomputer, the Digital Equipment Corporation PDP-8.

The PDP-8 has had a successful history. More than 100,000 units have been installed, many of which are still in use. The PDP-8 also has an extensive library of software and is a good machine for illustrating device interfacing.

The great advantage of the PDP-8 for our purpose is that it has a simple structure with only eight basic instructions. It exists in several models; each executes the same basic set of instructions, but they differ in minor ways. We will use the PDP-8I as the basis for our exercise. We will develop our design from first principles and make no reference to the Digital Equipment Corporation's design. The result will be functionally equivalent to the PDP-8I—for example, it will run PDP-8I software—but we will use top-down design techniques. The only detailed information we need about the PDP-8I is a description of the action of each instruction. We shall call our design the LD20.†

The statement of the problem is brief: build a computer that will execute the PDP-8I instruction set.

PDP-8I SPECIFICATIONS

The first step is the obvious one of studying the PDP-8I to see what we must emulate. The major characteristics of the PDP-8I are:

(a) *A 12-bit word size.* This is quite small and will cause memory-addressing limitations. If a memory word is used to hold an address, it can refer to only 4096 (2^{12}) different locations. Therefore, the standard PDP-8 is limited to 4096 words of addressable memory.

(b) *A single accumulator.* Several instructions refer to an accumulator (*AC*), used to store intermediate results for later manipulation. Having only one accessible register forces a programmer to use care in saving and restoring vital data in the *AC*, for example upon subroutine entry and exit. Many computers have several registers, which can speed the execution of programs but which expose the programmer to subtle bugs if the data in all registers is not properly handled. In many applications the single *AC* is a blessing!

(c) *A 3-bit operation code.* Each instruction occupies a 12-bit word, of which 3 bits are devoted to the operation code. This provides eight basic commands—an adequate but hardly abundant number. Only 9 bits remain in the instruction for such purposes as addressing memory, whereas the 4096-word memory requires a full 12-bit address.

(d) *Paging.* Addressing limitations in minicomputers and microcomputers have forced computer architects to find a number of ingenious solutions. The PDP-8's method is based on memory pages of 128 (2^7) words. The 4096-word address space is divided into 32 pages, and each memory-referencing instruction has 7 bits to address a word within a page. The missing 5 bits of the address are not a part of the instruction, but are derived implicitly from the context. Without some trick of this sort there would be no way to pack a 3-bit command and an address into a 12-bit word. Maneuvers such as this are common features of minicomputers. The paging mechanism of the PDP-8 is perhaps the simplest technique and serves as a foundation for studying more complicated schemes used in other computers.

Throughout this design exercise, we will use the octal numbering system to specify particular values of the PDP-8's instructions, addresses, and so on.

---

† The LD20 design developed in this book is used in instructional laboratories for digital design. The equipment to support this design and the LD30 microprogrammed version (developed in Chapter 10) is produced by Logic Design, Inc. A laboratory manual for the LD20 and LD30 is available. See Readings and Sources at the end of this chapter.

Any such numbers not in octal will have an explicitly designated base. Thus 305 is 305 octal, 1011_2 is 1011 binary, and 42_{10} is 42 decimal.

PDP-8 Memory Addressing

In many memory addressing schemes for small instructions, the location of the current instruction is used to specify part of the operand address. For example, assume a program with five instructions stored sequentially, starting at location 300. Call these instructions CM0 (command 0) through CM4 (command 4). A memory map of this program would be:

Location	Contents
300	CM0
301	CM1
302	CM2
303	CM3
304	CM4

If instruction CM3 is being executed, we know that it is located at address 303, since that is where we placed it. Instruction CM3 can employ a subset of the 12 bits in its word to reference data located close to location 303. In the PDP-8, "close to" means in the same page.

The PDP-8 splits 4096 words of memory into 32 pages of 128 words each, as shown in Fig. 7–1. Instruction CM3 is in page 1; 7 bits are sufficient for that instruction to access any word in that page.

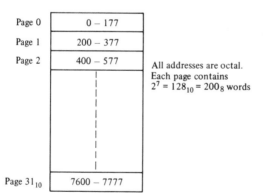

All addresses are octal.
Each page contains
$2^7 = 128_{10} = 200_8$ words

Figure 7–1 Page structure of the memory of the PDP-8.

We now have a mechanism such that an instruction needs only 7 bits to access a memory cell in one particular page. Let us call these 7 bits the *page offset*, and let the page offset occupy the rightmost 7 bits of a PDP-8 instruction:

Op code 3 bits	Uncommitted 2 bits	Page offset 7 bits

Suppose location 301 contains the 12 bits $001\,XY1\,000\,101_2$, (for the moment we will ignore the 2 bits X and Y). The operation code is 001_2, which means an addition of the AC and the contents of a memory location. Which location? The 7 page-offset bits are $1\,000\,101_2 = 105_8$. The instruction is to add the contents of location 105 in this page (the page containing the add instruction) to the accumulator. We know that the instruction is at location 301, and since the instruction is in page 1, the page offset is referring to page 1. Thus we will get the contents of word 105 in page 1 and add it to the AC.

What if instructions in different pages require the same data? It would be nice if some common page could be accessed by instructions in any page. In the PDP-8, page 0 has this function. We have two precious unused bits in the instruction, and we need one of them to tell if we want word 105 in the current page (page 1 in our example) or word 105 in the common page (page 0). In the PDP-8, the Y bit is used for this page selection; we call it the *page bit*.

If the page bit is 1, the page address of the current instruction is concatenated with the 7-bit offset in the instruction to form a full 12-bit address, which is sufficient to identify any word of the 4096-word memory. If the page bit is 0, the reference will be to a word in page 0 of the memory.

If we execute an instruction at location 301 that contains $001\,011\,000\,101_2$, we will add the contents of location 105 in page 1 to the AC. Location 105 in page 1 is memory location 305

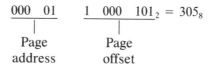

If location 301 contains $001\,001\,000\,101_2$, the instruction would mean to add the contents of location 105 in page 0 to the AC. Location 105 in page 0 is memory location 105.

Indirect addressing. We have shown how the page bit and the page offset combine to yield an address either in page 0 or in the current instruction page. What happens if a command in page 2 needs to access a location in page 7? We must use all 12 bits of a word as address bits. We can do this if the word accessed by an instruction is treated not as an operand but as the *address* of an operand. This extra step is called *indirect addressing*. The PDP-8 uses the remaining instruction bit X as the *indirect bit* to specify indirect addressing. The complete format of a memory referencing instruction is

Op code 3 bits	Indirect bit	Page bit	Page offset 7 bits

In the previous examples, the contents of locations 305 or 105 (for page bits 1 or 0) were treated as 12-bit *data* words. If the indirect bit is on, these contents are treated as 12-bit *addresses* of data. We require one extra memory cycle to access this final indirectly addressed data location.

Indirect addressing is a powerful concept since it provides a way to specify arbitrary 12-bit addresses. Into some memory word *IND* that is close to our instruction or in page 0, we load the address of the final location that we wish to access. We can then access the location by indirectly addressing it through *IND*.

It is useful to have a shorthand for the final memory location referenced in an instruction after all applicable paging and indirect addressing are invoked. We call this final location the *effective address, EA*. The contents of location *EA* is called the *contents of the effective address, CA*. (*CA* is sometimes called the *effective operand*.) Using *EA* and *CA*, we can compactly describe the memory references of any PDP-8 instruction.

Examples of memory addressing. Here are some examples of referencing memory using PDP-8 instructions. The addresses will have 12 bits, since the PDP-8 has 4096 words of memory. We refer to the contents of an addressed memory location by enclosing the address in parentheses: If location 0301 contains 0305, then (0301) = 0305. Note that (*EA*) = *CA*.

Now assume that the following memory locations have been loaded with the data shown:

$$
\begin{array}{ll}
(0301) = 1305 & (0305) = 1234 \\
(0302) = 1105 & (0105) = 4321 \\
(0303) = 1705 & (1234) = 5567 \\
(0304) = 1505 & (4321) = 7765
\end{array}
$$

(a) What are the *EA* and the *CA* for the instruction located at 0301?

$$\text{Command} = 001_2 = \text{TAD (Add)}$$

$$1305 = 001 \quad 01 \quad 1 \quad 000 \quad 101_2$$

Indirect bit = 0
Page bit = 1
Page offset = 105

$$EA = 000 \quad 01 \quad 1 \quad 000 \quad 101_2 = 0305$$

Page address   Page offset

$$CA = (0305) = 1234$$

This instruction would add the quantity 1234 to the contents of the *AC*.

(b) What are the *EA* and the *CA* for the instruction located at 0302?

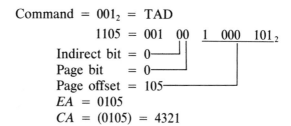

$$Command = 001_2 = TAD$$
$$1105 = 001 \quad 00 \quad 1 \quad 000 \quad 101_2$$

Indirect bit = 0
Page bit = 0
Page offset = 105
EA = 0105
CA = (0105) = 4321

This instruction would add the quantity 4321 to the contents of the *AC*.

(c) What are the *EA* and the *CA* for the instruction located at 0303?

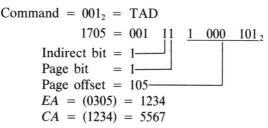

$$Command = 001_2 = TAD$$
$$1705 = 001 \quad 11 \quad 1 \quad 000 \quad 101_2$$

Indirect bit = 1
Page bit = 1
Page offset = 105
EA = (0305) = 1234
CA = (1234) = 5567

This instruction would add the quantity 5567 to the contents of the *AC*.

(d) What are the *EA* and the *CA* for the instruction located at 0304?

$$Command = 001_2 = TAD$$
$$1505 = 001 \quad 10 \quad 1 \quad 000 \quad 101_2$$

Indirect bit = 1
Page bit = 0
Page offset = 105
EA = (0105) = 4321
CA = (4321) = 7765

This instruction would add the quantity 7765 to the contents of the *AC*.

Auto indexing.   The PDP-8 has a feature called *auto indexing* that provides some flexibility in addressing. Most large computers have index registers to facilitate access to arrays of data. Unfortunately, specifying an index register takes 1 or more bits of the instruction and we have no bits left. The PDP-8's auto indexing is a primitive way to index without using bits in the instruction. An *auto index register* is a word in the memory that will be automatically incremented every time it is used as the source of an indirect address. The word is incremented before it is used as an address. Repeated use of the same auto index register will sequence the effective address *EA* throughout the full address space of the memory. There are 8 auto index registers in the PDP-8's main memory, locations 10_8 through 17_8. When not performing auto indexing, these locations behave like normal memory words.

Here are some examples of auto indexing. Assume that the following locations have the contents shown:

$$(0013) = 4102$$
$$(4102) = 1111$$
$$(4103) = 2000$$

(a) Instruction: $1013 = 001 \quad 00 \quad 0 \quad 001 \quad 011_2$

  Command $= 001_2 = $ TAD
  Indirect bit $= 0$
  Page bit $\quad = 0$
  $EA = 0013$
  $CA = (0013) = 4102$

Although location 0013 is the address, there is no auto indexing because the indirect bit is 0. This instruction adds the quantity 4102 to the contents of the AC.

(b) Instruction: $1413 = 001 \quad 10 \quad 0 \quad 001 \quad 011_2$

  Command $= 001_2 = $ TAD
  Indirect bit $= 1$
  Page bit $\quad = 0$

The initial address is 0013. This is an auto index location used as an indirect address. The auto indexing feature causes

$$(0013) (+) 1 \rightarrow (0013), \text{ or}$$
$$4102 (+) 1 \rightarrow (0013)$$

Then

$$EA = (0013) = 4103$$
$$CA = (4103) = 2000$$

The effect of executing this instruction is to increment the contents of location 0013 by 1, and to add the quantity 2000 to the contents of the AC.

PDP-8I Instructions

Its instruction set characterizes a computer, and therefore we must carefully study the PDP-8I's instructions. The effective address EA and contents of the effective address CA notations allow a compact description of the memory-referencing instructions.

AND (Twelve-bit logical AND). Operation code $000_2 = 0_8$.

$$AC \cdot CA \rightarrow AC$$

This is a bit-by-bit AND of the AC with the effective address contents. For example,

$$AC = 001 \quad 101 \quad 111 \quad 000_2$$

$$CA = \underline{110 \quad 111 \quad 101 \quad 100_2}$$

$$AC \cdot CA = 000 \quad 101 \quad 101 \quad 000_2$$

The value of $AC \cdot CA$ replaces the old contents of the AC.

TAD (Two's-complement add).   Operation code $001_2 = 1_8$.

$$AC \, (+) \, CA \rightarrow AC$$

The addition is performed in the two's-complement mode; that is, the instruction implies that the numbers are 12-bit signed quantities represented in the two's-complement notation. If an arithmetic overflow occurs, the CPU toggles (complements) a special flag called the *link bit* (*LINK*).

ISZ (Increment and skip if 0).   Operation code $010_2 = 2_8$.

$CA \, (+) \, 1 \rightarrow (EA)$; then, if $CA \, (+) \, 1 = 0$, skip the next instruction; otherwise execute the next instruction.

This instruction is useful in controlling loop execution.

DCA (Deposit and clear AC).   Operation code $011_2 = 3_8$.

$$AC \rightarrow (EA); \text{ then } 0 \rightarrow AC$$

The contents of the AC goes into the specified memory location, then the AC is set to 0.

JMP (Jump).   Operation code $101_2 = 5_8$.

Jump to location with address EA for the next instruction.

JMS (Jump to subroutine).   Operation code $100_2 = 4_8$.

Store the address of the word following the JMS instruction (i.e., the return location) in the memory word with address EA. Then jump to the location with address $EA \, (+) \, 1$ for the next instruction.

The return location is the word after the JMS instruction. This instruction stores the return address in the first word of the subroutine and then jumps to the second word, which must contain the starting instruction for the subroutine. The normal entry to a subroutine X is thus with a JMS X, which saves the return address in location X. The normal exit from the subroutine is with a JMP *X (indirect jump through location X).

OP (Operate).   Operation code $111_2 = 7_8$. This is by far the most complex command in the PDP-8. It does not reference memory, so the address field bits are available for other purposes. The Operate instruction permits the following basic actions:

Clear accumulator: $0 \rightarrow AC$

Clear link bit: $0 \rightarrow LINK$

Complement accumulator: $\overline{AC} \rightarrow AC$

Complement link bit: $\overline{LINK} \rightarrow LINK$

Increment accumulator: $AC\ (+)\ 1 \rightarrow AC$

Rotate the concatenated accumulator and link bit right or left, 1 or 2 bit positions.

OR console switches with AC: $SR + AC \rightarrow AC$

Skip on various conditions of the accumulator or link bit.

Halt the computer.

Each of these operations is controlled by 1 or more bits in the address field of the instruction. These are sometimes called *microcoded instructions* or *microinstructions*. The programmer may invoke combinations of these microinstructions within one Operate instruction. There is a huge number of possible combinations; about 20 of these are useful to the programmer. These combinations of microinstructions ease the pinch of having only eight basic instructions in the PDP-8.

The operation code 111_2 occupies bits 0 through 2, as usual. Instruction bits 3 through 11 have individual functions. The Operate instruction on the PDP-8I is split into two groups, group 1 (G1) and group 2 (G2). Bit 3 specifies the group: in group 1, bit 3 = 0; in group 2, bit 3 = 1.

The format for group 1 is

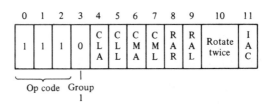

The meaning of the microcode bits in G1 is

Bit	Mnemonic	Name
4	CLA	Clear accumulator
5	CLL	Clear link
6	CMA	Complement accumulator
7	CML	Complement link
8	RAR	Rotate accumulator and link right
9	RAL	Rotate accumulator and link left
10	—	0 = 1-bit rotation; 1 = 2-bit rotation
11	IAC	Increment accumulator

The format for group 2 is

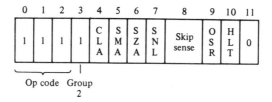

Op code   Group
2

The meaning of the microcode bits in G2 is

Bit	Mnemonic	Name
4	CLA	Clear accumulator
5	SMA	Skip on minus accumulator
6	SZA	Skip on zero accumulator
7	SNL	Skip on nonzero link
8	—	(specifies sense of skips; see discussion)
9	OSR	OR switch register into accumulator
10	HLT	Halt the computer

(In group 2 micro-operations, bit 11 is 0. On the PDP-8I, the condition of bit 11 is irrelevant, but some other models of the PDP-8 computer have another set of microinstructions, group 3, identified by bits 3 and 11, both of which are set to 1.)

To find the exact result of combining microinstructions, we must define the sequence in which the operations of each group occur. The PDP-8 describes the sequence in terms of *priorities*. There are four priority levels, 1 through 4: priority 1 operations occur before priority 2, and so on. The priority sequences of the micro-operations of G1 and G2 are

Priority	Group 1	Group 2
1	CLA CLL	Skips
2	CMA CML	CLA
3	IAC	OSR HLT
4	Rotates	

The group 2 "skip" microinstructions require further explanation. There are three conditions for skipping: SMA, SZA, and SNL. Bit 8 determines the skip mode. The operations are as follows:

If bit 8 is 0: a skip occurs if *any* of the chosen conditions is satisfied; otherwise, no skip occurs.

If bit 8 is 1: *no* skip occurs if any of the chosen conditions is satisfied; otherwise, a skip occurs.

IOT (input-output transfer).   The operation code is $110_2 = 6_8$. The PDP-8 has a primitive but adequate facility for the input and output of data. We will discuss the IOT instruction more thoroughly later; but now we will note how data enters and leaves the computer. Outgoing data (from the PDP-8 to the external world) comes from the AC. Incoming data reaches the AC by being ORed with the existing contents of the AC. There is a programmable facility for clearing the AC prior to accepting incoming data. Thus the basic input operations are

$$0 \rightarrow AC \quad \text{(optional)}$$
$$\text{Input.Data} + AC \rightarrow AC$$

The IOT instruction also permits the programmer to enable and disable the PDP-8's interrupt system. These IOT subcommands are ION (Interrupt System On) and IOF (Interrupt System Off), and have instruction bit patterns 6001_8 and 6002_8, respectively. The presence of interrupt commands alerts us to the need to investigate the interrupt mechanism.

Interrupts.   The PDP-8 specification requires that the machine be able to sense the presence of an external interrupt request. This request originates in some peripheral device and means that the device wishes to report an event of interest to the computer program. Any number of devices can request interrupt processing through this one external interrupt request line. When the PDP-8's interrupt system is activated, the computer monitors the interrupt request signal to see if any device needs servicing. If so, then at an appropriate time in the normal instruction processing cycle, the PDP-8 will force an automatic subroutine jump (JMS) to a fixed memory location (cell 0000). It is the programmer's responsibility to see that a valid subprogram for processing interrupts begins at location 0000. This subroutine is responsible for reading data from the peripheral device, writing data, or perhaps placing control information into the device. The characteristics of the device generating the interrupt determine what the interrupt subprogram must do. Therefore, the interrupt subprogram must determine which device is responsible for the interrupt and then perform actions tailored to that device. After servicing the interrupt, the subprogram will make a normal subroutine return through cell 0000 and processing of regular instructions will resume.

Interrupt requests originate from external devices running at their own pace, and may interrupt the program at any time. This is both a blessing and a curse to the programmer. Interrupt requests can occur whenever a peripheral device decides it needs service from the main computer. This is a potent programming tool, since the computer program need not waste time continually checking its peripheral devices to see if one needs service.

Interrupts are powerful; they are also tricky. The difficulty arises because interrupt requests originate from external devices and are therefore not reproducible. An interrupt may occur when the resident computer program is not prepared to handle it. For instance, suppose that the programmer has not established an interrupt service routine beginning at memory location 0000. Then the program will not run correctly if the computer recognizes an interrupt and jumps to location 0000. Even if the interrupt service program is present, it may not properly treat all the interrupt requests that may arise. These problems are difficult to diagnose, since the debugger of the program cannot reproduce the exact sequence of instructions that led to the difficulty. Interrupt programming requires much more foresight and care than conventional programming.

To allow more control over this difficult programming task, computers with interrupts always allow the programmer to *enable* (turn on) and *disable* (turn off) the computer's interrupt detection apparatus. The programmer may select those times when interrupt requests may result in the interruption of the program. Some computers permit the handling of several types of interrupts, each type having its own interrupt jump location. We will not pursue this subject, because our focus is on the PDP-8's interrupt capabilities.

The PDP-8 programmer may enable or disable the recognition of interrupts by using the ION (Interrupt System On) and IOF (Interrupt System Off) sub-commands of the IOT instruction. The PDP-8's hardware will automatically disable the interrupt system whenever an interrupt causes a jump to location 0000. This action is needed to give the programmer's interrupt service routine enough time to react to one interrupt without the danger of another interrupt occurring in the middle of the processing of the first interrupt. It is the programmer's responsibility to enable the interrupt system again at the proper time, to permit the detection of further interrupts. This gives rise to a subtle problem. The interrupt subroutine will normally leave the interrupt system disabled until it is time to return to the main (interrupted) program. At this time the interrupt subprogram must enable the interrupt system and return. The last two instructions of the subprogram are:

```
    .
    .
    .

ION        (Turn on interrupt system)
JMP *0     (Indirect jump to point of interruption)
```

We must make sure that we can execute the return jump to get back to the main program. Consider what would happen if an interrupt request is pending at the time the ION command is executed. The ION would reenable the interrupt system and the computer would immediately jump again to location 0000 *without* executing the jump instruction after the ION. The interrupt-forced JMS 0 causes cell 0000 to receive the address of the point of interruption—the address following the ION in this example. This act destroys the old return address in location 0000 which the unexecuted JMP *0 instruction wanted to use. The PDP-8's solution to this dilemma is to inhibit the recognition of interrupt requests for one instruction following an ION command, thus allowing the program the time to execute the crucial JMP *0 to return to the interrupted program before the computer recognizes any additional interrupt requests.

Interrupts are a complex feature of computers, and they place a heavy responsibility on the programmer. Whether or not the programmer does the job correctly, the computer must faithfully perform its assigned duty of detecting interrupt requests and forcing subroutine jumps to location 0000 whenever the interrupt system is enabled.

THE ARCHITECTURE OF THE LD20

We will now choose the main building blocks and data paths for the LD20. In order to be consistent with the principles set forth in Chapter 5, we plan for our design to be synchronous and static. We may make a few basic observations about the architecture:

(a) All major building blocks should be "out in the open." We wish to use building blocks of the right scale for our example. We could build everything from transistors or from AND, OR, and NOT circuit elements, as designers were forced to do a few years ago, but such approaches are far too detailed and would not teach you how to use higher-level building blocks. At the other end of the scale, it is possible to buy a complete PDP-8 processor in a single integrated circuit chip, and we would expect you to use such a chip in appropriate designs. However, a one-chip processor would be a poor vehicle for studying digital logic design, since all the interesting structure is buried inaccessibly within the chip.

(b) The control panel should continuously display all the major registers and control signals, so that the observer may see at a glance what is happening in the machine. With lamp drivers and LEDs (light-emitting diodes), the cost of displaying signals is minimal. Now is the time for the beginning designer to think about making designs that are easy to understand and debug.

(c) The control panel should provide ample mechanisms for the operator to control the progress of the machine. Synchronous, static design implies that we have ways to control the speed of the master clock, halt the clock entirely, and deliver manually produced clock pulses. The ability to reset or restart the machine is crucial in debugging. Since the basic unit of computer operation is the instruction, the ability to allow only one complete instruction at a time to be executed (single-stepping) is a powerful debugging aid. As creative designers, we will be eager to include such useful features in the control panel.

The Principal Elements of the Architecture

Our knowledge of the PDP-8 lets us specify certain architectural structures immediately.

Accumulator (AC). We will need a 12-bit register to store intermediate results.

Link. *LINK* is a 1-bit storage register used primarily to handle arithmetic overflow during addition to the *AC*. It also participates with the *AC* in shifting operations.

Memory (*MEM*). The PDP-8I has 4096 words of 12-bit memory. We will use standard RAM chips. The memory system requires two data items for its operation: a memory address and a memory buffer.

Memory address (*MA*). This is a natural part of any RAM, since the memory system must know where in its memory to perform a read or write operation. We will provide a register to hold the address—the memory address register *MA*.

Memory buffer (*MB*). Our selected memory presents read data on a set of output lines (*MEM*) for use in other parts of the computer. The memory must receive write data at a closely specified time in its cycle; this data is generated by other parts of the computer at earlier times. We need to hold the data until it is needed for the write operation, so we store it in a memory buffer register *MB*.

Control panel switches. Every PDP-8 contains a switch register *SR* of 12 toggle switches that allows the user to enter data into the computer manually. Calling these switches a register is unusual, since the data is stored in mechanical switches instead of flip-flops. Nonetheless, the switches are a memory device, and so we retain the nomenclature. The control panel also contains control switches such as *START, STOP,* and *CLEAR,* which will be described more fully in a later section.

Arithmetic logic unit (*ALU*). We must develop a building block to handle the various logical and arithmetic operations required by the PDP-8's instructions. The description of the PDP-8's instructions shows that the LD20 must perform the following operations on 12-bit data words:

Logical AND: from the AND instruction.

Logical OR: from one of the Operate microinstructions.

Logical NOT: from the Complement AC microinstruction.

Clear to zero: There are a number of times when a register must be cleared. The best way is to synchronously load the register with 12 bits of zero generated by the *ALU*.

Add: from the TAD instruction.

Increment: from the Increment AC microinstruction.

The *ALU* must be able to perform all these operations. We plan to use the 74LS181 Four-Bit ALU chip for this function, since it can perform all 16 logical operations as well as addition, subtraction, and incrementing. The 74LS181 is the only integrated-circuit chip that we specify at this time. Most of the actual selection of chips will be made much later in the design process. The only reason to fix on the 74LS181 now is to identify the likely features and limitations of our *ALU*. We care only about its functional characteristics, not about the details of how it is controlled.

Program counter (PC).   Every computer has a program counter that specifies the location of the next instruction to be executed.

Instruction register (IR).   The *IR* is another register found in every computer; it holds the current instruction for the duration of its execution.

Data Paths

We have identified major elements of the LD20's architecture by looking at the PDP-8's functions. Now that we have this set of elements, how do we put them together? In true top-down spirit, we will leave them scattered about on the desk and back off far enough to ask a question. How does the PDP-8 instruction set say they should be connected? Contemplating this question will lead us surprisingly close to the final architecture. We assume that the *ALU* has two 12-bit input paths and that it will handle all logical and arithmetic operations. Let's see how the PDP-8's instructions guide us to a model of the data paths among the building blocks.

(a) *IAC Increment Accumulator: AC (+) 1 → AC.* If the *ALU* does the addition, the *AC* and the *ALU* must be connected in a manner similar to Fig. 7–2. (In this and subsequent figures, the data paths are all 12 bits wide, and we call the output of the *ALU* the *ALUBUS*.) Execution of the IAC microinstruction must result in setting the *ALU* control lines to force the *ALU* to increment the input and place the result on the *ALUBUS*.

(b) *TAD Two's-Complement Add: AC (+) CA → AC.* This instruction requires an architecture like that in Fig. 7–3. Here the *ALU*'s control lines must make the *ALU* add its two data input quantities and place the result on the *ALUBUS*.

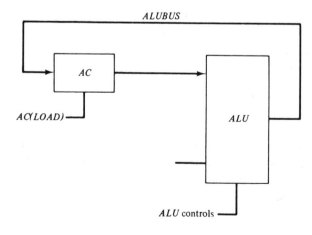

Figure 7–2 Data flow for incrementing the *AC*.

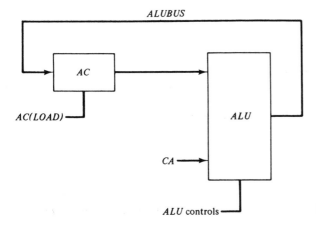

Figure 7-3 Data flow for addition.

(c) *AND: AC•CA → AC.* This instruction again leads to Fig. 7–3, where this time the *ALU* must perform the bit-by-bit logical AND of its data inputs.

(d) *OSR OR Switch Register: AC + SR → AC.* Both the switch register and the *AC* must be inputs to the *ALU*, as shown in Fig. 7–4.

(e) *Increment the program counter: PC (+) 1 → PC.* This operation is implied in any computer, since after one instruction is completed the next one will be executed. This *normal sequencing* will continue, instruction after instruction, until a programmed branch operation causes the *PC* to be set to a branch address. Normal (nonbranch) sequencing leads to Fig. 7–5, in which the *ALU* performs the increment operation.

(f) *Load the program counter:* This results from the branching case mentioned above and requires the operation *EA → PC.* At this point we could establish a private data path into the *PC* so that *EA → PC*; but look at the requirements of operations (a) through (e) that are leading to a common architecture. Examining those cases, we may reason as follows.

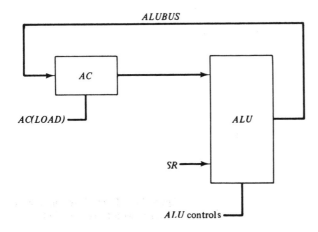

Figure 7-4 Data flow for the OR Switch Register (OSR) instruction.

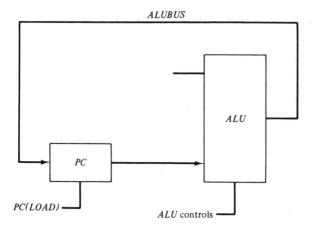

Figure 7–5 Data flow for incrementing the program counter (*PC*).

Developing the main bus structure. The AC seems to have pinned down one of the two data inputs that are a part of the *ALU* building block. The other input comes from a variety of sources: the contents of the effective address *CA*, the switch register *SR*, and the program counter *PC*, so we require a way of selecting the proper input to the *ALU*. Recall the discussions of data busing in Chapters 3 and 4. One good way to route several data sources to an output is to have three-state buffers on each source output. Enabling one of the source buffers lets that source "talk" to the *ALU*, as shown in Fig. 7–6. (Remember, all data paths are 12 bits wide.) This is an economical way to route the data. Some register integrated circuits include three-state output control; using such chips might eliminate some of the separate buffers in Fig. 7–6.

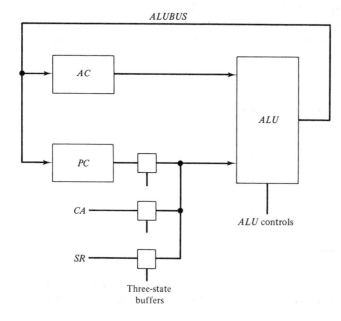

Figure 7–6 Controlling the input to the *ALU* with three-state buffers.

Another approach to the selection of the *ALU*'s input is to use the multiplexer building block to *select* the appropriate input (see Fig. 7–7). There is one multiplexer for each bit of the 12-bit data path. We have two good choices for busing the *ALU* input. To design the LD20, we have chosen the mux method of Fig. 7–7, primarily for its ease of debugging.

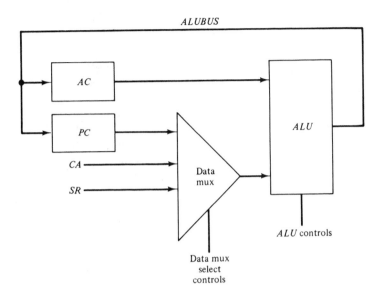

Figure 7–7 Selecting the input to the *ALU* with a multiplexer.

Our study of PDP-8 operations has led to this architecture. We may ask if the structure could be generalized to handle the case $EA \rightarrow PC$, which arises when the *PC* register is loaded with a branch address. The answer is, of course, yes: we add an *EA* input to the data multiplexer. To execute the branch operation, we must set the mux control inputs to select *EA*, and must set the *ALU* controls to pass this *ALU* input unchanged to the *ALUBUS*. If a load signal to the *PC* accompanies these operations, we would perform $EA \rightarrow PC$. It is significant that we have accomplished the goal by generalization rather than by specialization.

This is a discovery that we should try to carry as far as we can. We now have a mechanism for moving data from one register to any other, as well as a way of performing logical or arithmetic operations on the *ALU* data inputs. These capabilities are nearly all the structure we need to build a computer. Again, let us note that we were led to the architecture in a direct manner by a careful consideration of the fundamental requirements of the PDP-8's instruction set.

This is not the way some designers proceed. They start with a structure (a guess, really) based on a certain set of integrated circuit chips, and then try to bend the design to fit that guess. Sometimes the bending can require enormous leverage, when it could be side-stepped by redefining the architecture. Unfortunately, at that point it is too late; the designer's mind is already in a groove and it is difficult to jolt it out. If you let your target lead you to the architecture, you can avoid this groove until the last possible moment and save yourself anguish.

Adding memory. Now it is a simple matter to expand the tentative architecture of the LD20 to handle the remaining data transfers. We need to take care of memory and its interaction with the memory address and memory buffer registers *MA* and *MB*. The *MA* register gives memory the address at which to perform a read or write operation; therefore, *MA* must be wired directly to the memory. *MA* must be loaded with address data from various sources (e.g., *EA*), so we make *MA* a destination on the *ALUBUS*. So that data from the memory can reach other destinations, the output of memory connects to the data mux. Memory write operations require input from *MB* in addition to *MA*. The *MB* holds the write data and must be connected directly to the memory. The memory buffer register's input may come from a variety of sources yet to be specified, so we will make *MB* a destination on the *ALUBUS*. Figure 7–8 shows the proposed routing of the memory data.

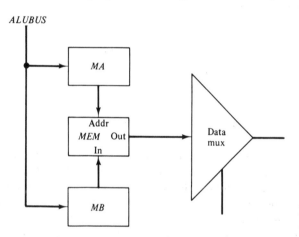

Figure 7–8 The data flow in the memory system.

Handling *CA*. This is a good time to investigate *CA*, the contents of address *EA*. *CA* is not a register; rather, it is a concept that we introduced to aid our understanding of the operations of PDP-8 instructions. Where do values for *CA* come from? Remember that *CA* = (*EA*): *CA* is the contents of the location referenced by *EA*. Thus *CA* comes from the memory. We have just proposed an architecture for reading and writing memory data. If we route an address *EA* into *MA* over the existing paths, then a subsequent memory read command will give us *CA*. Therefore, we may eliminate *CA* from our data mux inputs, realizing that our recent addition of the memory system incorporates the data movement for *CA*.

Fetching instructions. Can we use our basic data routing scheme to acquire the next instruction from memory and move it to the instruction register *IR*? The memory is already a source on the data bus, so we simply make the

IR a destination on the *ALUBUS*. At the time of instruction fetch, our design must select *MEM* through the data mux, cause the *ALU* to pass its input without modification, and then cause the *IR* to load the result from the *ALUBUS*.

Handling shift operations. Last, we must lay a plan for the data movements for the left and right shift instructions, which involve the *AC* and the link bit. One approach would be to include the shifting capability in the *ALU* building block. This would be acceptable design practice. On the other hand, one of our basic MSI building blocks is the parallel-in parallel-out shift register, which will shift or retain its value or load a new value upon command. Since the shift operations in the PDP-8 involve only the *AC* and *LINK*, let's make the *AC* a shift register in addition to its earlier assignment as a holding register. In this event it appears unlikely that the shift operations will affect the basic routing of the data.

The LD20 data bus. The initial proposal for the architecture of the data path is complete, and appears in Fig. 7–9. We have derived the structure from first principles based on the requirements of the PDP-8's instructions, and although we have used our knowledge of good building blocks we have not committed ourselves to any particular chip other than the 74LS181 ALU. This initial architecture turns out to be very close to the final requirements that will emerge from more detailed study.

It is now time to consider the LD20's control algorithm. We use our tentative architecture as a framework for developing the control; understanding the control will, in turn, lead us to a refined architecture. Throughout all this, we remain aloof from the actual hardware until we thoroughly understand both the architecture and the control.

A PRELIMINARY SKETCH OF THE LD20'S CONTROL

You have seen how a study of the PDP-8's instruction set led naturally to a tentative specification of the LD20's data storage and data paths. In this section we will write a control algorithm that will make the architecture execute the PDP-8's instructions. If we have chosen the tentative architecture well, we can write the control flowchart with only minor changes to the architecture. If we must make any major change to our structure because of the algorithm, we should start over and reconsider the architecture on the basis of our increased understanding of the problem. As you will see, our preliminary structure is a good one, requiring little revision.

First, we draw a preliminary flowchart of the control process. This broad treatment, although not detailed enough to serve as an implementation specification, nevertheless can help us understand the complex flow of operations involved in processing the instructions. Then, guided by our general description, we will develop a detailed ASM chart. It is at this point that we will find the ASM chart forcing small changes in the architecture. A complete ASM chart is the key to a good design. By its very nature, it requires us to think through our problem

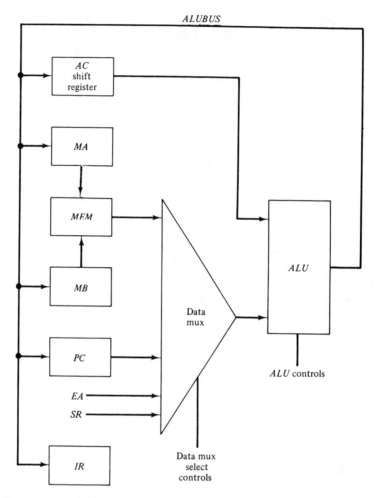

Figure 7-9 An initial proposal for the architecture of the main data path in the LD20.

completely before we touch a gate or a soldering iron. This is the time to avoid problems, and a good ASM chart is worth all the effort expended on it.

We have proposed a preliminary structure that can move data among registers, perform arithmetic and logical operations on the *AC*, and load the memory control registers *MA* and *MB* with data to support a memory read or write. The major commands required to accomplish these results are:

(a) *Data multiplexer select controls*. These signals select one of the data multiplexer input registers and present the data to the *ALU*.

(b) *ALU controls*. These signals tell the *ALU* what logical or arithmetic operation to perform on its inputs. The result appears as output on the data bus *ALUBUS*, for distribution throughout the machine.

(c) *Register load commands.* The *ALUBUS* distributes data to the registers of the machine. At any time, only certain selected registers must load this data. The loading operations are controlled by a load signal for each register.

(d) *Memory control.* The memory unit requires read and write commands, delivered in accordance with a standard memory control protocol.

The function of the control hardware is to generate a properly sequenced set of these commands to carry out the control algorithm. Before we plunge into an ASM chart formulation, we will express the control algorithm as a simplified diagram analogous to a programmer's flowchart, one which does not carry the detailed timing and signal information that will appear later in the ASM chart.

Fetch and Execute

At this stage, we are looking for ways to subdivide the control algorithm into manageable parts. In computer design, a common practice is to split the control flow into units for instruction fetch and instruction execute phases. *Fetch* is that portion of the control algorithm responsible for fetching the next instruction and getting the operands or operand addresses required to execute that instruction. *Execute* is that portion of the algorithm that carries out the operations required by a given instruction. Fetch calls Execute, which calls Fetch, and so on until the computer is halted. Figure 7–10 shows the gross flow of the control.

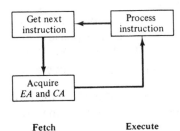

Fetch   Execute

Figure 7–10 The fetch and execute phases of the LD20.

Investigating the Fetch Phase

From our discussion of PDP-8 instruction codes, we may divide the LD20's instructions into three categories:

(a) Those that require neither *EA* nor *CA*. For example, a CMA (Complement AC) microinstruction needs no information other than the instruction word.

(b) Those that need an effective address *EA* but do not require *CA*. An example is a JMP instruction. We jump to the memory word at the *EA* but we do not care about the contents of that word.

(c) Those that require the contents *CA* of the effective address. An example is the AND instruction, which ANDS the *CA* and the *AC*.

The fetch unit in Fig. 7–10 is rather crude, since it always obtains an *EA* and a *CA*, whereas only a few instructions require *CA* and some instructions do not even need an *EA*. Can we expand the fetch phase to take advantage of these facts? If the instruction obtained from memory is CMA, for example, we could go to the execute phase as soon as we know that the instruction code did not require *EA* or *CA*. We modify Fig. 7–10 into the more efficient Fig. 7–11.

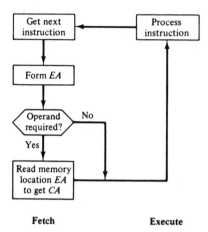

Figure 7–11 Saving time in the fetch phase.

The CMA microinstruction takes the "No" branch in Fig. 7–11. Another example is the DCA instruction. The location at the effective address *EA* will receive the contents of the *AC*. The contents of location *EA* (in other words, *CA*) will be destroyed, so the execution phase will not need *CA* from Fetch. Examples of instructions that will take the "Yes" branch in Fig. 7–11 are AND and TAD; these instructions operate on the *contents* of the *EA*, which is *CA*.

Forming *EA* and *CA*.   We must explore the effect of the PDP-8's memory accessing arrangements on the formation of the effective address *EA*. Recall that a PDP-8 memory referencing instruction has, in addition to the operation code, a page bit, a 7-bit page offset, and an indirect addressing bit. When the indirect addressing bit is 0, we may generate the *EA* as soon as we know what the operation code is. If the indirect addressing bit is 1, the formation of *EA* requires an additional memory access.

If the page bit is 1, the page address is the page that contains the instruction. If the page bit is 0, the page address is page 0. We expect to derive knowledge of the current instruction's location from the program counter *PC*. We may specify the 5-bit address as

$$PAGE.ADDRESS = (PC0-PC4) \cdot PAGE.BIT \qquad (7-1)$$

The high-order 5 bits of *PC* specify the page of the current instruction. If the page bit is 1, we wish to use these 5 bits of *PC* as our page address. If the page bit is 0, we need 5 zero bits for the page address. Equation (7–1) specifies that the page address is the logical AND of the page bit with each bit $PC0$ through $PC4$.

For direct addressing, the 12-bit effective address EA is the 5-bit page address concatenated with the 7-bit page offset:

$$EA = PAGE.ADDRESS \quad PAGE.OFFSET \qquad (7-2)$$

If the instruction requires an operand, we must access the memory at location EA to get CA. Instructions with indirect addressing require another iteration to obtain the final EA and CA. Thus, if the instruction requires an operand or if it specifies indirect addressing, we must obtain the contents of the memory location EA. We may think of the contents of EA as a tentative CA. For direct addressing, this is actually the CA we seek. For indirect addressing, this tentative CA becomes the new EA, so that the indirect process may continue. Figure 7–12 shows the result of this analysis.

Auto indexing requires a further addition to our general fetch flowchart. If the acquisition of an indirect EA makes use of one of the auto index locations

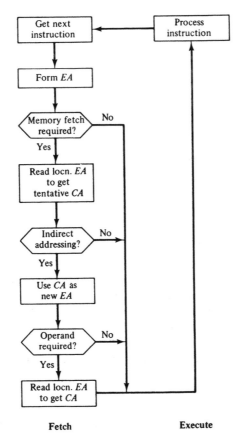

Figure 7–12 An LD20 with indirect addressing.

10_8 through 17_8, we must increment the contents of that location before we use it as an indirect address. This activity is shown in Fig. 7–13.

You can see that, little by little, as our understanding of the control process grows, the general flowchart becomes more complex and our notations become more succinct. This is characteristic of work on both hardware and software: as understanding grows, the detail increases and the notation is compressed.

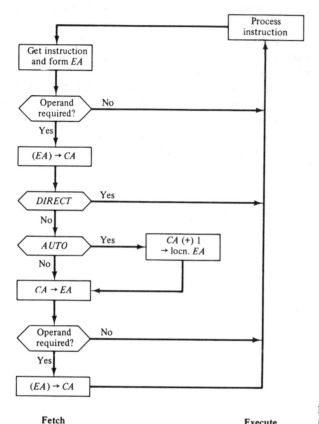

Fetch　　　　　　　　　　　　　　　Execute

Figure 7–13 An LD20 with auto indexing.

Halts and interrupts. The LD20's flowchart must make provision for halting and for detection of interrupts in a systematic way. A halt can result either from manual intervention or from the execution of the HLT microinstruction. In either case, the machine must move to an idle state whose purpose is to monitor the manual start switch. Halt processing must wait until the machine has finished the execution of the current instruction; the natural place to look for a halt is therefore at the start of the next fetch cycle.

This is also the natural place to examine interrupt requests. If the LD20 were to process an interrupt request at the instant it arose, chaos would result, since the LD20 would no doubt be in the middle of processing some instruction of a program. The proper time to handle interrupts is after an instruction is executed but before the computer fetches the next instruction from memory. Thus we will add a section for handling interrupt requests to our flowchart. Figure 7–14 shows these additions.

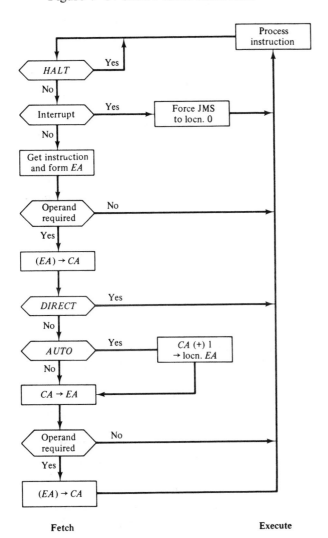

Fetch   Execute

Figure 7–14 A preliminary flowchart for the LD20.

The fetch portion of our preliminary flowchart for the LD20 is now complete. Since Fetch involves some rather complex steps (to produce the proper *EA* and *CA*), it was useful to undertake this preliminary study of the control algorithm. The execute phase of the LD20's control algorithm consists primarily of well-separated sections for each instruction, and we will be able to handle these when we produce the actual ASM chart, without needing a detailed preliminary analysis.

DEVELOPING AN ASM CHART FOR THE LD20

Figure 7–14 is a high-level flowchart of the algorithm for emulating the PDP-8. We must reduce it to a more detailed flowchart that recognizes the architecture developed earlier in this chapter as well as the timing of the individual operations of the hardware. The ASM chart is the appropriate way to show these relationships.

In this section we will develop the ASM chart for the LD20 in detail since it is the most important step in reaching a good final design. We cannot over-emphasize the importance of a good control algorithm. This portion of the design should occupy at least half of the time devoted to developing the digital system; every minute will be well spent. The design of the LD20 has been through at least five iterations and has had several hundred hours of thought devoted to it. You may think this excessive, but it isn't, for design errors in the hardware can be very hard to correct after a system is completed. It is far better to produce the best possible design before construction begins.

We will develop each segment of the LD20's ASM systematically and many of the remaining figures in this chapter show fragments of the complete algorithm. The entire ASM for the LD20 appears in three charts (Figs. 7–42, 7–43, and 7–44) toward the end of the chapter. You may wish to refer to these diagrams as you study the details in order to enhance your understanding of the inter-relationship of the parts.

Conventions

We will adopt the conventions developed in Chapter 6 for naming states and any conditional output terms associated with them. The symbolic names for the states of the fetch phase of the LD20 are **F1** through **F7**, and for the execute phase are **E0** through **E5**. The first fetch state of each instruction is always **F1**, and the first execute state is always **E0**. Some of the states have many conditional outputs. We would like the names of the conditional output terms to reflect the parent state of the term. Our convention is simple. We give each conditional output term within a state a number and use a period to concatenate this number to the parent state name. Thus E0.15 is conditional output term 15 of state **E0**. A conditional output term may produce several output signals.

Occasionally, we will need to label other points or sections of the ASM chart, so that we can identify them in different figures. Our convention is to give the state's name, then a period followed by a nonnumeric name. For instance, the execute phase is divided early into two parts, one for manual operations, and one for normal processing of the instruction register; we call these parts E0.MAN and E0.IR, respectively.

The main purposes of the fetch phase are to get an instruction from memory, to prepare the effective address *EA* and its contents *CA* as required by the instruction, and then to go to the execute phase to process the instruction. We need a uniform mechanism for passing the instruction word, the *EA*, and the *CA* from the fetch phase to the execute phase. Obviously, the instruction should go into the instruction register *IR*. The memory address register *MA* is a good spot to place the address *EA*. What about *CA*?

From the discussion of the LD20's tentative architecture, you recall that *CA* comes from the memory word at location *EA*—the word whose address is in the *MA* register. During the fetch phase, we will perform the memory read operation to get *CA* for later use in the execute phase. Where can we put *CA* until it is needed? We could add another register to the data architecture for this purpose, but perhaps we already have a satisfactory intermediate register. The memory buffer register *MB* is promising, and we adopt the standard convention that *CA* appears in register *MB* when we enter the execute phase. Notice that in Fig. 7–9 the *MB* is not connected to the data mux. If the *MB* is to be the source of *CA*, we must add *MB* to the data mux inputs. Our general data-routing scheme allows us to make this modification of the architecture immediately, without elaborate analysis.

Memory Control

It will be useful to define a protocol for memory control before we embark on the detailed development of an ASM for the LD20. We want to treat memory as a black box that will behave in a reliable and well-defined way when it is presented with proper input commands and data.

In Chapter 4 you saw how typical memory building blocks behave. Memory integrated circuit chips operate at their own pace, asynchronous to outside events. Our design goals for the LD20 allow the system clock to operate at any reasonable rate. Depending on the clock speed and on the requirements of the particular memory device, the memory may be slower or faster than the LD20 clock. We must develop a set of rules—a protocol—according to which the LD20 will use the memory. We hope that these rules will be general enough to allow the use of memory independently of clock speeds and particular memory chips. What control signals and data are necessary to deal with a memory system? As far as the memory user (the LD20) is concerned, the memory box should have three elements for handling data:

    (a) A memory address input.

    (b) An input for memory write data.

    (c) An output for memory read data.

To control the memory's activities, the box should have:

    (d) An input command to start read operations.

    (e) An input command to start write operations.

    (f) An output that announces when the memory operation is finished.

In addition to these specific signals, the system clock is available, if it is needed. Figure 7–15 shows the black box for the memory.

At this point, we *assume* that the box exists, and we establish the protocol for using it in the ASM for our LD20. We will worry about the actual design of the box later. This modular approach to the problem of controlling the memory is an important aid to clear design, since we may put aside the intricacies of

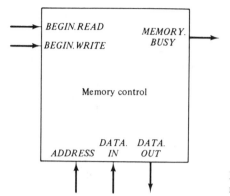

Figure 7–15 A control unit for the memory of the LD20.

manipulating particular chips, knowing that the memory controller's interface with the outside world must conform to the established protocol.

In Fig. 7–9 we specified the data interconnections between the LD20 and its memory. The address input to the memory box comes from *MA*, the *MB* register provides memory-write data, and the memory's data output feeds into the data mux. There remains to specify the behavior of the control and status lines of the memory box. We will use *START.READ*, *START.WRITE*, and *BUSY* for the LD20 signals interfacing with the corresponding inputs and outputs of the memory box. Figure 7–16 shows the resulting interface.

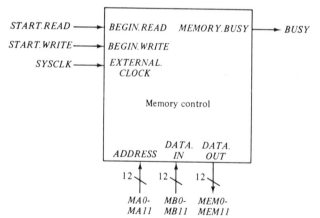

Figure 7–16 The LD20's interface with its memory control unit.

We initiate a memory read by asserting *START.READ* during some state. The memory controller senses this signal and begins the read operation on the next LD20 clock pulse. Thus read begins at the transition from the "start read" state. When the read operation begins, the memory–busy status signal *BUSY* must become true, and the controller must keep the signal true at least until the operation is complete. For convenience, we will require that changes in the *BUSY* signal be synchronous with the LD20 clock. The transition from *BUSY*

= T to *BUSY* = F will occur at the next clock pulse after the memory box determines that the memory cycle is complete. The write operation, initiated by *START.WRITE*, behaves analogously.

The completion of a write operation implies that the memory box no longer needs the address and data inputs. Completion of a read operation means that the memory has had time to produce valid data, based on the stable address supplied to the box. As long as the address remains the same (or until another memory operation begins), the output data is valid.

Standard memory read and write sequences. The typical memory read sequence based on the memory protocol appears in the ASM chart of Fig. 7–17. The first state prepares to form a stable memory address in *MA*, and prepares to initiate the read operation. The second state loops until *BUSY* becomes false. We then exit that state while loading the now-valid memory data into a register. The third state is the earliest time that the contents of the register are available for use. This is a perfectly valid algorithm for reading a memory, and we could use it unhesitatingly in the LD20 ASM. It has the disadvantage that the memory activity is spread out over three states. If the design requires many accesses to memory, our ASM will contain quite a few states and will appear cluttered and unwieldy. Can we conveniently collapse the presentation of the algorithm into a more compact form? If we could arrange to wait upon, load, and then use the memory's contents in a single state, we may achieve some improvement. The following reasoning is useful.

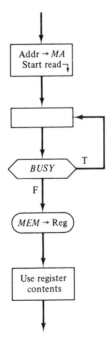

Figure 7–17 An ASM for a memory read.

We introduce a signal *BUSY.HELD* that behaves like *BUSY* except it remains true one clock cycle longer. The purpose of this signal is to help identify the first state time after *BUSY* becomes false; only then is *BUSY* = F while *BUSY.HELD* = T. This is the time when we wish to load the stable memory output into a register. Any time after this we may use the register contents. This fancy footwork allows us to express the memory read sequence in only two states, as illustrated in Fig. 7–18.

We have reduced the number of states but have not improved the readability of the ASM sequence. On the contrary, it is more obscure than before. If we notice that it does no harm to load invalid memory data into the register as long as we finish by loading valid data, we may entirely eliminate the test on *BUSY*, leaving the single test signal *BUSY.HELD*, as shown in Fig. 7–19.

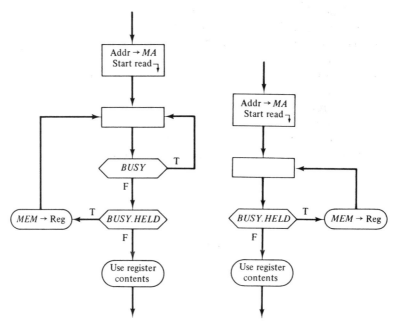

Figure 7–18 Reducing states in the ASM for memory read.

Figure 7–19 Improving the ASM for memory read.

Last, as a matter of personal style, we choose to cast the remaining test in terms of the concept of Cycle Complete (*CC*), which in this treatment is the inverse of *BUSY.HELD*. Figure 7–20a is the final form of a memory read sequence. The somewhat simpler sequence for memory write appears in Fig. 7–20b.

This is a general read–write protocol. The requirement on *CC* is that during the *entire last clock cycle* before *CC* became true, the memory operation was complete. We may now use the protocol with confidence, with all the detail packaged in the memory box and the associated *CC* circuit.

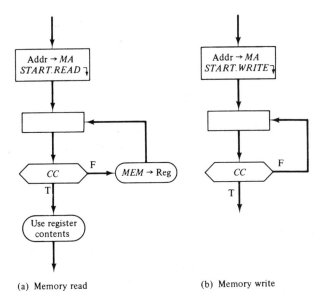

(a) Memory read   (b) Memory write

Figure 7–20 ASMs for the LD20's memory.

State F1 (Fig. 7–21)

Halt detection. State **F1** is the first state of every fetch cycle, and is a good place to check for a halt condition. The LD20 must remember halt requests that arose during the processing of the last instruction, so we will add a halt-status flip-flop *HALTFF* to the architecture. (At this time we do not care exactly what type of chip we use for the flip-flop, although we suspect it will be a JK since the ASM will set and clear the halt status at various points in its processing.) If a halt request occurred, the ASM will have set *HALTFF* to T, and we can

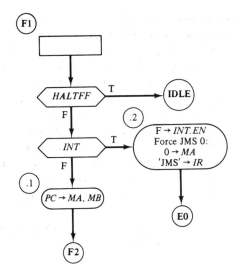

Figure 7–21 ASM of state **F1**.

test it in state **F1**. If *HALTFF* is true at this time, the ASM should move to an idle state **IDLE**, which will inhibit further processing of instructions. If *HALTFF* is false, the fetch phase proceeds.

Detecting an interrupt. State **F1** is the proper place to test the status of the interrupt request system. Using the commands ION (6001_8) and IOF (6002_8), the programmer may enable or disable the interrupt system. To record the current enabled or disabled condition, we need a memory element; this addition to the LD20 architecture is the *INT.EN* (Interrupt Enabled) flip-flop. Let *INTREQ* be the interrupt request signal coming into the LD20, and let *INT* represent a detected interrupt signal. Then

$$INT = INTREQ \cdot INT.EN \cdot \overline{ION}$$

The term $\overline{ION}$ ensures that interrupt recognition will be inhibited immediately after an ION command, in conformity with the PDP-8's instructions. At the time the ASM enters state **F1** to start processing a new instruction, the instruction register *IR* still contains the previous instruction. If that instruction was an ION, we must ignore any pending interrupt request and instead allow an additional instruction to complete its execution. The equation for interrupt recognition means that *INT* is true only if *INTREQ* is true and *INT.EN* is true and *ION* is false. *INT* is interrogated in state **F1**, and if it has the value T at that time, interrupt processing takes place.

If *INT* is true, we must force a subroutine jump to location 0000 (JMS 0) and disable the interrupt system. Jump commands require an effective address *EA*, and the execute portion of the ASM expects to find the *EA* for a JMS in the memory address register and the operation code in the instruction register. Therefore, the proper response to interrupt recognition is shown in the conditional output term F1.2 in Fig. 7–21.

A style-motivated digression. Term F1.2 changes the *MA* and *IR* registers. We ask ourselves if there is any condition which would make us regret destroying the old values of *MA* and *IR*. As long as the machine is running at full speed we would never notice, since the overall processing of the interrupt would be correct—the changes in console displays for *MA* and *IR* would last only a few microseconds. The only time we might notice a change is when the machine is stopped, when its clock is operating at slow or manual speed, or when the single instruction switch is on and we are executing individual instructions. Suppose an interrupt request arose during the execution of a halt instruction. In that case *HALTFF* would be true upon the next entry to state **F1**. In Fig. 7–21 we see that the *HALTFF* test occurs before the *INT* test in state **F1**, so the machine will go to **IDLE** without interrupting, with the old instruction in *IR*; there is no problem with this behavior.

In manual clock mode, or at slow clock speeds, we are presumably checking for correct machine operation or debugging a newly constructed machine. Then we would want to see all the detailed micro-operations of the hardware, including any interrupt processing, so our ASM is appropriate for this situation also.

The remaining condition arises when the single instruction switch is on and we are single–stepping through the program when the interrupt occurs. But again, the *HALTFF* flip-flop would be true, causing the machine to return to **IDLE** after each instruction, with the old values of *MA* and *IR* preserved. Thus we find that we may jam 0 into *MA* and the JMS operation code into *IR* bits 0–2 without the loss of useful information. This digression points up the desirability of thorough study of a design in its conceptual stages; now is the easiest time to detect and correct design flaws.

Normal processing of instructions. If *HALTFF* and *INT* are both false, normal instruction processing will proceed and the ASM must prepare to get the next instruction from memory. In state **F1**, the program counter *PC* contains the address of the next instruction. Since we have to perform a read from memory at that location to get the next instruction, in F1.1 we transfer the contents of the program counter to the memory address register, in preparation for the read. Later in the fetch phase the memory data becomes the current instruction. The effective address *EA* (which may be needed by this new instruction) is formed by concatenating a 5-bit page address with the 7-bit page offset. For this purpose, the *EA* computation may require *PC* bits 0 through 4, so we must preserve *PC* until *EA* is formed. But the formation of *EA* must await the arrival of the new instruction from the memory. Normal instruction sequencing dictates that *PC* be incremented by 1, and we wish to do this in the fetch phase so as not to waste time in the execute phase. (The execution of a branch instruction will later replace the program counter's value with a branch address.) Now we have a subtle problem: We must hasten to increment *PC*, yet we must save *PC* for possible use in forming *EA*. *PC* bits 0–4 may be different in the incremented *PC* than in the current *PC*. (Why?) Our solution is to look for a temporary place to hold the current value of the *PC* until *EA* is formed. The memory buffer register *MB* will serve. Until this point, *MB*'s only role has been to hold data to write into memory and to hold *CA* for instruction execution. Neither of these is occurring now, so *MB* may hold the old *PC* awhile. F1.1 moves *PC* to *MB*.

State F2 (Fig. 7–22)

Incrementing the program counter is an operation common to all one-address computers. Such machines assume that the next instruction will come from the memory location immediately following the current instruction, unless there is a jump instruction. We can achieve the normal sequential case by incrementing *PC* in state **F2**. We also initiate a memory read cycle to obtain the next instruction. *MA* contains the address, placed there in state **F1**. The downward arrow in Fig. 7–22 implies that the start-read operation is to begin at the *end* of the state (like a register load or a flip-flop set).

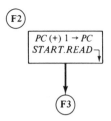

Figure 7–22 ASM of state **F2**.

State F3 (Fig. 7–23)

This state serves two main purposes:

(a) To wait for the memory to finish the read process started at the end of state **F2**, and load the memory data into the instruction register.

(b) To generate the effective address *EA* from the instruction newly placed in *IR*.

We use the standard protocol for controlling the memory read operation: The ASM loops in state **F3** until the cycle complete signal *CC* becomes true. During the last loop with *CC* = F, the memory data copied into *IR* at term F3.2 comprises the new instruction.

Once the new instruction is in IR, control will pass through term F3.1. Here we generate *EA* for the new instruction by concatenating the 5-bit page address with the 7-bit page offset, according to Eq. (7–2). The program counter

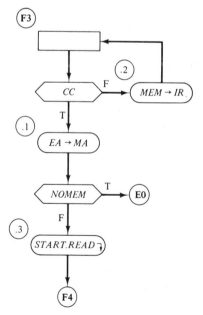

Figure 7–23 ASM of state **F3**.

The Art of Digital Design   Part II

PC no longer points to the current instruction, since state **F2** incremented the *PC* register. However, we saved the old value of *PC* in the *MB* register, so we can calculate *EA* by

$$EA = [(MB0{-}MB4)\cdot IR4] \overset{\frown}{[IR5{-}IR11]} \qquad (7\text{--}3)$$

(Remember that if *IR4* is 0, the first 5 bits of *EA* are 0.) Since in the execute phase the effective address is to be found in the memory address register, we move *EA* to *MA* in F3.1.

Early exits to Execute. We now have an opportunity to be clever by noticing that certain instructions need no more fetch-phase processing. Two such instructions are IOT and OP, which do not reference the memory. The instruction bits that would normally be treated as a part of a memory address have a different meaning. All the information that Execute needs is already in the instruction register, so we may immediately branch to the execute phase.

Three other instructions are possible candidates for this early exit to Execute: the directly addressed versions of DCA, HMP, and JMS. These instructions do something *to* or *at* a memory location, but they do not require the *contents* of that location. For example, a JMP X instruction will jump to location X regardless of the value stored in X. Remember the meanings of *EA* and *CA*:

EA: The target address of the operand after any indexing or indirect addressing is done.

CA: The contents of the word at the target address: $CA = (EA)$.

The PDP-8 does not have an index register. If the instruction calls for direct addressing (bit *IR3* is 0), then we are already moving the final *EA* into the *MA* register. For the directly addressed DCA, JMP, and JMS instructions, we may exit immediately to state **E0**. Since the condition for branching to **E0** involves several logic terms, we find it convenient to define a new variable *NOMEM* (No Memory Reference) to express the opportunities to exit to **E0** at this point:

$$NOMEM = OP + IOT + DIRECT\cdot(DCA + JMS + JMP)$$

where

$$DIRECT = \overline{IR3}$$

We may use *NOMEM* in the ASM chart for state **F3** as a compact representation of the **E0** branching condition.

If *NOMEM* is true, the ASM goes to state **E0**. If *NOMEM* is false, the instruction in *IR* requires further memory accesses, so we may initiate a read from the memory in term F3.3 as we exit from state **F3**. The memory read will use the newly established value of the memory address register.

State F4 (Fig. 7-24)

Since state **F3** initiated a read operation, state **F4** must observe the memory-read protocol and loop on *CC*, capturing the memory's output. We are reading an operand at this point, so the memory buffer register *MB* is the correct place to put the memory data, in anticipation that this operand will be required by Execute. This activity occurs in term F4.1 of Fig. 7-24.

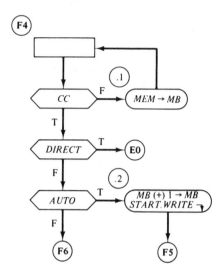

Figure 7-24 ASM of state **F4**.

If the instruction involves direct addressing, we have just acquired the necessary operand *CA* and we may go immediately to state **E0**. This case arises when there are directly addressed AND, TAD, and ISZ instructions. We have already defined the variable *DIRECT* as $\overline{IR3}$; if *DIRECT* is true, we now exit to state **E0**. In order to continue, all remaining fetch instructions have *DIRECT* = F (*IR3* = 1).

We are now on a path for indirect addressing, since *IR3* = 1. We must test to see if we are referencing one of the auto index memory locations, which have addresses 10_8 through 17_8. If a PDP-8 instruction accesses one of these locations for use as an indirect address, the location is automatically incremented and put back into the memory before it is used for indirect addressing. To keep the ASM chart compact, we introduce a variable *AUTO* that is true if the effective address *EA* specifies one of the locations 10_8 through 17_8. These addresses have bits 0-8 in common, whereas bits 9-11 select the particular address. Since the effective address is in *MA* at this time, the equation for *AUTO* is

$$AUTO = \overline{MA0} \cdot \overline{MA1} \cdot \overline{MA2} \cdot \overline{MA3} \cdot \overline{MA4} \cdot \overline{MA5} \cdot \overline{MA6} \cdot \overline{MA7} \cdot MA8$$

If *AUTO* is true in state **F4**, auto indexing must occur. The ASM must increment *MB*, which holds the contents of the auto index location, and also initiate a memory-write operation. The *MA* register still points to the auto index location and remains unchanged during the write operation. The auto indexing preparation takes place in term F4.2; the ASM must then move to a state (**F5**) that will wait for the completion of the write.

If no auto indexing was associated with the indirect addressing, the ASM exits from state **F4** into state **F6**, where indirect processing continues.

State F5 (Fig. 7-25)

This state completes the auto indexing operation and then merges into the main fetch stream at state **F6**. Auto indexing appeared as a "side operation" from state **F4**, and it has no further influence on indirect addressing.

State F6 (Fig. 7-26)

We enter this state only in the presence of a memory referencing instruction involving indirect addressing. Indirect addressing requires that we use the current Contents of Effective Address *CA* (held in *MB*) as the final Effective Address *EA*. Thus state **F6** moves *MB* into *MA*. If the instruction is of the type DCA, JMS, or JMP that does not use the contents of *EA*, we may again exit immediately to state **E0**. The surviving instructions are indirectly-addressed AND, TAD, and ISZ. They require memory contents as operand data, and so we must start one last read operation and move to a final fetch state **F7** to complete the read.

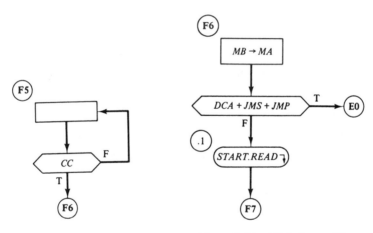

Figure 7-25 ASM of state F5.   Figure 7-26   ASM of state F6.

State F7 (Fig. 7-27)

In state **F7** we complete the acquisition of the necessary *CA* data for indirectly addressed instructions. The contents of the memory go into the *MB* register, as required by our conventions for the entry into the execute phase of the LD20's ASM.

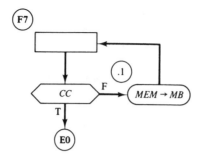

Figure 7-27 ASM of state **F7**.

State IDLE

In many respects, **IDLE** is the most important state in the machine; without it we would have no way to halt or start the computer. The function of **IDLE** is to scan the control panel switches for activity. When it occurs, **IDLE** must pass control to an appropriate section of the ASM. When the machine is in **IDLE**, the LD20 will loop in that state as long as no switch is depressed.

Control panel switches. Let's consider what control switches we desire on the control panel.

CONT (Continue): This is the start switch. When the **IDLE** ASM senses that *CONT* = T, it will branch to the fetch phase to continue routine processing of instructions.

Loading the registers: Our design calls for the contents of each main register to be continuously on display. We also desire to be able to load the contents of each register from the control panel's switch register *SR*. We provide six switches for this purpose:

LDMA (Load memory address register from *SR*)

LDMB (Load memory buffer register from *SR*)

LDPC (Load program counter register from *SR*)

LDIR (Load instruction register from *SR*)

LDAC (Load accumulator register from *SR*)

LDMEM (Load the memory word at address *MA* from *SR*)

These load switches are not part of the commercial PDP-8 but are a great convenience to the LD20 designer during implementation and to the LD20 programmer who is debugging programs and operating the computer.

Two more switches have counterparts in the PDP-8:

DEP (Deposit) is used to load the panel switch register *SR* into sequential memory locations. Since it advances *MA* after each memory load, *DEP* is useful for entering programs or other data into consecutive locations.

EXAM (Examine): The currently addressed word is always on display in the memory's panel lights. *EXAM* advances the memory address register *MA*, allowing the operator to view the next word. This is of value for examining the contents of consecutive memory locations.

Three more switches complete the panel controls:

CLR (Clear) fulfills a PDP-8 requirement that the operator be able to clear the link bit and the interrupt enable flag from the control panel.

RESET is the master reset switch. Its function is to force the LD20 into the **IDLE** state, regardless of the present condition of the machine. This operation is useful for starting the LD20 and, during debugging, for escaping from any hangup that may arise from improper design or implementation, or from malfunctioning components.

SING.INST (Single Instruction): In its normal operating mode, when *SING.INST* is false, the LD20 processes a continuous sequence of instructions. When *SING.INST* is true, the LD20 halts after processing each instruction. Whereas the other control switches are "momentary–on" pushbuttons which return to the false condition after the button is released, Single Instruction is a toggle switch that has stable false and true positions.

The **IDLE** state examines the condition of all the LD20 control switches except *RESET* and *SING.INST*. State **E0** processes *SING.INST*; the action of the *RESET* switch is asynchronous to the LD20's clock.

The other switches are also asynchronous to the LD20's system clock, but we will insist that our design handle them within the synchronous ASM structure that we are now developing. The output of each of the switches will have the ∗ appended to its name to indicate its asynchronous nature.

IDLE operations. To specify the operation of the **IDLE** state, it is useful to define a signal *MANSW*∗ that is a composite of 10 manual pushbutton signals:

$$MANSW* = LDMA* + LDMB* + LDPC* + LDIR* + LDAC*$$

$$+ LDMEM* + DEP* + EXAM* + CLR* + CONT*$$

In the **IDLE** state *MANSW*∗ can assist in detecting a depressed switch. By now you should be sensitive to our two old problems: testing asynchronous signals and processing a signal from a pushbutton more than once. We can coordinate **IDLE**'s scan of the pushbuttons with the operator's push of the button by synchronizing *MANSW*∗. This will synchronize the pushbutton operation with the LD20. If the operator keeps the button down until the operation is completed, we will not need to synchronize each individual switch signal.

We must also arrange to handle each depression of a switch just once. For practice, you might consider what would happen if depressions of the *DEP*, *EXAM*, or *CONT* pushbuttons were acted upon more than once. Of the several schemes for dealing with the problem, we choose to add a single pulser to the architecture to isolate the first whole clock period in which *MANSW*∗ is true. As we showed in Chapter 6, the single pulser provides for synchronizing the input and delivering truth on the output for only the first clock period after the input becomes true. In the present case, the input to the single pulser is *MANSW*∗

and the output is *MANPULSE*. **IDLE** can test *MANPULSE*, and the operation appears in the ASM in a compact form.

With *MANPULSE* true, **IDLE** will go to state **F1** if *CONT∗* is also true, and will go to state **E0** otherwise. The execute phase therefore not only controls the execution of instructions, but also provides for manual operations. At various points, the execute phase will depend on whether entry to state **E0** was from Fetch (normal instruction processing) or from **IDLE** (manual operations). We provide a storage element *MANEX* (Manual Execution) in the architecture that will be true in Execute only if state **E0** was entered from **IDLE**. *MANEX* protects the execute phase from improper pushbutton depressions, since Execute will respond to manual signals only if *MANEX* is true.

The **IDLE** state performs one further function: In the normal process of halting the LD20, state **F1** finds *HALTFF* = T and branches to **IDLE**. Since *HALTFF* has served its purpose, **IDLE** must reset the flip-flop to false.

Figure 7–28 shows the ASM for state **IDLE**.

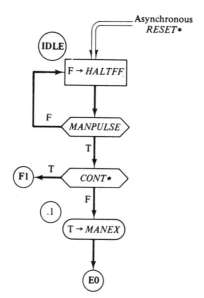

Figure 7–28   ASM of the **IDLE** state.

Execute-Phase Processing

The execute phase processes program instructions and manual operations. Within each category, we must perform a many-way branch to accomplish the processing for each type of command. In Fig. 7–29 we show the structure of the execute phase. Each square box represents the sequence of actions required for that operation. Our task is to convert this general flow diagram into an ASM chart.

State E0: The Single–Instruction Switch

The main state in Execute is **E0**; it is rather complex because each of the 17 different operations begins (and many also end) in this state. In addition to the

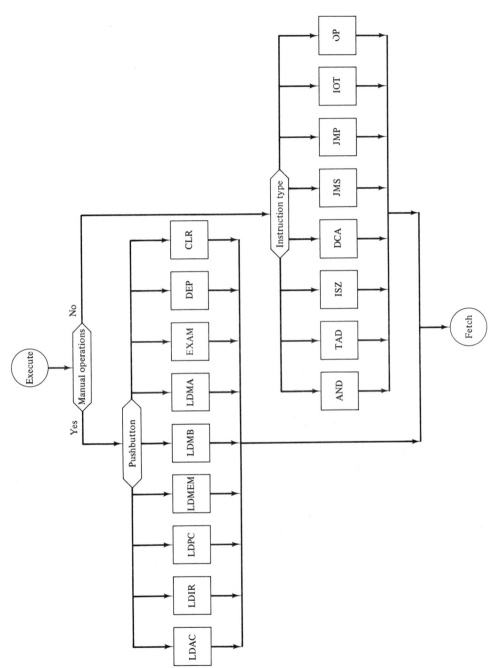

Figure 7-29 The execute phase of the LD20. (This is not an ASM chart.)

individual operations, some Execute activities are common to all the operations. The principal common activity is the examination of the Single Instruction Switch, a switch that allows the operator to select either continuing instruction processing or a halt after the execution of each instruction. The signal emerging from the debounced toggle switch is *SING.INST∗*. Before testing this signal, we must synchronize it with the LD20 system clock. We use a clocked D flip-flop with input *SING.INST∗* and output *SING.INST*. If *SING.INST* is true, then state **E0** sets the halt flip-flop *HALTFF* to true; otherwise, state **E0** will set the halt flip-flop to false. In other words, *HALTFF* receives the value of *SING.INST* during state **E0**. Setting *HALTFF* to true will not stop the machine immediately, since the ASM does not examine the value of *HALTFF* until state **F1** of the next fetch phase. If *HALTFF* is true at that time, the ASM will jump to **IDLE** rather than continuing to process instructions routinely. There is no need for a separate stop switch on the LD20, since it would duplicate the effect of the single instruction switch.

State E0: Detection of Manual Operations

The execute phase is entered from a fetch state or from **IDLE**. Entering **E0** from Fetch implies normal instruction processing; entering from **IDLE** must cause processing of the relevant manual switch. For calls from **IDLE**, the manual execute flip-flop *MANEX* is true, having been set in IDLE.1 so that Execute can determine the correct type of processing. Experience has shown us that it is advantageous to elaborate slightly on the simple testing of *MANEX* in **E0**. In debugging, we often wish to execute the instruction in the instruction register *IR* without going through the normal fetch process. This is especially true in the early stages of debugging, when the memory system is not yet working. During normal operations the system clock will be running at microsecond speeds. A debounced depression of a manual switch lasts for at least 100 msec, which would be on the order of 100,000 clock cycles. Thus, during normal operation, a manual switch that causes a call to Execute will be depressed for the entire duration of the execute phase. However, during debugging, the operator will advance the system clock at a slow rate, probably manually. The operator can press a button for one clock period to escape from the **IDLE** state, and then *release* the button before the next clock edge arrives. The LD20 is then in state **E0**, deciding either to process a manual pushbutton or to process the contents of the instruction register. When no button is depressed, the composite pushbutton signal *MANSW∗* is false. By insisting that *MANSW∗* be true in order to process manual operations, we have a way to force the machine to process an instruction, even though we did not reach Execute by way of Fetch.

We now have two requirements for manual operations: *MANEX* must be true to assure that any manual activity originated from **IDLE**, and *MANSW∗* must be true to show that a button is still depressed. We define the relevant test signal as

$$MANEX.SW∗ = MANEX \cdot MANSW∗$$

If *MANEX.SW*∗ is false, the execute phase will process the contents of IR. The ASM chart for the beginning stages of **E0** is shown in Fig. 7–30.

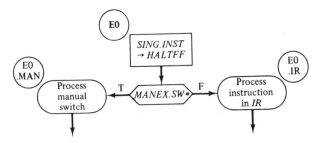

Figure 7–30 Initial processing in state **E0**.

Here is what the operator does when debugging a given instruction while bypassing the fetch phase:

(a) Loads *IR* from *SR* with the desired command, using a high-speed clock and the *LDIR* button.

(b) Loads the desired effective address into *MA*, using the *LDMA* switch.

(c) Loads the desired operand into *MB*, using the *LDMB* switch.

(d) Switches to a low-speed (or manual) clock.

(e) Holds any manual pushbutton down for the duration of one clock edge. This will cause a transition from **IDLE** to **E0**.

(f) Releases the pushbutton before the next clock edge. The ASM will now be in state **E0** on the false path in Fig. 7–30.

(g) Proceeds through the execute ASM chart.

When *MANEX* is true, we may thus traverse either the manual or instruction parts of the execute phase. We must clear *MANEX* to false before we leave the execute phase. It is tempting to put this clearing operation into the unconditional part of state **E0**, right up front. This will not work, because throughout its duration the execute phase needs a knowledge of the status of *MANEX* to control the flow. We must remember to insert the term F → *MANEX* just before any exits from the execute phase back to Fetch or **IDLE**. This is a subtle point, and in one of the exercises at the end of this chapter we ask you to consider what would happen if *MANEX* were reset unconditionally in **E0**.

As we develop the ASM for the execute phase, we will insert the F → *MANEX* term at the appropriate places, without further comment.

Execute Phase: Manual Commands

Most of the LD20's manual pushbuttons are not a part of the standard PDP-8 computer. We have incorporated them into this design for the convenience of the operator. Six pushbutton signals load a given register from the *SR*: *LDMA*∗, *LDMEM*∗, *LDMB*∗, *LDPC*∗, *LDAC*∗, and *LDIR*∗. *LDMEM*∗ requires that the *MA* be set appropriately. The *EXAM* pushbutton supports a sequential dump to the memory lights on the display panel: each depression of the *EXAM* switch will increment *MA*, and thus the contents of the next memory location will appear in the lights. Recall the design of the memory box. As long as *MA* is stable, the memory box produces good output data after the short memory read cycle. The *DEP* pushbutton enables a sequential memory load from *SR*. It acts like *LDMEM* except that, after loading the memory, it increments *MA*, so that the next word appears in the memory lights and is available for changing.

The ASM description of most of these pushbutton operations is straightforward. *EXAM* does not require us to perform an ASM memory read because as long as *MA* is stable, the read occurs automatically at a speed thousands of times faster than human reactions. According to the memory protocol, *LDMEM* and *DEP* activities require the execution of a state *after* the state that initiates the write operation. Also, looking ahead to instruction processing, we know that at least DCA (Deposit and Clear Accumulator) requires a memory write. We could declare a separate state for each path from **E0** (for *LDMEM*, *DEP*, *DCA*, etc.), but the memory write operations have so much in common that we should see if it is convenient to combine all these second execute states into a single state. Let's call the new state **E1**. If any of the users of state **E1** require special processing during or after **E1**, we must be prepared to introduce a many-way branch, similar to but smaller than the one in **E0**. With this in mind, we derive the ASM descriptions of the manual operations (see Fig. 7–31).

The *EXAM* and *DEP* operations require that the *MA* register be incremented. According to our plan for the LD20's architecture, we perform the incrementing by routing the *MA* through the *ALU*, and routing the incremented value, on the *ALUBUS*, back into the *MA*. This action requires that *MA* be attached to the data multiplexer as a source. Our design allows us to add this register immediately, without difficulty.

Execute Phase: Memory–Accessing Instructions

Our convention upon entering state **E0** from Fetch is that *MA* contains the effective address *EA*, and *MB* contains the contents of that address, whenever these values are relevant to the instruction. The ASM descriptions of the memory-accessing instructions are simple and are based directly on the descriptions of the PDP-8's instructions earlier in this chapter. Figure 7–32 is the ASM for AND, TAD, ISZ, DCA, JMS, and JMP.

Execute Phase: The IOT Instruction

This instruction performs input and output operations to or from the LD20 accumulator. It is an important command, and PDP-8 input–output procedures require careful study. We defer a detailed discussion of PDP-8 input–output programming and interfacing until Chapter 9; at this time we treat only the LD20's responsibilities for the execution of IOT instructions. The IOT instruction has three fields: the operation code (110_2), a 6-bit device address field, and a 3-bit control field. The format of the instruction word is

0	1	2	3	8	9	10	11
Op code 110			Device address		IOP4 request	IOP2 request	IOP1 request

The interface between the LD20 and the external devices consists of the following signals:

(a) Six device address lines from the LD20: *IR3–IR8*.

(b) Twelve data lines from the LD20: *AC0–AC11*.

(c) Three control lines *IOP1*, *IOP2*, and *IOP4* from the LD20, derived as requested by *IR11*, *IR10*, and *IR9*, respectively.

(d) Twelve external data lines into the LD20: *INPUT0–INPUT11*.

(e) Three control lines into the LD20: *IOSKIP*, *ACCLR*, and *ORAC*.

IOT protocol. The protocol requires the following. With the device address stable on the outgoing device address lines, the LD20 issues a sequence of 0 to 3 *IOP* signals, as specified by 1-bits in *IR11*, *IR10*, and *IR9*. *IOP1* will be issued first, if requested by *IR11*, followed by *IOP2* and then *IOP4*, if requested. During the lifetime of each of these three *IOP* signals, the LD20 must interrogate the three incoming control signals, in the order *IOSKIP*, *ACCLR*, and *ORAC*. If *IOSKIP* is true, the LD20 must skip an instruction. If *ACCLR* is true, the LD20 must clear the *AC*, and if *ORAC* is true, the LD20 must perform a bit-by-bit OR operation of incoming data into the *AC*:

$$AC + INPUT \rightarrow AC$$

Figure 7–33 shows the general structure of IOT processing in the LD20.

IOT architecture and control algorithm. To support the IOT protocol, the LD20 needs:

(a) A way to assert the requested *IOP* signals, in order.

(b) Within each *IOP* time, a way to interrogate the incoming control signals, in the specified order.

We have chosen a simple yet effective design. During the execution of an IOT instruction, state **E0** is responsible for creating each requested *IOP* signal. An asserted *IOP* signal stays true throughout three more states, **E3**, **E4**, and **E5**, which perform a sequence of tests on *IOSKIP*, *ACCLR*, and *ORAC*. State **E5**, in addition to testing the *ORAC* signal, determines if three passes through the state sequence **E0**, **E3**, **E4**, **E5** have elapsed. If fewer than three passes have occurred, state **E5** will return control to state **E0** for another pass; otherwise, control passes to state **F1**, ending the execution of the IOT instruction.

The architecture must contain an element to produce the requested *IOP*

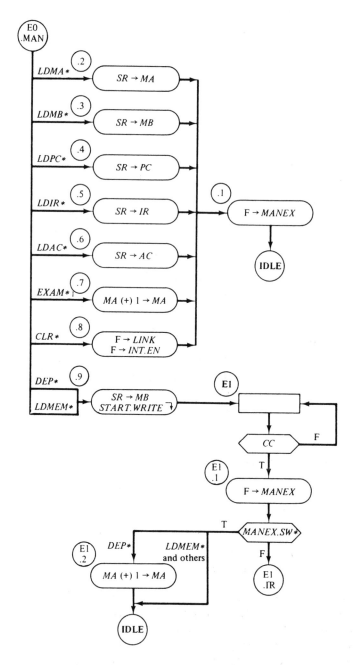

Figure 7-31 ASM of manual operations in the execute phase.

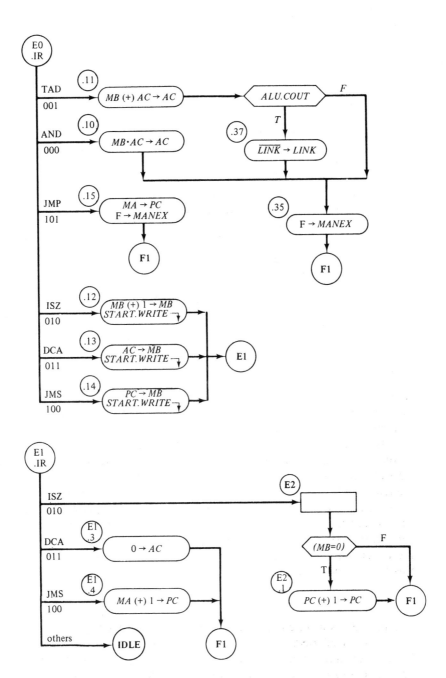

Figure 7–32 ASM of instructions that access the memory in the execute phase.

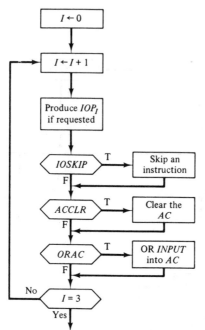

Figure 7-33 General flow of an IOT instruction. (This is not an ASM chart.)

signals in the proper sequence and hold them throughout the execution of the **E3**, **E4**, and **E5** states. A 4-bit parallel-in parallel-out shift register, one of our standard building blocks, serves nicely to provide the sequencing of the *IOP* signals. We initialize the shift register with the binary bit pattern TFFF prior to reaching the execute phase. For an IOT instruction, passage through state **E0** shifts the bits one position to the right. Think of the bit positions as being named 0, 1, 2, and 4. Output signals from bits 1, 2, and 4 represent the enabling conditions for the respective *IOP* signals, say *IOP1.EN*, *IOP2.EN*, and *IOP4.EN*. For instance, after two right shifts, the bit pattern in the *IOP* enabler is FFTF. This corresponds to *IOP2* time and *IOP2.EN* is true; if the programmer has requested an *IOP2* signal (with instruction bit $IR10 = 1$), it will be issued at this time. The logic equivalent of this operation is

$$IOP2 = IOP2.EN \cdot IR10$$

Figure 7-34 is a functional sketch of the *IOP* enabler system.

The LD20 ASM of the execute phase of IOT processing is shown in Fig. 7-35. Our architecture for the LD20 already can support the $PC \,(+)\, 1 \rightarrow PC$ and the $0 \rightarrow AC$ operations required in response to the incoming control signals. To accomplish $AC + INPUT \rightarrow AC$, we must provide a data path for the 12 bits of *INPUT* into the *ALU*, where the OR operation will occur. Since we have adopted a readily expansible data routing scheme, we can easily add *INPUT* to the list of items on the data multiplexer inputs.

State **F1** is a convenient place to initialize the *IOP* enabler; see Fig. 7-39a.

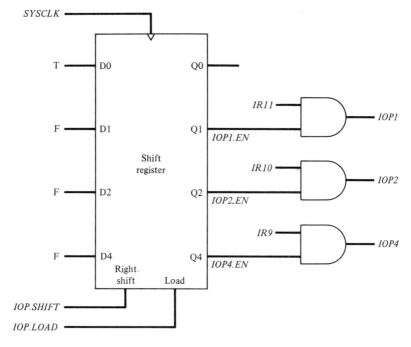

Figure 7-34 An enabler for the *IOP* signals.

The initialization to TFFF occurs with each instruction fetch; this does no harm since the only shifting of the *IOP* enabler occurs with IOT instructions.

Enabling and disabling the interrupt system.   The IOT instruction contains the ION (Interrupt System On) and IOF (Interrupt System Off) commands, which allow the programmer to enable or disable the interrupt system. The instruction format for these operations is 6001_8 for ION and 6002_8 for IOF; in both cases the device address is 0. Execution of these commands requires that the ASM set the interrupt enable flip-flop *INT.EN* to T or F, respectively. Figure 7-35 shows actions in state **E0** that handle these activities.

State E0: Operate Microinstructions

Operate priority system.   From the programmer's point of view, the PDP-8I's Operate instruction is divided into two groups, G1 and G2. In our hardware, the important structural feature of the Operate instruction is the priority system for specifying the sequence of micro-operations. The simplest way to process the Operate instruction is to devote four states to it, one for each priority level, as shown in Fig. 7-36.

Frequently, a programmer will use an Operate instruction that specifies only one micro-operation. For example, CLA (7200_8), a priority-1 operation, has only the CLA bit set. The ASM chart in Fig. 7-36 would perform the CLA in the first state and then do nothing in the next three states except waste time.

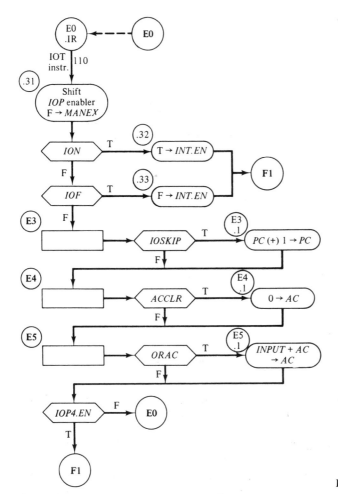

Figure 7-35 ASM of the IOT instruction.

We prefer to use only one clock period to process the CLA bit and then go immediately to the fetch phase for the next instruction. Similarly, an RAR instruction involves a single priority-4 operation. We prefer to bypass the first three priority states and go directly to the fourth state. In general, we wish to process the highest-priority micro-operation, followed in order by lower-priority operations, without wasting clock cycles on unpopulated priorities.

We may formalize this priority–resolution problem by assuming that there exist four *requests for service, RQST1* through *RQST4,* each representing one priority level. The requests are initialized prior to the execute phase of the Operate instruction, based on the pattern of bits in the instruction word. For example, if the instruction specifies IAC, a priority-3 operation, then *RQST3* would be true after the initialization. If the highest-priority request (*RQST1*) is true, the LD20 must perform the specified priority-1 operations and set *RQST1* false; then if any priority requests remain, the LD20 must return for another

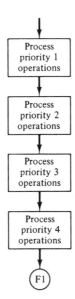

Figure 7-36 A primitive ASM of the Operate instruction.

pass through the execute phase of the Operate instruction. If *RQST1* is false (which is possible the first time through, and will definitely be so the second time), then *RQST2* is examined; and so on. Figure 7–37 is a flowchart of this process.

The hardware for this scheme involves storage elements for the *RQST* terms; four flip-flops will serve. Also, we need the ability to determine if any priority requests remain after a pass, so that control may return for more processing of the instruction. We will define a variable *MOREOP* to be true when more processing is needed and false otherwise. We may test this variable at the end of each pass and act accordingly.

The ASM chart for managing Operate instruction priorities is almost within our grasp. One final point needs study. The shift microinstructions (the "rotates") generate either a 1-bit or a 2-bit shift. The shift register building blocks described in Chapter 4 perform only 1-bit shifts per clock cycle, leading us to implement the 2-bit rotate operations as two 1-bit rotates. This will mean looping twice through the priority-4 part of the Operate ASM. We need a flag to tell the ASM when a second pass through priority 4 is needed, so we will add to the architecture a flip-flop with output *DOUBLE*, to serve as a signal to "shift again." When *DOUBLE* is false, no more shifts are needed. When *DOUBLE* is true, another shift must occur, and *DOUBLE*, having served its purpose, must be reset to false. Bit *IR10* = 1 specifies a double shift, and *IR10* = 0 denotes a single shift. We may initialize *DOUBLE* with the value of bit *IR10*, once the instruction has been acquired but before the execute phase begins.

With these additions to the LD20's architecture, we may specify the ASM chart for the priority control of the Operate instruction. The execute phase for this instruction occurs entirely within state **E0**. Figure 7–38 is the ASM counterpart of the flowchart in Fig. 7–37.

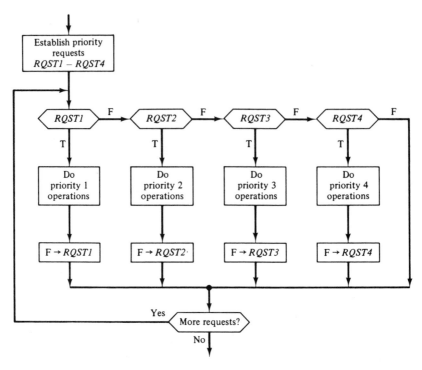

Figure 7–37 The general flow of an efficient execution of the Operate instruction. (This is not an ASM chart.)

Initializing the priority request flip-flops. To complete the processing of the Operate instruction, we must arrange to initialize the priority request flip-flops prior to entering state **E0**. This must occur during Fetch, after the instruction reaches the instruction register *IR*. Examining the fetch-phase ASM, we see that there is only one place to do this—in term F3.1 of state **F3**. We will set or reset the request flip-flops (and also the *DOUBLE* flip-flop for shifts) at this point, according to the values of the bits in the instruction. These additions to the fetch-phase ASM appear in Fig. 7–39b; Fig. 7–39a contains the addition to state **F1** for initializing the *IOP* enabler.

Individual microinstructions. We have conquered the Operate instruction priority system in an elegant way. There remains the specification of the individual micro-operations at each of the four priority levels. Except for the skip and rotate operations, the ASMs are simple. We have already developed a structure that will handle both single and double rotates. Let's examine the skip instructions.

Bit 8 of the Operate instruction specifies the sense of the skip logic. We may define a variable *SKIP* that will tell the LD20 whether the condition for a skip is met. In a group 2 Operate instruction, *SKIP* = T will mean that the LD20 must skip an instruction; *SKIP* = F will mean no skip. To aid the description of the logic, we require variables that indicate the three skip criteria: a zero accumulator, a minus accumulator, and a nonzero link. For the last

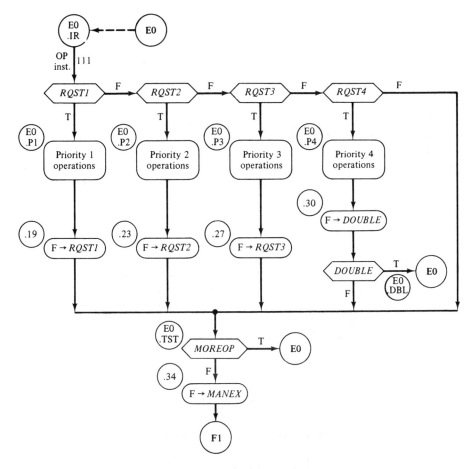

Figure 7–38 ASM of the Operate instruction's priority system.

condition, the value of the link bit suffices. In the PDP-8's two's-complement representation of numbers, bit $AC0$ gives the sign of the number in the accumulator: $AC0 = 0$ for positive numbers and $AC0 = 1$ for negative numbers. We use a new variable $(AC=0)$ to denote a zero accumulator; its implementation in terms of the bits of the accumulator is given in Chapter 8.

Assume that a group 2 Operate instruction is in the instruction register and that the skip-sense bit (bit $IR8$) is 0. Then we may describe the PDP-8's skip

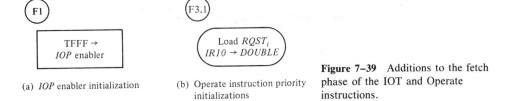

(a) *IOP* enabler initialization   (b) Operate instruction priority initializations

Figure 7–39 Additions to the fetch phase of the IOT and Operate instructions.

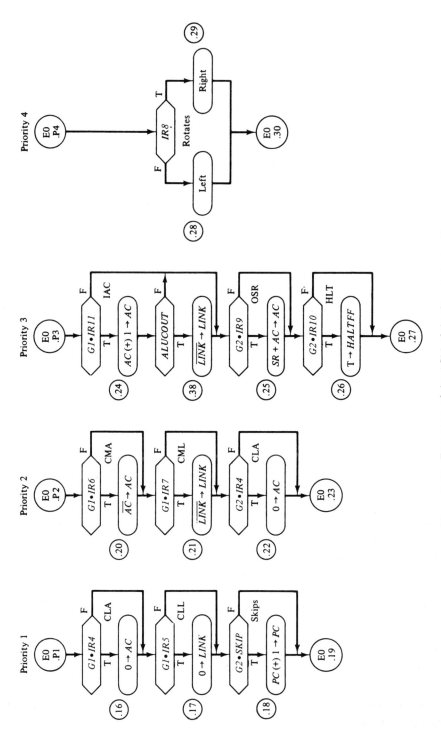

Figure 7-40 Fragments of the ASM of the Operate instruction.

actions for this situation by a variable *SKIP0:*

$$SKIP0 = SMA \cdot AC0 + SZA \cdot (AC=0) + SNL \cdot LINK$$

If bit 8 is 1, then we may declare another variable *SKIP1* to express this condition:

$$SKIP1 = \overline{SMA \cdot AC0 + SZA \cdot (AC=0) + SNL \cdot LINK}$$

The definition of the PDP-8's skip micro-operation expresses *SMA* as bit *IR5*, *SZA* as bit *IR6*, and *SNL* as bit *IR7*. Noticing that $SKIP1 = \overline{SKIP0}$, we may combine these two conditions into a single equation for the skip operation:

$$
\begin{aligned}
SKIP &= \overline{IR8} \cdot SKIP0 + IR8 \cdot SKIP1 \\
&= \overline{IR8} \cdot SKIP0 + IR8 \cdot \overline{SKIP0} \\
&= IR8 \oplus SKIP0 \\
&= IR8 \oplus (IR5 \cdot AC0 + IR6 \cdot (AC=0) + IR7 \cdot LINK)
\end{aligned}
$$

We now have all the tools for describing the individual Operate micro-operations. In Fig. 7–40 we show the ASMs of each of the four priority levels.

CONCLUSION

Now that we have derived the details of the LD20's architecture and control, it is desirable to collect the results into a more compact form. Figure 7–41

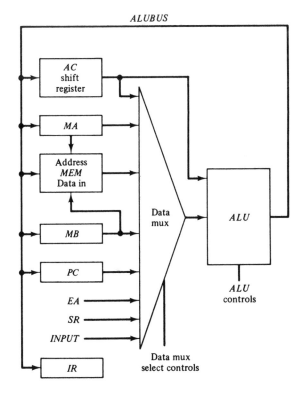

Figure 7–41 Architecture of the LD20: main data paths; one of twelve 1-bit slices.

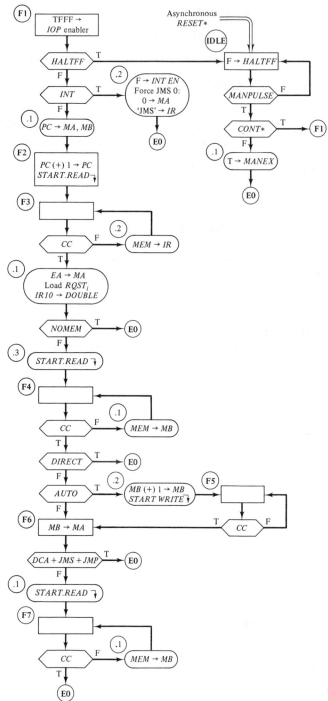

Figure 7–42 ASM of the LD20's fetch phase.

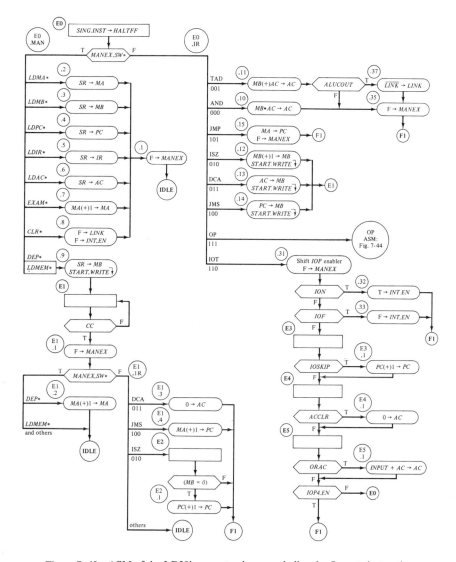

Figure 7–43   ASM of the LD20's execute phase, excluding the Operate instruction.

shows the architecture of the main data paths in the LD20; each of the 12 bits has a similar diagram. (In Chapter 8 we will make one further refinement to this architecture to support possible expansions of the system.) In addition to this main data structure, the LD20's architecture at this point includes simple storage elements for the following variables:

LINK: Link bit
HALTFF: Halt status
INT.EN: Interrupt enable

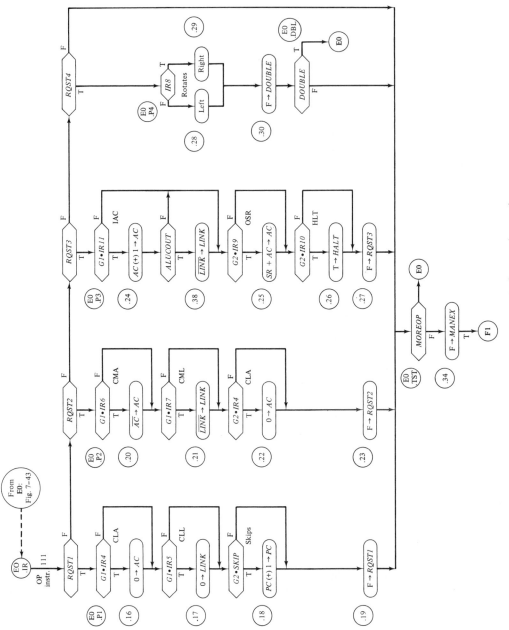

Figure 7-44 ASM of the LD20's Operate instruction.

The Art of Digital Design   Part II

MANEX: Denotes that the execute phase was reached from **IDLE**

SING.INST: Synchronized single-instruction switch signal

$RQST_i$:   Four priority request flip-flops for the Operate instruction

DOUBLE: Number of 1-bit rotate micro-operations specified

The architecture also has a single-pulser unit with output *MANPULSE*, to denote the first full clock cycle after the depression of a manual pushbutton. The *IOP* signal enabler, consisting primarily of a shift register, completes the list.

Figures 7–42, 7–43, and 7–44 present the LD20's ASM in three charts: the fetch phase, the execute phase, and the complex Operate instruction. These three ASM charts will be of great use in Chapter 8.

This completes the specification of the architecture and the control for the LD20. The implementation lies before us! This will involve:

(a) Deriving the logic equations for the ASM outputs.

(b) Developing the state generator.

(c) Specifying the hardware for the architecture and converting the ASM output equations to fit the exact integrated circuits.

(d) Specifying hardware for the control logic and the ASM outputs.

READINGS AND SOURCES

BELL, C. GORDON, J. CRAIG MUDGE, and JOHN E. MCNAMARA, *Computer Engineering— A DEC View of Hardware Systems Design.* Digital Press, Bedford, Mass., 1978. Chapter 8 discusses the history and structure of the PDP-8 family.

KUCK, DAVID J., *The Structure of Computers and Computations,* Vol. 1. John Wiley & Sons, New York, 1978.

MANO, M. MORRIS, *Computer System Architecture,* 2nd ed. Prentice-Hall, Englewood Cliffs, N.J., 1982.

PDP8/I and PDP8/L Small Computer Handbook. Digital Equipment Corp., Maynard, Mass., 1972. User's manual for the PDP-8I minicomputer.

PROSSER, FRANKLIN, and ROBERT WEHRMEISTER, *C421-C422 Advanced Computer Organization Laboratory Manual.* Computer Science Department, Indiana University, Bloomington, Ind. 47405, 1985. Laboratory manual to support the construction, debugging, and study of the LD20 and LD30 implementations of the PDP-8I. The laboratory project uses the Logic Engine Development System, manufactured by Logic Design, Inc. Ask the authors of this book for information.

EXERCISES

7–1. What determines the maximum number of words in the PDP-8I's memory? What determines the size of a memory page? What two pages of the PDP-8I are directly accessible from an instruction? What memory locations are accessible using indirect addressing?

7–2. Here are the contents of six words of PDP-8 memory. Treat each word as an instruction and give the command, the effective address *EA*, and, where appropriate, the contents of the effective address *CA*. All numbers are in octal.

$$(0061) = 0261$$
$$(0071) = 0061$$
$$(0342) = 3245$$
$$(1000) = 5725$$
$$(1015) = 2600$$
$$(2410) = 1471$$

7–3. Some computers allow multilevel indirect addressing, such that each word used as an indirect address can specify an additional level of indirect addressing. Why is this not possible in the PDP-8?

7–4. Explain the PDP-8's auto indexing. Here are the contents of three memory locations (all numbers are in octal):

$$(0016) = 0016$$
$$(0017) = 0016$$
$$(0020) = 0016$$

Starting always from this point, show the results of executing each of the following instructions:

(a) 0417
(b) 0420
(c) 0416
(d) 2016
(e) 2416
(f) 2420

7–5. Describe the PDP-8's interrupt mechanism. How is an interrupt request conveyed to the PDP-8? What does the machine do if it detects an interrupt request? How can the programmer control the issuing of interrupts? What are the programmer's responsibilities for processing an interrupt?

7–6. Explain why the PDP-8's ION (Interrupt System On) command does not take effect until after the execution of one additional instruction.

7–7. What is meant by a static design? Why is this such a valuable attribute of a digital system?

7–8. List the principal elements of the LD20's architecture. Which of these elements may the LD20 programmer directly manipulate?

7–9. Why does the LD20's architecture have the accumulator register as the sole source for one of the two *ALU* inputs?

7–10. Why is the instruction register *IR* not attached to the LD20 data multiplexer as a source?

7–11. Design a three-state data bus for the LD20, to replace the multiplexer bus control system used in the text.

7–12. What are the principal types of commands needed to control the LD20 architecture?

7-13. What is the principal function performed by the fetch ASM? The idle ASM? The execute ASM?

7-14. This exercise takes you through most of the fetch ASM and parts of the execute ASM. Assume the following (numbers are in octal):

$$HALTFF = INT = F$$

(0010) = 2221	(0200) = 2222	(2222) = 1111
(AC) = 1234	(PC) = 0305	

Starting from state **F1**, trace the LD20's ASM operations in each of the following cases.

(a) (0305) = 5200

(b) (0305) = 3200

(c) (0305) = 0200

(d) (0305) = 0600

(e) (0305) = 0410

7-15. This exercise takes you through the major sections of the execute ASM. Assume that (MA) = 0200 and (MB) = 2222. Trace the ASM flow from state **E0** to state **F1** for each of the following instructions. (X means "irrelevant." All numbers are in octal.)

(a) (IR) = 1XXX

(b) (IR) = 2XXX

(c) (IR) = 4XXX

(d) (IR) = 6XXX

7-16. This exercise takes you through the operate section of the execute ASM. Trace each of the following instructions from state **E0** to state **F1**. The numbers are in binary.

(a) (IR) = 111 011 000 001

(b) (IR) = 111 000 100 001

(c) (IR) = 111 000 010 111

(d) (IR) = 111 101 001 010

7-17. Make a table showing the sequence of fetch-phase states for each distinct form of the eight PDP-8I instructions. Also list the total number of memory references required in each case (including memory references in the execution phase).

7-18. Why does the LD20 not process an interrupt request immediately upon its assertion?

7-19. Explain carefully the significance of the three terms in the interrupt generation signal:

$$INT = INTREQ \cdot INT.EN \cdot \overline{ION}$$

7-20. In the LD20 ASM, is it feasible to move the test for interrupt from its present position in state **F1** into state **F2**?

7-21. Give a specific illustration to show why we must save the LD20 program counter (in state **F1**) for later use in computing the effective address *EA*. Why not use the value in *PC* at the time *EA* is generated?

7-22. What is the latest point in the fetch ASM that we may place the $PC (+) 1 \rightarrow PC$ operation?

7-23. Refer to the memory control interface in Fig. 7-16. Carefully state the difference between the similar labels inside and outside the box.

7-24. Explain the memory accessing protocol of the LD20 as displayed in Fig. 7-20. What are the conditions under which this protocol works?

7-25. Redraft the LD20's fetch phase using the basic read protocol of Fig. 7-17. Can you see a strong argument against using this memory control ASM?

7-26. State two ways in which the LD20 user may cause the machine to halt.

7-27. Why is it desirable that the *RESET** pushbutton signal perform an *asynchronous* reset of the LD20?

7-28. Show why the asynchronous *CONT** input in the **IDLE** state does not need to be synchronized with the LD20 system clock.

7-29. The most complex processing in the fetch phase involves a memory-referencing instruction that employs indirect addressing and auto indexing to fetch an operand. The minimum number of fetch-phase clock cycles for such an instruction is 11.
  (a) Give an example of a PDP-8 instruction that requires at least this number of LD20 clock cycles.
  (b) Verify from Fig. 7-42 that 11 clock cycles are minimal for this type of instruction.
  (c) Can you find a way to shorten the number of clock cycles by modifying the fetch ASM? (*Hint:* Look at state **F6**.)

7-30. With the present architecture of the LD20, why can't we eliminate state **F2** by including **F2**'s actions in a conditional output of state **F1**?

7-31. Suppose you wish to reduce the number of states in the LD20's fetch phase (Fig. 7-42).
  (a) Collapse states **F2** and **F3** into a single state that performs actions equivalent to the former two states.
  (b) What are the disadvantages of the collapse in part (a)?

7-32. Suppose that all the $F \rightarrow MANEX$ commands in the execute phase were replaced by a single unconditional $F \rightarrow MANEX$ in state **E0**. What would be the effect on the execution of the LD20 ASM?

7-33. (a) With the present LD20, show what happens if an operator depresses a manual pushbutton such as *LDAC* while the LD20 is not in the **IDLE** state.
  (b) Answer part (a) assuming that the entire *MANEX* signal is deleted from the ASM.

7-34. The LD20's manual operations are performed in the execute phase, like regular instruction processing. An attractive alternative would be to perform manual pushbutton executions in a manual phase, using states separate from Execute.
  (a) Develop an LD20 ASM with such a manual phase.
  (b) What simplifications does this form of LD20 ASM introduce?
  (c) What disadvantage of this form of ASM can you see?

7-35. State **E2** may be eliminated from the LD20 ASM. Eliminate it. What are the advantages and disadvantages of this elimination?

7–36. Design a detailed ASM for the LD20's Operate instruction along the pattern of Fig. 7–36. Discuss the advantages and disadvantages of this primitive approach.

7–37. Modify the ASM of the LD20's Operate instruction to replace the two-position rotate flag *DOUBLE* by a fifth priority level, *RQST5*.

7–38. The LD20's algorithm for the IOT instruction always makes three passes through states **E3**, **E4**, and **E5**, regardless of the number of *IOP* signals requested by the instruction. Modify the LD20's ASM so that **E3**, **E4**, and **E5** (or their equivalent) are executed only when the LD20 asserts one of the *IOP* signals.

7–39. Suppose the PDP-8 IOT instruction protocol is altered so that the computer issues all requested *IOP* signals at one time. How will this affect the LD20's ASM?

7–40. Combine states **E3**, **E4**, and **E5** for the IOT instruction into a single state that performs activities equivalent to the former three states.

8

Building the Minicomputer

In Chapter 7, we designed the architecture and control algorithm for the LD20 minicomputer—the architecture in terms of functional units, and the control as an ASM chart. In this chapter, we will sketch the implementation of the LD20 from these starting points. We will not specify each building block and logic equation in detail, because we use the standard methods illustrated in Chapters 5 and 6, and you will be able to work out the details with only general guidance.

Good design style dictates that we defer the actual decisions about hardware as long as we can. Up to this point, the design of the LD20 has required us to commit ourselves to only a few specific pieces of hardware. In this chapter, we will continue to defer hardware decisions until we have dealt with the system on the logical level to the fullest useful extent.

We use the ASM chart extensively to derive the control signals. The ASM chart for the LD20 was shown in Figs. 7–42, 7–43, and 7–44.

PRELIMINARIES

Auxiliary Variables

The ASM chart uses a number of convenient variables that we have constructed from primary signals. The definition of many of these appears in Chapter 7, but we list them here for reference. In general, the outputs of elements of the architecture are primary signals and do not appear in this list.

$$AUTO = \overline{MA0} \cdot \overline{MA1} \cdot \overline{MA2} \cdot \overline{MA3} \cdot \overline{MA4} \cdot \overline{MA5} \cdot \overline{MA6} \cdot \overline{MA7} \cdot MA8$$

$$DIRECT = \overline{IR3}$$

$$G1 = \overline{IR3}$$

$$G2 = IR3$$

$$INT = INTREQ \cdot INT.EN \cdot \overline{ION}$$

$$IOF = IOT \cdot \overline{IR3} \cdot \overline{IR4} \cdot \overline{IR5} \cdot \overline{IR6} \cdot \overline{IR7} \cdot \overline{IR8} \cdot \overline{IR9} \cdot IR10 \cdot \overline{IR11}$$

$$ION = IOT \cdot \overline{IR3} \cdot \overline{IR4} \cdot \overline{IR5} \cdot \overline{IR6} \cdot \overline{IR7} \cdot \overline{IR8} \cdot \overline{IR9} \cdot \overline{IR10} \cdot IR11$$

$$IOP1 = IOP1.EN \cdot IR11$$

$$IOP2 = IOP2.EN \cdot IR10$$

$$IOP4 = IOP4.EN \cdot IR9$$

$$MANEX.SW* = MANEX \cdot MANSW*$$

$$MANSW* = LDMA* + LDMB* + LDPC* + LDIR* + LDAC*$$
$$+ LDMEM* + DEP* + EXAM* + CLR* + CONT*$$

MOREOP (see implementation of operate priority circuit)

$$NOMEM = OP + IOT + DIRECT \cdot (DCA + JMS + JMP)$$

$$SKIP = IR8 \oplus (IR5 \cdot AC0 + IR6 \cdot (AC = 0) + IR7 \cdot LINK)$$

$$(AC = 0) = \overline{AC0} \cdot \overline{AC1} \cdot \overline{AC2} \cdot \overline{AC3} \cdot \overline{AC4} \cdot \overline{AC5} \cdot \overline{AC6} \cdot \overline{AC7} \cdot \overline{AC8}$$
$$\cdot \overline{AC9} \cdot \overline{AC10} \cdot \overline{AC11}$$

$$(MB = 0) = \overline{MB0} \cdot \overline{MB1} \cdot \overline{MB2} \cdot \overline{MB3} \cdot \overline{MB4} \cdot \overline{MB5} \cdot \overline{MB6} \cdot \overline{MB7} \cdot \overline{MB8}$$
$$\cdot \overline{MB9} \cdot \overline{MB10} \cdot \overline{MB11}$$

Effective address *EA*: According to Eq. (7–3), the effective address is

$$EA_i = MB_i \cdot IR4 \qquad (i = 0\text{–}4)$$
$$EA_i = IR_i \qquad (i = 5\text{–}11)$$

Instruction variables: We may most easily form the individual instruction variables as the outputs of a decoder. Technically, they are not auxiliary variables, but it is useful to emphasize their derivation here.

$$0: \quad AND = \overline{IR0} \cdot \overline{IR1} \cdot \overline{IR2}$$
$$1: \quad TAD = \overline{IR0} \cdot \overline{IR1} \cdot IR2$$
$$2: \quad ISZ = \overline{IR0} \cdot IR1 \cdot \overline{IR2}$$
$$3: \quad DCA = \overline{IR0} \cdot IR1 \cdot IR2$$
$$4: \quad JMS = IR0 \cdot \overline{IR1} \cdot \overline{IR2}$$
$$5: \quad JMP = IR0 \cdot \overline{IR1} \cdot IR2$$
$$6: \quad IOT = IR0 \cdot IR1 \cdot \overline{IR2}$$
$$7: \quad OP = IR0 \cdot IR1 \cdot IR2$$

Labels for ASM Chart Locations

These variables are useful parameters for forming output equations and control routings in a systematic manner. The logic equations come immediately from the ASM chart. We have two types of labels: the conditional output terms, which are the basic parameters for expressing the output equations, and the other position markers, not directly associated with conditional outputs but used to label various components of the states of the execute phase.

ASM chart position labels. You should verify each of these equations.

$$DBL.IS.T = \overline{RQST1} \cdot \overline{RQST2} \cdot \overline{RQST3} \cdot RQST4 \cdot DOUBLE$$

(this is an auxiliary variable used in $E0.DBL$ and $E0.TST$ below)

$$E0.DBL = E0.IR \cdot OP \cdot DBL.IS.T$$
$$E0.IR = E0 \cdot \overline{MANEX.SW*}$$
$$E0.MAN = E0 \cdot MANEX.SW*$$
$$E0.P1 = E0.IR \cdot OP \cdot RQST1$$
$$E0.P2 = E0.IR \cdot OP \cdot \overline{RQST1} \cdot RQST2$$
$$E0.P3 = E0.IR \cdot OP \cdot \overline{RQST1} \cdot \overline{RQST2} \cdot RQST3$$
$$E0.P4 = E0.IR \cdot OP \cdot \overline{RQST1} \cdot \overline{RQST2} \cdot \overline{RQST3} \cdot RQST4$$
$$E0.TST = E0.IR \cdot OP \cdot \overline{DBL.IS.T}$$
$$E1.IR = E1 \cdot \overline{MANEX.SW*}$$

Conditional output terms. Here are a few of these terms. You should verify them and derive the rest.

$$IDLE.1 = IDLE \cdot MANPULSE \cdot \overline{CONT*}$$
$$F1.2 = F1 \cdot \overline{HALTFF} \cdot INT$$
$$F4.2 = F4 \cdot CC \cdot \overline{DIRECT} \cdot AUTO$$
$$E0.1 = E0.2 + E0.3 + E0.4 + E0.5 + E0.6 + E0.7 + E0.8$$
$$E0.11 = E0.IR \cdot TAD$$
$$E0.18 = E0.P1 \cdot G2 \cdot SKIP$$
$$E0.37 = E0.11 \cdot ALU.COUT$$

The most difficult conditional output term to derive is $E0.34$, at the final exit to **F1** from the execution of an Operate instruction (see Fig. 7–38). To assist the synthesis of this equation and the equations for moving from state **E0** to **E0** and from state **E0** to **F1**, we have defined ASM labels **E0.DBL** and **E0.TST** and the auxiliary definition $DBL.IS.T$. With these judiciously chosen variables, we have

$$E0.34 = E0.TST \cdot \overline{MOREOP}$$

We will encounter *E0.TST* and *E0.DBL* again in the development of state transitions. You should derive the defining equations for *E0.DBL* and *E0.TST*, using the ASM in Fig. 7–44. The exercise will increase your intuitive feeling about implementing the flow of control and also will be excellent practice in manipulating Boolean expressions.

THE DATA-ROUTING SYSTEM

Inputs to the Data Multiplexer

The LD20's 12-bit data path uses 12 multiplexers to control access to the *A* input of the *ALU*. In the initial specification, in Fig. 7–9, we had four sources in the data mux system. As the design progressed, we found that we needed to add *MB*, *MA*, and *INPUT* to the data mux. Looking ahead, we find that the implementation of *ALU* operations will require that the *AC* register be available at the *ALU*'s *A* input, in addition to its position as the sole source of the *ALU*'s *B* input.

The data path now has eight sources of 12-bit data, a convenient number because 8-input multiplexers are available. However, as good designers we consider the possibility that later we might wish to extend the LD20's basic design to accept additional instructions. New instructions are likely to require additional registers on the main data path. Is there an easy way to leave open an input to the data multiplexers, to allow for such a possibility? Yes: We choose to combine *INPUT* and *EA* into a single data multiplexer input, thereby reducing to 7 the data mux inputs committed to our basic design. In the treatment of the input-output interface in Chapter 9, you will see that the *INPUT* lines from the external device are required to have three-state control. Data is presented on these lines only when required by an appropriate IOT instruction. On the other hand, *EA* is used only in ASM state **F3**. If we use three-state control on the 12 bits of *EA*, we may create a "minibus" consisting of *INPUT* and *EA*. This minibus will use only a single input on the data multiplexer system. *INPUT*'s access to this data mux input is controlled by the IOT protocol. The signal to control the three-state outputs of *EA* is

$$SELEA = F3.1$$

Although we may use AND gates for the first 5 bits of *EA*, we find it convenient to use three-state multiplexers for these bits and three-state buffers for the remaining bits, as shown in Fig. 8–1.

The Select Signals of the Data Multiplexer

In our final design the data multiplexer system has seven 12-bit sources: *AC*, *MA*, *MB*, *MEM*, *PC*, *SR*, and the combined *INPUT/EA*. A set of twelve 8-input multiplexers such as the 74LS151 will provide efficient control of these inputs. The muxes require three select inputs, *B4*, *B2*, and *B1*. Our task is to develop the logic equations that provide the proper inputs to these three control

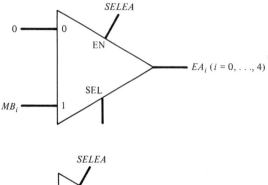

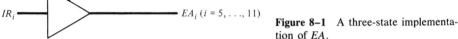

Figure 8–1 A three-state implementation of *EA*.

pins on each mux chip. We first make an assignment of the registers to input positions on the multiplexers. The order is arbitrary; Fig. 8–2 shows our choice. For each source register, we may write the contributing state terms from the ASM chart. Remember, we are looking for registers used as a *source* of data in the chart. Table 8–1 contains the result. Now we can derive the equations for *B4*, *B2*, and *B1*.

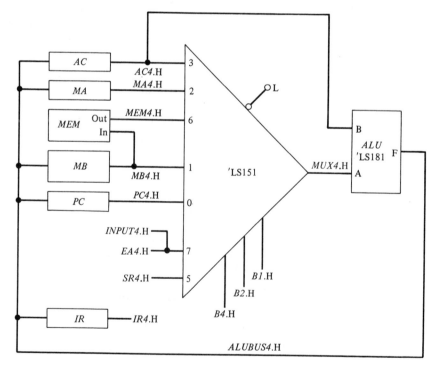

Figure 8–2 Bit slice 4 of the main data-routing system, showing the data multiplexer and register assignments to the data multiplexer inputs.

TABLE 8-1 DATA MULTIPLEXER CONTROLS

Source	Mux input	Mux controls B4	B2	B1	ASM terms
PC	0	0	0	0	F1.1 + F2 + E0.14 + E0.18 + E2.1 + E3.1
MB	1	0	0	1	F4.2 + F6 + E0.10 + E0.11 + E0.12
MA	2	0	1	0	E0.7 + E0.15 + E1.2 + E1.4
AC	3	0	1	1	E0.13 + E0.20 + E0.24
SR	5	1	0	1	E0.2 + E0.3 + E0.4 + E0.5 + E0.6 + E0.9 + E0.25
MEM	6	1	1	0	F3.2 + F4.1 + F7.1
EA	7	1	1	1	F3.1
INPUT	7	1	1	1	E5.1

$$B4 = SR + MEM + EA + INPUT$$
$$= F3.1 + F3.2 + F4.1 + F7.1 + E0.2 + E0.3 + E0.4 + E0.5 + E0.6$$
$$+ E0.9 + E0.25 + E5.1$$
$$B2 = MA + AC + MEM + EA + INPUT$$
$$= F3.1 + F3.2 + F4.1 + F7.1 + E0.7 + E0.13 + E0.15$$
$$+ E0.20 + E0.24 + E1.2 + E1.4 + E5.1$$
$$B1 = MB + AC + SR + EA + INPUT$$
$$= F3.1 + F4.2 + F6 + E0.2 + E0.3 + E0.4 + E0.5 + E0.6 + E0.9$$
$$+ E0.10 + E0.11 + E0.12 + E0.13 + E0.20 + E0.24 + E0.25 + E5.1$$

ALU Operations

Although we have always supposed that we would use the 74LS181 ALU chips as the basis for our *ALU*, we have done nothing to this point that requires these particular chips. Rather, we specified only that the *ALU* black box must be able to perform certain logic and arithmetic functions on its inputs. We may carry this generality one step further and develop logic equations for each *ALU* operation. A scan of the LD20's ASM chart shows that we require the following operations: INCREMENT, PLUS, AND, OR, and NOT. In Chapter 7, we decided to use the *ALU* as a source of zeros to clear registers, so the ZERO operation emerges. Since the *ALU* also serves as a transparent box for moving data from a source to a destination, the last *ALU* operation is PASS.

Our *ALU* black box has two sets of 12-bit inputs *A* and *B*. The *AC* feeds into the *B* input, and the data multiplexer output goes into the *A* input. We may now tabulate the conditions under which the *ALU* must perform each operation. Table 8–2 shows this information; the logic equations for each *ALU* operation can be drawn from the presentation.

Now at last we specify that the *ALU* unit contains three 74LS181 Four-Bit ALU chips, linked together to perform 12-bit arithmetic. All the chips share 5 control inputs: *S3*, *S2*, *S1*, *S0*, and *M*. The least significant chip has a carry-in input that we must control, and the most significant chip has a carry-out

TABLE 8–2 *ALU* OPERATIONS

ASM notation[a]	*ALU* operation	ASM terms
R (+) $1 \rightarrow$	INCREMENT A	$F2 + F4.2 + E0.7 + E0.12 + E0.18 +$ $E0.24 + E1.2 + E1.4 + E2.1 + E3.1$
$R \cdot AC \rightarrow$	A AND B	$E0.10$
$R + AC \rightarrow$	A OR B	$E0.25 + E5.1$
$\overline{R} \rightarrow$	NOT A	$E0.20$
R (+) $AC \rightarrow$	A PLUS B	$E0.11$
$0 \rightarrow$	ZERO	$F1.2 + E0.16 + E0.22 + E1.3 + E4.1$
$R \rightarrow$	PASS A	All other conditions

[a]R is any source register on the data multiplexer system.

output, which we may use if necessary. Figure 8–3 shows the arrangement, with the proper LD20 logic variables as inputs and outputs. From a TTL data book, we find the 74LS181 control input values for our seven arithmetic and logical operations. Table 8–3 contains the information; $S3$–$S0$ and M are T = H, while CIN is T = L. Now, using Tables 8–2 and 8–3, we may write the sum-of-products form of each control input equation:

$$ALUS3 = \text{AND} + \text{PLUS}$$
$$= E0.10 + E0.11$$
$$ALUS2 = \text{F}$$
$$ALUS1 = \text{AND} + \text{ZERO}$$
$$= F1.2 + E0.10 + E0.16 + E0.22 + E1.3 + E4.1$$
$$ALUS0 = \text{AND} + \text{OR} + \text{PLUS} + \text{ZERO}$$
$$= F1.2 + E0.10 + E0.11 + E0.16 + E0.22 + E0.25 + E1.3$$
$$+ E4.1 + E5.1$$

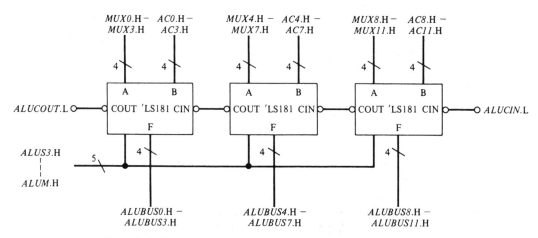

Figure 8–3 Implementation of the *ALU*. The most significant bits are on the left.

TABLE 8–3 74LS181 ALU CONTROL INPUTS[a]

S3	S2	S1	S0	M	CIN	Function
F	F	F	F	F	T	INCREMENT
T	F	T	T	F	T	AND
F	F	F	T	F	F	OR
F	F	F	F	T	F	NOT
T	F	F	T	F	F	PLUS
F	F	T	T	T	F	ZERO
F	F	F	F	F	F	PASS

[a]Adapted from 74LS181 data sheet. *S0, S1, S2, S3,* and *M* are T = H; *CIN* is T = L.

$$ALUM \quad = \text{NOT} + \text{ZERO}$$
$$= F1.2 + E0.16 + E0.20 + E0.22 + E1.3 + E4.1$$
$$ALUCIN = \text{INCREMENT} + \text{AND}$$
$$= F2 + F4.2 + E0.7 + E0.10 + E0.12 + E0.18 + E0.24 + E1.2$$
$$+ E1.4 + E2.1 + E3.1$$

Register–Load Signals

The output of the *ALU* serves as the 12-bit data bus *ALUBUS* that connects to the input of several of the registers. We need equations to specify when each register should load the *ALUBUS* data. Such loading activity shows up in the ASM charts as a register-transfer notation of the form "→ register". We may read the relevant terms for each register load from the ASM charts. Of the five equations—for *MA(LD), MB(LD), PC(LD), IR(LD),* and *AC(LD)*—we show two below. You should develop the remaining ones.

$$IR(LD) = F1.2 + F3.2 + E0.5$$
$$AC(LD) = E0.6 + E0.10 + E0.11 + E0.16 + E0.20 + E0.22 + E0.24$$
$$+ E0.25 + E1.3 + E4.1 + E5.1$$

We will implement four of the registers (*MA, MB, PC,* and *IR*) with enabled D-register chips. The 74LS378 Six-Bit Enabled D-Register and 74LS379 Four-Bit Enabled D-Register are good choices. (The 74LS378 presents its outputs with one voltage polarity; the 74LS379 provides both voltage polarities for its outputs.) These registers have a load enable input, so the register load equations derived above apply directly to the enable inputs of these registers. The *AC* register must also serve as a shift register to implement the rotate micro-operation. The shift register building block, with responsibilities for shifting left, shifting right, loading, and holding data unchanged, is more complex to control than is the simple enabled D register. If we choose the 74LS194 as our shifter, two control lines *S1* and *S0* govern the operations. The serial inputs for left and right shifts must come from the *LINK* bit, since the PDP-8 rotate operations treat the *AC* and *LINK* as a combined 13-bit circular register. In Fig. 8–4 we show the structure of the *AC* circuit. Table 8–4 contains the relationship of accumulator controls to the ASM chart logic. Let *ACS1* and *ACS0* be the LD20 signals that control the *S1* and *S0* inputs. Then the logic equations follow from the table:

$$ACS0 = \text{LOAD} + \text{RIGHT}$$
$$= AC(LD) + E0.29$$
$$ACS1 = \text{LOAD} + \text{LEFT}$$
$$= AC(LD) + E0.28$$

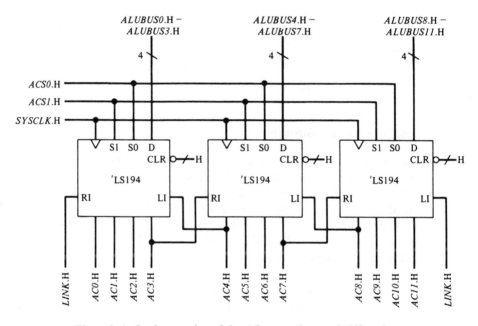

Figure 8–4 Implementation of the *AC* accumulator and shift register.

TABLE 8–4 ACCUMULATOR CONTROLS

ASM output	74LS194 operation	74LS194 control inputs[a]		ASM terms
		S1	*S0*	
→ *AC*	LOAD	T	T	= *AC(LD)*
ROTATE RIGHT	RIGHT	F	F	*E0.29*
ROTATE LEFT	LEFT	T	T	*E0.28*
No Change	HOLD	F	F	All others

[a]Adapted from 74LS194 data sheet. *S1* and *S0* are T = H.

332   The Art of Digital Design   Part II

Architecture and Control of the Link Bit

The PDP-8 uses the link bit in several ways. There are specific instructions to clear and complement the link. Overflows from arithmetic operations cause *LINK* to be complemented. The rotate micro-operations in the Operate instruction use the combined *AC–LINK* as a 13-bit circular register.

We could use a JK flip-flop for *LINK* storage, and derive the control equations for setting and clearing the flip-flop from the ASM chart. Since there are so many different operations on *LINK*, we look for a more orderly way to proceed. Suppose that we implement *LINK* as a simple data storage element such as a D flip-flop. What is the input to *LINK* in each of the necessary operations? For clearing and complementing, the inputs are zero and $\overline{LINK}$, respectively. For left rotates, *AC0* is the input to *LINK*, and for right rotates, *AC11*. Is this all? If we use a D flip-flop, we must also provide for the frequent occasions when nothing happens to *LINK*. The "do–nothing" situation requires that the input to *LINK* be *LINK* itself. If an enabled D flip-flop were available, we could use the enable feature to manage the constructive actions, leaving the flip-flop disabled during the do-nothing periods. However, such individual enabled D flip-flops are not common, so we choose to implement *LINK* with a simple D flip-flop. Therefore, at all times we must select one of the five inputs to the link register: *LINK*, $\overline{LINK}$, zero, *AC0*, or *AC11*. That magic word *select* shows us the way: use the multiplexer building block. We dodged the link architecture in Chapter 7 because at that time we did not understand the complete role of the link bit. Now that we know what is required, we can add a *LINK* circuit to our architecture. We need a 5-input multiplexer, but an 8-input mux is the smallest useful and commonly available chip. Choose a 74LS151; we will control its three select inputs with LD20 signals *LINKMB4*, *LINKMB2*, and *LINKMB1*. Figure 8–5 shows the link unit. Table 8–5 is a summary of the control conditions

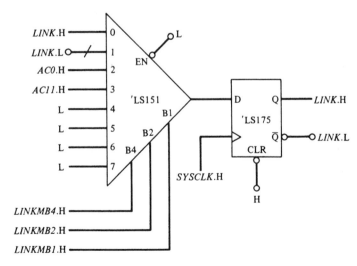

Figure 8–5 Circuit for the link bit.

TABLE 8-5 MULTIPLEXER CONTROLS FOR LINK

| ASM output | Link mux controls | | | ASM terms |
	$B4$	$B2$	$B1$	
$0 \rightarrow LINK$ (Clear)	1	X	X	$E0.8 + E0.17$
SHIFT RIGHT	0	1	1	$E0.29$
SHIFT LEFT	0	1	0	$E0.28$
$\overline{LINK} \rightarrow LINK$ (Complement)	0	0	1	$E0.21 + E0.37 + E0.38$
No Change (Hold)	0	0	0	All others

for the link multiplexer. The resulting equations for the control inputs are

$$LINKMB4 = \text{CLEAR}$$
$$= E0.8 + E0.17$$
$$LINKMB2 = \text{SHIFT.RIGHT} + \text{SHIFT.LEFT}$$
$$= E0.28 + E0.29$$
$$LINKMB1 = \text{SHIFT.RIGHT} + \text{COMPLEMENT}$$
$$= E0.21 + E0.29 + E0.37 + E0.38$$

STATE SEQUENCING SYSTEM

The State Generator

To produce the proper sequence of states, we will use the multiplexer table-lookup technique. There are fourteen states in the LD20's ASM, so 16-input multiplexer chips are suitable. Four 16-input control multiplexers will provide inputs to the four state flip-flops. The 4 flip-flop outputs specify a 4-bit code for the current state. This code serves as the select input to each of the four control muxes. The state generator's architecture is illustrated in Fig. 8-6, which shows one of the four state-variable circuits.

The state decoder. The encoded form of the current state is fine for driving the control muxes, which require an encoded select input, but the output terms in the ASM require individual signals for each state. To provide these, we decode the state code using a decoder building block. Since we have fourteen states, we choose the 74154 Four-Line-to-Sixteen-Line Decoder.

State assignment and state transitions. With the multiplexer implementation of a synchronous ASM, we have little reason to favor one state assignment over another. An orderly and easily remembered assignment is **IDLE** = 0000_2, **F1** through **F7** = 0001_2 through 0111_2 (1_8 through 7_8), and **E0** through **E5** = 1000_2 through 1101_2 (10_8 through 15_8). Using this assignment and the ASM charts, Table 8-6 shows the conditions for state transitions. Whenever

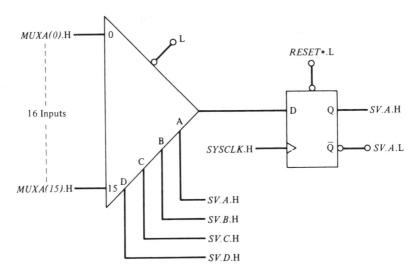

Figure 8–6 Implementation of one of the four state variables.

it is profitable, we use conditional output terms, since we must implement them anyway to provide the appropriate output signals.

There are two unused states, with assignments 1110_2 and 1111_2. If for any reason the machine is in either of these states, we wish to go next to the **IDLE** state. Also, for any state, an asynchronous *RESET** signal from the control panel must force an immediate (unclocked) jump to **IDLE**. This apocalyptic master reset is the only asynchronous state change in the design. The 74LS175 Quad D Register that serves as our state-variable memory has an asynchronous clear input. Therefore, if we choose positive logic (T = H) for the state variables, we may use this clearing feature with input *RESET** to force the LD20 immediately to **IDLE** (0000_2). Although in general we are wary of asynchronous circuit elements, the feature is convenient here.

State multiplexer inputs. Using Table 8–6, we may derive the inputs to the four state multiplexers, such as the one shown in Fig. 8–6. Here is a selection of the input equations; you should verify them and derive the remainder.

$$MUXA.E0 = MUXA(8) = E0.9 + E0.12 + E0.13 + E0.14 + E0.15$$
$$+ E0.31 \cdot \overline{ION} \cdot \overline{IOF} + E0.32 + E0.33$$
$$+ E0.34 + E0.35$$
$$MUXB.F1 = MUXB(1) = F1.1$$
$$MUXC.F5 = MUXC(5) = \overline{CC} + CC = T$$
$$MUXC.E1 = MUXC(9) = F$$
$$MUXD.E1 = MUXD(9) = \overline{CC} + E1.IR \cdot ISZ$$
$$MUXD.E4 = MUXD(12) = T$$

TABLE 8–6 STATE TRANSITIONS IN THE LD20

Present state	Next state	D	C	B	A	Conditions for transition
IDLE	IDLE	0	0	0	0	$\overline{MANPULSE}$
	F1	0	0	0	1	$MANPULSE \cdot CONT*$
	E0	1	0	0	0	$IDLE.1$
F1	IDLE	0	0	0	0	$HALTFF$
	F2	0	0	1	0	$F1.1$
	E0	1	0	0	0	$F1.2$
F2	F3	0	0	1	1	T
F3	F3	0	0	1	1	$\overline{CC}$
	F4	0	1	0	0	$F3.3$
	E0	1	0	0	0	$CC \cdot NOMEM$
F4	F4	0	1	0	0	$\overline{CC}$
	F5	0	1	0	1	$F4.2$
	F6	0	1	1	0	$CC \cdot \overline{DIRECT} \cdot \overline{AUTO}$
	E0	1	0	0	0	$CC \cdot DIRECT$
F5	F5	0	1	0	1	$\overline{CC}$
	F6	0	1	1	0	CC
F6	F7	0	1	1	1	$F6.1$
	E0	1	0	0	0	$(DCA + JMS + JMP)$
F7	F7	0	1	1	1	$\overline{CC}$
	E0	1	0	0	0	CC
E0	IDLE	0	0	0	0	$E0.1$
	F1	0	0	0	1	$E0.15 + E0.32 + E0.33 + E0.34 + E0.35$
	E0	1	0	0	0	$E0.DBL + E0.TST \cdot MOREOP$ $= E0.1R \cdot OP \cdot (DBL.IS.T + MOREOP)$
	E1	1	0	0	1	$E0.9 + E0.12 + E0.13 + E0.14$
	E3	1	0	1	1	$E0.31 \cdot \overline{ION} \cdot \overline{IOF}$
E1	IDLE	0	0	0	0	$CC \cdot MANEX.SW*$
	F1	0	0	0	1	$E1.3 + E1.4$
	E1	1	0	0	1	$\overline{CC}$
	E2	1	0	1	0	$E1.IR \cdot ISZ$
E2	F1	0	0	0	1	T
E3	E4	1	1	0	0	T
E4	E5	1	1	0	1	T
E5	F1	0	0	0	1	$\overline{IOP4.EN}$
	E0	1	0	0	0	$\overline{IOP4.EN}$
(1110)	IDLE	0	0	0	0	T
(1111)	IDLE	0	0	0	0	T
(Any)	IDLE	0	0	0	0	$RESET*$ (asynchronous)

Priority Control in the Operate Instruction

Perhaps the most sophisticated aspect of the LD20's design is the priority system for the Operate instruction. We must develop logic equations for controlling the priority-level memory element *RQST1–RQST4* and the double-shift flag *DOUBLE*, and deriving the value of the Operate priority looping variable *MOREOP*.

Table 8–7 is a summary of the priority assignments for the PDP-8I's Operate instruction and the instruction register's bit patterns for the Operate micro-operations. From this table we derive logic equations for each priority level *PRI1–PRI4*:

$$PRI1 = \overline{IR3} \cdot (IR4 + IR5) + IR3 \cdot (IR5 + IR6 + IR7 + IR8)$$
$$PRI2 = \overline{IR3} \cdot (IR6 + IR7) + IR3 \cdot IR4$$
$$PRI3 = \overline{IR3} \cdot IR11 + IR3 \cdot (IR9 + IR10)$$
$$PRI4 = \overline{IR3} \cdot (IR8 \oplus IR9)$$

Once the LD20's *IR* is loaded, these equations establish the priority levels requested by the Operate instruction.

TABLE 8–7 PRIORITY ASSIGNMENTS FOR THE PDP-8I'S OPERATE INSTRUCTION

Priority	Group	PDP-8I Operate micro-operations	Definition
1	1	CLA, CLL	$\overline{IR3} \cdot (IR4 + IR5)$
	2	Skips	$IR3 \cdot (IR5 + IR6 + IR7 + IR8)$
2	1	CMA, CML	$\overline{IR3} \cdot (IR6 + IR7)$
	2	CLA	$IR3 \cdot IR4$
3	1	IAC	$\overline{IR3} \cdot IR11$
	2	OSR, HLT	$IR3 \cdot (IR9 + IR10)$
4	1	Rotates	$\overline{IR3} \cdot (IR8 \oplus IR9)$
	2	None	$IR3 \cdot F$

Controlling the priority flip-flops. The priority flip-flops designate those priority levels of the Operate instruction not yet acted upon. The priority flip-flops *RQST1–RQST4* are initialized in state **F3** at ASM term F3.1 (see Fig. 7–42). According to the ASM chart of the Operate instruction, Fig. 7–44, *RQST1*, *RQST2*, and *RQST3* are cleared in state **E0** at ASM terms E0.19, E0.23, and E0.27, respectively; *RQFF4* does not require clearing in state **E0**. We may use JK flip-flops for *RQSTi*. We have the following equations for controlling the priority flip-flops:

$$RQST1(SET) = F3.1 \cdot PRI1$$
$$RQST2(SET) = F3.1 \cdot PRI2$$

$$RQST3(SET) = F3.1 \cdot PRI3$$
$$RQST4(SET) = F3.1 \cdot PRI4$$
$$RQST1(CLR) = F3.1 \cdot \overline{PRI1} + E0.19$$
$$RQST2(CLR) = F3.1 \cdot \overline{PRI2} + E0.23$$
$$RQST3(CLR) = F3.1 \cdot \overline{PRI3} + E0.27$$
$$RQST4(CLR) = F3.1 \cdot \overline{PRI4}$$

The double-rotate flag *DOUBLE*. We implement *DOUBLE* with a JK flip-flop, which requires set and clear control inputs. In F3.1, *DOUBLE* receives the value of bit *IR10* of the instruction register; term E0.30 resets *DOUBLE* to F. The equations are

$$DOUBLE(SET) = F3.1 \cdot IR10$$
$$DOUBLE(CLR) = F3.1 \cdot \overline{IR10} + E0.30$$

Using our mixed-logic notations for the JK flip-flop, we may use either *J* or *K* as the set input for *DOUBLE*, adopting the most convenient voltage polarities for the set and clear terms above.

The *MOREOP* loop test variable. Look now at the *MOREOP* test at the bottom of the Operate instruction's ASM in Fig. 7–44. *MOREOP* is a derived variable that specifies when it is necessary to return to state **E0** to process another priority level of the Operate instruction. How is the value of *MOREOP* related to the condition of the request flip-flops? *MOREOP* must be true if and only if there is at least one more priority level to process. In any pass through the *MOREOP* test, the request flip-flop presently being honored is still true, since it will not be reset until the end of the current **E0** state. Thus *MOREOP* must be true whenever *two or more* request flip-flops are true. Table 8–8 explicitly displays this relation.

Once again we have several possible implementations. Using SSI gates to derive *MOREOP* is straightforward but messy. Another implementation of this "two-or-more-of-four" circuit is to use a 16-input multiplexer with *MOREOP* as its output. *RQST1*, *RQST2*, *RQST3*, and *RQST4* serve as select inputs to the mux, and each data input is permanently true or permanently false, according to Table 8–8.

We obtain a more compact implementation using an 8-input multiplexer, again with *MOREOP* as its output. Look at Table 8–8 and view the data as arranged, not in sixteen rows but in eight groups of 2 rows each, based on the values of *RQST1*, *RQST2*, and *RQST3*. The two entries for each pair are distinguished by the value of *RQST4*, and the entries for each pair have outputs that behave in one of three ways:

a. Both outputs are true;
b. Both outputs are false; or
c. The outputs have the same value as *RQST4*.

TABLE 8-8 *MOREOP* AS A TWO-OR-MORE-OF-FOUR IMPLEMENTER

RQST1	*RQST2*	*RQST3*	*RQST4*	*MOREOP*
F	F	F	F	F
F	F	F	T	F
F	F	T	F	F
F	F	T	T	T
F	T	F	F	F
F	T	F	T	T
F	T	T	F	T
F	T	T	T	T
T	F	F	F	F
T	F	F	T	T
T	F	T	F	T
T	F	T	T	T
T	T	F	F	T
T	T	F	T	T
T	T	T	F	T
T	T	T	T	T

This leads us to the elegant implementation of *MOREOP* shown in Fig. 8-7. If by now you suspect that the multiplexer is a flexible and powerful building block, you are quite right.

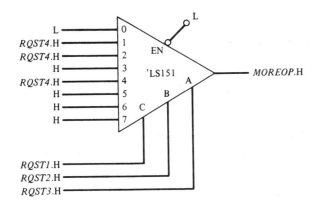

Figure 8-7 Implementation of *MOREOP*.

The *IOP* Signal Enabler

In Chapter 7, we developed a shift register architecture for issuing the *IOP* signals for the IOT instruction. To select the proper time for enabling each *IOP* signal, we have outputs *IOP1.EN*, *IOP2.EN*, and *IOP4.EN* from the shift register. The equations for issuing each *IOP* signal are

$$IOP1 = IOP1.EN \cdot IR11$$

$$IOP2 = IOP2.EN \cdot IR10$$
$$IOP4 = IOP4.EN \cdot IR9$$

State **E5** tests the signal *IOP4.EN* to see if the third pass through the IOT states is in progress. If *IOP4.EN* is true at that time, the execution of the instruction is complete and the next state will be **F1**; otherwise, the next state will again be **E0**.

We select a 74LS194 Four-Bit Universal Shift Register as the *IOP* enabler. This chip has two control inputs *S1* and *S0*. Table 8–9 shows the conditions for each operation required of the *IOP* enabler. We may immediately develop logic variables *IOPS1* and *IOPS0* to control the 74LS194's *S1* and *S0* inputs:

$$IOPS0 = LOAD + RIGHT.SHIFT = F1 + E0.31$$
$$IOPS1 = LOAD + LEFT.SHIFT \ = F1$$

TABLE 8–9 *IOP* ENABLE CONTROL

ASM output	74LS194 operation	74LS194 control[a] *S1* *S0*		ASM terms
TFFF → *IOP* Enabler	LOAD	T	T	*F1*
IOP Enabler Right Shift	RIGHT	F	T	*E0.31*
(Not used)	LEFT	T	F	F
(No change)	HOLD	F	F	All others

[a]Adapted from 74LS194 data sheet. *S1* and *S0* are T = H.

Control of the Interrupt System

Sensing an interrupt request. Since an interrupt request signal *EXTINT*∗ may arrive at any time, we must synchronize this signal to the system clock to produce the LD20's *INTREQ*, used in the formation of the auxiliary signal *INT*. We may synchronize *EXTINT*∗ with a D flip-flop; the input is *EXTINT*∗, the output is *INTREQ*.

Interrupt system enable control. The interrupt system's enable flip-flop *INT.EN* requires set and clear commands to control its *J* and *K* inputs. According to the ASM chart, the equations are

$$INT.EN(SET) \ = E0.32$$
$$INT.EN(CLR) = F1.2 + E0.8 + E0.33$$

Forcing the interrupt. State term F1.2 generates an interrupt by forcing a JMS operation code into the first 3 bits of the instruction register, so that the execute phase will proceed as if it is processing a regular JMS instruction. The *MA* register is also set to zero at this time to provide the proper effective address for the JMS instruction. The *ALU* output *ALUBUS* is the normal data path for input to the instruction register *IR*. We must add the necessary logic to the inputs of *IR0*, *IR1*, and *IR2* to create the JMS operation code (100_2) in term F1.2, but to pass normal *ALUBUS* data at other times.

One way to do this is to add a "JMS" (4000_8) source to the data mux system and select this source into *IR* during F1.2. However, we are reluctant to use our last, carefully preserved data mux input. But there are only three data bits involved in this transaction. Therefore, rather than enlarging the data muxes to 16-input varieties, we will modify the inputs to instruction register bits *IR0*, *IR1*, and *IR2* to accommodate the JMS jam at term F1.2. The equations are

$$IR0 = \overline{F1.2} \cdot ALUBUS0 + F1.2 \cdot T = ALUBUS0 + F1.2$$
$$IR1 = \overline{F1.2} \cdot ALUBUS1 + F1.2 \cdot F = \overline{F1.2} \cdot ALUBUS1$$
$$IR2 = \overline{F1.2} \cdot ALUBUS2 + F1.2 \cdot F = \overline{F1.2} \cdot ALUBUS2$$

The inputs to bits *IR3* through *IR11* are unchanged by this activity and remain directly connected to the corresponding *ALUBUS* lines.

The Manual System

MANPULSE. According to our plan, a depression of a manual switch must cause the assertion of a signal *MANPULSE* for only one clock cycle following the depression. *MANPULSE* is the output of a single-pulser circuit whose input is *MANSW**.

MANEX. In the ASM for the execute phase we use the *MANEX* flip-flop's output to help decide if manual operation or regular instruction processing is appropriate. If you implement *MANEX* using a JK flip-flop, you should be able to derive the equations for *MANEX(SET)* and *MANEX(CLR)* from the ASM charts.

Logic Equations

We have derived some of the logic equations in this chapter, but have shown actual circuits for only a few. Figure 8–8 shows some typical circuits for LD20 logic equations.

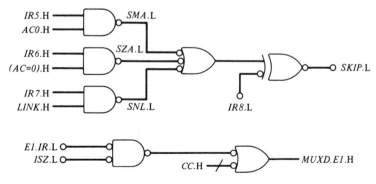

Figure 8–8 Some typical circuits for logic equations in the LD20.

With your knowledge of mixed logic, you should be able to draft clear and correct circuits for realizing the logic equations needed for controlling the LD20. All the control signals can be constructed with SSI or MSI chips. But don't forget PALs as candidates for creating some of these signals. For instance, several control signals are functions of the bits of the *IR*: *ION*, *IOF*, *NOMEM*, the Operate instruction priorities *PRI1–PRI4*, and so on. Perhaps many or all of these signals can be generated by a suitable PAL, with fewer chips.

THE MEMORY SYSTEM

Access to the LD20's Memory

Read and write signals.   In the LD20, *START.READ* and *START.WRITE* initiate memory reads and writes, according to the memory protocol developed in Chapter 7. From the ASM, we may derive the equations for these variables:

$$START.READ\ \ = F2 + F3.3 + F6.1$$
$$START.WRITE = F4.2 + E0.9 + E0.12 + E0.13 + E0.14$$

The LD20's memory interface.   The LD20 signals that interface with the memory control unit appear in Fig. 7–16. The correspondence between the standard memory unit signals and our LD20 variables is

$$
\begin{aligned}
BEGIN.READ &= START.READ \\
BEGIN.WRITE &= START.WRITE \\
MEM.BUSY &= BUSY \\
ADDRESS_i &= MA_i \\
DATA.IN_i &= MB_i \\
DATA.OUT_i &= MEM_i \\
EXTERNAL.CLOCK &= SYSCLK
\end{aligned}
$$

The LD20 can supply all these inputs directly from signals available in its architecture or control circuits. Data from the memory unit also goes directly into the LD20's architecture. The remaining LD20 signal from the memory control unit, *BUSY*, does not appear directly in our LD20 ASM charts. To achieve a compact ASM representation of LD20 memory reads and writes, we introduced in Chapter 7 a signal *CC* (Cycle Complete), which has the property of becoming false when a memory read or write begins, and becoming true one clock cycle after the *BUSY* signal from the memory controller becomes false. We must fabricate this LD20 extension to the standard memory controller. If we use a D flip-flop for *CC*, we may most readily specify the input *CC(D)* by describing when *CC(D)* is *false:*

$$\overline{CC(D)} = BUSY + START.READ + START.WRITE$$

$CC(D)$ must be false whenever $BUSY$ is true, or whenever we are about to begin a read or write (i.e., when $START.READ$ or $START.WRITE$ is true). This assures that the output CC is false during the first clock cycle of the memory operation, because either $START.READ$ or $START.WRITE$ was true when the clock pulse that began the memory operation arrived. After that, CC remains false during all remaining cycles in which the memory operation is in progress, because $BUSY$ is true. $BUSY$ will not become false until the first clock pulse after the memory operation is finished; the output CC will not respond to the change in $BUSY$ until the following clock pulse. Figure 8–9 shows this circuit.

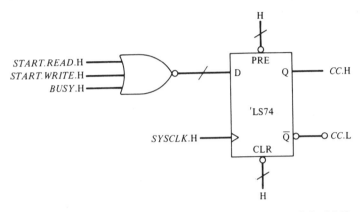

Figure 8–9 Extension of the standard memory interface of the LD20.

The Memory Unit

It is now time to specify the memory chips we wish to use and to design the control unit to use these memory chips in conformity with the protocol developed in Chapter 7. The memory unit has the following responsibilities:

(a) Use the 12 input address lines to access the proper word in the memory.
(b) Perform a write operation upon command, using the information on the data input lines.
(c) Perform a read operation upon command, and maintain the information on the data output lines thereafter, until another memory operation commences.
(d) Provide a timing signal $BUSY$ to indicate to the user when the memory unit is occupied with a user-originated read or write operation.

We have chosen the 6147 4K × 1 Static RAM as the building block for the LD20 memory. Here is a brief description of its characteristics.

The 6147 memory chip. The 6147 is a 4K × 1 static RAM having 12 address pins, 1 data input pin, 1 three-state data output pin, a write control pin, and a chip–enable pin. The write control and chip enable pins are low-active;

the rest are high-active. The data-out line is in its high-impedance state during write operations and whenever chip enable is false.

We will use a version of the 6147 having a 70-nsec cycle time: the minimum read and write cycles require 70 nsec. The maximum guaranteed read access time is 70 nsec. Since our design will stabilize the address, chip enable, and input data signals at about the same time (shortly after the initiating clock pulse), we can simplify the rather complex timings for read and write. There are no additional critical timings for the read operation. For the write operation, the address-write-setup time (time that the address lines must be stable before making the write signal true) is 0, the address-hold time (time that the address must be stable after write becomes false) is 15 nsec, and the data-hold time (time that the data must be stable after write becomes false) is 10 nsec.

Memory organization and addressing. The LD20 requires 4096 words of 12 bits. We will use twelve 6147 RAM chips, one for each bit. The RAM chips will be controlled identically.

Memory timing. The 6147 memory chip has no mechanism to announce when its operations are complete. We must determine the guaranteed upper limits of its timing from the data sheet, and we must establish our own means of providing the proper intervals for its operation. In the discussion above, you saw that with stable inputs our major timing constraint is the memory cycle time, which is 70 nsec. How can we produce a signal that reflects the memory cycle time? Perhaps the simplest way is to use a single-shot, for instance the 96L02 Dual Single Shot discussed in Chapter 4. With the proper combination of external resistor and capacitor, the single-shot will produce a true output for the required time whenever it receives a triggering signal. Call the single-shot output *TIMER*. Since the single-shot is an asynchronous edge-triggered device, we must be careful in our design to avoid spurious edges (glitches) on the single-shot inputs. We must now consider how to use this *TIMER* signal in our memory control unit.

The busy status signal. The memory protocol defines *MEM.BUSY* as a signal that changes synchronously with the external clock signal. By the nature of the protocol, a memory operation begins synchronous with the clock, so *MEM.BUSY* and its LD20 counterpart *BUSY* become true after a clock edge. Since the termination of a memory operation is an event that is asynchronous to the external clock, we deduce that we will need a synchronizing element to produce a valid *MEM.BUSY* signal. We have three sources of control of this element—the *BEGIN.READ* and *BEGIN.WRITE* inputs and the internal signal *TIMER*, which indicates the completion of a memory operation. A controlled (JK) flip-flop is the proper building block; we will use two, one to control read and one to control write, each clocked by the input signal *EXTERNAL.CLOCK*. The input equations for the read-control flip-flop are

$$READ.BUSY(SET) \quad = BEGIN.READ$$
$$READ.BUSY(CLEAR) = \overline{TIMER}$$

The flip-flop's output is *READ.BUSY*. The circuit for the write-control flip-flop is similar, with output *WRITE.BUSY*. *MEM.BUSY* is the OR of these two outputs:

$$MEM.BUSY = READ.BUSY + WRITE.BUSY$$

You can see this structure in Fig. 8–10, which shows the entire memory control circuit.

We must investigate the behavior of the flip-flop outputs to verify that they do indeed represent the synchronized busy conditions for read and write operations. Table 8–10 is a summary of the *READ.BUSY* flip-flop's output during the sequence of operations dictated by the memory protocol. Under the legitimate input conditions we see that the circuit performs correctly. There is an equivalent formulation for *WRITE.BUSY*.

MEM.BUSY is a clean signal, composed of the OR of the output of two flip-flops, only one of which will be changing at any time. Since *MEM.BUSY* is free of glitches, we may confidently use it to trigger the *TIMER* single-shot used to time the memory operations.

The *WRITE* signal. The 6147 has a low-active control pin for write; the memory chips will be in the read mode whenever we are not writing. The only time we wish to send a write signal to the memory is when *WRITE.BUSY* is true. We may use the *WRITE.BUSY* flip-flop's output as a source of the 6147 *WRITE* input. The address setup time is 0, but the hold times require that we release the 6147's write signal at least 15 nsec before any changes in address or input data. Fifteen nsec is on the order of one 74LS gate delay—much shorter than our LD20 clock cycle. The memory protocol for our LD20 assures that

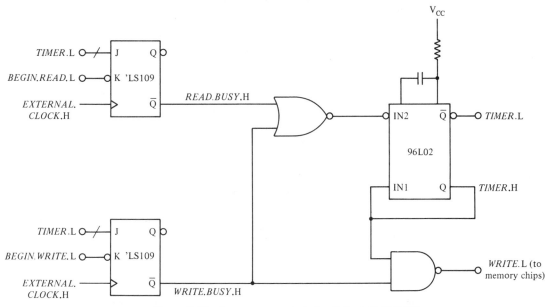

Figure 8–10 A memory control circuit for a 6147 RAM.

Sequence of input changes	Inputs *BEGIN.READ* $J_{(n)}$	$\overline{TIMER}$ $K_{(n)}$	Output *READ.BUSY* $Q_{(n)}$ $Q_{(n+1)}$	Interpretation
1	F	T	X → F	Normal quiescent condition. Flip-flop cleared.
2	T	T	F → T	*BEGIN.READ* toggles the flip-flop.
3	F	F	T → T	Holding: both inputs are false.
.				
.				
.				
4	F	T	T → F	Time out. Flip-flop clears to the quiescent condition.
—	T	F	X → T	Illegal activity.

the LD20 ASM will remain in its read or write loop for one full clock cycle after the single-shot timer has forced *CC* to become true. Thus if we control the 6147's write signal with the single-shot timer, we are assured of an ample hold time. As shown in Fig. 8–10, the 6147's write signal is true whenever *TIMER* is true and we are performing a write operation.

This completes the design and implementation of the memory control unit of the LD20. In Fig. 8–11 we show the timings of the events in a typical memory read sequence of the LD20.

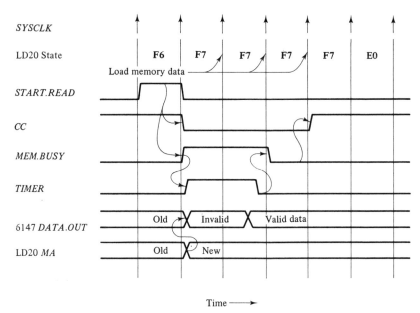

Time ⟶

Figure 8–11   Timing of a typical memory read operation of the LD20, beginning at state **F6**.

FINISHED!

The specification of the LD20 is complete. The implementation of this chapter was easier and more mechanical than the design of architecture and control in Chapter 7. This is typical of good design. Because of the care and work put into the abstract design, the implementation is straightforward. When we have drafted the remaining circuit diagrams, and added pin numbers, chip locations, and so on, to each element, we will be ready to commence actual construction. The LD20 is a real computer. Many people have constructed their own LD20 (or one of its predecessors), and it really works.

Designing the LD20 was a major project. In the course of the design, you have learned in considerable detail how a typical computer is organized, how data flows among its components, and how the control system provides for orderly processing of instructions. Most important, you have had a chance to observe the design process in its entirety. It would be wrong to think that designing the LD20 is easy or obvious. It is neither. Much thought and work went into its structure. So do not be overconfident of your abilities but conversely, do not be dismayed if you cannot produce a quick solution of a complex problem. The main theme of this book is that by adhering to good top-down design methodologies—good style—you can deal with the inherent complexity of a problem at the abstract logical level, where it is easier to manage, without creating additional complexities and restrictions by the premature introduction of hardware. Only when you have mastered the problem at the abstract level should you introduce the hardware.

In order to use the LD20 fully as a real PDP-8 emulator, we must now consider how to interface peripheral devices to the computer. We examine this problem in Chapter 9.

READINGS AND SOURCES

KUCK, DAVID J., *The Structure of Computers and Computations,* Vol. 1. John Wiley & Sons, New York, 1978.

MANO, M. MORRIS, *Computer System Architecture,* 2nd ed. Prentice-Hall, Englewood Cliffs, N.J., 1982.

PDP8/I and PDP8/L Small Computer Handbook. Digital Equipment Corp., Maynard, Mass., 1972. User's manual for the PDP-8I minicomputer.

PROSSER, FRANKLIN, and ROBERT WEHRMEISTER, *C421-C422 Advanced Computer Organization Laboratory Manual.* Computer Science Department, Indiana University, Bloomington, Ind. 47405, 1985. Laboratory manual to support the construction, debugging, and study of the LD20 and LD30 implementations of the PDP-8I. The laboratory project uses the Logic Engine Development System, manufactured by Logic Design, Inc. Ask the authors of this book for information.

EXERCISES

8-1. Why do we sometimes define auxiliary variables for use in the ASM chart?

8-2. Verify the logic equations for the position labels in the LD20's ASM chart. Pay particular attention to the derivation of *E0.DBL* and *E0.TST*.

8-3. Verify the logic equations for the LD20 ASM conditional output terms given in the text. Derive the remaining terms.

8-4. Derive the equation for *E0.34* directly from Fig. 7-44, without using the auxiliary variables *E0.DBL*, *E0.TST*, or *DBL.IS.T*.

8-5. The equations for the data multiplexer select control signals *B4*, *B2*, and *B1* depend on the mux input position assigned to each register. Although the assignment is a designer's choice, some choices may result in simpler implementation logic than others. Show that the register attached to mux position 0 does not enter into any of the sum-of-product equations for *B4*, *B2*, or *B1*. Is the register used in position 0 in the LD20 the optimal choice? If not, which register is optimal?

8-6. Suppose we elect to assign sources to the 8 data mux inputs in the following order: *AC*, *MA*, *MEM*, *PC*, *EA*, *SR*, *MB*, *INPUT*. Derive appropriate logic equations for the data mux select controls *B4*, *B2*, and *B1*. Does such an assignment change any other equations in the LD20?

8-7. In Fig. 8-1, the instruction register *IR* is not an input to the data mux. Why?

8-8. Suppose the multiplexers in the main data path were replaced with three-state circuits to control access to the ALU. We then would have three-state output-enable signals on each register: *AC(OE)*, *MB(OE)*, etc. Draw a circuit diagram of such a data bus, and derive logic equations for each of the output-enable signals.

8-9. For several paths in the LD20's ASM, no data mux activity is required, yet the main data mux in Fig. 8-1 is always enabled, and the *ALU* controls are performing a PASS operation. Examples are in state **F5** and conditional term E0.26. What information is present on the *ALUBUS* at such times? Why is this phenomenon not harmful to the activities of the LD20?

8-10. Why is *MEM* not a destination on the *ALUBUS*? Why are *EA*, *SR*, and *INPUT* not destinations?

8-11. In the LD20, the *AC* forms the *B* input to the 74LS181 ALU chips, and the data mux output is the *A* input. Consult a data sheet for the 74LS181, and show why we cannot successfully *reverse* these two inputs to the *ALU*.

8-12. Derive the LD20's logic equations for the 74LS181 ALU control inputs.

8-13. Table 8-3 presents the logic for the 74LS181 ALU control inputs, under our assumption that *CIN* is T = L. Show why this is a wise choice for *CIN*. (*Hint:* Consider the expression for the PASS operation in Table 8-2.)

8-14. Derive equations for the register load signals *MA(LD)*, *MB(LD)*, and *PC(LD)*.

8-15. Verify Table 8-4. Derive the accumulator register control inputs *ACS1* and *ACS0*.

8-16. Implement the link bit as a JK flip-flop. Derive logic equations for setting and clearing the link flip-flop. Compare the clarity and compactness of this approach with the method used in the LD20.

8-17. Verify Table 8-5. Why do we assign input position 0 on the link mux to the hold condition? Derive logic equations for the link mux select inputs.

8-18. Implement a one-hot state generator for the LD20. Compare the complexity with that of the multiplexer controller method. What effect must the *RESET** signal have on the one-hot system?

8-19. In Fig. 8–5, the mux select code is implemented as T = H. What changes would be required in Fig. 8–5 if the code were:

    (a) Implemented with T = L?

    (b) Implemented with the state variables in reverse order; in other words, with *SV.A*.H going into mux select input *D*, and so on? Are such variations on Fig. 8–5 useful or wise?

8-20. The 74154 Four-Line-to-Sixteen-Line Decoder, used as a state decoder in the LD20, is a large 24-pin chip that requires more power than chips of the low-power Schottky family. Show a circuit for the state decoder with 74LS42 Four-Line-to-Ten-Line Decoder chips, which have 16 pins and require less power.

8-21. Working directly from the LD20's ASM, derive Table 8–6.

8-22. Derive logic equations for all inputs to the state generator multiplexers in the LD20.

8-23. The 74LS163 Programmable Binary Counter chip can serve as a 4-bit D register with a synchronous clear. Why do we not use this chip to implement the state generator flip-flops, and thereby support a *synchronous* resetting of the LD20?

8-24. Show why the variable *MOREOP*, which controls looping in the LD20 Operate instruction, is a two-or-more-of-four implementer rather than a one-or-more-of-four.

8-25. Show that the Operate instruction variable *MOREOP*, described in Table 8–8, has three more implementations equivalent to Fig. 8–7.

8-26. Derive equations for the control inputs to the *MANEX* 74LS109 flip-flop.

8-27. The halt flip-flop *HALTFF* may be implemented with a JK or an enabled D flip-flop.

    (a) Using a JK flip-flop, write logic equations and draw circuit diagrams for inputs *HALTFF(SET)* and *HALTFF(CLR)*.

    (b) Using an enabled D flip-flop, write logic equations and draw circuit diagrams for inputs *HALTFF(D)* and *HALTFF(EN)*.

8-28. Draft circuit diagrams for each auxiliary variable in the LD20.

8-29. Draft circuit diagrams for the LD20 ASM chart's position labels and conditional output terms.

8-30. Draft circuit diagrams for the LD20 data multiplexer's select inputs, the *ALU* control inputs, and the register load signals.

8-31. Draft complete circuit diagrams for the LD20 *ALU* system in Fig. 8–2. Show pin numbers.

8-32. Draft circuit diagrams for the LD20 registers *MA*, *MB*, *PC*, and *IR*. Show pin numbers.

8-33. Draft circuit diagrams for the LD20 *AC* and *LINK* circuits. Also draft circuits for the control signals for these systems. Show pin numbers.

8-34. Draft circuit diagrams for the LD20 state generator system, including the instruction decoder. Show pin numbers.

8-35. Implement each input to the state-generator multiplexers.

8-36. Suppose the LD20 ASM were implemented with a one-hot state generator, using D-registers. Write logic equations and draw a circuit diagram for the D-inputs to the flip-flops. How will you implement the RESET operations in the ASM?

8-37. Implement the LD20's multiplexer-based state generator, using PALs. For the inputs to the multiplexers, use the logic equations you developed in earlier exercises.

8-38. Implement the LD20's one-hot state generator of Exercise 8–36, using PALs.

8-39. Draft circuit diagrams for implementing the priority system of the LD20's Operate instruction using MSI-level technology. Include the priority request flip-flops and their inputs, the *DOUBLE* flip-flop and its inputs, and the *MOREOP* function.

8-40. Show two different implementations of the circuit for the Operate instruction variable *DOUBLE*, using a 74LS109 flip-flop. In one, assume that *DOUBLE(SET)*.H and *DOUBLE(CLR)*.L are conveniently available; in the other, assume the availability of *DOUBLE(SET)*.L and *DOUBLE(CLR)*.H.

8-41. The LD20's *SKIP* signal may be implemented with a PAL20L10. With this PAL, you can also implement the intermediate term *(AC=0)* that appears in the equation for *SKIP*.

    (a) The most straightforward implementation with this PAL yields *SKIP*.L. Show this implementation of *SKIP*.

    (b) By producing the inverse of *SKIP* within the PAL, you can achieve an efficient implementation of *SKIP*.H. Show this implementation.

8-42. Using PALs, implement the Operate instruction's priority request system. This can be done, for instance, with a PAL20L10 and a PAL16R4. The PAL16R4 provides four D flip-flops, which can serve as the request flip-flops. The 20L10 can generate the PRI_i signals, which are functions only of the bits in the instruction register. (There is room left over in the 29L10 to generate several other auxiliary signals that are functions of the instruction register bits.) You will have to recast the equations in the text for the inputs to the request flip-flops so that they apply to the D flip-flop implementation.

8-43. Draft circuit diagrams for the LD20 *IOP* signal enabler. Include the logic for the control inputs.

8-44. What changes would be required in the LD20 memory unit if 1024×1 RAM chips were used instead of 4096×1 chips? What if 1024×4 RAM chips were used?

8-45. Why is it necessary that the memory unit's output signal *MEM.BUSY* be synchronized with the LD20 clock?

8-46. Develop and explain a table similar to Table 8–10 for the behavior of the *WRITE.BUSY* flip-flop in the memory unit of the LD20.

8-47. The 96L02 Dual Single Shot has two equivalent and independent sections. Figure 8–10 uses only one of these. Devise an alternative memory control circuit that uses both sections of the 96L02 and avoids the necessity of using the OR gate as a part of the sensitive single-shot trigger input.

8-48. Complete the memory control circuit diagram of Fig. 8–10. Include pin numbers. By referring to a data sheet for the 96L02 single shot, estimate values of the single-shot timing resistor and capacitor required to produce a delay of 200 nsec.

8-49. In Exercise 7–34 we proposed a "manual phase" of the LD20's ASM to perform manual pushbutton operations separately from the execute phase. Implement a modification of the LD20 that incorporates this manual phase.

8-50. To provide console control of the link bit, we wish to modify the LD20 so that manually loading the accumulator (using the LOAD AC pushbutton) will also toggle the link bit. Make this modification, showing any alterations or additions to the LD20's ASM chart, and any alterations or additions to the LD20's logic equations and architecture.

8–51. Suppose we have decided to add a new PDP-8 instruction to the LD20, as a part of group 3 of the Operate instruction. The new instruction is: if the interrupt request input is true, then *skip two instructions and set the link bit true*. Show any changes in the LD20's architecture and ASM required by this new micro-instruction.

8–52. The PDP-8E, another model of the PDP-8 minicomputer, has an additional programmer-accessible element—a 12-bit *MQ* (multiplier-quotient) register. The *MQ* register is manipulated by an extension of the Operate instruction that includes a set of group 3 microinstructions, designated by Operate instruction bits 3 and 11 equal to 1. Group 3 has three microinstructions, which fall into two priority classes:

> Priority 1: CLA (instruction bit 4 is 1)
>
> Priority 2: MQA (instruction bit 5 is 1)
>
> MQL (instruction bit 7 is 1)

The PDP-8E defines these microinstructions as follows:

> CLA (Clear accumulator): $0 \rightarrow AC$
>
> MQL (*MQ* load): $AC \rightarrow MQ; 0 \rightarrow AC$
>
> MQA (*MQ* into *AC*): $MQ + AC \rightarrow AC$

In this exercise you will add this set of microinstructions to the LD20. There is one point that makes this modification nontrivial. The PDP-8E manual states that specifying both MQL and MQA in a microinstruction will cause the *AC* and *MQ* register contents to be swapped. The defined operations of MQL and MQA will not support this swap (try it!), so you must be artful in your solution. Your task: make suitable modifications to the architecture and ASM of the LD20 to support the group 3 microinstructions.

8–53. A customer insists that she needs a computer like the LD20 but in which the computer's AND instruction is replaced by a new instruction that performs the customary AND operation and then replaces the contents of the referenced memory location by its one's complement. Show all the modifications to the LD20's architecture and ASM, and implement the modified instruction.

8–54. Assume you are testing the LD20, with the clock in its manual mode so that you can easily deliver clock pulses and observe the effects on the display panel lights. The LD20 is in its **IDLE** state and the Single Instruction Switch is off. The present contents (in octal) of some of the registers are

> *MA*: 1017
>
> *IR*: 4212
>
> *AC*: 0013
>
> *MB*: 2222
>
> *SWR*: 4444

What is the sequence of states executed by a properly functioning LD20 if you hold down the CONT (Continue) button and issue a sequence of manual clock pulses?

9

Interfacing with the Minicomputer

To make the LD20 truly useful, we need to be able to attach input and output devices to it. In this chapter, we will design an asynchronous interface between a keyboard-display terminal and the LD20. The interface will support the execution of PDP-8 IOT instructions designed for controlling the terminal. We will first describe the terminal device and its communication mechanism. Then after reviewing the PDP-8 input-output protocol, we will describe the specific IOT instructions for the terminal. With this background, we will design the interface. The design will follow the usual pattern of separating architecture and control, but we will develop an asynchronous ASM for the control section of the interface.

TERMINAL COMMUNICATIONS

A typical terminal consists of a keyboard and a display which communicate with other devices over two pairs of signal wires. One pair of wires transmits data from the terminal; the other pair receives data into the terminal. The terminal transmits and receives data serially, one character at a time. Each character is itself sent serially, 1 bit at a time, at a set rate, typically 300 to 9600 bits per second. The terminal transmits and accepts characters in an 8-bit code: usually the 7-bit ASCII character code plus an eighth bit that sometimes represents character parity. Data is transferred using the standard *asynchronous communication discipline*, which sends each character as a self-contained entity. In this discipline, a signal line has two valid levels: *mark*, corresponding to a data 1-bit, and *space*, corresponding to a data 0-bit. When no data is being communicated, the signal line is in the mark level. To begin transmitting a character, the transmitting device changes the signal level from mark to space, and holds the space level for one bit-time. This creates a *start* bit for the character. After the start bit, the 7 ASCII character code bits and the eighth (parity) bit follow, each occupying one bit-time. Following the eighth bit are one or two *stop* bits, in the mark level. Figure 9–1 shows the waveform for the asynchronous transmissions of the ASCII character "P," which has the code 120_8. We assume that the parity bit records even parity for the 8 data bits; in the example we use one stop bit. A character requires 7 data bits, 1 parity bit, 1 start bit, and 1 stop bit, for a total of 10 bits per transmitted character, making the maximum rate of character transmission one-tenth of the transmission bit rate.

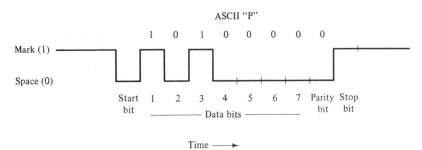

Figure 9–1 Asynchronous transmission of the ASCII character "P." Even parity, 1 stop bit.

The receiving device detects the start bit by observing the transition from mark to space on its incoming data line, and then senses each bit as it arrives, at the agreed-upon data rate. Each character forms a packet of information independent of other characters.

THE PDP-8'S INPUT-OUTPUT PROTOCOL

The PDP-8's input-output protocol is specified by the IOT instruction, which is described in Chapter 7. The programmer may request that any of three *IOP* signals be sent to the external world, along with an address to identify the particular device. The PDP-8 (and our LD20) receives control information from the addressed device over the three lines *IOSKIP, ACCLR,* and *ORAC.* Twelve bits of data leave the computer from the *AC,* and 12 bits enter the computer over the LD20's *INPUT* lines. The external device must respond only when addressed, and then only according to the protocol established between that device and the PDP-8. The PDP-8's protocol specifies the sequence for delivering *IOP* signals, and for processing the incoming control signals.

Our principal concern in this chapter is to design an interface between a keyboard–display terminal and the LD20. The interface must reconcile the vast difference between the behavior of the terminal and the behavior of the LD20. The terminal will transmit and receive characters according to the asynchronous communication discipline, whereas the LD20 uses the rather curious input-output protocol implied by the PDP-8's IOT instruction.

The PDP-8I computer manual contains a set of elementary instructions for communicating with a simple keyboard–display terminal. Since we wish to use PDP-8I software, we must adopt the standard format of PDP-8 terminal instructions. The PDP-8 interface has two device addresses: 03 (*DA03*) for receiving characters from the terminal, and 04 (*DA04*) for transmitting characters to the terminal. Each of the three *IOP* signals plays a role in both transmitting and receiving data. Here are the IOT instructions for the PDP-8's standard interface with a terminal.

From Terminal to PDP-8 (*DA03*)

IOP1 (Instruction 6031_8). If there is a character from the terminal ready to be sent into the PDP-8, the interface must respond to the *DA03 IOP1* signal with *IOSKIP* = T; at all other times, *IOSKIP* = F. This instruction allows the programmer to skip an instruction if incoming data is ready. If no new data is present, no instruction is skipped.

IOP2 (Instruction 6032_8). In response to the *DA03 IOP2* signal, the interface clears its internal indication that a character is ready and sends *ACCLR* = T to the PDP-8. At other times, *ACCLR* = F. Asserting *ACCLR* causes the PDP-8 to clear the accumulator, anticipating the *IOP4* exchange to follow.

IOP4 (Instruction 6034_8). When *IOP4* = T, the *DA03* interface places its incoming data (12 bits) on the external input lines *INPUT* to the PDP-8. The interface also issues *ORAC* = T to the PDP-8, which initiates the loading of incoming data into the previously cleared *AC*. Of the 12 bits of data supplied by the interface, the leftmost 4 are zeros, and the rightmost 8 contain the terminal data.

The programmer will usually combine *IOP2* and *IOP4* into one instruction (6036_8). The following is a simple input sequence for receiving one character from the terminal; the code begins at address 1004_8.

Location	Instruction		
1004	6031	IOT	1,DA03: Skip if character ready
1005	5204	JMP	$-1: Wait for character
1006	6036	IOT	2 + 4,DA03: Clear flag and load character
1007	3---	DCA	---: Store data from accumulator

From PDP-8 to Terminal (*DA04*)

IOP1 (Instruction 6041_8). If the *DA04* interface can accept a character from the PDP-8, the interface must respond to the *DA04 IOP1* signal with *IOSKIP* = T; otherwise, *IOSKIP* = F. This gives the programmer the ability to skip an instruction if it is acceptable to send a character to the terminal.

IOP2 (**Instruction 6042₈).** The transmit interface must clear its internal indication that the transmit stage is free, thus marking *DA04* as busy. The interface remains in this condition until a character has passed through the transmit stage and the transmit stage can again accept a character.

IOP4 (**Instruction 6044₈).** This signal causes the *DA04* interface to load a character of data from the lower 8 bits of the PDP-8's accumulator. The interface transmits these 8 data bits to the terminal, and having done so, marks the *DA04* internal flag as free to accept another character.

The programmer will ordinarily issue *IOP2* and *IOP4* together in one IOT instruction. Here is a sequence for transmitting one character to the terminal; the code begins at location 1004:

Location	Instruction		
1004	- - - -	----	Load character data into accumulator
1005	6041	IOT	1,DA04: Skip if transmit unit is available
1006	5205	JMP	$-1: Wait until transmit unit is free
1007	6046	IOT	2 + 4,DA04: Clear transmit flag, and send character

REQUIREMENTS OF THE LD20-TERMINAL INTERFACE

We have described the terminal and the PDP-8 instructions for reading and writing terminal information. Now we can specify our interface between the LD20 and the terminal. The interface has three responsibilities: to support input from the terminal to the LD20, to support output from the LD20 to the terminal, and to provide proper interrupt signals to the LD20.

Interface Flags and Interrupts

The interface must maintain a status flag for each device address. *DA03FLG* becomes true when the interface has a new character for the PDP-8; *DA04FLG* becomes true when the interface is able to accept a new character from the PDP-8. The PDP-8 programmer is responsible for resetting each flag to false, using appropriate IOT commands.

Whenever either flag is true, the PDP-8's external interrupt signal *EXTINT*∗ must be true. For a single terminal device, this means

$$EXTINT* = DA03FLG + DA04FLG$$

These two flags not only affect the PDP-8's external interrupt system; they are also the key to controlling the terminal operations using the IOT instructions.

The Receive Section (*DA03*)

The receiving section of the interface:

(a) Receives serial data from the terminal according to the asynchronous communication discipline, converts the data to an 8-bit parallel byte, and holds the byte for possible transmission to the LD20.

(b) Keeps an internal receive status flag (*DA03FLG*) to inform the LD20 of the status of the receive station. The interface sets *DA03FLG* to true upon receipt of a complete character from the terminal. The LD20 programmer may set the flag to false with an IOT instruction that issues *IOP2* with device address *DA03*.

(c) Recognizes *DA03*, and when *DA03* is present responds to the LD20's signals *IOP1*, *IOP2*, and *IOP4* with the actions described earlier. Device address 03 is

$$DA03 = \overline{IR3} \cdot \overline{IR4} \cdot \overline{IR5} \cdot \overline{IR6} \cdot IR7 \cdot IR8$$

(d) Issues T on the external interrupt line *EXTINT** whenever *DA03FLG* is true.

The Transmit Section (*DA04*)

The transmitting section of the interface:

(a) Recognizes *DA04*, and when *DA04* is present responds to the LD20's signals *IOP1*, *IOP2*, and *IOP4* with the actions described earlier. Device address 04 is

$$DA04 = \overline{IR3} \cdot \overline{IR4} \cdot \overline{IR5} \cdot IR6 \cdot \overline{IR7} \cdot \overline{IR8}$$

(b) Upon recognizing *IOP4* with *DA04*, receives 8 bits in parallel from the LD20 and transmits these bits serially to the terminal, using the asynchronous communication discipline.

(c) Keeps an internal transmit status flag (*DA04FLG*) to inform the LD20 of the status of the transmit section. This flag becomes false when the LD20 program issues *IOP2* with device address *DA04*. The flag goes true after the LD20 has written a character to the terminal and the character has passed through the interface sufficiently for the interface to be able to accept another character from the LD20.

(d) Issues T on the external interrupt line *EXTINT** whenever *DA04FLG* is true.

PRELIMINARY ARCHITECTURE OF THE INTERFACE

In Fig. 9–2, we show the signals between the LD20 and the interface, and between the terminal and the interface.

To build the interface black box, we could design a complete MSI serial-to-parallel and parallel-to-serial system. However, 8-bit serial data transmission is common, and manufacturers have produced a device to assist in the conversion between serial and parallel data.

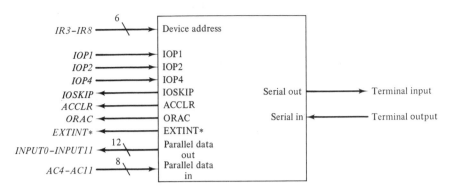

Figure 9–2 Interface signals in the LD20-terminal interface.

The UART

The *universal asynchronous receiver-transmitter* (UART) is a complex LSI circuit in a 40-pin integrated circuit package. Several manufacturers produce functionally equivalent chips. The purposes of the UART are to receive data according to the asynchronous communication protocol and render it into parallel form, and to accept parallel data and transmit it serially according to the asynchronous protocol. The UART has inputs to specify the number of bits per character (from 5 to 8), the parity mode, and the number of stop bits; hence the name "universal." In our design these are fixed parameters, so we will not consider them further. Design Examples 3 and 4 in Chapter 6 illustrate the internal operations of a UART.

The UART has independent serial receive and serial transmit sections. Here is a description of the important UART signals. All signals are high-active (T = H) unless otherwise specified. (UART nomenclature is not standardized; check the appropriate data sheet for the names of signals, the direction of bit numbering in registers, and so forth.)

Signals of the UART's receive section

Receive Serial In (*RSI*): The serial data line from the terminal.

Receive Buffer Register (*RBR8–RBR1*): Outputs from an 8-bit register that holds an assembled character received from the terminal (bit 1 is on the right). These outputs are three-state.

Data Received (*DR*): A status flag that becomes true when a valid terminal character enters *RBR*. *DR* becomes false upon the F → T transition of the *DRR* signal, described next.

Data Received Reset (*DRR*): A F → T transition of this signal clears *DR* to false. This signal is low-active (T = L).

Receive Register Disable (*RRD*): The receive buffer register has three-state outputs that are disabled as long as *RRD* is true. When *RRD* = F, the outputs *RBR8–RBR1* are available for use.

Receive Clock: The UART expects a receiver clock input of 16 times the serial bit rate. A typical clock frequency is $1200 \times 16 = 19{,}200$ Hz.

Signals of the UART's transmit section

Transmit Serial Out (*TSO*): The serial data line to the terminal.

Transmit Buffer Register (*TBR8–TBR1*): Inputs to an 8-bit register that holds a character prior to its being serialized to the terminal (bit 1 is on the right).

Transmit Buffer Register Load (*TBRL*): Upon a F → T transition on this line, the UART accepts an 8-bit character at inputs *TBR8–TBR1*. When the transmit data serializing section of the UART is free, the UART will then move the new character into the serializing section, thus freeing the *TBR*. This signal is low-active (T = L).

Transmit Buffer Register Empty (*TBRE*): This signal is true whenever the UART can accept a character into the *TBR*. It is false when the *TBR* is full, awaiting access to the transmit data serializing section.

Transmit Clock: Provides clocking for the UART's transmit section. It has the same characteristics as the receive clock and will usually originate from the same external timing source.

Incorporating the UART into the Design

Using the UART will save us much design work. With it, we have transformed our main problem for interfacing the LD20 to the terminal into interfacing the LD20 to the UART—a much simpler task. We must now determine how to reconcile the LD20's input-output protocol with the UART's status and command signals. Before we can further specify the architecture of the interface, we must investigate the control algorithm for the interface between the LD20 and the UART.

THE INTERFACE CONTROL ALGORITHM

Clocking Events in the Interface

So far, the designs in this book have been synchronous, driven by a master clock whose effect was implied in the ASM chart but which did not explicitly appear. What is the clock for the LD20-terminal interface? We must supply a UART clock, but it is too slow (perhaps 19,200 Hz) to interface with the LD20, which runs a hundred times faster. Also, the UART's fixed clock rate violates our variable-clock-speed design criterion. The LD20 system clock would serve nicely, especially since most of the interface control information consists of *IOP* signals that are synchronized with the LD20's clock. However, the PDP-8's input-output protocol does not specify that the computer's clock is available to its peripheral devices.

Should we devise an asynchronous design? In Chapter 5, we sidestepped asynchronous methods in favor of synchronous design, which is easier to manage and has fewer pitfalls. Providing the interface with the LD20 system clock would allow a synchronous design—the easiest solution to our problem. However, to present the flavor of unclocked circuits, we will develop an asynchronous interface.

In asynchronous design, with no master clock to drive the system, states can change only when the test inputs change. What are the test inputs for our interface? The main tasks are recognizing the two terminal device addresses and then responding to the LD20's *IOP* signals. The start of an event in our algorithm will be the recognition of *DA03* or *DA04* and the assertion of an *IOP* signal. The event will last until the *IOP* signal becomes false or the device address changes. Figure 9–3 is a partial ASM chart for the interface events.

In synchronous design, fluctuations in input signals cause no trouble provided that the inputs are stable at the clock transition. In an asynchronous circuit, the inputs act as clocks; any change in an input can produce ASM transitions, intended or accidental. This is the great difficulty with asynchronous design; the circuit is sensitive to its inputs at all times. Unclocked circuits lack the systematic periods of calm that are characteristic of clocked systems. In our interface, hazards in the device address or the *IOP* signals can cause spurious asynchronous transitions, but by careful planning we can eliminate hazards without elaborate techniques.

Figure 9–3 shows that interface events depend on *pseudo-clocks* such as *DA03•IOP1*, formed from a combination of a specific device address and an *IOP*

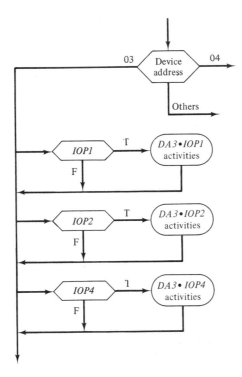

Figure 9–3 An asynchronous ASM for *IOP* signals.

signal. Bits *IR3–IR8* of the LD20's instruction register present a "device address" at all times; spurious device addresses appear as the LD20 processes assorted instructions. Figure 9–3 shows that fluctuations of the device address lines produce no algorithmic actions unless they occur while an *IOP* signal is true. *IOP* signals are true only during the execution of IOT instructions. Since the instruction register's outputs change only during the instruction fetch in LD20 state **F3**, the device address is stable whenever an *IOP* signal is asserted.

We must also verify that the *IOP* signals themselves are free of hazards. *IOP* signals arise in the LD20 from AND-gate implementations of equations like

$$IOP1 = IOP1.EN \cdot IR11$$

Since *IOP1.EN* and *IR11* are outputs of flip-flops, their transitions are smooth and free of hazards. Transitions of these two signals are well separated in the ASM; *IR11* changes only during instruction fetch in state **F3**, and *IOP1.EN* changes only at the end of states **E0** or **F1**. Therefore, the composite *IOP1* signal experiences hazard-free F → T and T → F transitions. Similarly, *IOP2* and *IOP4* are free of hazards.

We have shown that the interface pseudo-clocks are free of hazards. We emphasize that in synchronous design we avoid this type of analysis, freeing our minds to construct our algorithm rather than worrying about the behavior of the hardware.

The LD20 will deliver only one *IOP* signal at a time; we need not worry about two or more of these asynchronous events occurring simultaneously.

Synchronizing Signals to the LD20 and the UART

The LD20 expects replies from its *IOP* signals to be synchronous with its clock. How do we achieve this if the LD20's clock is not available to the device interface? We must use the pseudo-clocks developed from the combination of device address and *IOP* signal, since these are themselves synchronous with the LD20. To satisfy the synchronizing requirement, a selected device must issue proper replies for the duration of the *IOP* signals. The ASM notation in Fig. 9–3 is intended to convey this behavior.

To satisfy our LD20 synchronizing requirement, the combinational interface circuits must use only LD20-synchronized inputs to generate the *IOSKIP*, *ACCLR*, and *ORAC* replies. This means that we must synchronize certain UART outputs to the LD20 pseudo-clocks.

The UART's control inputs *DRR* (Data Receive Reset) and *TBRL* (Transmit Buffer Register Load) are both edge-triggered. Any F → T edge will trigger the UART's actions; the UART has its own internal synchronizing circuitry to avoid difficulties caused by input transitions that are asynchronous to its clock.

We next specify the interface's detailed reactions to each *IOP* signal.

Algorithm for the Receive Circuit

IOP1 **for device address *DA03*.** To satisfy the PDP-8's protocol, *IOSKIP* must assume the value of *DA03FLG* during this event. *DA03FLG* must become true when a character is available for input, and must become false when the program issues an *IOP2* signal. The UART signal *DR* (Data Received) parallels *DA03FLG*: it becomes true when a character is ready, and it becomes false when the UART input *DRR* (Data Received Reset) is asserted. We use a terminal * on names of signals that change independently of the LD20 pseudo-clocks. Thus we have

$$DA03FLG* = DR*$$

DR∗ (and thus *DA03FLG*∗) may change from F to T at any time, depending on the activities of the terminal and the UART. Before it can serve as a source for *IOSKIP*, *DA03FLG*∗ must be synchronized to the LD20's timings. We care if *DA03FLG*∗ is synchronized only when *IOP1* is true, since only then is the LD20 looking at the signal. To achieve synchronization, we will clock a D flip-flop with the hazard-free *DA03·IOP1* signal. The flip-flop's input is *DA03FLG*∗; the output is *DA03FLG.SYNC*. *IOSKIP* must have the value *DA03FLG.SYNC* for the duration of *IOP1*. In our ASM, these activities appear as conditional outputs of *IOP1*:

> *DA03FLG* ∗ → *DA03FLG.SYNC* ⬏
> *IOSKIP* = *DA03FLG.SYNC*

The upward-pointing arrow indicates that the flip-flop activity occurs on the rising (F → T) edge of the *DA03·IOP1* pseudo-clock. The equals sign in the second expression implies that the item on the left has the value of the item on the right for the duration of the pseudo-clock. The synchronizing D flip-flop becomes a part of the interface's architecture.

IOP2 for device address DA03. During this event, we must reset *DA03FLG*∗ and must assert the LD20's input signal *ACCLR*. Clearing *DA03FLG*∗ is the PDP-8 programmer's way of clearing the terminal receiver's contribution to the external interrupt signal *EXTINT*∗. Since *DA03FLG*∗ comes directly from the UART signal *DR*∗, we may clear *DA03FLG*∗ by asserting the UART's control input *DRR* (Data Received Reset). *DR*∗, and thus *DA03FLG*∗, will not return to T until the UART has received another character from the terminal. The ASM conditional output term for *IOP2* is

> *DRR*
> *ACCLR*

IOP4 for device address DA03. During this event (and only during this event), we must place the received data on the LD20's external input lines *INPUT0–INPUT11*, and assert the PDP-8's input signal *ORAC*. We gain access to the UART's 8-bit data character by enabling the three-state data outputs *RBR8–RBR1*, using the control signal *RRD* (Receive Register Disable). We need to assert $\overline{RRD}$. However, we have a small problem with notation. In the ASM charts, signals are considered false except when they are specifically asserted.

Here we have a signal *RRD* that is normally true, but must be false only when *IOP4* is true. We avoid this situation by defining another signal *RRE* (Receive Register Enable) to use in the ASM chart:

$$RRE = \overline{RRD}$$

During *IOP4* time, we assert *RRE*; it is otherwise false.

The UART provides 8 bits of three-state data for the LD20, but the LD20 expects 12 bits. Our interface must therefore contain 4 bits of three-state zero, which is enabled at the same time as the UART data bits. Adding a three-state zero box to our architecture, we have this ASM for *IOP4*:

This completes the algorithm for the receive section of our interface.

Algorithm for the Transmit Circuit

***IOP1* for device address *DA04*.** During this event, the interface must supply the value of *DA04FLG** to *IOSKIP*. It is tempting to let *DA04FLG** be the UART signal *TBRE** (Transmit Buffer Register Empty), and for the moment we will make this assumption. Since we must synchronize the signal, we will use a D flip-flop clocked by the *DA04·IOP1* pseudo-clock. This addition to the architecture has input *DA04FLG** and output *DA04FLG.SYNC*, and our ASM conditional output term is

$$\left(\begin{array}{l} DA04FLG* \rightarrow DA04FLG.SYNC \\ IOSKIP = DA04FLG.SYNC \end{array}\right)$$

***IOP2* for device address *DA04*.** For this event, the interface's only activity is to clear the output status flag *DA04FLG**. Now we have a tricky problem, typical of the kind of irritation found in device interfacing. Using the assumption we made in the previous paragraph, we should clear *DA04FLG** by clearing *TBRE**, but the only way to clear *TBRE** is to load an outgoing character into the UART, using signal *TBRL* (Transmit Buffer Register Load). The PDP-8's protocol does not allow the loading of a character during *IOP2* time, so we must seek an alternative solution.

To hold *DA04FLG**, we introduce a flip-flop that we may clear independently of the UART. Now we must determine exactly how to set the *DA04FLG** flip-flop. *DA04FLG** must record the availability of the UART's Transmit Buffer Register (*TBR*) until the LD20 program clears the flag. Once *DA04FLG** is cleared, it must remain false until a *new event* occurs to make *TBR* freshly available; we must avoid recording an old buffer availability status. The typical program activity involves testing the flag (*IOP1*), clearing the flag (*IOP2*), and writing a character to the interface (*IOP4*). Once cleared, *DA04FLG** should remain false until *TBR* again becomes available after the transmission of the character. To set *DA04FLG∗*, we must record the occurrence of a F → T *edge* of the *TBRE∗* signal. A clock input is the only readily available edge-sensitive element in our design repertoire; *TBRE∗* becomes a pseudo-clock in our ASM, as follows:

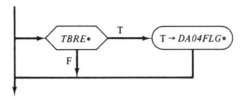

This implies a D flip-flop to hold *DA04FLG∗*, as proposed earlier. The flip-flop will have T as its input and *DA04FLG∗* as its output. The F → T transition of *TBRE∗* will act as an active clock edge, which will transfer the true input into the flip-flop, where it will remain until an *IOP2* command clears the flag. After clearing, only a subsequent F → T transition of *TBRE∗* will clock truth into *DA04FLG∗* again.

The clearing operation originates with the LD20 and is not synchronized with the UART's *TBRE∗* activity. To permit clearing at *IOP2* time, we must use a D flip-flop with an asynchronous clear input, such as the 74LS74. Figure 9–4 shows this architecture and the controls of the *DA04FLG∗* circuit. The *IOP2* conditional output of the ASM chart looks as follows:

$$\left(\text{F} \rightarrow DA04FLG* \right)$$

IOP4 for device address DA04.   After the tortuous *IOP2* exercise, *IOP4* is easy! At this time, the LD20 is placing a character on the external output data lines, and our interface must accept it and initiate serial transmission to the

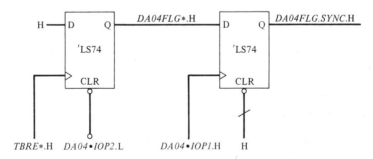

Figure 9–4   A circuit for *DA04FLG∗*.

terminal. The UART signal *TBRL* (Transmit Buffer Register Load) will do this if data from the LD20 feeds into the *TBR8–TBR1* UART inputs. The ASM segment for *IOP4* is

$$\left(\text{TBRL} \right)$$

Interrupt Generation

The LD20's external interrupt signal *EXTINT*∗ is a composite of the device flag signals. *EXTINT*∗ is true if any of the device flags is true. Programmers may clear an interrupt request by resetting the appropriate flag. The terminal interface must transmit the status of its two flags onto the LD20's external interrupt input. This activity is continuous, independent of the ASM pseudo-clocks. Therefore, the ASM unconditional output for generating an interrupt is

$$EXTINT* = DA03FLG* + DA04FLG*$$

This completes the development of the algorithm for the terminal interface. Figure 9–5 shows the full ASM.

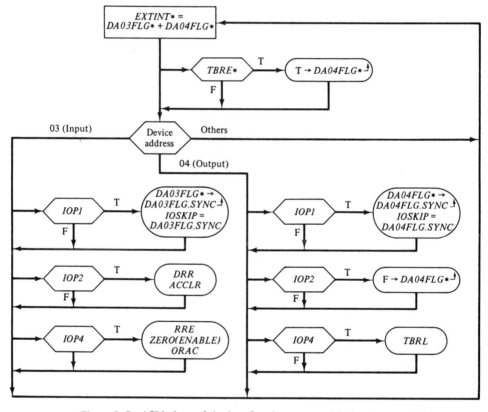

Figure 9–5 ASM chart of the interface between an LD20 and a terminal.

IMPLEMENTING THE TERMINAL INTERFACE

With the foregoing discussion, specifying the circuit for the interface is simple. Figure 9–6 shows the architecture and a sketch of the control circuits. You should find it straightforward to complete the details of the circuit.

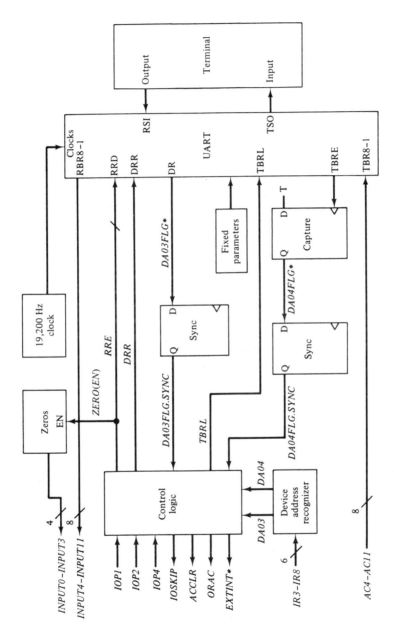

Figure 9–6 Architecture of the interface between the LD20 and a terminal.

We can provide the clock for the UART's clock inputs with a standard 555 timer circuit, as described in Chapter 12. The UART is connected with the terminal through a nondigital circuit that matches the TTL-compatible voltage signals of the UART with signal-level requirements of the terminal. The circuit contains a *line driver* and *line receiver* such as the 1488 and 1489.

POLISHING THE LD20'S INPUT-OUTPUT PROCESSES

Maintaining the LD20's Control of the Status Inputs

Suppose an LD20 program is trying to read from the terminal. Once the LD20 issues an IOT instruction with a nonzero device address, the algorithm executes states **E0**, **E3**, **E4**, and **E5** three times. This occurs whether or not the IOT instruction specifies any *IOP* signals. For proper hardware execution, we must have valid logic levels (T or F) on the test inputs *IOSKIP*, *ACCLR*, and *ORAC*. This is an important but subtle point: whenever the LD20 executes an IOT instruction, the test signals must be valid. Furthermore, the input signal *EXTINT*∗ must be correct at all times.

If we could be sure that the terminal and its interface were attached and operating, our problem would be less serious, since we could require that the device interface respond with false on the status inputs whenever it is not actively responding with truth. But we cannot guarantee that the terminal interface is plugged in whenever an LD20 program issues an IOT instruction. We cannot weasel out of this predicament by rationalizing that no self-respecting computer operator would run a terminal program if the terminal were not turned on. The terminal might be located in another building or another city, outside the control of the operator. In any event, we do not want our hardware to be dependent on an operator's intervention. If the terminal interface were turned off, there would be no logic signals at all on the *IOSKIP*, *ACCLR*, *ORAC*, and *EXTINT*∗ lines from the terminal. Obviously, then, the LD20 must *protect itself* from hardware failures caused by missing control signals.

When will IOT instructions arrive? The LD20 does not know, since IOT instructions are a function of the particular program in the memory. So, we may extend our concept of when the LD20 should protect itself: Whenever its power is on, the LD20 must protect itself against spurious external signals. Since the peripheral power supply is independent of the LD20's power supply, this protection must be in the LD20 hardware itself.

How to do this? Consider gating each input status line with the device address so that the LD20 holds *IOSKIP* and its counterparts false except when the device is being addressed. This won't work, for three reasons:

(a) Just because the LD20 addresses a device does not assure that the device will respond.

(b) Assertion of the *EXTINT*∗ signal occurs independently of device addressing.

(c) Making the LD20 hardware dependent on particular device addresses is abominable design.

We need a soft but constant source of false for the status lines *IOSKIP*, *ACCLR*, *ORAC*, and *EXTINT∗*. By "soft" we mean that any valid true signal will override the LD20's protective false. The solution is to require each peripheral device to transmit its status signals using open-collector outputs. Recall from Chapter 2 that the open-collector circuit allows and requires a pull-up resistor. This resistor connects the open-collector output to a source of V_{CC}, which in the TTL family is equivalent to connecting the output to the high-voltage level. If the resistor is located in the LD20, then the LD20's power can provide V_{CC}. In this way, whenever the LD20 is functioning, the signal it uses will be H unless a peripheral device pulls it to L. By adopting the convention T = L for each of the status signals, we have solved our problem.

This analysis causes us to modify the LD20's design to include pull-up resistors on the inputs *IOSKIP*, *ACCLR*, *ORAC*, and *EXTINT∗*. We also require that peripheral devices provide these signals as open-collector outputs, using the T = L convention.

The open-collector solution works for our LD20 status inputs even if no peripheral device sends a signal, since the pull-up resistor would hold the input high (false). In Chapter 12, we describe how to determine a suitable value of the resistor.

Attaching Several Devices

In our design of the LD20, we assume the presence of input-output data and control signals as required by the PDP-8's input-output protocol. We may place these lines into three categories:

(a) Signals from the LD20 (device address, *IOP*, and data out).
(b) Data lines to the LD20 (*INPUT*).
(c) Device status responses to the LD20 (*IOSKIP*, *ACCLR*, *ORAC*, and *EXTINT∗*).

Let's investigate how several devices can share these lines. When the LD20 has only one peripheral device, there is little difficulty, since we may attach the interface wires directly to the LD20. When there is more than one device, this crude method fails. Let's examine each of the three categories of signals.

Signals from the LD20. All devices share access to the device address, data, and *IOP* lines. The individual interface must interpret the message on these lines and act appropriately (or refrain from acting). The LD20 must produce output signals of adequate power, but the computer has no further responsibilities for getting signals to their proper destinations.

Data lines to the LD20. Input devices must share access to the input data lines. A device must take control at the proper point in its input sequence and at that time must present valid data on each of the 12 lines. At other times, the device must relinquish control. Treating the *INPUT* lines as a data bus is a convenient way to manage this activity. Of the methods in Chapter 3 for controlling bus access, the most convenient is to use three-state output controls. We may formalize the data input system for more than one device by requiring each input device to have three-state control of its data lines to the LD20. Each device must keep its three-state outputs disabled unless an IOT instruction requires the device to place data on the *INPUT* lines. Since data input is accompanied by the *ORAC* signal, the LD20 knows exactly when to accept the data.

Device status responses to the LD20.   We have already shown that the status input signals will reach the LD20 from open-collector outputs. Since open-collector technology provides a wired OR under the T = L convention, peripheral devices should have the open-collector sources of each status signal wired together at the input to the LD20. There are four wired ORs, for *IOSKIP, ACCLR, ORAC*, and *EXTINT**. This method will work well for our LD20 input-output system.

A FINAL WORD ABOUT PART II

The goal of the LD20 project was to extend your design skills using techniques and components at the MSI level of complexity. As a study project in digital design, the hardwired LD20 minicomputer is nearly ideal. This scale of problem is about at the upper limit of practicality for design with hardwired control. The PDP-8 is a simple computer—that is why we chose it for our example. Although the fundamental concepts of register architecture, memory, and sequential control apply to more complex computers as well as to our teaching example, more complex designs require additional components and techniques. More powerful VLSI circuits combining the basic elements of computer control or arithmetic operations are available. In Part III we will describe some of these tools and introduce you to microprogramming—a powerful method of digital control that forms part of the bridge between hardware and software.

The methods and materials of Part II of this book are fundamental components of digital design. With your work thus far, you are well on the way to acquiring a solid proficiency in design.

READINGS AND SOURCES

ARTWICK, BRUCE A., *Microcomputer Interfacing*. Prentice-Hall, Englewood Cliffs, N.J., 1980.

FLETCHER, WILLIAM I., *An Engineering Approach to Digital Design*. Prentice-Hall, Englewood Cliffs, N.J., 1980. Asynchronous ASMs.

KUCK, DAVID J., *The Structure of Computers and Computations*. Vol. 1. John Wiley & Sons, New York, 1978.

PDP8/I and PDP8/L Small Computer Handbook. Digital Equipment Corp., Maynard, Mass., 1972. User's manual for the PDP-8I minicomputer.

PROSSER, FRANKLIN, and ROBERT WEHRMEISTER, *C421-C422 Advanced Computer Organization*

Laboratory Manual. Computer Science Department, Indiana University, Bloomington, Ind. 47405, 1985. Laboratory manual to support the construction, debugging, and study of the LD20 and LD30 versions of the PDP-8I. The laboratory project uses the Logic Engine Development System, manufactured by Logic Design, Inc. Ask the authors of this book for information.

WIATROWSKI, CLAUDE A., and CHARLES H. HOUSE, *Logic Circuits and Microcomputer Systems.* McGraw-Hill Book Co., New York, 1980. Asynchronous ASMs.

EXERCISES

9–1. Describe the asynchronous communication discipline used in terminals and other low-speed serial devices.

9–2. What type of clocking is necessary in the asynchronous communication discipline? Design Example 4 in Chapter 6 deals with serial-parallel conversions involving a synchronous stream of data bits. Explain the difference between the two data communication disciplines.

9–3. Describe the PDP-8's IOT protocol for the terminal, including the expected responses of the terminal interface to the *IOP* signals.

9–4. What is the purpose of an input-output interface? Why do we need an interface between the terminal and the LD20? Why does the LD20 not process the terminal's signals directly?

9–5. Why must the LD20 process the external interrupt signal *EXTINT** independently of any IOT instructions?

9–6. Why must the LD20-terminal interface maintain internal transmit and receive flags?

9–7. What type of event causes state transitions in an asynchronous ASM? How is this behavior different from a synchronous system?

9–8. Why must the test inputs in an asynchronous ASM be free of hazards? Explain how we assure that the asynchronous test inputs in the LD20-terminal interface are hazard-free.

9–9. Describe the behavior of the receive section of the LD20-terminal interface ASM.

9–10. Describe the behavior of the transmit section of the LD20-terminal interface ASM. Pay particular attention to the circuit in Fig. 9–4.

9–11. Describe the behavior of a UART.

9–12. Why do we not need to present the UART with input signals that are synchronized to its clock?

9–13. Develop a complete circuit diagram for the LD20-terminal interface suitable for use as a construction guide.

9–14. Assume that the LD20's system clock is available to the interface. Design a synchronous LD20-terminal interface (using a UART) that is driven by the LD20's clock. Compare this interface with the asynchronous one.

9–15. The LD20 delivers only one *IOP* signal at a time. Suppose that the *IOP* signals for a given device address had no coherent interrelationship; in other words, suppose that they were asynchronous *to each other*. What effects might this situation have on the proper functioning of the LD20-terminal interface?

9–16. Expand the LD20-terminal interface to allow its use with other devices similar to a terminal. To do this, add an additional device address to the interface. The new address should allow the LD20 programmer to specify the following UART parameters with suitable IOT instructions:

Number of bits per character (from 5 to 8, using a 2-bit code to the UART).
Parity mode (odd or even, using 1 UART input bit).
Parity inhibit (no parity checking or production, using 1 UART input bit).
Number of stop bits (1 or 2, using 1 UART input bit).

Also, the new device address should permit the programmer to determine the various error conditions in the UART:

Parity error on input (1 UART output bit).
Framing error on input (1 UART output bit).
Data overrun error on input (1 UART output bit).

Design appropriate sequences of IOT instructions to set the UART control parameters and to read the UART status information. Develop an ASM for the interface. Implement the interface. (You may wish to consult the complete data-sheet description of a UART.)

9–17. Explain why a computer must maintain hardware control over all incoming status signals. Show how the LD20 handles this problem. Why did we choose an open-collector structure? Why not three-state?

9–18. In the open-collector structure by which the LD20 protects itself against spurious activity of external devices, why do we insist on the convention T = L?

9–19. Discuss the peripheral device's responsibilities for assuring orderly operation of the LD20's input-output busses when several devices are attached.

Microprogrammed Design

10

At the heart of our development of digital design is the algorithm. It is our basic tool for organizing our thoughts, and we use it to guide the design process. The actual implementation of the control algorithm is less important than the algorithm itself. We will accept any reasonable implementation scheme that conforms to our demands for clarity, simplicity, and regularity. In Part II, we developed systematic methods of realizing algorithmic state machines using building blocks of the scale of MSI integrated cicuits. Are there other ways to transform ASM charts into circuits?

From the earliest days of computers, programmers have regarded computers as machines for executing algorithms. In 1951, Maurice Wilkes* proposed building a special "computer" for executing algorithmic state machines, with the logic of the algorithm residing in a special program callled a microprogram. Wilkes's concept, microprogramming, was well ahead of the state of digital technology. Microprogramming was not used commercially until 1964, when IBM employed it extensively in the construction of the System/360 series of computers.

The year 1964 also saw the beginnings of the small, inexpensive computer. In that year the Digital Equipment Corporation introduced the PDP-8 minicomputer, which was the first CPU inexpensive enough to be dedicated to running algorithms to control a particular device. The PDP-8 and its successors and imitators

---

* M. V. Wilkes, "The best way to design an automatic calculating machine," *Manchester University Computer Inaugural Conference,* 1951, p. 16. A more accessible reference is M. V. Wilkes, "The growth of interest in microprogramming: a literature survey," *Computing Surveys,* Vol. 1, September 1969, pp. 139–145.

enjoyed wide use for more than ten years in sophisticated digital control applications, although engineers looked upon these uses as simply an extended form of conventional programming. In 1974 another wave of technology produced the dramatically less expensive CPUs that we call microprocessors and microcomputers. The ensuing explosion of applications will continue in the foreseeable future. Unfortunately, many people equate the microprocessor with the concept of microprogramming, a serious misconception. Today's microprocessors and microcomputers are inexpensive, small, conventional computers, programmed in a conventional way. Microprogramming represents a different approach to programming, and although many computers are *constructed* with the aid of microprogramming techniques, the conventional software programmer normally does not use the technique. To separate these ideas more clearly, we refrain from using the sadly diluted "microcomputer" and "microprocessor" names when referring to microprogrammed devices. Instead, we will speak of a "microprogrammed controller," or "microcontroller."

CLASSICAL MICROPROGRAMMING

Wilkes recognized the fundamental separation between controller and architecture and was able to contemplate new and systematic ways of implementing the control function. Although formal ASM charts had not yet been invented, Wilkes proposed a machine whose fundamental operation was the execution of the ASM state in Fig. 10–1. This standard state has at most one test variable, and may have none.

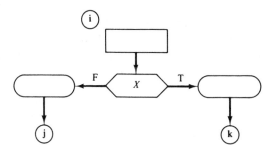

Figure 10–1 The ASM state executed by Wilkes's microprogramming machine.

Wilkes proposed to use diodes for his machine. At that time, diodes were used to construct logic AND and OR functions. Although we no longer use diodes for this purpose, to appreciate Wilkes's proposal you should understand that diode construction provides a wired-OR capability similar to that of the modern open-collector gate.

We will describe Wilkes's method by means of an example. Consider the simple ASM in Fig. 10–2. Let's proceed with a clocked implementation of the state generator using an encoded state assignment of the usual sort; Fig. 10–2 shows an arbitrary assignment of state variables B and A for the three states. Implementing the algorithm calls for constructing the new values of the state variables, $B(NEW)$ and $A(NEW)$, which serve as inputs to the clocked state flip-

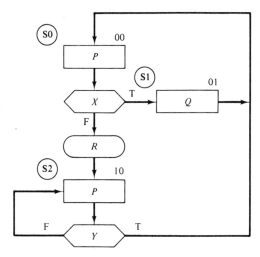

State variables: *BA*

Figure 10–2 An ASM with three states, to illustrate classical microprogrammed control.

flops. If we decode the current state variables, we may produce logic signals *S0*, *S1*, and *S2* for the individual states. A routine examination of the ASM leads us to the following equations for the various outputs required in the implementation:

$$B(NEW) = S0 \cdot \overline{X} + S2 \cdot \overline{Y}$$
$$A(NEW) = S0 \cdot X$$
$$P = S0 + S2$$
$$Q = S1$$
$$R = S0 \cdot \overline{X}$$

Wilkes proposed a systematic way of implementing these equations, which we will model as follows. Arrange the test inputs on vertical wires, with individual state signals emerging from the decoder on horizontal wires. Then, with diode AND gates, produce the branch path terms required for each state and send these horizontally to the right for use in constructing the command and next state functions. For instance, in the first step, the scheme accepts state signal *S0* and test input *X*, and produces $S0 \cdot X$ and $S0 \cdot \overline{X}$. Rather than use notations for diode or simple AND gates, let us suppress the detail in order to emphasize the standard, systematic structure of the design. We use a special square symbol with two outputs:

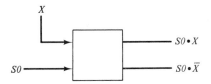

One such operation in each branching state will produce all the product terms required by the logic equations. If a state has no test input, we omit the square symbol, sending the raw state signal on to the right.

To merge the terms into final expressions, we must have a systematic way of performing the logic OR operation of the appropriate product terms. The product terms all appear on horizontal lines, so we will arrange the output signals on a second set of vertical wires, to the right of the input lines. Taking advantage of the wired OR capability of diodes, we add a diode at wire crossings where we require the OR operation. For simplicity, we show this as a dot ✦. The hardware will produce the OR of all the dotted horizontal signals on a vertical wire. Figure 10–3 shows the implementation for the ASM of Fig. 10–2, using our square and dot notations for the logic operations.

Wilkes's idea was masterful. Here is a systematic method of building up the random logic needed to implement any ASM with no more than one branch per state. There is another powerful way of viewing the diode matrix, which shows its close connection to conventional programming and which led to the term *microprogramming*. The orderly arrangement of the matrix of wires in Fig. 10–3 suggests a lookup table. From the discussion of ROMs in Chapter 4 you saw that we could represent any Boolean expression as a table in which we look up the result for a particular set of values of the input variables. This was what Wilkes did, using the technology of his day; he implemented the next-state computation as a "table" of diodes. Each row at the right in Fig. 10–3 corresponds to a table entry for a particular branch path of the algorithm. The table entry describes the values of each output element in the system: the new values of state variables and values for each command variable. We can interpret the

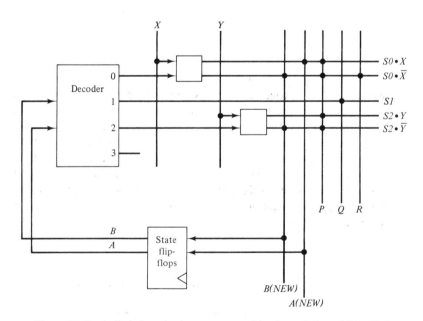

Figure 10–3 A diode-based microprogrammed implementation of Fig. 10–2.

state machine as a primitive computer whose instruction set executes only the ASM branch operation, and whose instructions come from a special bit pattern program, a *microprogram*. Each row in Fig. 10–3 is an instruction in the microprogram; in each row, the OR dots represent 1-bits, and the absence of a dot represents a 0-bit.

Why is this viewpoint powerful? We have substituted bits of memory for a random collection of gates. Since the bits of the diode array have a close correspondence to the ASM structure, and are out in the open, we may readily understand and manipulate them. Changing a Wilkes microprogram merely involves changing some diodes in a systematic manner.

CLASSICAL MICROPROGRAMMING WITH MODERN TECHNOLOGY

Let us now explore the design of microprogrammed controllers with modern devices. We will find that this causes a few changes in Wilkes's scheme, but only in the details. At the conceptual level, microprogramming remains the realization of controllers by means of tables rather than gates, an idea that survives from Wilkes. We may reduce any ASM chart to a table with inputs of current state and status variables and outputs of next-state variables and commands. If we translate an ASM chart into a table and then implement the table directly, using hardware lookup techniques, we are microprogramming in the broadest sense. The essential step is the direct implementation of the table, bypassing gates, Boolean algebra, and so on.

To simplify our treatment, we adopt for the present a uniform representation of logic truth in the microcode. The customary convention is positive logic, with T = H; we adhere to this convention in this section. In software programming, the choice of convention is of no consequence to the programmer, who is not dealing with voltage. In microprogramming, where we remain close to the hardware, this choice of positive logic will create problems, and we will return to discuss how to reinsert the full power of mixed logic into the microcode.

If our microcontroller is to implement really large algorithms, comparable to computer programs, we may need a sizable memory to hold the microinstructions for all the branch paths of the ASM. At this stage, ROM is a good choice since its contents remain intact even when the power is off. This means that the algorithm is instantly available when the power goes on, which seems quite desirable. The absence of inexpensive, fast ROM was the stumbling block in implementing microprogrammed control after its introduction, and many years of research ensued before the development of practical devices; it is no longer a problem.

In the testing of status variables newer technology has forced some changes in Wilkes's scheme. In the jargon of microprogramming, ASM variables are called *qualifiers*. In Fig. 10–3, we supply the current 2-bit state address B,A that we decode into individual signals for each state. After this decoding of the state address, we incorporate the qualifier tests using AND gates, to create one line for each decision path in the ASM. There are more branch paths than states, and each branch path results in one microinstruction.

As long as we implement the micromemory bit by bit with diodes, we can dive into the hardware following the address-decoding stage and insert or remove AND gates as needed. But ROMs and RAMs come as indivisible integrated circuits—the designer has access to address inputs and memory outputs, but to none of the interior circuitry. With RAM or ROM, the decoder of Fig. 10–3 is inside the device, and we have no way to get in to insert the AND gates. We need some other way to test the status variables in any given state, while preserving the table-driven nature of microcoded design.

Microprogramming with Multiple Qualifiers per State

The only way to access microinstructions stored in RAM or ROM is through the address inputs. The important elements in identifying a microinstruction are the current state and the test inputs or qualifiers. We might construct the ROM address from these two sources of bits: the current state code, obtained from the state flip-flops, and the individual qualifier signals, using one address bit for each qualifier. The size of the address field is the sum of the number of state flip-flops and the number of test inputs used in the design. For n address bits, a ROM contains 2^n words. Since the ROM's size grows exponentially with the number of qualifiers, this method rapidly gets out of hand.

In this approach, the value of every qualifier contributes to each next instruction address. This is highly redundant addressing, since most of the combinations of qualifiers are of no interest. For ASMs built from such states as are shown in Fig. 10–1, at most one qualifier is needed in each state, yet the microinstruction address includes all the qualifiers. In general, every address is possible, so each word in the ROM must contain a valid microinstruction. Each microinstruction must supply the value of all command outputs and the value of the next ROM address:

| Next-state address | Command outputs |

To illustrate this approach, let's again implement the small algorithm in Fig. 10–2, this time basing it on a ROM. (In practice, we would choose PROM, EPROM, or RAM, to permit more ready modification of the microcode during the development phase of the design. After the design has stabilized, we could then find a manufacturer to produce the ROMs if our production volume warranted this step. Let's use PROM in this example.) There are two state variables and two qualifiers, we we must have at least 4 ROM address bits, resulting in sixteen microinstructions. We begin by exhaustively enumerating the outputs required for each of the sixteen instructions:

	Address			Contents				
B	A	X	Y	B(NEW)	A(NEW)	P	Q	R
0	0	0	0	1	0	1	0	1
0	0	0	1	1	0	1	0	1
0	0	1	0	0	1	1	0	0
0	0	1	1	0	1	1	0	0
0	1	0	0	0	0	0	1	0
0	1	0	1	0	0	0	1	0
0	1	1	0	0	0	0	1	0
0	1	1	1	0	0	0	1	0
1	0	0	0	1	0	1	0	0
1	0	0	1	0	0	1	0	0
1	0	1	0	1	0	1	0	0
1	0	1	1	0	0	1	0	0
1	1	0	0	0	0	0	0	0
1	1	0	1	0	0	0	0	0
1	1	1	0	0	0	0	0	0
1	1	1	1	0	0	0	0	0

Next, we implement the table directly in a system of sixteen 5-bit words. Figure 10–4 is a sketch of the circuit. All lines into the PROM are address inputs; of the 5 output bits from the PROM, 2 are inputs to the state flip-flops, and 3 are command outputs to the architecture.

This approach to microprogrammed design is conceptually straightforward but requires enormous ROMs as the algorithm becomes more complex. There is an added benefit, however. Our original treatment based on Wilkes's work allowed us to implement ASMs containing at most one test variable per state. Since the present approach requires us to create a microinstruction for every possible combination of qualifier values for each state, we are automatically able to implement an ASM of arbitrary complexity; hence the name "multiple qualifier." No matter how complicated the branch path through a state, it corresponds to some microinstruction in this ROM-based design. Note the strong similarity to the ROM-based implementation of logic circuits discussed in Chapter 4.

As an exercise in using the multiple qualifier method, you might wish to implement the Black Jack Dealer machine of Chapter 6, using microprogrammed control. Suppose you use the same architecture as in the hardwired solution presented in Chapter 6. In Fig. 6–32, you can identify two state variables (say B and A) and eight qualifiers (CARD.RDY.SYNC, CARD.RDY.DELAYED, STAND, BROKE, ACECARD, ACE11FLAG, SCOREGT16, and SCOREGT21), so the ROM address field will contain 10 bits. The number of microinstructions is $2^{10} = 1024$! The eleven command signals (including the adder select signals) together with the two inputs to the state-variable flip-flops require that each ROM word have 13 bits. You will need a system containing a 1K × 13 ROM.

You may wish to write down the contents of some of the 1024 microinstructions to solidify your grasp of the concepts in the multiple qualifier method. You

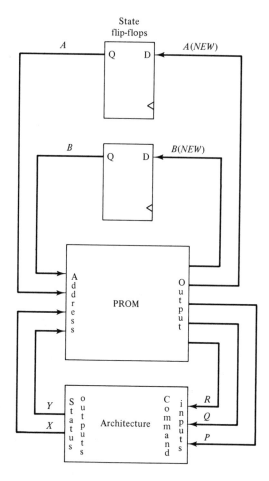

State
flip-flops

Figure 10–4 A PROM-based micro-programmed implementation of Fig. 10–2.

can appreciate the tedium of using this "simple" method to implement a complex algorithm manually.

What is your reaction to this approach? Ours is:

(a) It is a straightforward but tedious implementation of a general ASM.
(b) The tables are very large, even for relatively small problems.
(c) The method would be feasible only with inexpensive ROMs or PROMs.

The problem is that the address for the ROM is a concatenation of a small number of encoded state-variable bits and a large number of individual qualifier bits. The state variables are important at all times in the execution of the algorithm, but each qualifier appears only occasionally in the ASM. Most of the time, the algorithm is indifferent to the value of most of the qualifiers, yet we must enumerate each combination. We are forced to use a canonical form of truth table rather than the compact form allowed by the typical ASM. This ROM-based method is feasible if you have a "smart" PROM programmer that can accept your logic equations and expand them into a canonical truth table. But in most applications another approach might be better.

One Qualifier per State

The Wilkes scheme tests only one qualifier per state. The address is formed from the state variables alone, requiring the decoding of only a small number of variables. The scheme implements the qualifier tests with AND gates inserted in an orderly manner *inside* the circuit, following the decoding of the address. You have seen that using a ROM precludes this method, and our first attempt was to move all the qualifier signals out into the address field. This allowed us to implement general ASMs, but at a severe penalty. We would like to remove the qualifier variables from the ROM address field so that we can eliminate redundant microinstructions.

Providing properly sequenced command outputs to the architecture is the purpose of an ASM. In microprogramming, the command outputs arise from bits in the microcode. In addition to these command bits, microprogramming instructions also provide the new values of the state variables. This gives us a clue: our microcode has two components, an external one (command outputs) and an internal one (the next microinstruction address). Perhaps by enlarging the internal portion of the microinstruction we can shrink the number of instructions. We are striving to develop a method of handling large problems, and we may have to compromise the generality of the ASM structures that our method will handle.

Let's start instead with the most elementary useful ASM operation; one qualifier per state with no conditional outputs. Further, let's try to realize each state with only one microinstruction. In such a scheme, the ROM address would consist solely of the state variables, which would select the proper microinstruction. This instruction must contain sufficient information to guide the development of the next-state address. In particular, the instruction itself must specify *which* qualifier this particular state is testing.

If we organize all the qualifiers in a list, we may designate any qualifier by its index n in the list. The ASM structure we are trying to realize is

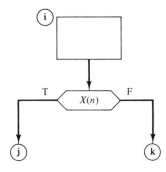

where $X(n)$ is one of the qualifiers X. Let's include this index n—the qualifier index—in the microinstruction. Then, in our single microinstruction for state i, we must also include the next-state addresses **j** and **k** for the true and false branches. Each microinstruction in our ROM will now look like:

Qualifier index n	True address TA	False address FA	Command outputs

Each word is now wider than in the multiple qualifier method, since it includes the index field and an extra address field, but our microinstruction table is reduced to one row per state.

To execute a microinstruction, our primitive microcontroller must be able to select the proper $X(n)$ using the value of n in the instruction. Based on the present value of the selected test input $X(n)$ the processor must choose one of the two address fields as input to the state flip-flops. Let's construct this selection hardware. Index n is an address in a table, so we may use a multiplexer building block with n as the code for the test input selection. The output of this multiplexer is a variable $X(n)$, whose value must specify either the true or the false address. We can perform this last selection with a set of two-input multiplexers, using $X(n)$ as the select input. Figure 10–5 is the circuit for our processor.

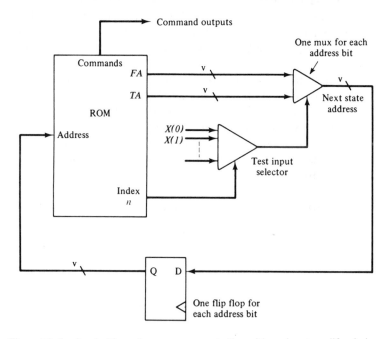

Figure 10–5 A primitive microprogram controller with an input qualifier index and both true and false jump addresses.

To illustrate the use of this microprogrammable machine to execute an algorithm, we will write the microcode for the ASM of Fig. 10–6, which is a

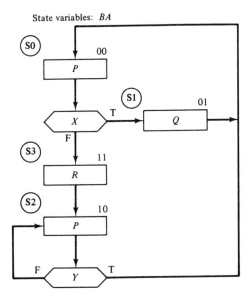
State variables: BA

S0
00
P

S1
01
X — T → Q

F

S3
11
R

S2
10
P

F — Y — T

Figure 10–6 Figure 10–2 redesigned to have no conditional outputs.

variant of Fig. 10–2 modified to eliminate the conditional output. Assigning indices 0 and 1 to the qualifiers X and Y, respectively, yields the following four microinstructions:

Address	n	TA	FA	P	Q	R
			Code			
0	0	1	3	1	0	0
1	0	0	0	0	1	0
2	1	0	2	1	0	0
3	0	2	2	0	0	1

TA and FA each require 2 bits, n has 1 bit, and there are three command outputs, so this design would require four words of 8-bit ROM.

Unconditional state transitions occur in instructions 1 and 3. Since each microinstruction must specify the index n of some test variable, we simply choose any index and make both TA and FA point to the same next instruction.

Why did we eliminate the ASM's conditional output from our single-qualifier scheme? Conditional outputs arose naturally in Wilkes's scheme and in the multiple-qualifier approach. But here we have exactly one microinstruction per ASM state, and all the information for the execution of that state must reside in that microinstruction. If we were to permit conditional outputs, we would need a way to designate, for *each command output bit*, whether it is to be asserted unconditionally, or only on the true branch, or on the false branch, or not at all. This would require 2 microinstruction bits per command output—a considerable burden on the hardware. One of the virtues of the present method is its simplicity of form. Another reason for eliminating conditional outputs will surface later.

Single-Qualifier, Single-Address Microcode

The preceding single-qualifier structure is feasible, but the two address fields can consume considerable space in the microinstruction. We can eliminate one address field if we adopt a rule for inferring that address from the present address. The obvious choice, which conforms closely to the practice used in conventional computers, is to insist that one of the branch addresses be the next sequential address. In the single-qualifier method, state assignments correspond to microinstruction addresses, and since the state assignments are at our disposal, we may use normal sequencing to save bits in the microinstruction. The ASM operation reflecting this modification of the single-qualifier scheme is shown in Fig. 10–7a.

The microinstruction now contains one jump address, the qualifier index, and the command output bits. In the version of ASM in Fig. 10–7a the jump address is always the true path. We can enhance the versatility of this approach by adding one more bit to the microinstruction, to allow the microprogrammer to specify which path, true or false, the jump address refers to. Now the format for the microinstruction is

Index n	TFBIT	Jump address JA	Command outputs

This form of microprogramming implements the basic ASM operation in Fig. 10–7b.

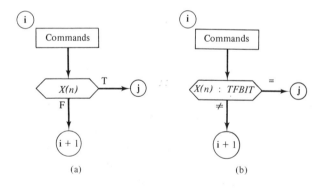

Figure 10–7 ASMs for single-qualifier, single-jump address microinstructions. (a) Microinstruction branches on a true qualifier. (b) Microinstruction allows selection of the jump condition.

Now consider how we might build the processor for this method. Since we have eliminated one of the address fields, we have also removed the need for the two-input multiplexers on the state flip-flop inputs in Fig. 10–5. Instead, we need to be able to *increment* the current address whenever the test variable value is opposite of *TFBIT* in the microinstruction. We may incorporate this operation into our state flip-flop assembly by replacing the simple flip-flops with a programmable binary counter building block. Our microcontroller processes a microinstruction in each clock cycle, so we are always either branching, which corresponds to landing a new value into the counter, or sequencing the counter.

With the aid of the ASM in Fig. 10–7b, we may derive the condition for loading the counter. We load the counter whenever the next address is to be the microinstruction jump address JA:

$$(NEXT.ADDRESS = JA) = X(n) \cdot TFBIT + \overline{X(n)} \cdot \overline{TFBIT}$$
$$= X(n) \odot TFBIT$$

We sequence the counter whenever we do not jump. An implementation of this microprogrammable controller is shown in Fig. 10–8.

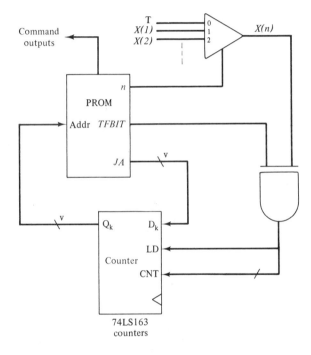

Figure 10–8 A counter-based implementation of the microinstruction control in Fig. 10–7b.

Another convenience shown in Fig. 10–8 is making test input position 0 a permanent true signal. The microprogrammer may then execute an unconditional branch (no test variable) by designating $n = 0$ and $TFBIT = 1$ in the microinstruction. Conversely, unconditional sequencing to the next instruction occurs when $n = 0$ and $TFBIT = 0$.

Having constructed this sophisticated single-qualifier controller, let's use it for the simple ASM in Fig. 10–6. Earlier, when we used the single-qualifier method with two address fields for this algorithm, our exact choice of state assignment was unimportant. (Why?) In the present method, we are required to make one of the exits from each state sequential. We encounter difficulty with the assignment in Fig. 10–6 since, as it happens, neither branch from state S2 is sequential. To avoid this problem, we may renumber the states in Fig. 10–6 so that S0 = 2, S1 = 0, S2 = 1, and S3 = 3. If we assign indices 1 and 2 to qualifiers X and Y, respectively, the microcode for the program is

State	Address	Index n	TFBIT	Jump address JA	Command outputs P	Q	R
S1	0	0	1	2	0	1	0
S2	1	2	0	1	1	0	0
S0	2	1	1	0	1	0	0
S3	3	0	1	1	0	0	1

As another illustration, we could implement the Black Jack Dealer of Chapter 6 using the sophisticated single-qualifier scheme. The ASM in Fig. 6–32 is not suitable, since it contains conditional outputs and states with multiple tests. Converting this ASM to an appropriate form is a useful exercise. In most cases, we may convert a conditional output to an unconditional one by creating a new state for the output. This simple method will not work when the exact timing of the conditional output is crucial; an example is in the GET state of Fig. 6–32. If we create a separate state for the *HIT* output, the hit light will blink on and off as the ASM loops around the two-state loop. We will deal with this particular problem presently.

States with multiple tests also require modification before we can use the single-qualifier method. Where the timings are not critical, we may create new states to perform each individual test. This approach would handle all cases in Fig. 6–32 except the first two tests in the GET state. You will recall that this structure is a manifestation of the familiar single pulser; it requires that *CARD.RDY.SYNC* and *CARD.RDY.DELAYED* be tested simultaneously. For the Black Jack Dealer algorithm, the solution, which also solves the problem of the *HIT* output, is to use one of the other forms for describing the action of the single pulser. (Notice how valuable was the knowledge that this troublesome structure represented a standard design element. How much more difficult the analysis would have been without this knowledge!)

Figure 10–9, the result of the ASM transformation, is a version of the Black Jack Dealer algorithm that we may implement as a single-qualifier microprogram. Figure 10–9 contains an address (state) assignment that makes good use of the requirement that one branch of each state must lead to the next sequential state. We have made an arbitrary choice of the qualifier index, and have shown the index in brackets in each test in the ASM.

Here is the microcode for the Black Jack Dealer of Fig. 10–9, without the command outputs. You should find it easy to include command bits in each microinstruction. What is the size of the ROM required by this microprogram?

Address	Index n	TFBIT	Jump address JA
0	1	1	0
1	1	0	1
2	2	1	11
3	3	1	11
4	4	1	12
5	6	0	0
6	7	0	10
7	5	0	9
8	0	1	5
9	0	1	0
10	0	1	0
11	0	1	4
12	5	1	5
13	0	1	5

Comparison of the Microprogramming Approaches

We have considered microprogramming from two viewpoints. How does the single-qualifier approach compare with the multiple-qualifier method? Some facts to consider are:

(a) The single-qualifier method has a one-to-one correspondence with an ASM chart; the multiple-qualifier method does not.

(b) The single-qualifier method requires much less microcode.

(c) The single-qualifier method will handle large problems easily; the multiple-qualifier method cannot tolerate many qualifiers before it becomes unmanageable.

(d) The multiple-qualifier method handles a completely general ASM; the single-qualifier method handles only a special case.

(e) The single-qualifier method requires more special hardware in the micro-controller, although the multiple-qualifier method requires the larger ROM.

From these points we may draw some conclusions. The single-qualifier method is clearer and easier to manage. It is more likely to result in correct microcode the first time. Field-service personnel are more likely to understand single-qualifier programs and they are much easier to document.

Item (c) is probably decisive, but even so, our sense of style leads us to recommend the single-qualifier method for most applications.

MOVING TOWARD PROGRAMMING

As we have developed the concepts of microprogramming in this chapter, our language and our emphasis have come ever closer to those of the software programmer. We began by emphasizing hardware, looking for ways to systematize

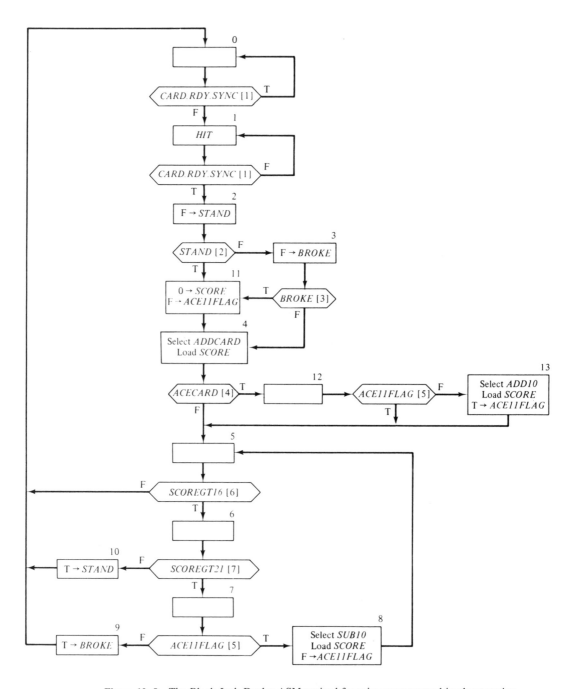

Figure 10–9 The Black Jack Dealer ASM revised for microprogrammed implementation.

the design process to handle large problems. We followed Wilkes through his discovery of table-driven hardware controllers. We finally arrived at a design of a machine that executes only a restricted form of ASM operation but that executes all such operations in a systematic manner. We call the specification of each standard ASM state a "microinstruction," and we refer to the state-variable values as a "memory address." We think in terms of writing a set of instructions—a microprogram—to describe an algorithm for this specialized machine, and we place the program into a memory.

It sounds like programming, but how far have we gone? Microprogramming is a middle ground between hardwired design and conventional programming, drawing advantages from each. From the viewpoint of the designer of hardware, microprogramming offers a way to tackle large and complex control problems. It retains much but not all of the speed and capability of parallel action that is characteristic of hardwired ASM implementations. The single-qualifier approach to microprogramming permits the designer simultaneously to receive status information, control the flow of the algorithm, and issue detailed commands to the controlled device. We lose multiple branches and conditional outputs and also a bit of speed, but we gain a compact, highly structured, easily modified method of formulating and implementing algorithms. At the other end of the microprogramming spectrum, the multiple-qualifier approach costs nothing in speed or in the flexibility of the algorithm, but tends to overpower the designer with its exponentially increasing size of microcode.

The software programmer sees microprogramming as an entry into the field of hardware. The serial, one-step-at-a-time nature of conventional programming gives way to a more parallel but still program-oriented approach. The micro-controller has a more primitive command structure than a conventional computer, yet the simple but parallel nature of its operations makes for much greater speed and versatility.

Why do we stress the programming aspect? Why has microprogramming come to be considered a conceptual breakthrough in hardware design? The answers lie in the great store of experience that computer science has gained in using programs to emulate algorithms. Conventional programmers have a host of software tools and strategies that we can use in microprogramming. By transforming a hardware problem into the programming domain, we may look forward to using editors, assemblers, language translators, and debugging aids in support of the development of our microcode.

Let us try, then, to borrow from programming concepts to expand the usefulness of microprogrammable controllers, without detracting from their inherent power as emulators of hardware algorithms.

Cleaning Up the Outputs

Throughout our development of microprogramming, the microinstruction memory—ROM, PROM, or RAM—has been the source of command outputs to the architecture and control signals to the next-state controller. Unfortunately, these memories undergo relatively long periods of instability when their address inputs change, and so the signals emerging from the memory outputs have undesirable voltage characteristics. Our circuits for sequencing microinstructions have worked despite this drawback, since the changes in the memory addresses have been synchronized with the system clock. The architecture is exposed to the impurities of the command outputs, but since it is driven by the same clock as the control unit, the designer may use these outputs reliably to feed the clocked architectural elements. The command outputs are unsatisfactory for nonclocked uses, such as serving as control signals to the world outside the clocked design. In these important cases, the designer must purify the command outputs, usually by passing them through a clocked flip-flop to assure a clean output.

The microinstruction memory has served as an instruction register, but in recent practice the role of instruction register is removed from the microprogram memory and is assumed by a true clocked register. This register, known as the *microinstruction register* or *pipeline register*, receives the full output of the microinstruction memory and delivers clean, reliable signals to the architecture and to the circuits that produce the next microinstruction address. The designer then has a uniform and reliable interface with the microprogram control unit.

Enhancing the Control Unit

Just as Fig. 10–8 grew out of consideration of more primitive controllers, we may generalize it to produce a more powerful controller. In Fig. 10–8, the counter and the coincidence gate perform a control function that responds to inputs and produces an output. The inputs are the test input signal from the external architecture and the *TFBIT* and jump address from the present microinstruction. The output is the address of the next microinstruction. We may view the *TFBIT* and jump address as components of an elementary computer branch instruction, and the counter and coincidence gate as a primitive computer control unit. In this view, Fig. 10–8 implements a computer with one flow-of-control instruction, a simple branch. But ordinary computers have much more sophisticated branching than this, so why should we not incorporate some of this sophistication into the next-instruction-address evaluator within our microinstruction control unit?

We will view a microinstruction as consisting of two components, a microinstruction sequencing part and a command output part:

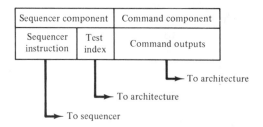

The sequencing component of the microinstruction becomes an instruction to be processed by a *microprogram sequencer*, contained within the microprogram control unit. The sequencer and its instruction input may be relatively simple, as in Fig. 10–8 and earlier figures, or it may be quite sophisticated. Our view of the microinstruction control unit has been transformed into Fig. 10–10. With these moves, we have a structure that is close indeed to conventional computers, yet retains much of the power of hardware. Our microprogram control unit has a memory, an instruction register, and is capable of determining the microprogram's flow based on the present state of the system and an external input.

With these expanded capabilities, we can express quite complex control algorithms. Our new model of a microprogrammable controller (Fig. 10–10) still implements an ASM similar to Fig. 10–7 in which all command outputs are unconditional and control of the microprogram either moves to the next sequential state or branches to a new state in response to the test of a single input signal. However, the possibilities for branching are considerably enlarged. We will see that we may use subprogram calls and returns, loops, and other useful constructs from the programming world. The opportunity to express highly complex algorithms as microprograms means that the designer will need sophisticated aids to support the development, debugging, and maintenance of the microprogram. At the same time, the architectures to be managed by these more complex algorithms become larger and more complex, requiring additional aids for hardware development.

Several microprogram sequencers are available as integrated circuit chips.

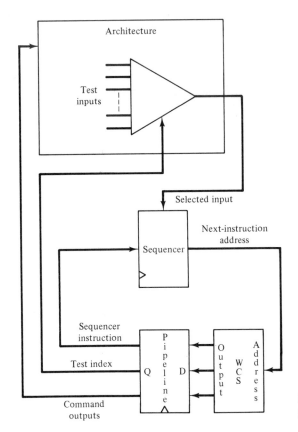

Figure 10–10 A sophisticated microprogram sequencer.

The first was the 2909, introduced by Advanced Micro Devices in 1975. The 2909 accepts a 2-bit operation code and provides 4 bits of next-instruction address; several 2909s can be cascaded to produce larger addresses. The 2909 supports conditional branches and subprogram calls and returns. As the technology advanced, more address bits were provided within a single chip and more complex operations were introduced. For instance, the 2910 integrated circuit produces 12 bits of next-instruction address and executes 32 instructions. The Texas Instruments 74AS890 supports 64 instructions and has 14 address bits.

The 2910 Microprogram Sequencer

Microprogram sequencers are sophisticated devices with many features. We will limit our discussion to the 2910 and those characteristics that support our study of microprogramming. If another microsequencer is used, it will have similar characteristics. Figure 10–11 shows the principal signals entering and leaving the 2910. The 2910 is designed to produce the address of the next microinstruction to be loaded into the pipeline register. It accepts a 4-bit operation code I, a test-input signal CC (Condition Code), a control signal $CCEN$ (Condition Code Enable) to guide the use of the test-input signal, and a 12-bit data-input field D. The D-field usually provides a microprogram branch address (our familiar jump address) from the pipeline register, although it has other uses in the 2910. The output of the 2910 is a 12-bit next-instruction address Y. The 2910 can support microprograms containing up to 4096 instructions. Since the 2910 contains internal registers, we must supply a system clock signal CP.

Figure 10–12 shows the internal architecture of the 2910. The next-instruction address Y originates from one of four sources, selected by a 4-input multiplexer based on the operation code and the value of the test input signal. Two of the four sources are already familiar to us: the next sequential address (current microinstruction address + 1), and a branch address derived from the current microinstruction. The 2910 contains a microprogram counter-register (μPC) which records $Y + 1$, the next sequential address, in case it is needed later.

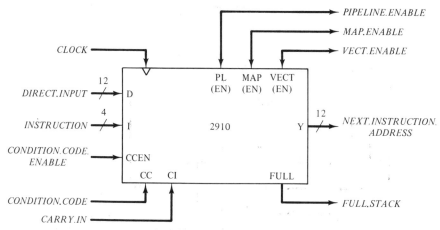

Figure 10–11 The 2910 microprogram sequencer.

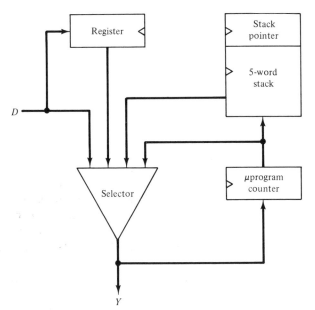

Figure 10–12 Twelve-bit data paths in the 2910 microprogram sequencer.

The output of the μPC forms one input to the Y multiplexer. The 12-bit external data input D forms another multiplexer input. Usually, D comes from the jump-address field of the pipeline register.

The two remaining multiplexer inputs support subprogram calls and program looping. The 2910 contains a five-word stack. In a subprogram call, the 2910 must save the return address on the top of the stack; a subprogram return must supply the return address from the top of the stack. The 2910's internal stack allows calls to microprogram subprograms to be nested five deep. Each subprogram call results in a stack push operation, and each subprogram return causes a stack pop operation. When a microinstruction executes a subprogram call, the required return point is the address of the control store word following the subprogram call instruction. This address is exactly the quantity that is currently stored in the 2910's μPC; Fig. 10–12 shows a data path from the μPC to the stack that supports subprogram calls.

The fourth input to the Y multiplexer is from an internal register R that can hold a loop counter. The R-register can be loaded from the 2910's D-input. Several 2910 instructions support the loading, testing, and decrementing of the value in the R-register.

The 2910's 4-bit operation code supports 16 basic instructions. Each instruction has a "pass" and a "fail" option, generating a total of 32 possible operations. The selection of pass or fail is controlled by the values of the 2910 inputs CC and $CCEN$, according to the following prescription: if the enable signal $CCEN$ is true and the test input (condition code) CC is false, then fail; otherwise pass. Viewed another way, this structure allows the execution of the pass version of an instruction when we are not testing the input, or when the input is true while we are testing it. The following table specifies the conditions for selecting the option:

CCEN	CC	Result
F	F	Pass
F	T	Pass
T	F	Fail
T	T	Pass

CJP (conditional jump) is a typical 2910 instruction. In its fail mode, this instruction selects the μPC as the Y-output, accomplishing normal sequencing. In its pass mode, CJP selects the D-input as the Y-output, thus performing a branch. At the next system clock edge, the pipeline register will receive the appropriate instruction and the 2910's μPC will capture the address + 1 of this instruction.

Another example is CJS (conditional jump to subprogram). In its fail mode, CJS sequences to the next microinstruction address, with no effect on the 2910's internal stack. In its pass mode, CJS performs a subprogram jump, which selects the branch address in the D-input as the value of Y, and, when the system clock fires, causes the contents of μPC to be pushed onto the internal stack. (As usual, μPC will receive the new $Y + 1$ when the clock transition occurs.)

In the fail mode, the instruction CRTN (conditional subprogram return) performs normal sequencing, with no effect on the 2910's stack. In the pass mode, CRTN delivers the top-of-stack element to Y, thereby supplying the return address to the previous subprogram as the address of the next microinstruction. When the clock transition occurs, the 2910 pops its stack, and μPC receives $Y + 1$.

When the 2910 is used as the sequencing element in Fig. 10–12, an appropriate form for the flow-of-control portion of the microinstruction is:

		Sequencer component	
I	CCEN	D	Test index

Each microinstruction provides the 2910 with the I, $CCEN$, and D fields. The index field goes to the architecture to guide the selection of the appropriate test input, which becomes the 2910's CC input.

Thus far, the 2910's D-field arises from the corresponding field in the microinstruction pipeline register. Although this is by far the most common and useful mode of operation, the 2910 also permits an alternative source of the D-input. With each instruction, the 2910 asserts one of three D-field selection signals. In the instructions described above, the 2910 asserts its Pipeline-Enable signal $PL(EN)$. On the other hand, the 2910's JMAP (Jump on Map Address) instruction, which causes an unconditional jump to the D-field address, asserts the Map-Enable signal *MAP(EN)* instead of *PL(EN)*. This feature provides a limited yet useful capability to select the *D*-field input from a source in the designer's architecture, under control of the *MAP(EN)* signal. We will use this feature of the 2910 in a subsequent design example. One other 2910 instruction has similar characteristics; all other instructions cause the assertion of *PL(EN)*.

If *MAP(EN)* is chosen, all inputs to the 2910's *D*-field must have three-state characteristics and the designer must use the 2910 *PL(EN)* and *MAP(EN)* signals to select the proper input.

Table 10–1 is a summary of the 2910's instructions. In this chapter we use about half of these instructions, and will explain each new instruction at the time of use. Consult an AM2910 data sheet for additional information, if you desire. The instructions CJP, CJS, CRTN, and CONT (Continue) are by far the most commonly used 2910 instructions. In this chapter, we use the alternative mnemonics JUMP, CALL, and RTN in place of CJP, CJS, and CRTN.

Choosing a Microprogram Memory

With the realization that our microprogramming methods are capable of describing and executing quite complex algorithms, we begin to see the need for sophisticated equipment to help the designer to manage the complexity. We have assumed that the microprogram storage was a read-only memory—ROM, PROM, or EPROM. In accordance with good programming practice, our microprograms do not change; all the "data storage" is in the architecture. Even when sophisticated microprogram sequencers such as the 2910 are used, the microprogram remains fixed during execution—the sequencer itself contains storage for sub-program return points and loop control. For such an environment, ROM seems the natural choice. Many designers initially discarded RAM for this purpose because of the volatility of its contents when power drops. But as the size and complexity of modern microprograms have increased, this choice has been reversed. For debugging complex microprograms, and when the microcode may be modified in the field, RAM is essential. In microprogramming jargon, the microprogram storage is the *control store*. If the control store is easily alterable, as is RAM, it is called *writable control store* (WCS). If we use RAM, we must load it frequently, and we need powerful microprogramming aids. In the next section we describe a microprogrammable development system that provides the designer with the hardware and software tools required to manage the design and development process.

THE LOGIC ENGINE—A DEVELOPMENT SYSTEM FOR MICROPROGRAMMING*

Microprogramming permits the designer to tackle complex control tasks, but this ability to deal conceptually with complex designs entails numerous practical

* This section is modified from Franklin Prosser and David Winkel, "The Logic Engine Development System—Support for Microprogrammed Bit-Slice Development," *Proceedings of MICRO-16,* October 1983, pages 84–91.

TABLE 10-1 INSTRUCTIONS OF THE 2910 MICROPROGRAM SEQUENCER

HEX I_3-I_0	MNEMONIC	NAME	REG/ CNTR CON- TENTS	FAIL $\overline{CCEN}$ = LOW and $\overline{CC}$ = HIGH		PASS $\overline{CCEN}$ = HIGH or $\overline{CC}$ = LOW		REG/ CNTR	ENABLE
				Y	STACK	Y	STACK		
0	JZ	JUMP ZERO	X	0	CLEAR	0	CLEAR	HOLD	PL
1	CJS	COND JSB PL	X	PC	HOLD	D	PUSH	HOLD	PL
2	JMAP	JUMP MAP	X	D	HOLD	D	HOLD	HOLD	MAP
3	CJP	COND JUMP PL	X	PC	HOLD	D	HOLD	HOLD	PL
4	PUSH	PUSH/COND LD CNTR	X	PC	PUSH	PC	PUSH	Note 1	PL
5	JSRP	COND JSB R/PL	X	R	PUSH	D	PUSH	HOLD	PL
6	CJV	COND JUMP VECTOR	X	PC	HOLD	D	HOLD	HOLD	VECT
7	JRP	COND JUMP R/PL	X	R	HOLD	D	HOLD	HOLD	PL
8	RFCT	REPEAT LOOP, CNTR $\neq$ 0	$\neq$ 0	F	HOLD	F	HOLD	DEC	PL
			= 0	PC	POP	PC	POP	HOLD	PL
9	RPCT	REPEAT PL, CNTR $\neq$ 0	$\neq$ 0	D	HOLD	D	HOLD	DEC	PL
			= 0	PC	HOLD	PC	HOLD	HOLD	PL
A	CRTN	COND RTN	X	PC	HOLD	F	POP	HOLD	PL
B	CJPP	COND JUMP PL & POP	X	PC	HOLD	D	POP	HOLD	PL
C	LDCT	LD CNTR & CONTINUE	X	PC	HOLD	PC	HOLD	LOAD	PL
D	LOOP	TEST END LOOP	X	F	HOLD	PC	POP	HOLD	PL
E	CONT	CONTINUE	X	PC	HOLD	PC	HOLD	HOLD	PL
F	TWB	THREE-WAY BRANCH	$\neq$ 0	F	HOLD	PC	POP	DEC	PL
			= 0	D	POP	PC	POP	HOLD	PL

Note 1: If $\overline{CCEN}$ = LOW and $\overline{CC}$ = HIGH, hold; else load.   X = Don't Care

I-field Value	Mnemonic	Function
$0	JZ	Jump to location 0
$1	CJS	Conditional jump to subroutine at pipeline address
$2	JMAP	Jump to map address
$3	CJP	Conditional jump to pipeline address
$4	PUSH	Push with conditional load of counter
$5	JSRP	Conditional jump to subroutine at R address or at pipeline address
$6	CJV	Conditional jump to vector address
$7	JRP	Conditional jump to R address or pipeline address
$8	RFCT	Repeat loop if counter is non-zero
$9	RPCT	Jump to pipeline address if counter is non-zero
$A	CRTN	Conditional return from subroutine
$B	CJPP	Conditional jump to pipeline address with stack pop
$C	LDCT	Load counter from D input
$D	LOOP	Test end of loop
$E	CONT	Continue
$F	TWB	Three-way branch

Alternative instruction mnemonics:

CALL	Equivalent to CJS
RTN	Equivalent to CRTN
JUMP	Equivalent to CJP

problems. On what type of breadboard should we construct the architecture? How do we debug the architecture? How do we produce the microcode? How do we load the microcode into a control store? How do we design and build the microinstruction sequencer? How do we debug the microcode? How do we modify the microcode?

These questions imply that designers need a powerful support system to allow them to manage microprogrammed control. The control unit is itself only one part of a good development system. The system must also support the development and debugging of the architecture and the control algorithm. It should minimize the usual headaches of design and the subtleties of constructing the hardware. It should provide for convenient wire-wrapping for initial testing, and for lights and switches for displaying and controlling individual signals during the testing of the design, as well as lend powerful support to the development, debugging, and modification of the control program.

Several commercial microprogrammable development systems have appeared. We will describe one of these, the Logic Engine, which we designed and built. Our goal is to reach a position from which we may easily produce and manage complex hardware projects using microprogrammed control. The goal is ambitious, and to achieve it requires an understanding of the design principles and practices presented in this section.

Figure 10–13 shows the parts of the Logic Engine Development System. The base unit houses the microprogrammable controller, a microcomputer-based

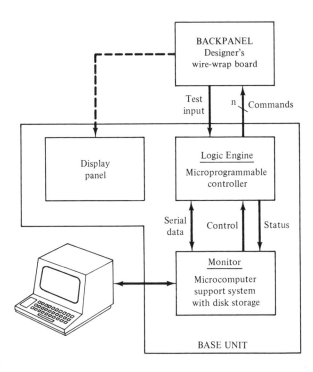

Figure 10–13   The Logic Engine.

debugging support system, and a debugging display panel. Attached to the base unit is a large, detachable backpanel for wire-wrap of the hardware architecture. A terminal provides for convenient interaction between the development system and its user.

The Base Unit

The Logic Engine's microprogrammable control unit contains a 2910 microprogram sequencer, a writable control store of up to 4K words, a microinstruction pipeline register to deliver command signals to the designer's architecture, and a buffer register to support communication between the controller and the microcomputer-based monitor. Figure 10–14 shows the structure of the Logic Engine's controller. The task of the controller is to present a properly sequenced set of signal voltages (commands) to the designer's architecture. Therefore, a Logic Engine micro-instruction has two primary fields: a fixed-format sequencing field to direct the 2910 in the production of the address of the next microinstruction and an open-ended field for specifying command signals to the designer's circuit. The sequencing field consists primarily of the items required to direct the 2910: I, $CCEN$, and D. The length of the command-bit field is determined by the particular project and may exceed 100 bits.

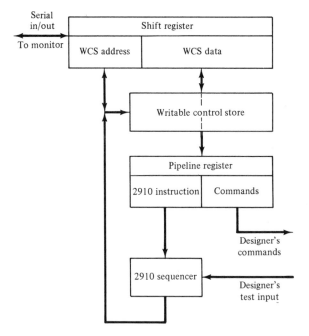

Figure 10–14 The architecture of the Logic Engine controller.

The designer may choose two ways of executing the microinstructions. In the automatic mode, the controller loads microinstructions from the writable control store (WCS) into the pipeline register under the control of the 2910 sequencer. In the debugging mode, the designer can influence the delivery of command signals to the architecture in several ways, using features of the Logic Engine's debugging monitor.

The Logic Engine's support system consists of software running on a microcomputer inside the base unit. The microcomputer has dual floppy-disk drives, two serial input-output ports, and one parallel input-output port. The parallel port provides the interface between the support system and the Logic Engine's controller. One serial port is dedicated to the designer's display terminal; the other serial port is available for connecting a serial printer, remote computer, or other device. The Logic Engine's support software is organized around a debugging monitor. Additional software includes a text editor, a microprogram assembler, and various utility programs.

The Logic Engine's display panel provides about 100 LEDs for displaying data, over two dozen pushbuttons and toggle switches for entering data, and a variable-speed clock that includes a manual mode. The designer has access to these whenever the backpanel is attached to the base unit. The base unit contains a power supply adequate to operate the Logic Engine and the designer's circuit.

The Backpanel

The Logic Engine's backpanel is large: 16 in. wide and 20 in. high. It has a general-purpose work area to handle integrated circuit chips of 8 to 64 pins. For a typical design, the backpanel can accommodate several hundred chips. The designer has access to both sides of the board at all times. Ground and +5 V appear as power grids on opposite sides of the board and there are extensive provisions for attaching power-bypass capacitors (see Chapter 12).

Along one side of the backpanel is an area committed to the microinstruction pipeline register and WCS for the designer's command signals. This permits easy wire-wrapping of the command signals to the architecture, and allows the designer to employ as many command signals as the design requires.

The Supporting Software

The Logic Engine's development and debugging monitor supports the detailed control of the WCS and of the operations of the microprogram sequencer and pipeline register. The designer may load the WCS from a floppy-disk file—an example of *downloading*. The designer may read and modify any word in the WCS, modify any word without disturbing the remainder, and display the contents of a block of WCS. Since the 2910 microprogram sequencer is an integral part of the Logic Engine, the monitor knows its characteristics in detail and thus can support the display and modification of all of the sequencer's internal registers. The designer may display the microinstruction pipeline register and modify any portion of it. The monitor also permits the designer to specify whether, with each manual change of the pipeline register, a designer's clock signal is to be issued. These features give the designer an important debugging tool: the manual entry of microinstructions into the pipeline register without modifying the writable control store. Since the pipeline register's command field is wired to the designer's architecture, the designer may exert detailed manual control of the circuit.

Table 10–2 summarizes some of the functions of the Logic Engine's monitor that are available to the designer. In executing microcode from the WCS, the most powerful debugging features are *single-step* and *breakpoint*. Single-step permits the designer to execute one instruction at a time from the WCS, with a complete Logic Engine register dump accompanying each instruction. The breakpoint feature is used when the designer is running microcode at high speed. The designer announces a particular WCS address that, if it becomes the candidate for next microinstruction, will cause the controller to halt. Breakpoints permit the designer to stop the execution of instructions at any address and then observe the status of the system.

TABLE 10–2 FUNCTIONS OF THE LOGIC ENGINE'S MONITOR

M	Display and modify the WCS
E	Examine a block of the WCS
R	Display and modify the Logic Engine's registers
P	Load the pipeline and execute an instruction
C	Clear the 2910
B	Set or clear a breakpoint
G	Go! (Run microcode from the WCS)
I	Idle the Logic Engine
S	Execute a single instruction
H	Help!
L	Load the WCS from a disk file
U	Unload the WCS to a disk file

The Logic Engine's microprogram assembler provides powerful development features within a structured microprogramming language. The microassembly language encourages the designer to express the control algorithm in high-level terms and provides for transforming the high-level specification into microcode. The assembler supports the symbolic naming of single command bits and fields of bits, and there is a convenient syntax for invoking the desired values of the command bits. The assembler provides full mixed-logic capabilities, giving the designer the freedom to specify signal values as voltages or as logic levels and to describe the voltage convention for truth for each signal. The designer may specify default values for command signals, so that in writing microcode only command signals that deviate from the default values need be described. In the sequencing portion of the microinstruction, the assembler supports the 2910's instruction set and has a convenient syntax for specifying the designer's test inputs.

DESIGNING AND DEBUGGING WITH THE LOGIC ENGINE

To illustrate the process of design using microprogrammed control, let's study the design of a small portion of a machine that can directly execute the Forth language in hardware. (It is not necessary to know the Forth language to follow the example.)

The Initial Design

Our first step is to work out the architecture—the registers, busses, and data paths. Forth is a stack-oriented language, and several of its important operations involve manipulations of the elements on the stack. In this example, we focus on the stack operations. We wish the several top elements of the stack to be available for direct use; the deeper elements will be kept in a RAM. Figure 10–15 is a portion of the architecture, showing the top three elements of the stack. (The stack elements may contain as many bits as required by the problem, but this decision does not concern us here.) The input to each stack element is through a set of multiplexers. Each of the potential sources of a given element of the stack becomes an input to that element's multiplexer. (Notice the similarity of this data-routing design to that used in the LD20.) For testing purposes, in

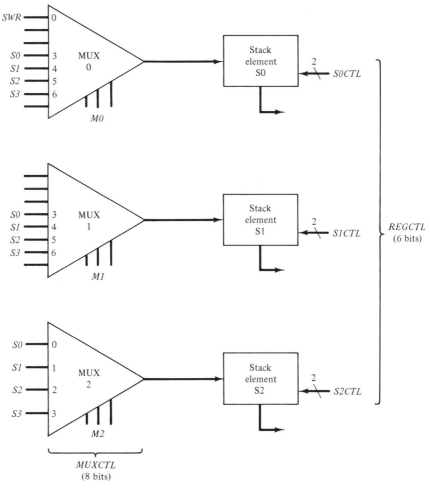

Figure 10–15 The architecture of a microprogrammed design.

addition to the elements of the stack, inputs include the switch register on the display panel. We will call the select signals for the three multiplexers *M0*, *M1*, and *M2*, and refer to the entire collection of multiplexer select signals as *MUXCTL*. In addition to holding its contents and loading new information, each element of the stack may perform internal bit-shifting operations. To support these activities, each stack element requires 2 control inputs; for stack elements, *S0*, *S1*, and *S2*, we call the stack controls *S0CTL*, *S1CTL*, and *S2CTL*, and we call the collection of six stack-element controls *REGCTL*.

The next step in the design is to work out, in rough form, the algorithms to control the architecture. The thought put into this step will often lead to modifications of the architecture. The design process involves moving between increasingly refined sketches of the architecture and control until we feel reasonably confident that we understand our problem thoroughly. Then there is hope that when we build our machine, it might actually work!

At this stage in the design of the algorithm, we can see what signals from the architecture we must test in our microcode in order to direct the flow of the microinstructions. The 2910 sequencer accepts a single signal as a test input, and the 2910's instructions may act upon the value of this signal. Consistent with our earlier development of microprogrammed controllers, we place a test-input multiplexer in the Forth Machine's architecture to deliver the single test signal to the 2910. Since we know which, if any, signal is required for testing in each microinstruction, we may use some of the microinstruction's command bits as the select code for the multiplexer. The full Forth Machine design requires about a dozen test inputs, so a 16-input multiplexer with a 4-bit select code is appropriate. Figure 10–16 shows the structure of the apparatus for selecting test inputs, including the two that we use in this design. We have allocated the first 4 microinstruction command bits (bits 0–3) to the control of the test multiplexer; this is an arbitrary choice.

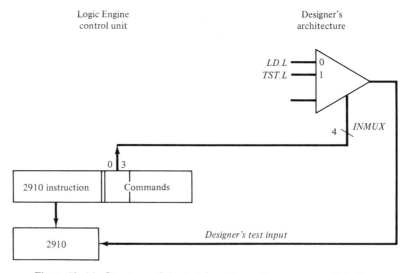

Figure 10–16 Structure of the test input in a microprogrammed design.

The Initial Testing of the Architecture

Now it is time to construct and test the architecture. On the Logic Engine's backpanel, we lay out the chips required by the architecture, assemble the appropriate sockets and chips, and wire-wrap the design. The size of the backpanel permits us to develop and debug the architecture without partitioning the components among small printed circuit boards.

At this point, we usually make some preliminary tests of the registers and the data paths. The Logic Engine's display panel has numerous lights and switches to assist us. Using wire-wraps or jumpers, we connect the important outputs to any of the display panel's LEDs and connect switches to the inputs. A disposable cardboard overlay for the display panel allows us to label the lights and switches. We may use the display panel's variable-speed clock to provide clocking signals. Manual clocking permits us to debug statically—we deliver clock transitions only when we wish. This is a powerful debugging technique.

Now we exercise the architecture with the display panel's switches, and observe the results on the lights. In effect, we are manually delivering rudimentary control to the architecture prior to developing the actual control program, thus allowing early detection of gross errors in the wiring or design.

Developing the Control Program

Once we are satisfied that the architecture is working properly, we turn to the detailed development of the control algorithm. We rely on the Logic Engine to help us develop the control in two ways: by providing a standard environment for developing and executing microprograms, and by aiding us to program and test the code.

The Logic Engine's microprogram assembler, LEASMB, has two parts. In the *declaration phase* we specify symbolic names for all the variables and quantities of interest, and we describe the structure of the microinstruction. The *program phase* contains the microcode itself, in symbolic form. The use of symbolic notations is of great value because of their descriptive power and because changes in the design may usually be made with little disturbance to the program. During the earlier phases of the design, natural names will emerge for the important signals that control the architecture. It is convenient to use these names in the microcode.

Figure 10–17 is a microprogram for our Forth Machine design. We will use this code to introduce the elements of the microassembly language. (Later, you will study the process of generating a microprogram for a more complex microprogrammed design. Our treatment of the microassembly language will be informal.)

The microcode in Fig. 10–17 supports a small portion of the testing of the Forth Machine—the manual loading of data from the switch register on the display panel and the exercising of the Forth language's rotate instruction. This code includes declarations and microinstructions that illustrate a variety of features of the microassembler.

```
LOGIC ENGINE DEVELOPMENT SYSTEM MICROPROGRAM ASSEMBLER
FORTH_TEST: LOGIC ENGINE DEMONSTRATION PROGRAM

                    ID    FORTH_TEST

         * FORTH ENGINE
         * SAMPLE DECLARATIONS AND SAMPLE MICROCODE
                    SIZE 18; Number of command bits
                    MODE LOGIC
         * TEST MUX CONFIGURATION
         INMUX       COM  (0:3),T=%HHHH,D=0
         LD.L        INV  INMUX=0,T=%L
         TST.L       INV  INMUX=1,T=%L
         * COMMAND FIELD DECLARATIONS
         MUXCTL(7:0) COM  (4:11),T=$FF,D=%TTTTTTTT
         M0          EQU  MUXCTL(7:5); Mux 0 select signals
         M1          EQU  MUXCTL(4:2); Mux 1 select signals
         M2          EQU  MUXCTL(1:0); Mux 2 select signals
         M0S2        INV  M0=5; Select Reg S2 thru Mux 0
         M0SWR       INV  M0=0; Select Switch Reg thru Mux 0
         M1S0        INV  M1=3; Select Reg S0 thru Mux 1
         M2S1        INV  M2=1; Select Reg S1 thru Mux 2
         REGCTL      COM  (12:17), T=%HHHHH,D=%FFFFFF
         LOAD3       EQU  %111111; Load S0,S1,S2
         ROTATE      INV  M0S2,M1S0,M2S1,REGCTL=LOAD3; Rotate stack
                     PROG

LOC XDDDI CCCC C
         000        BEGIN    ORG  0
         000                 EQU  *
000 10033 OFF0 0    LOAD     JUMP TEST IF LD.L=%F
001 50013 OFF0 0             JUMP *    IF LD.L=%T
                             JUMP BEGIN;M0SWR,M1S0,M2S1,
002 30003 00DF C                 REGCTL=LOAD3; **Push switches onto stack
003 10003 1FF0 0    TEST     JUMP LOAD IF TST.L=%F
004 50043 1FF0 0             JUMP *    IF TST.L=%T
005 30003 0ADF C    ROT      JUMP BEGIN;ROTATE; **Rotate top 3 stack elements
                             END

0 ERROR(S) DETECTED
```

Figure 10-17 Microcode for the design example.

LEASMB microprogram statements have an optional label field, an operation field, and an optional operand field, in that order. Within the operand field, the required subfields are separated by semicolons; comments may follow the operand field, if preceded by a semicolon. Lines beginning with an asterisk are comments. An LEASMB output listing, such as in Fig. 10–17, shows the source program and, to the left of the program phase, the object code in hexadecimal notation.

An LEASMB program begins with the directive ID and ends with the directive END. The ID is the name assigned to the object code produced by the assembler, END marks the physical end of the program text. The declarations precede the program; the directive PROG separates the two. In the declaration phase, the directive SIZE specifies how many command bits are used in the design. The directive MODE, which takes an argument LOGIC or VOLTAGE, announces how the assembler is to interpret numeric data that could describe either logical or voltage values. As mixed logicians, we choose LOGIC mode.

Unless otherwise specified, numbers are in decimal notation. Binary constants are preceded by %, and may contain numerals 1 and 0, logical T and F, or voltage H and L.

The declaration phase has three main directives: COM, INV, and EQU. The program phase has, besides mnemonics for each of the 2910 sequencer's instructions, two directives: ORG and EQU. COM specifies the nature of the command bits in the microinstruction; INV allows a symbolic name to invoke complex operations on command bits; EQU allows the equivalencing of names to values or to other names. ORG declares the origin of the microinstructions in the control store.

Command bits, presented to the architecture by the command field of the pipeline register, are defined by the COM directive. In Fig. 10–17, the definition of REGCTL provides the following information: REGCTL is a field of 6 bits which occupy bits 12 through 17 of the command field of the microinstruction. [If we choose, we may refer to individual bits or groups of bits of REGCTL using a default set of indices, 0 for the leftmost bit through 5 for the rightmost bit. Thus REGCTL(0:3) would reference the leftmost 4 bits of the field. This notation frees the programmer from a dependence on the positions of particular command bits.] For each bit of REGCTL, truth is represented by a high voltage level (T = %HHHHHH). Whenever any bits of REGCTL are not mentioned in a microinstruction, the default values will be false (D = %FFFFFF). (A microinstruction will usually deal explicitly with only a few command bits; the default declarations of command bits are therefore of great importance in freeing the programmer from unnecessary details. In this example, the default values hold the current contents of the stack registers.)

The MUXCTL(7:0) declaration specifies that MUXCTL occupies bits 4 through 11 of the command field. In our program we may refer to individual bits or groups of bits using indices 7 (for the leftmost bit) to 0 (for the rightmost). This explicit indexing notation overrides the normal default notation. For each bit of MUXCTL, truth is represented by a high voltage (T = %HHHHHHHH). By convention, their unspecified default values are false.

In our illustration, we usually wish to deal with the group of command signals that controls a particular multiplexer in Fig. 10–15. So, for convenience, we define three variables M0, M1, and M2 in the next three lines of the program. M0 is declared to be a field of 3 bits, equivalent to bits 7 to 5 of MUXCTL. Similarly, M1 and M2 are declared to be equivalent to bits 4 to 2 and bits 1 to 0 of MUXCTL. With these definitions, we may refer to the field MUXCTL as a whole or to subfields M0, M1, and M2, or to any bit or group of bits of MUXCTL.

These uses of the EQU directive equate names with fields. EQU may also be used to equate names with values or with addresses. LOAD3 is defined with an EQU as equivalent to the binary pattern %111111. BEGIN is defined in the program phase as equivalent to ∗. Since ∗ in this context designates the current assignable control-store address, this notation equates the symbol BEGIN to the address 000.

In examining Fig. 10–15, you will see that, in order to select stack element *S0* as the output of multiplexer 1, we must present the code 3 (binary %011) to the *MUX1* select inputs. For convenience, we define a symbol M1S0 that will invoke (INV) the value 3 on the field M1. If we wish to pass element *S0* through *MUX1*, we write M1S0, thus assigning the value 3 (%011) to the field M1 in the microinstruction. In the microcode in Fig. 10–17, the instruction at location 002 illustrates this usage. The symbol ROTATE illustrates how we easily may develop complex invocations. The use of ROTATE in the instruction at location 005 invokes the previously defined invocations M0S2, M1S0, and M2S1, and invokes the value LOAD3 in the command bits defined for REGCTL. Invocations are the most important concept in LEASMB; they are the key to achieving high-level specifications of our microinstruction operations.

The microprogram performs two operations: loading the contents of the display panel's switch register into stack element *S0* (and simultaneously pushing the stack down), and performing a cyclic rotation of the top 3 elements of the stack. Two pushbuttons on the display panel, LD and TST, control the actions. When LD is pressed and released, the load-and-push operation will occur; pressing and releasing TST will cause the execution of the rotate operation. It is necessary to assure that the microcode for loading and rotating will be executed only once for each push of the button. The code at locations 000 and 001 performs a single-pulser function for the LD button; the code at location 003 and 004 performs a similar function for the TST button. From the discussion in Chapter 6, you can see that in our microcode we have developed pure-control versions of the single-pulsers. We return for a closer examination of this structure later in this chapter.

When the instruction at location 003, JUMP LOAD IF TST.L = %F, is executed, test input signal *TST*.L must be selected through the designer's test multiplexer, and a false value of *TST*.L must cause a branch to microinstruction LOAD. In the declaration phase, the command variable INMUX describes the select signals for the test multiplexer. The declaration of the variable TST.L specifies that TST.L will invoke the value 1 for INMUX. This agrees with Fig. 10–16, in which *TST*.L appears at input position 1 of the test multiplexer.

Test signals entering the test multiplexer may be represented as T = H or

T = L, like *LD*.H and *TST*.L. A jump may be desired for either a true or a false value of a test signal, as in the microinstructions at locations 000 and 001, or at 003 and 004. The LEASMB microassembler arranges for the correct voltage to be presented to the 2910's *CC* input, in accordance with the specifications in the program's instructions and declarations. The designer declares the voltage conventions once and may thereafter deal with signals as logical entities. In our example, the declaration of TST.L has a term $T = \%L$ appended to an otherwise normal invocation. This declares that if a microinstruction ever specifies *TST*.L as the source for the 2910's *CC* input, then truth is represented as a low voltage level. The microprogram sequencer portion of an LEASMB microinstruction contains a bit, set by the assembler, that specifies whether the incoming test voltage should be inverted before it enters the 2910. (In exercises at the end of the chapter, you are asked to ascertain from Fig. 10–17 which bit of the micro-instruction corresponds to this *CC*-inversion flag, and what voltage transformations are induced by the instructions in the microcode.)

Use this informal description of the language to follow the test program. In the Logic Engine's microassembly language, we have tried to encourage the use of high-level structured coding within a simple syntax. Let us now return to the design of our Forth Machine circuit.

Testing the System

Using the command-bit structure declared in the microprogram, we wire the appropriate bits of the Logic Engine's pipeline register to our architecture. We assemble the test microprogram and load it into the Logic Engine's WCS, making the appropriate monitor commands. We clear the Logic Engine, thereby causing the next microinstruction to be taken from location 000. Feeling bold, we enter the automatic run mode, causing the Logic Engine repeatedly to execute the instructions at locations 000 and 003, waiting for us to press the LD or TST pushbuttons on the display panel. We put a desired value into the display panel's switch register, and press and release the LD button. Assuming that we have wired the outputs of the stack elements to the LEDs on the display panel, we may observe the results directly on the panel.

If, as is likely, there is some problem with the behavior of the load-and-push operation, we must debug the architecture and the code. The single-step mode is useful at this point; it allows us to execute our microcode yet freeze the system after each instruction so that we may observe the status of any signals in the architecture. Breakpoints are also useful, to suspend the execution of microinstruction. In debugging larger microprograms, a combination of breakpoint and single-stepping is a powerful aid to the designer.

If we isolate a problem and wish to supply a particular set of commands to the architecture, we may use the monitor's manual-execution mode, in which we present manually generated command-bit patterns for loading into the pipeline register. During manual execution, we may (in fact, must) specify whether we wish the Logic Engine to issue a clocking signal to our architecture. We have detailed control of the entire debugging process. If we determine that a particular microinstruction is incorrect and the error does not warrant a reassembly of the program at this time, we may manually change the microcode in the WCS. Without all this assistance from the development system, the designer would find it very difficult to express the control algorithm and debug the design.

Now we are ready to study a more complex microprogrammed design.

DESIGNING A MICROPROGRAMMED MINICOMPUTER

With the aid of powerful development facilities such as those provided by the Logic Engine, we are in a position to tackle a more difficult problem. You have already studied the LD20 hardwired minicomputer in Chapters 7, 8, and 9; in this chapter we will borrow its architecture and will develop a microprogrammed version of the control algorithm. We call the new design the LD30. Although the execution will be slower, the development of the LD30's control algorithm will be vastly simpler than the hardwired algorithm of the LD20. Nevertheless, to accomplish a microprogrammed implementation of this magnitude will require the use of all the sophisticated development aids offered by the Logic Engine.

Developing a Microprogram

The following prescription is useful for developing the microprogram:

1. Develop the main architecture. For our LD30, we would follow almost identically the path taken in Chapter 7, and we will adopt the LD20's main architecture.
2. Specify the obvious command bits for controlling the architecture. We need not worry about details at this stage, but our knowledge of the architecture will give us much insight into the commands needed to control it. We may modify our list of commands later.
3. Write high-level microinstructions for the control algorithm. We will use the Logic Engine's microassembly language, which will allow us to express complex operations in symbolic terms without going into details. Since we have already studied the elements of the control process for the LD20, we can draw on this experience, using notations that are as close as reasonable to those used in the LD20's ASM chart.
4. Develop the declaration phase of the microcode. This is primarily the specification of invocation variables to expand in ever-increasing detail the high-level notations used in the microinstructions.
5. Tidy up. Complete the details of command-bit representations, test inputs, minor architectural elements, and definitions of the behavior of specific chips used in the architecture.

The Architecture of the LD30

We adopt almost intact the architecture developed for the LD20. Figure 10–18 shows one of the twelve bit-slices in the main data structure. We see the principal registers, the *ALU*, and the main bussing system for the 12-bit data of the LD30.

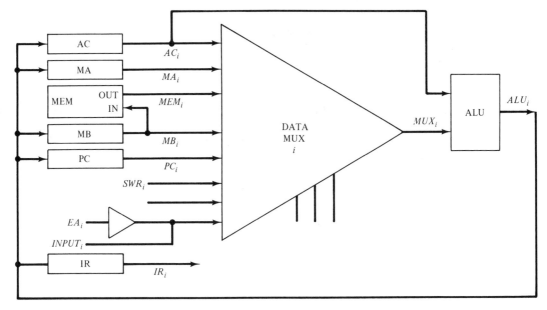

Figure 10–18 The main data paths of the LD30.

We will use the LD20's structure for the accumulator and link; this is shown in Fig. 10–19. Also, we will use the LD20's structure for the memory and its controller. We anticipate that the LD20's state generator will be completely missing from our microprogrammed implementation, since the microprogram controller in the Logic Engine performs the next-state selection. We expect our LD30 to contain at least one new architectural element: the test multiplexer that supports the delivery of the specified test signal to the microprogram controller. We will specify other architectural elements as we encounter them in developing the control algorithm for the LD30.

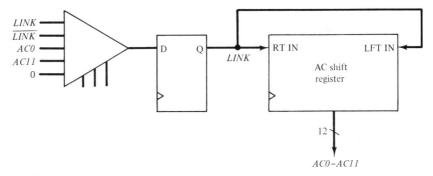

Figure 10–19 The accumulator and link of the LD30.

A First Approximation of the Command Bits

To provide a concrete point of reference for our later work, it is useful to specify as many of the architecture's control signals (the microinstruction commands) as are evident from the LD30's present architecture. This step could be deferred, since at this point it is just an approximation, but we have learned to appreciate having this early anchor to reality. As we pursue a top-down design, this knowledge can keep us from generating interesting but unproductive microcode.

Our first attempt to define control signals for our LD30 might take the form of Table 10–3. At a rough guess, we expect to have to test around 30 input or status signals; 5 or 6 bits will probably be required for the test multiplexer's select code. The controls for the data mux, the *ALU*, and the principal data registers are the same as in the LD20. The link multiplexer and the memory controls are also carried over from the earlier work on the LD20.

TABLE 10–3 PRELIMINARY
COMMAND SIGNALS FOR THE LD30

TESTMUXCTL	COM	5 or 6 bits
DATAMUXCTL	COM	3 bits
ALUCTL	COM	6 bits
MALD	COM	1 bit
MBLD	COM	1 bit
PCLD	COM	1 bit
IRLD	COM	1 bit
ACCTL	COM	2 bits
LINKMUXCTL	COM	3 bits
MEMORYCTL	COM	2 bits
ENABLE.EA	COM	1 bit

We expect to need additional architectural elements for our LD30's interrupt control, input-output control, Operate instruction priority control, and the halting operations, but we are not yet clear about their nature.

Writing the High-Level Microcode

We will use our earlier analysis of the LD20's control algorithm (see Chapter 7) to help develop the LD30's microcode. The hard work—understanding the PDP-8's specifications well enough to describe a correct control algorithm—is the same for the microprogrammed LD30 as for the hardwired LD20. In the LD20, we expressed the algorithm as an ASM chart. For the LD30, we will use the Logic Engine's LEASMB microassembly language. The two versions differ dramatically in appearance—the ASM chart is a two-dimensional flow diagram, whereas the microassembly language is linear, like a computer program. The two versions also differ in detail, since the LD20's ASM contains many states with multiple tests (multiple qualifiers) and conditional outputs, which are not permitted in the single-qualifier microinstructions supported by the Logic Engine. And the implementations of the two control algorithms are wildly different. However, the overall sequencing should remain the same. We do not need an ASM for our work on the LD30, but to tie this work more closely with the LD20, we will use LD20 nomenclature when practical. Figure 10–20 is a high-level diagram of the LD30's control algorithm. The algorithm consists of four

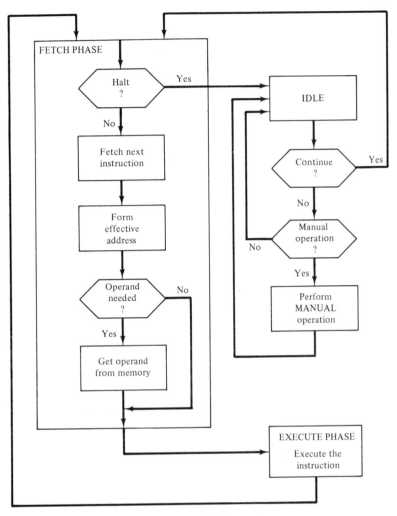

Figure 10–20   LD30 control.

principal blocks: the fetch phase, the execute phase, the idle phase, and the manual phase. We must decompose each block into a sequence of microinstructions. As always, we will strive for a top-down development.

The Idle Phase of the LD30

The purpose of the idle phase is to detect an operator's action at the LD30 display panel. As in the LD20, we must take care to process only synchronized input signals in our algorithm, and we must provide a means of processing each depression of a pushbutton on the display panel only once.

We may contemplate the use of a flip-flop *HALTFF* to record the status of halt requests, as we did in designing the LD20. A request to halt will eventually cause the algorithm to reach the idle phase; we should be prepared to clear the halt request at this time. The addition of a *HALTFF* flip-flop to our architecture will require appending two command bits to our list (for controlling J and K in a JK flip-flop or D and its enable in an enabled D flip-flop).

In the LD20 a composite signal *MANSW** indicates the assertion of at least one pushbutton signal. *MANSW** is the input to a single-pulser whose output is *MANPULSE*. *MANPULSE* is synchronized with our system clock and assures that a depression of a pushbutton is responded to only once. Following the discussion of single-pulsers in Chapter 6, we can envision several implementations of the *MANPULSE* signal.

Figure 10–21a shows an architectural implementation of *MANPULSE*, involving the two flip-flops and an AND gate that we have come to expect of this circuit. In this circuit, the microcode for the LD30's idle phase might be:

```
IDLE    EQU    *
        JUMP   * IF MANPULSE=%F; CLEAR.HALTFF
        JUMP   FETCH IF CONT.SW=%T
        CALL   MANUAL
        JUMP   IDLE
```

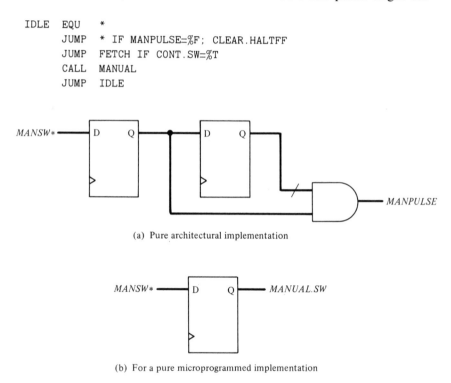

(a) Pure architectural implementation

(b) For a pure microprogrammed implementation

Figure 10–21 Two versions of the hardware for the LD30's single-pulser.

From Chapter 6, we learned that the single-pulser could be expressed as a pure algorithm rather than as an architectural element. In hardwired design, both views result in the same circuit. In microprogrammed design, the algorithm is expressed as microcode, and is never reduced to a hardwired circuit. Thus, we may implement the LD30's idle phase using a single flip-flop to synchronize the asynchronous signal *MANSW**, as shown in Fig. 10–21b and in the following microcode:

```
IDLE    EQU   *
        JUMP  * IF MANUAL.SW=%T; CLEAR.HALTFF
        JUMP  * IF MANUAL.SW=%F
        JUMP  FETCH IF CONT.SW=%T
        CALL  MANUAL
        JUMP  IDLE
```

The first instruction jumps in place until all buttons are released, thus assuring that any previous depression of a pushbutton is processed only once. The next step is to hang up until the operator depresses a pushbutton, which then allows the idle phase to proceed. This second procedure is superior, since it requires less hardware and only one additional microinstruction. This form of single-pulser was used in our Forth Machine design example.

As we did in the LD20, we will insert into the architecture the large OR gate necessary to produce *MANSW** from the collection of individual pushbutton signals on the display panel. We could dispense with this OR gate by writing microcode that serially single-pulses through the individual pushbutton signals, but this would require synchronizing each signal separately and would generate much additional microcode.

The Manual Phase of the LD30

The processing of manual operations in the LD20 was rather tedious, since the manual operations were merged with the execute phase. In the LD30, the manual phase is much more straightforward, as the following microcode shows:

```
MANUAL    EQU   *
          CALL  LDMA IF LDMA.SW=%T
          CALL  LDMB IF LDMB.SW=%T
          CALL  LDPC IF LDPC.SW=%T
          CALL  LDIR IF LDIR.SW=%T
          CALL  LDAC IF LDAC.SW=%T
          CALL  CLEAR IF CLEAR.SW=%T
          CALL  EXAMINE IF EXAMINE.SW=%T
          CALL  LDMEM IF LDMEM.SW=%T
          CALL  DEPOSIT IF DEPOSIT.SW=%T
          RTN

LDMA      RTN   ; SWR.TO.MA ; (the first ";" marks a null sequencer
                                    field)
LDMB      RTN   ; SWR.TO.MB
LDPC      RTN   ; SWR.TO.PC
LDIR      RTN   ; SWR.TO.IR
LDAC      RTN   ; SWR.TO.AC
CLEAR     RTN   ; CLEAR.LINK, CLEAR.INT.ENABLE

LDMEM     CALL  WRITE; SWR.TO.MB
          RTN
DEPOSIT   CALL  WRITE; SWR.TO.MB
          RTN   ; INCREMENT.MA
```

This microcode employs a sequential test of each pushbutton; if one is depressed, a subprogram is called to process it. A typical operation such as SWR.TO.MA must cause the transfer of the switch register to the selected register, in this case the memory address register. At this stage, we are unconcerned with the details of this transfer, other than to realize that, in our architecture, it can be performed in one clock cycle. Later, we will expand SWR.TO.MA and its cousins into the proper command-bit assertions, but now we do not permit these details to interrupt our thoughts.

Writing into memory requires the presentation to the memory buffer register of the data to be written, followed by an execution of a standard memory-write subprogram. We may specify the details of the WRITE subprogram at our convenience, but to satisfy your curiosity, we will do so now:

```
WRITE   EQU   *
        CONT  ; START.WRITE
        JUMP  * IF CC=%F
        RTN
```

Using the standard memory protocol of the LD20, we issue a *START.WRITE* signal (one of our two memory command signals) and wait until the cycle-complete response *CC* becomes true.

We've completed the microcode for the idle and manual phases. Next comes the instruction-fetch phase.

The Fetch Phase of the LD30

The conversion of the fetch phase of the LD20's ASM to our microprogrammed version is straightforward:

```
FETCH       EQU   *
F1          JUMP  IDLE IF HALTFF=%T
            CALL  INTERRUPT.TEST
F1.1        CONT  ; PC.TO.MA, PC.TO.MB, LOAD.IOP.ENABLER

F2          CONT  ; INCREMENT.PC

F3          CALL  READ.TO.IR
F3.1        JUMP  FETCH.DONE IF NO.MEMORY; EA.TO.MA

F4          CALL  READ.TO.MB
            JUMP  FETCH.DONE IF DIRECT.ADDRESSING
            CALL  AUTO IF AUTO.INDEXING

F6          JUMP  FETCH.DONE IF NO.INDIRECT.OPERAND; MB.TO.MA

F7          CALL  READ.TO.MB

FETCH.DONE  JUMP  EXECUTE;; operand, if any, is in MB
*                          EA, if any, is in MA
```

```
AUTO        EQU   * ; auto-indexing
            CALL  WRITE; INCREMENT.MB
            RTN
```

This microcode calls on several subprograms that are as yet unspecified: INTERRUPT.TEST, READ.TO.IR, READ.TO.MB. The memory-read subprograms are short, and we will elaborate them here; the treatment of INTERRUPT.TEST will appear later.

```
READ.TO.MB  EQU   * ; read memory(MA) into MB
            CALL  READ
            RTN   ; MEM.TO.MB

READ.TO.IR  EQU   * ; read memory(MA) into IR
            CALL  READ
            RTN   ; MEM.TO.IR

READ        EQU   * ; initiate and synchronize memory reads
            CONT  ; START.READ
            JUMP  * IF CC=%F
            RTN
```

In Chapter 7, developing the fetch phase was rather complex but now that we understand it, rendering our understanding into microcode is simple. As always, we have made good use of our ability to describe each microinstruction in high-level terms.

The Execute Phase of the LD30

At the start of the execute phase, *IR* will contain the current instruction, *MB* will contain any needed operand from memory, and *MA* will contain the effective address *EA*, if required. For the execute phase, we must develop microcode to decode the instruction and perform each type of PDP-8 operation. Given our work on the LD20, the execution of most of the instructions is straightforward, and we might proceed as follows; some of the details are left for you to complete.

```
EXECUTE     EQU   * ; main entry for execute phase
            JUMP  *+2 IF SING.INST.SW=%F
            CONT  ; SET.HALTFF ; only if single instruction switch is on
            JUMP  DECODE.INST ;; decode the instruction, somehow

EXEC.DONE   JUMP  FETCH ;; all done. return to fetch phase

* PDP-8 instruction-execution code. The LD30 instruction
* decoding step must branch to the proper code.

AND.CODE    EQU   * ; for PDP-8 AND instruction
* You complete this microcode.
```

```
TAD.CODE    EQU    * ; for PDP-8 TAD instruction
            JUMP   TAD.END IF ALU.COUT=%F; ADD.MB.TO.AC
            CONT   ; COMPLEMENT.LINK; only for a carry-out
TAD.END     JUMP   EXEC.DONE


ISZ.CODE    EQU    * ; for PDP-8 ISZ instruction
            CALL   WRITE; INCREMENT.MB
            JUMP   ISZ.END IF MB.IS.ZERO=%F
            CONT   ; INCREMENT.PC ; only if result is zero
ISZ.END     JUMP   EXEC.DONE


DCA.CODE    EQU    * ; for PDP-8 DCA instruction
* You complete this microcode.


JMS.CODE    EQU    * ; for PDP-8 CALL instruction
            CALL   WRITE; PC.TO.MB
            JUMP   EXEC.DONE; INCR.MA.TO.PC


JMP.CODE    EQU    * ; for PDP-8 JUMP instruction
* You complete this microcode.


IOT.CODE    EQU    * ; for PDP-8 IOT instruction
* This one is more complex. Defer it until later.


OP.CODE     EQU    * ; for PDP-8 Operate instruction
* This one is more complex. Defer it until later.
```

In the TAD.CODE block, the development of the arithmetic sum is caused by the execution of the ADD.MB.TO.AC invocation in the first microinstruction. The result of the addition, both the sum and the carry-out status, are available during the instruction. At the end of the instruction, the sum is loaded into the *AC*. During the instruction, we test the carry-out signal *ALU.COUT*, and if it is true, we arrange to complement the link bit.

The microcode for the execution of three of the PDP-8's instructions is left to you to complete. Of the algorithm for the LD30's execute phase, there remain only to specify the code for the Operate and IOT instructions, deal with interrupts, settle on how we will decode instructions, and clean up a few loose ends.

The Operate instruction. In designing the LD20, we developed a sophisticated method of saving clock cycles in the execution of the PDP-8's Operate instructions. Microprogrammed control, although allowing much parallel activity in each microinstruction, is inherently more serial than hardwired control. The elaborate priority-detection system of the LD20 usually saved a few cycles (and served as a good illustration of priority circuits), but attempting a similar scheme in our inherently slower LD30 would be misplaced effort. We resign ourselves to plodding serially through all the possible operations within each group of the PDP-8's operate instruction. Let's assume that the decoding of the instruction, in addition to detecting the Operate instruction, has also identified the subgroup.

This is an arbitrary but reasonable choice, and it will fit right in with the elegant scheme to be discussed later.

A reasonable microcode for a part of the PDP-8's Operate instruction might then be as follows:

```
OP.G1.CODE  EQU   * ; for PDP-8 Operate instruction, group 1

G1.P1       JUMP  *+2 IF IR4=0
            CONT  ; CLEAR.AC ; CLA operation

            JUMP  *+2 IF IR5=0
            CONT  ; CLEAR.LINK ; CLL operation

G1.P2       JUMP  *+2 IF IR6=0
            CONT  ; COMPLEMENT.AC ; CMA operation
```

In problems at the end of the chapter, you are asked to complete the microcode for the Operate instruction.

IOT Instruction. The PDP-8's IOT instruction performs input and output operations. As in the LD20, our LD30 must arrange for the signals *IOP1*, *IOP2*, and *IOP4* to be asserted, if requested, long enough to examine the values of the incoming status signals *IOSKIP*, *ACCLR*, and *ORAC*. This examination will require several microinstructions, during which the appropriate *IOP* signal must remain solidly asserted. We could write brute-force microcode for this problem, with code to test the status of each of the three *IOP* signals. However, our choice is to introduce an element into the LD30's architecture to generate and maintain the *IOP* signals as required. We have already seen such an element in the LD20: the *IOP* signal enabler. This circuit has a 4-bit shift register that allows the enabling of each *IOP* signal in turn. We will use this circuit, shown in Fig. 7–34, in our LD30. The microcode for the IOT instruction must then loop three times over the status tests—once for each of the possible *IOP* signals. We use the 2910 sequencer's internal counting instructions to manage the looping.

```
IOT.CODE  EQU   * ; for PDP-8 input-output operations
          LDCT  2 ;; load the 2910 R-register with (loop-count - 1)
IOT.LOOP  CONT  ; SHIFT.IOP.ENABLER
E3        JUMP  *+2 IF IOSKIP=%F
          CONT  ; INCREMENT.PC ; if IOSKIP is asserted
E4        JUMP  *+2 IF ACCLR=%F
          CONT  ; CLEAR.AC ; if ACCLR is asserted
E5        JUMP  *+2 IF ORAC=%F
          CONT  ; OR.INPUT.TO.AC ; if ORAC is asserted

          RPCT  IOT.LOOP; decrement & test R-reg, branch if non-zero
          JUMP  EXEC.DONE
```

The IOT instruction provided two suboperations ION and IOF for enabling and disabling the interrupt system. The suboperations are distinguished from regular IOT instructions by a zero device address in *IR3–IR8*. The microcode for these operations is:

```
IOT.ION.CODE  EQU  * ; turn on (enable) interrupt system
              JUMP EXEC.DONE ; SET.INT.ENABLE

IOT.IOF.CODE  EQU  * ; turn off (disable) interrupt system
              JUMP EXEC.DONE ; CLR.INT.ENABLE
```

Interrupt Processing in the LD30

The LD30 must be able to detect an interrupt request and present a synchronized version of the request to the microprogrammed control algorithm. The PDP-8's interrupt protocol also requires that we be able to determine if interrupts are enabled. If an interrupt is to occur, we must be able to force the execution of a CALL 0 instruction. For these functions, we adopt the same architectural elements used in the LD20. We show the architecture for the interrupt system in Fig. 10–22.

Handling an interrupt request began early in the fetch phase with a subprogram call to INTERRUPT.TEST. At the time we wrote that call, we had no clear

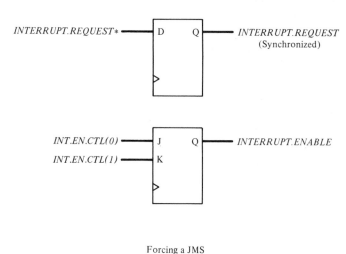

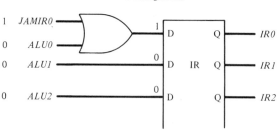

Figure 10–22 Architecture of the interrupt system in the LD30.

idea how interrupts would be handled. Now, having made some architectural commitments to generate the *INTERRUPT.REQUEST* and *INTERRUPT.ENABLE* signals, and to finesse the awkward problem of generating a CALL 0, we are able to specify the microcode fully.

```
INTERRUPT.TEST  EQU  * ; determine if an interrupt should occur
* no interrupt is possible if the previous instruction was an
* IOT instruction that enabled the interrupt system.
                RTN   IF IOT.ION=%T ; no interrupt if just enabled
                RTN   IF INTERRUPT.ENABLE=%F; no int if disabled
                RTN   IF INTERRUPT.REQUEST=%F; no int if no request

* Create an interrupt! Disable interrupts and force a ''CALL 0''.
                CJPP  EXECUTE,PASS; CLR.INT.ENABLE, FORCE.JMS.TO.ZERO
```

We called the interrupt-test code as a subprogram. If no interrupt occurs, the INTERRUPT.TEST microcode returns to the subprogram caller. If an interrupt occurs, the algorithm enters the regular execute-phase microcode, which is not a subprogram. Then, the code never makes a normal subprogram return back to the fetch phase, so we must take care to adjust the 2910 stack to remove the unused return point. The 2910 operation CJPP in the pass mode accomplishes both a jump and a stack pop.

Instruction Decoding in the LD30

The microcode for the execute phase calls a subprogram to decode the PDP-8 instruction residing in the *IR*. The operation code for most instructions is specified by the three bits in *IR0*, *IR1*, and *IR2*. Since our microinstructions can examine only one signal at a time, a pure microcode solution to instruction decoding is clumsy (although perfectly feasible). Here is a brute-force way:

```
DECODE.INST  EQU  * ; pure microcode instruction decoding
             JUMP  CODE.1XX IF IR0=1
CODE.0XX     JUMP  CODE.01X IF IR1=1
CODE.00X     JUMP  TAD.CODE IF IR2=1
             JUMP  AND.CODE
CODE.01X     JUMP  DCA.CODE IF IR2=1
             JUMP  ISZ.CODE
CODE.1XX     JUMP  CODE.11X IF IR1=1
CODE.10X     JUMP  JMP.CODE IF IR2=1
             JUMP  JMS.CODE
CODE.11X     JUMP  OP.CODE IF IR2=1
             JUMP  IOT.CODE
```

In developing microcode for the execution phase, we decided that the instruction-decoding step would distinguish the three groups of Operate instructions and the two special cases of the IOT instruction. We could complete this decoding with further tests of additional bits of the LD30's *IR*, although the test for the IOT-instruction special cases of the IOT instruction is complicated by having to identify a 6-bit field of zeroes in bits *IR3–IR8*. All of this microcode is a serialized version of a basically parallel function—the decoding of a several-bit code. We already know that combinational logic, implemented with gates, decoders, or ROMs, can achieve rapid decoding.

We may simplify the decoding of LD30 instructions and illustrate some powerful control techniques by developing a solution that mixes architecture and algorithm. The 2910, at the heart of the Logic Engine's microprogram controller, normally expects to receive its *D* input from the microinstruction pipeline register; 14 of the 2910's 16 operation codes assume this. However, for the 2910's JMAP instruction, the *D* input comes from a different source, usually from a "mapping ROM." Let's use a ROM to assist the decoding.

Our goal is to examine bits of the LD30's *IR* and branch within our control microcode to the proper code for executing each type of PDP-8 instruction. The decoding of the PDP-8's basic instructions and our special cases requires examining all 12 bits of the *IR*. A pure ROM solution would use ROMs with 12 address inputs to produce about a dozen microcode addresses. However, many of the bits of the *IR* are needed only to detect the two special cases of the IOT instruction, ION and IOF. Our plan is to incorporate two signals in hardware: *IOT.ION* will be true for the interrupt-enable suboperation; *IOT.IOF* will be true for the interrupt-disable suboperation. For now, we assume the existence of these two signals; we will discuss how to create them later.

Now the inputs to our instruction-decoding ROM are considerably simplified. We require *IR0*, *IR1*, and *IR2* to decode the main operation code; *IR3* and *IR11* to distinguish the Operate instruction groups; and *IOT.ION* and *IOT.IOF* to identify the IOT subinstructions and to distinguish ordinary input-output IOT instructions from the special cases. These 7 signals form the inputs to the ROM; the ROM's output must produce microcode branch addresses that we can use to transfer to sections of the microprogram dealing with the PDP-8's 12 instructions and subinstructions. What are the 12 microcode branch addresses? From our microcode for the execute phase we see examples such as AND.CODE, IOT.CODE, and OP.G2.CODE. But the experienced microprogrammer will recognize a pitfall: the actual addresses of these locations in the microprogram memory depend heavily on the details of the microcode. If we modify the microcode, the likelihood is that a reassembly of the microcode will change the addresses of the subprograms. We certainly do not desire to reprogram our ROM every time we reassemble the LD30's control microprogram! To avoid this difficulty, we introduce a jump table into our microprogram. This table will be in a fixed location in the microinstruction memory, and we will guarantee not to alter its location. The ROM's outputs will point to entries in this jump table, and the table entries, from their fixed locations, will jump to the subprogram addresses. The advantages are great: the contents of the ROM remain fixed even though we modify the microcode, and our microassembler will handle the drudgery of locating the subprograms. Table 10–4 shows the structure of the mapping ROM and the microcoded jump table. If speed of execution is sufficiently important, after the design is ready for production you may program a new

TABLE 10-4 DECODING THE LD30'S INSTRUCTIONS

CONTENTS OF THE MAPPING ROM

IR0	IR1	IR2	IR3	IR11	IOT.ION	IOT.IOF	Jump address	
0	0	0	X	X	X	X	$10	AND.INST
0	0	1	X	X	X	X	$11	TAD.INST
0	1	0	X	X	X	X	$12	ISZ.INST
0	1	1	X	X	X	X	$13	DCA.INST
1	0	0	X	X	X	X	$14	JMS.INST
1	0	1	X	X	X	X	$15	JMP.INST
1	1	0	X	X	1	0	$16	IOT.ION.INST
1	1	0	X	X	0	1	$17	IOT.IOF.INST
1	1	0	X	X	0	0	$18	IOT.INST
1	1	1	0	X	X	X	$19	OP.G1.INST
1	1	1	1	0	X	X	$1A	OP.G2.INST
1	1	1	1	1	X	X	$1B	OP.G3.INST

The Inputs span columns IR0 through IR11; IOT.ION and IOT.IOF; Outputs span Jump address and the instruction name.

MICROCODED JUMP TABLE

```
              ORG   $10; opcode jump table
AND.INST      JUMP  AND.CODE
TAD.INST      JUMP  TAD.CODE
ISZ.INST      JUMP  ISZ.CODE
DCA.INST      JUMP  DCA.CODE
JMS.INST      JUMP  JMS.CODE
IOT.ION.INST  JUMP  IOT.ION.CODE
IOT.IOF.INST  JUMP  IOT.IOF.CODE
OP.G1.INST    JUMP  OP.G1.CODE
OP.G2.INST    JUMP  OP.G2.CODE
OP.G3.INST    JUMP  OP.G3.CODE
```

mapping ROM to point directly to the instruction execution code, thereby eliminating the microcoded jump table.

With the addition of the mapping ROM for decoding instructions, we may now write the DECODE.INST microcode whose existence we assumed when we developed the original microcode for the LD30's execute phase:

```
DECODE.INST  EQU   * ; instruction decoding
             JMAP  ;; jump to correct instruction processing code
```

Declarations for the LD30 Control Microprogram

We have developed the main elements of the LD30's architecture and specified the control algorithm in high-level terms. Now we will develop the declaration phase of the microcode, in which we will expand our high-level invocations into actual command signals directed to the LD30 architecture.

Consider operations involving the main data paths. If we scan the executable instructions in the microprogram, we can catalog a large number of operations on the main data path. PC.TO.MA, CLEAR.AC, ADD.MB.TO.AC, and SWR.TO.IR are examples. In our version of the microcode, there are 27 different invocations that refer to the main data path architecture. These require selecting an operand through the data multiplexer, performing an appropriate *ALU* operation, and loading the result into some register. Still resisting the temptation to jump into the details of the hardware too soon, we break down each of these high-level invocations into more detailed specifications:

```
PC.TO.MA       INV  SEL.PC, ALU.PASS, LOAD.MA
CLEAR.AC       INV  ALU.ZERO, LOAD.AC
ADD.MB.TO.AC   INV  SEL.MB, ALU.PLUS, LOAD.AC
SWR.TO.IR      INV  SEL.SWR, ALU.PASS, LOAD.IR
```

We have tried to choose obvious names for the intermediate operations. For instance, SEL.MB means "select the *MB* register as the input to the *ALU*," ALU.PLUS means "cause the *ALU* to add its operands," and LOAD.AC means "load the *AC* with whatever is at its data inputs." At this stage, if we were to change our minds about some details of our main data path architecture, it is likely that these declarations would not require alteration.

Next, we must expand the specifications for each of the intermediate-level commands. Eventually, we will end up with detailed assignments of elementary signals to microinstruction command bits, but we need not hurry this process.

We may describe the data multiplexer selection codes as follows, using the same ordering of the multiplexer inputs as in the LD20.

```
* DATA MULTIPLEXER SELECTIONS
SEL.PC      INV  DATAMUXCTL=0
SEL.MB      INV  DATAMUXCTL=1
SEL.MA      INV  DATAMUXCTL=2
SEL.AC      INV  DATAMUXCTL=3
SEL.SWR     INV  DATAMUXCTL=5
SEL.MEM     INV  DATAMUXCTL=6
SEL.INPUT   INV  DATAMUXCTL=7
SEL.EA      INV  DATAMUXCTL=7, ENABLE.EA
```

We have described the selection of each input to the multiplexer in terms of its multiplexer select code. (As in the LD20, we have left data multiplexer input 4 unused, to allow for expansion. We have routed *INPUT* and *EA*, both controlled by three-state signals, into a single data multiplexer input. *EA*'s control signal is ENABLE.EA, which will become one of the LD30's microinstruction command signals; *INPUT* needs no microinstruction command signal for its three-state control since that is governed by the PDP-8's input-output protocol.)

Now we can finally specify the details of DATAMUXCTL: it is a microinstruction command-bit field of 3 bits, as suggested in Table 10-3. Until we begin the wiring, the actual locations of the three bits are unimportant, even distracting.

If we later decide to change their position in the microinstruction, *the only change to the microprogram is in the COM statement itself.*

A suitable final version of this command statement might be:

```
DATAMUXCTL  COM  (6:8),T=%HHH
```

where each select bit is declared as $T = H$.

We may unravel the specifications of the *ALU* operations in a similar way. Here are specifications for two of the seven *ALU* operations needed in the LD30:

```
* ALU OPERATIONS (using the 74LS181)
ALU.PASS  INV  ALUCTL=LS181.PASS
ALU.PLUS  INV  ALUCTL=LS181.PLUS
```

To declare the operation of the 74LS181 ALU chip, we go to the integrated circuit data book and extract the voltage behavior of the chip. For instance:

```
* 74LS181 arithmetic logic unit
* Order of signals is S3,S2,S1,S0,M,CIN
LS181.PASS  EQU  %LLLLLH
LS181.PLUS  EQU  %HLLHLH
```

The final specification of the command-bit field ALUCTL is:

```
ALUCTL  COM  (9:14),D=LS181.PASS
```

The field occupies 6 bits. We choose to specify the default assignment so that whenever we are not specifically controlling the *ALU*, it will be performing a PASS operation.

Specifying the control of most of the data registers is simple, since only a single control signal is involved, but the *AC*, with its ability to shift and load, requires two control signals. Here is a representative sample of the declarations:

```
* SOME DATA REGISTER OPERATIONS
LOAD.MB     INV  MBLD
LOAD.PC     INV  PCLD


* ACCUMULATOR OPERATIONS (using 74LS194)
LOAD.AC     INV  ACCTL=LS194.LOAD
AC.RIGHT    INV  ACCTL=LS194.RIGHT
AC.LEFT     INV  ACCTL=LS194.LEFT


* 74LS194 SHIFT REGISTER OPERATION
* Order of signals is S1,S0
LS194.HOLD   EQU  %LL
LS194.RIGHT  EQU  %HL
LS194.LEFT   EQU  %LH
LS194.LOAD   EQU  %HH
```

Finally, the specification of the command bits for the *MB*, *PC*, and *AC* registers is:

```
MBLD    COM  (16),T=%L,D=%F
PCLD    COM  (17),T=%L,D=%F
ACCTL   COM  (19:20),T=%HH,D=LS194.HOLD
```

We are nearly done. We may now tabulate the test input signals needed to drive the microcode, and prepare the necessary declaration statements. Table 10–5 contains the declarations. Each invocation describes the voltage polarity of the signal and its position within the test multiplexer.

TABLE 10–5 TEST INPUT
DECLARATIONS OF THE LD30

```
*TEST MULTIPLEXER INPUTS
TM               EQU  TESTMUXCTL
MANUAL.SW        INV  TM=2,T=%H
CONT.SW          INV  TM=12,T=%L
LDMA.SW          INV  TM=11,T=%L
LDMB.SW          INV  TM=10,T=%L
LDPC.SW          INV  TM=9,T=%L
LDIR.SW          INV  TM=8,T=%L
LDAC.SW          INV  TM=7,T=%L
CLEAR.SW         INV  TM=6,T=%L
EXAMINE.SW       INV  TM=5,T=%L
LDMEM.SW         INV  TM=4,T=%L
DEPOSIT.SW       INV  TM=3,T=%L
SING.INST.SW     INV  TM=13,T=%L
IR4              INV  TM=26,T=%H
IR5              INV  TM=27,T=%H
IR6              INV  TM=28,T=%H
IR7              INV  TM=29,T=%H
IR8              INV  TM=30,T=%H
IR9              INV  TM=31,T=%H
IR10             INV  TM=32,T=%H
IR11             INV  TM=33,T=%H
INTERRUPT.REQUEST INV TM=1,T=%H
INTERRUPT.ENABLE INV TM=0,T=%H
HALTFF           INV  TM=15,T=%L
CC               INV  TM=14,T=%H
NO.MEMORY        INV  TM=23,T=%L
DIRECT.ADDRESSING INV TM=25,T=%L
NO.INDIRECT.OPERAND INV TM=22,T=%L
AUTO.INDEXING INV TM=21,T=%L
IOT.ION          INV  TM=24,T=%L
IOSKIP           INV  TM=18,T=%L
ACCLR            INV  TM=17,T=%L
ORAC             INV  TM=16,T=%L
ALU.COUT         INV  TM=34,T=%L
MB.IS.ZERO       INV  TM=20,T=%L
SKIP             INV  TM=19,T=%H
```

This concludes our presentation of the development of the LD30's microcode declarations. The declarations tend to be lengthy, but they contain information important both for the microprogram assembler and for the reader of the microcode. Properly specified, the declarations exhibit a gradual evolution of detail, from the high-level specifications in the microprogram algorithm down to the voltage behavior of each command line in the microinstruction. Later modification and maintenance of the design are greatly simplified by the existence of this orderly, well documented record. We said earlier that the LEASMB invocation was the key to top-down design. Now you can see the power of this concept.

Our Design of the LD30 is Complete!

Our treatment of the microprogram for the LD30 is finished. We have shown representative portions of the algorithm and of the declarations that describe the terms used in the algorithm. How different is the LD30 machine from the LD20?

The main elements of the architecture are the same in both the hardwired and microprogrammed designs. Several minor architectural features of the LD20 survive in the LD30, and the LD30 contains two major additions: the test input multiplexer and the jump-address EPROM used in instruction decoding. Figure 10-23 shows the elements of the LD30's architecture. Not much is left.

The LD30 requires about 70 chips; the LD20, 125. The dramatic change between the two is, of course, in the implementation of the control algorithm. In moving from the LD20's hardwired control to the LD30's microprogrammed control, we eliminate about 55 integrated circuit chips and innumerable wires. The LD30's control algorithm looks like a program, and it is one. It contains about 120 microprogram instructions and several hundred lines of declarations. Comments are often useful in the microprogram, but with well-chosen nomenclature and systematic top-down development, the microcode is largely self-documenting.

In the senior-level hardware laboratory at Indiana University, our students study, construct, debug, and extend the LD20 hardwired and the LD30 micro-programmed versions of the PDP-8. With the aid of the Logic Engine, our students accomplish this in one semester, using wire-wrap technology. The proof of their performance is to download and execute actual PDP-8 programs—a source of immense satisfaction for the students and the instructors.

SUMMING UP

Starting in 1964 with IBM's use of microprogramming in their System/360 digital computers, manufacturers have increasingly adopted microprogramming methods for the control of computers. It is fair to say that most computer designs now have at least some microcode in their control—a fact often unknown to the programmer, since the microcode is not visible. The conventional programmer works with the computer's machine or assembly language, or with higher-level languages, without being aware that there is really another layer of programming—the microcode—buried in the hardware.

The great advantages of microcoding are its uniformity and ease of mod-

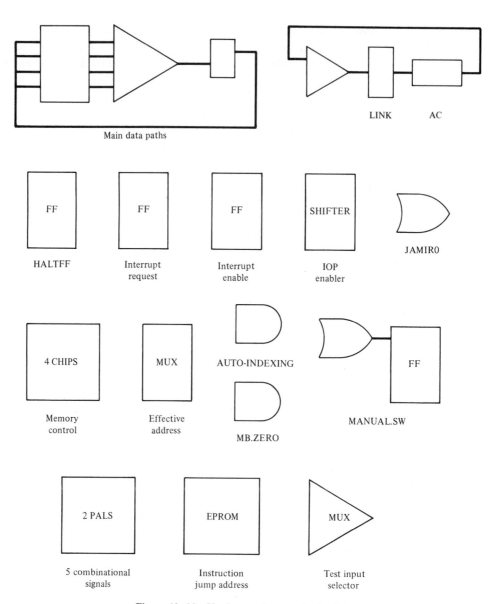

Figure 10–23 Hardware elements in the LD30.

ification. Carrying the notion further, we could control *all* digital tasks with a single type of controller, such as a Logic Engine. Using microcode, we make the controller perform a specialized task for each type of device, for instance executing the PDP-8's instructions in the LD30. With identical copies of the basic microprogrammable processor, we could control computers, line printers, card readers, floppy disks, terminals, and other devices having suitable speed requirements. Each of these would have its own architecture, which would include the device itself and the logic needed to interface with the microcode. Each controller would have its own microcode to control the architecture.

This has great potential advantages for both the designer and the maintenance engineer. The designer develops control algorithms with a uniform style, in a programming mode, using support systems derived from experience with software. The maintenance engineer would have to master only one type of control hardware. Each device's specialized control algorithm could be dealt with as a microprogram, a form that is generally easier to understand than a hardwired design. Diagnostic and documentation aids based on software principles can greatly speed both the development and the maintenance processes.

Unfortunately, although computer manufacturers make extensive use of proprietary microprogrammable processors, few general-purpose microprogramming systems such as the Logic Engine are available commercially.

In this chapter, we have discussed some basic principles of microprogramming and have shown how the concepts may become a powerful tool for design, requiring sophisticated support systems. We have not attempted to discuss the wide variety of microprogramming techniques; for this you may consult more specialized tests, such as those listed in the Readings and Sources.

Microprogramming is a great stride toward bridging the gap between hardware and software. Let us now move all the way, to the use of small conventional computers, microcomputers, as digital controllers.

READINGS AND SOURCES

ANDREWS, M., *Principles of Firmware Engineering in Microprogram Control.* Computer Science Press, Woodland Hills, Calif., 1980. Good discussions of classical microprogrammed architectures. Techniques for reducing the width and length of a ROM. Uses ASM notation for microprogramming.

Bipolar Microprocessor Logic and Interface. Advanced Micro Devices, 901 Thompson Place, P.O. Box 3453, Sunnyvale, Calif. 94088. Technical data and applications for the Am2900, Am29100, and Am29300 series.

BLAKESLEE, THOMAS R., *Digital Design with Standard MSI and LSI*, 2nd ed. John Wiley & Sons, New York, 1979. Good description of target versus host machines.

GLASSER, LANCE A., and DANIEL W. DOBBERPUHL, *The Design and Analysis of VLSI Circuits.* Addison-Wesley Publishing Co., Reading, Mass., 1985.

KLINGMAN, EDWIN E., *Microprocessor System Design*, Vol. 2. *Microcoding, Array Logic, and Architectural Design.* Prentice-Hall, Englewood Cliffs, N.J., 1982. Microprogramming.

LSI Databook. Monolithic Memories, 2175 Mission College Blvd., Santa Clara, Calif. 95954. PALs, memory products, arithmetic units, system building blocks.

MANO, M. MORRIS, *Computer System Architecture*, 2nd ed. Prentice-Hall, Englewood Cliffs, N.J., 1982.

MEAD, CARVER, and LYNN CONWAY, *Introduction to VLSI Systems.* Addison-Wesley Publishing Co., Reading, Mass., 1980. The first VLSI textbook.

MICK, JOHN, and JAMES BRICK, *Bit-Slice Microprocessor Design*. McGraw-Hill Book Co., New York, 1980. A collection of design notes for the Advanced Micro Devices 2900 bit-slice family. This book is useful far beyond the AM2900 chips.

MYERS, GLENFORD J., *Digital System Design with LSI Bit-Slice Logic*. John Wiley & Sons, New York, 1980.

PROSSER, FRANKLIN, *Logic Engine Development System: System Reference Manual*. Logic Design, Laramie, Wyo., 1983.

PROSSER, FRANKLIN, *Logic Engine Development System: LEASMB Microprogram Assembler Reference Manual*. Logic Design, Laramie, Wyo., 1983.

PROSSER, FRANKLIN, and ROBERT WEHRMEISTER, *C421-C422 Advanced Computer Organization Laboratory Manual*. Computer Science Department, Indiana University, Bloomington, Ind., 47405, 1985. Laboratory manual to support the construction, debugging, and study of the LD20 and LD30 implementations of the PDP-8I. The laboratory project uses the Logic Engine Development System, manufactured by Logic Design. Ask the authors of this book for information.

PROSSER, FRANKLIN, and DAVID WINKEL, "The Logic Engine Development System: Support for microprogrammed bit-slice development," *Proceedings MICRO 16,* October 1983, page 84.

System Design Handbook, 2nd ed. Monolithic Memories, 2175 Mission College Blvd., Santa Clara, Calif. 95054, 1985. Section 3 contains a shifter and pipeline architecture similar to that of the Logic Engine.

WESTE, NEIL, and KAMRAN ESHRAGHIAN, *Principles of CMOS VLSI Design: A Systems Perspective*. Addison-Wesley Publishing Co., Reading, Mass., 1985.

EXERCISES

10–1. Describe Wilkes's great contribution to the systematic development of control algorithms.

10–2. Produce a Wilkes implementation similar to Fig. 10–3 for the ASM in Fig. 5–13.

10–3. Produce a Wilkes implementation similar to Fig. 10–3 for the ASM in Fig. 5–20.

10–4. Explain the meaning of "microprogram."

10–5. How does a ROM implementation of an ASM differ from Wilkes's implementation? In what ways is the ROM method less flexible?

10–6. In microprogramming terminology, what is a "qualifier"?

10–7. Produce a ROM-based multiple-qualifier microprogrammed implementation of the ASM in Fig. 5–13.

10–8. Produce a ROM-based multiple-qualifier microprogrammed implementation of the ASM in Fig. 5–20.

10–9. Sketch a representative part of a ROM-based multiple-qualifier implementation of the Black Jack Dealer ASM in Fig. 6–32.

10–10. Discuss the advantages and disadvantages of the following methods of ROM-based microprogramming design. Consider the ease of implementation, generality of ASM structure, understandability of the result, ease of use, and so on.
(a) Multiple qualifier.

(b) Single qualifier, two addresses.

(c) Single qualifier, one address.

(d) Single qualifier in a Logic Engine environment.

10–11. Produce a single-qualifier, single-address microprogram implementation of an appropriate modification of the ASM in Fig. 5–13. For this task you must alter Fig. 5–13 to accommodate the requirements of the microprogram. State clearly any assumptions you make in producing the altered ASM.

10–12. Perform Exercise 10–11 for the ASM in Fig. 5–20.

10–13. Design a controller for single-qualifier, two-address microinstructions (after Fig. 10–5). The controller should support up to 256 microinstructions, up to 32 test inputs, and up to 40 command outputs. Specify the MSI chips that you use.

10–14. Perform Exercise 10–13 for a controller to execute microinstructions of the form of Fig. 10–8.

10–15. Why does the text's single-qualifier method not permit conditional outputs? Design a single-qualifier microinstruction interpreter that supports conditional outputs.

10–16. Complete the specification of the microcontroller in Fig. 10–10.

10–17. Characterize microprogrammed design as seen by:

(a) The hardware designer.

(b) The computer programmer.

10–18. Why is the hardwired control described in ASM charts not the best tool for the design of large, complex systems?

10–19. Propose a Logic Engine implementation of the Black Jack Dealer's control algorithm of Fig. 10–9. Express the microcode in the style of the Logic Engine's microassembly language described in the text.

10–20. Although RAM is appropriate for storing microcode in the Logic Engine during the development and debugging of an algorithm, why might RAM not be the best choice for a production version of a microprogrammed controller? On the other hand, what advantages might RAM-based microprogram storage have in a production version?

10–21. Compare the LD20 and LD30 as to speed, number of states, total design effort, and anticipated ease of maintenance.

10–22. The text mentions an early microprogram sequencer, the 2909 4-bit sequencer slice. To achieve a realistic number of microprogram address bits, several of these chips are to be cascaded together. What connections do you anticipate will be required *between the cascaded 2909 chips?*

10–23. Look at Fig. 10–17, which contains the microcode for the Forth Machine design example. The object code is presented in hexadecimal notation, with 4 bits per hexadecimal digit. In the object code, 1 represents a high voltage and 0 a low voltage. One of the bits in the X field of the object code is the 2910's *CCEN* input. The Logic Engine implements an unconditional jump by making *CCEN* false, thereby forcing a pass operation in the 2910. A conditional jump, which makes use of the 2910's *CC* input, requires *CCEN* to be true.

(a) By inspection of the object code in Fig. 10–17, determine which bit of the X-field is *CCEN*.

(b) State whether the 2910's *CCEN* is high-active or low-active.

10–24. Another of the X-field bits in the Logic Engine microcode of Fig. 10–17 is *CCINV*, the assembler's specification that the voltage of the incoming test signal must be inverted in order that the 2910 properly interpret the signal. The 2910's *CC* input is low-active.

(a) By inspection of the object code, determine which bit corresponds to *CCINV*.

(b) Using the declaration for LD.L and TST.L, determine the correct status of the *CCINV* bits (true or false) in the microinstructions at locations 000, 001, 003, and 005.

(c) Determine if *CCINV* is high-active or low-active.

10–25. From Fig. 10–17, expand the hexadecimal digits for the command bits in the object microinstructions at locations 001 and 002, so that you can observe the 18 command signals individually. For each of these microinstructions, verify that the bits in the object code are consistent with the declarations in the source program.

10–26. In Fig. 10–17, why is the curious term $T = \%L$ appended to the declarations for LD.L and TST.L?

10–27. Complete the LD30 microcode for the following PDP-8 instructions:

(a) AND.

(b) DCA.

(c) JMP.

10–28. Complete the LD30 microcode for the parts of the PDP-8's Operate instruction not elaborated in the text:

(a) Group 1.

(b) Group 2.

10–29. Exercise 8–52 requires the addition of the PDP-8's Group 3 instructions to the LD20's hardwired design. Repeat that exercise for the LD30, showing all modifications or extensions of the architecture and the microcode.

10–30. Consider the PDP-8's IOT instruction. Assume that the IOP signal enabler is eliminated from the LD30's architecture. Implement the IOT instruction solely in microcode. Remember to keep each required IOP signal asserted throughout the testing of the incoming status signals.

10–31. Explain clearly why we chose to implement the decoding of the LD30's instructions with a mapping ROM and a separate jump table rather than by using the faster method of having the mapping ROM jump directly to the appropriate microcode.

10–32. From the LD30 microcode given in the text, find ten examples of operations on the main data path, other than *PC.TO.MA*, *CLEAR.AC*, *ADD.MB.TO.AC*, and *SWR.TO.IR*. Write LD30 invocation declarations for your ten operations, following the examples in the text. Expand any intermediate operations until each element is finally reduced to command declarations.

10–33. Perform Exercise 8–50 using the LD30's microprogrammed design. Show modifications or additions to the architecture and microcode.

10–34. Perform Exercise 8–51 using the LD30's microprogrammed design. Show modifications or additions to the architecture and microcode.

10–35. Perform Exercise 8–52 using the LD30's microprogrammed design. Show modifications or additions to the architecture and microcode.

10–36. Perform Exercise 8–53 using the LD30's microprogrammed design. Show modifications or additions to the architecture and microcode.

10-37. Choose a computer of your choice, such as the PDP-11, M6809, M68000, or Intel 8080, and write a microprogram to implement the processor's instruction set. Use the style of the Logic Engine's microassembly language. The suggested processors are complex, and you may wish to consider a subset of the instructions. The user's manual for your chosen processor will probably not give much insight into the internal organization of the processor. As we did with the LD20 and LD30 versions of the PDP-8, start with the instruction set and the obvious registers referenced by the instructions. You are free to specify additional registers, data paths, and control signals required to achieve your implementation.

10-38. Propose a Logic Engine implementation of the Black Jack Dealer's control algorithm of Fig. 10–10. Express the microcode in the style of the Logic Engine's assembly language described in the text.

11

Microcomputers in Digital Design

In this chapter, we will complete the bridge from hardware to software by considering the role of microcomputers in digital design. We assume that you have some experience in programming a microcomputer at the assembly language or machine language level. From the programming viewpoint, microcomputers are small-scale versions of larger computers; given enough time and resources, any microcomputer can execute any algorithm. In this chapter, we will confine the discussion to the use of microcomputers to control digital devices.

In Chapter 10, we developed primitive microprogrammable computers specialized to emulate algorithms of a specific form. Their basic command implemented an ASM state with a conditional branch. In many cases, there is no need to use such a specialized computer, since any computer is a universal algorithm emulator when properly programmed. Of course, some computers may be a poor choice for certain design problems, because of an ill-adapted instruction set, inadequate speed, unsuitable input-output facilities, or high cost.

In digital design, the microcomputer can often serve as the controller—the executer of the control algorithm. In earlier design methods used in this book, we maintained direct and intimate contacts between the controller and the controlled devices; the processing of status signals and command outputs was flexible, with a high degree of parallelism. The relationship of a microcomputer controller to its environment is quite different and more restrictive, and the control algorithms must abandon much of the former parallelism. We will devote most of this chapter to these aspects of microcomputer-based design. But first, a bit of history will show how ordinary computers came to be used in digital design.

THE COMPUTER AS A DEVICE CONTROLLER

The first computer widely used as a controller of hardware was the PDP-8, primarily because it was the first computer whose price was less than astronomical. The history of the PDP-8 sheds insight on the evolution of microcomputer-based design. In the early 1960s, a group of engineers was assigned to build a controller for a nuclear reactor. The reactor supplied status signals such as neutron flux, core temperature, and power output. The primary controller commands were directed to motors that inserted and withdrew cadmium rods that absorbed neutrons and thereby controlled the reactor power. The initial design of the controller was hardwired. Unfortunately, the problem was incompletely specified, since the reactor was still under construction. The algorithm changed often, and each change required that the designers rip out gates and wires and start anew. The conviction gradually arose among the engineers that they must seek a general solution that could handle all the cases then known and the likely new ones to come.

What characteristics must such a device have? It must be an arbitrary algorithm emulator. The algorithm may involve computation as well as control. For example, the motion of the control rod may be a complicated mathematical function of power output and neutron flux. Speed was of no particular importance, since the inertia of the control rods would limit the mechanical responses to milliseconds, regardless of the speed of the controller.

There were two ways to meet these requirements:

a. Extend the controlled architecture (the reactor) to include primitive arithmetic units, registers, and memories to support the computational requirements. The control algorithm would then manipulate these elements to perform computations as well as control the cadmium rods. This is a hardwired approach, using more powerful and general architectural elements such as bit slices that allow the handling of more general classes of algorithms.

b. Use a conventional computer to emulate the algorithm. The reactor would be interfaced to the computer and treated as a peripheral device that received commands from the computer and supplied status signals in return. At that time, all computers were expensive, and were difficult or impossible to connect to so strange a device as a reactor.

One way to meet the design goals was to design and build a small, inexpensive computer with sufficient flexibility to allow easy interfacing of arbitrary peripheral devices. In an inspired decision, the engineers chose this approach, and the PDP-8 was the result. This early PDP-8 was crude by later standards, but it worked! A steady stream of new *minicomputers* followed, as manufacturers hastened to develop small computers priced to encourage their use in process control and automation applications.

Enter the Microcomputer

In the early days, minicomputers were made with integrated circuit logic of the MSI level of complexity. Minicomputers had lots of chips; the more elaborate minicomputers can have chip counts exceeding one thousand.

The VLSI technology that emerged in the late 1970s led to placing more and more circuitry on a single integrated circuit chip. With the introduction of the Intel 4004 microprocessor chip, the microcomputer revolution was under way. A *microprocessor* is a complete computer instruction processing unit on a single chip. Add a memory and a method for input and output of data, and you have a *microcomputer*.

These devices differ from their large computer and minicomputer counterparts in many details, but in their basic approach to problem solving they remain conventional computers. It is their incredibly low cost that sets them apart; price, not concept, is the reason for the explosion of uses of microcomputers. A microprocessor chip can cost only one dollar; a microcomputer with mounted microprocessor, memory, input-output components, power supply, and cabinetry can cost as little as a hundred dollars. Common abbreviations for the microprocessor are MPU and CPU.

THE MICROCOMPUTER IN DIGITAL DESIGN

The technical community now has much experience in using computers to solve a wide variety of problems. Our interest is in digital design: how do we use a conventional computer (a minicomputer or a microcomputer) as an element of hardware design? A computer is a general algorithm executer, so it would seem reasonable that computers may be able to perform the control portion of suitable designs.

This approach brings our view of the design of a controller much closer to conventional programming. We express our control algorithms in a computer assembly language or perhaps in a high-level language such as Fortran. The conventional computer greatly restricts the ways in which a designer can access the controlled device. The hardware view of testing continuously available variables gives way to a computer program that directs the computer to read status information from the controlled device into the computer for later testing. The conditional branch of the ASM becomes a step in a conventional software flowchart, or a high-level statement such as

```
IF (STATUS) THEN . . . ELSE . . .
```

The parallel activities in an ASM state become a sequence of steps in a computer program. Commands to a controlled device take the form of write statements. There remains little sense of the detailed hardware timings that characterized our earlier methods of control.

We retain the underlying model of a controller as directing a controlled device (the architecture), but our view of control is quite different when using the microcomputer than when using hardwired or microprogrammed control. Here is a comparison of the three approaches:

a. *Hardwired control.* The fastest emulation of an algorithm, with its speed limited by gate propagation delays. The ASM chart is the most convenient way to display algorithms. Multiway branches and conditional outputs can reduce the number of states, allowing many operations to proceed in parallel, further enhancing the speed of the circuit. In systems with up to about 16 states, this method is simple enough to be our choice. The cost of components is minimal, but fabrication and debugging are laborious, and errors in design or implementation result in tedious alterations of the hardware.

b. *Microprogrammed control.* Slightly slower than hardwired control, its speed is still limited by propagation delays. The slowdown results from the extra states required by multiple branches and conditional outputs. Microprogrammed control is most suited to complex algorithms, say with more than 16 states, and is limited by the address space of the control store and by the management of the complex algorithms. Microprogram designers need high-level design aids such as the Logic Engine. The cost of the production version of a microprogrammed control is small.

c. *Conventional microcomputer control.* The method of choice when slow speed and somewhat higher cost are acceptable. Speed is generally 10 to 100 times less than in the two previous methods, because of two characteristics of computers. The basic time unit is that of a computer instruction, which requires several clock cycles, whereas hardwired and microprogrammed controllers execute a state in one cycle, and the execution of a state requires several instructions. The reading of status variables or the writing of command outputs requires one or more separate instructions, as does the testing of status variables.

Offsetting these restrictions is the excellent human interface accompanying most computers, including microcomputers. The basic expression of the algorithm is the program, perhaps with the aid of a software flowchart. Complex computer programs, whether for control or for general problem solving, require considerable software support. Loaders, editors, assemblers, and operating systems are necessary, and we expect these facilities to accompany the computer hardware. Still, after a microcomputer-based control design is complete, you may often remove from the production version much of the equipment that supported the development.

Microcomputer control of a digital system can replace much of the control hardware with software, but the architecture remains "hard." The controlled device is frequently some existing piece of equipment, around which we construct sufficient hardware to adapt it to the requirements of our microcomputer system.

DATA FLOW IN A MICROCOMPUTER

The severe limitation on the number of pins in a microprocessor chip means that there is usually only a single data path into and out of the chip. Instructions and data from memory, output to peripheral devices, and all other transfers of data into or out of the MPU occur over this one path, called the *data bus*. Remember, a bus is a data path consisting of a set of signal paths or wires, one wire per bit, over which many components may send or receive information. The data bus is usually as wide as a memory word—8, 16, and 32 bits are typical sizes.

To direct the flow of information into and out of the microprocessor, two more busses emerge from the chip: an *address bus* and a *control bus*. The address bus designates the destination of a data transfer; since all data moves over the data bus, the microprocessor will use the address bus not only to describe memory locations but also to designate the peripheral devices attached to the system. This bus is usually 16 or 24 bits wide. The control bus determines the direction of information flow; input from the addressed device or output to it. The structure of the control bus varies. A simple-control bus might have just two lines, a read line for initiating inputs to the microprocessor, and a write line for starting output operations from the microprocessor. Most control busses have additional signals for managing timing and interrupts.

Memory references are the most frequent use of the microprocessor busses. To retrieve a word from the memory, the MPU must place the proper address on the address bus and then place a read signal on the control bus. The memory must have enough internal logic to recognize the read signal, accept the address, and start a read cycle that will eventually place memory data on the data bus so the MPU can accept it. The MPU performs a write operation by placing the address, the MPU data, and a write signal on the proper busses; the memory must accept all this information and perform the memory write operation.

The microcomputer's memory, terminal, and mass storage share the bus with other devices. Figure 11–1 is a generalized microcomputer bus. Each

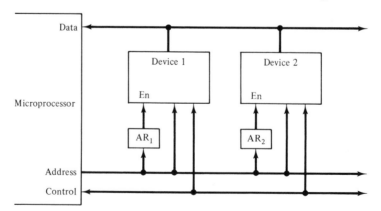

Figure 11–1. The bus structure of a microcomputer. Each device has an address recognizer *AR*.

device has an *address recognizer,* which scans the address bus and asserts an enabling signal whenever the bus contains an address associated with the device. The address recognizer is usually a simple combinational circuit, whose details vary with the microprocessor's addressing scheme and the addresses assigned to the device. Two addressing schemes are in common use. We may divide the activities of a microprocessor bus into memory accesses and input-output transfers. In a microcomputer system, these two bus activities may be separate concepts or manifestations of a single concept. In systems where memory and input-output are conceptually separated, we speak of the *separate input-output* approach; the concept of unified memory and input-output is called *memory-mapped input-output.*

Memory-Mapped Input-Output

In memory-mapped input-output, every device attached to the system behaves like a memory location with an address. In a microprocessor designed for this approach, we input a byte from a terminal keyboard by executing a load instruction, just as we would to obtain the contents of a memory location. The load instruction has an operand address that identifies the particular device. Outputs arise through store instructions; there are no read or write instructions. A data transfer causes the address to appear on the address bus, to be interrogated by the address recognizers in all devices (including the memory). The selected device recognizes its address; then, governed by the control bus signals, the device transmits or receives information over the data bus.

Memory-mapped input-output is elegant, simple, and flexible. The necessity of decoding the large address field is only a minor drawback. Many microcomputers use this approach: the Motorola 6809 and 68000 series are popular examples.

Separate Input-Output

The separate input-output strategy incorporates two bus systems, one for memory and a separate one for input-output devices. Peripheral devices are attached to the input-output bus, and the RAM goes on the memory bus. These microprocessors have read and write instructions for input and output. The Intel 8080 has a separate input-output system.

Since the microprocessor chip itself has only a single data, address, and control bus structure, there must be extra hardware outside the chip to route the microprocessor bus information onto the correct external busses. In this method, the microprocessor has a *select* signal in its control bus to specify which of the external bus systems to use. This signal allows the decoding hardware to enable the address, data, and control signals onto either the memory bus system or the input-output bus system.

Since the decoding hardware has separated the memory references from the input-output, the memory unit does not need to do further decoding to identify memory addresses. Similarly, the input-output address bus transmits only the addresses of peripheral devices, not memory addresses, so the address recognizers in the devices are usually simple. Because there are far fewer peripheral devices than memory words, the input-output address bus may be much smaller than the memory address bus: 8 bits allow for addressing an ample number of peripherals.

Separate input-output is less flexible than memory-mapped input-output, but the separation of memory and input-output functions seems more natural to some people. Either scheme is satisfactory, and the designer sees little fundamental difference between them.

Device addressing. To each device one or more unique addresses are assigned. They may be permanently assigned and built into the device or they may be selected with switches on the device's cabinet. A printer may have only a single address, whereas a RAM would respond to a host of addresses corresponding to each word in the memory. All devices monitor the address bus lines, and respond only to their particular addresses.

Few devices have suitable circuitry for recognizing addresses built into them. If you are using the microcomputer as a controller for a piece of "foreign" equipment, you will have to interface the controlled device to the microcomputer controller. By using the methods described in Chapters 2 and 3, you can design efficient MSI address detection circuits, using gates for a fixed address or arithmetic magnitude comparators for a selectable address.

Suppose the devices in Fig. 11–1 are assigned the following 16-bit addresses in a memory-mapped system. The addresses are given in hexadecimal notation; bit A_{15} is most significant and bit A_0 is least significant.

Device 1: Address F002
Device 2: Addresses F004–F007

In Device 1, the address recognizer AR_1 must implement the equation

$$AR_1 = A_{15} \cdot A_{14} \cdot A_{13} \cdot A_{12} \cdot \overline{A_{11}} \cdot \overline{A_{10}} \cdot \ldots \cdot \overline{A_2} \cdot A_1 \cdot \overline{A_0}$$

Device 2 has a range of four addresses. All combinations of the least significant 2 bits are valid, so address recognizer AR_2 must implement

$$AR_2 = A_{15} \cdot A_{14} \cdot A_{13} \cdot A_{12} \cdot \overline{A_{11}} \cdot \overline{A_{10}} \cdot \ldots \cdot \overline{A_3} \cdot A_2$$

Hardware for the Bus Interface

The interface of a component to a bus is called *bidirectional* if the device can both send and receive signals, and *unidirectional* otherwise. For instance, RAMs have a bidirectional interface with the data bus and a unidirectional (input only) interface with the address bus. An MPU has a bidirectional interface with the data bus and a unidirectional (send only) interface with the address bus.

Most busses in microcomputer systems rely on three-state control, as observed in Chapter 3. Microcomputer components are low-power devices, yet many components may be attached to the bus as receiving circuits. To provide adequate power for reliable bus operation, it is often necessary to *buffer* the outputs onto the bus. For unidirectional outputs, three-state buffer chips such as the 74LS244

Octal Buffer are useful "drivers." Most logic inputs may serve directly as unidirectional bus receivers.

For bidirectional bus attachments, in which a single pin of a device must carry both input and output data, we may use a three-state *bus transceiver,* shown in Fig. 11–2. To operate such a device, we send control signals to specify whether the transceiver is to transmit, receive, or do neither. This requires two control lines as in Fig. 11–2a. Often a single control line for the transceiver is sufficient, in which case the transceiver always passes data in one direction or the other, as in Fig. 11–2b. In some LSI chips in microcomputer systems, bus transceivers are built in; others require separate circuits.

Part of the task of interfacing a device with a computer is to manage the system bus's transactions. This requires detailed adherence to the timing and format of the particular bus system. Many manufacturers provide support hardware to make this task easier at the peripheral end. Typical of this are special LSI chips with names such as *input-output ports* and *peripheral interface adapters* (PIAs). These chips are versatile, often managing 16 bits of parallel data in elaborate configurations determined by programming. PIAs have registers and buffers for the input and output of data, status and control signals for handling bus activities and peripheral timings, some address decoding capability, flexible support of interrupts, and other useful features. The microcomputer program establishes the complex operating conditions of the PIA by programming the contents of the PIA's control registers.

Bus Protocols

If you are attaching a device to a microcomputer, you must be thoroughly familiar with the bus control protocol of that system, whether you design the protocol yourself or use an established one. Bus protocols are more complex than those shown here. There are many protocols in use, including quite a few promulgated by manufacturers or technical groups. Several such "standard" protocols have gained favor, although not all of them are well conceived, nor are they completely

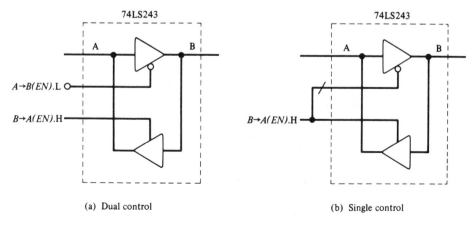

Figure 11–2. A bus transceiver in two configurations.

standardized. Still, when it is possible, use a standard protocol rather than creating your own.

MICROCOMPUTER INPUT-OUTPUT

To the digital designer, a microcomputer program represents a control algorithm; the program issues commands and receives status information by means of input and output operations directed to the controlled device. Usually the microcomputer is dedicated to the task of device control, and the microcomputer program controls every aspect of the architecture's interface with the system. The technique for this detailed control is called *programmed input-output,* which can be used either with or without interrupts.

Programmed Input-Output

Programmed input-output is a simple and versatile scheme supported by every microcomputer. The MPU program explicitly controls the transmission or receipt of each byte of data passed over the data bus; thus the name *programmed* input-output. Usually, the program uses one of the MPU's registers to hold a byte of outgoing data or to receive an incoming data byte. MPU input-output instructions cause transfers between the MPU's register and the system data bus; these instructions are the loads and stores of memory-mapped input-output or the reads and writes of separated input-output.

Since the only data connection between the controller (the MPU) and the controlled device is the system data bus, command outputs from the controller take the form of a byte of data sent out by an MPU output instruction to the addressed device. The device may interpret one or more of the bits in this byte as control signals or may use the information as data, as specified by the designer. If the device requires more data or control information than one byte can hold, the designer must assign additional addresses to the device, providing another "home" for command information, or to process bytes of control information serially.

The controlled device usually generates status information to guide the control algorithm in the MPU. These signals do not influence the MPU's activities the instant the device develops them, since the MPU must issue a programmed read before the device is allowed to place the status information on the data bus. After the status data is in the MPU, the program tests the status as a part of its execution of the control algorithm.

Here is a simple example of microcomputer control using programmed input-output. We wish to turn on a light in response to a switch. We use a PIA to sense the setting of the switch and to maintain the control signal that governs the light. Our model for this example is the Motorola 6809 system, which has an 8-bit data bus and a 16-bit address bus. By software, we specify an 8-bit input port and an 8-bit output port in the PIA, as shown in Fig. 11–3. The switch is connected to the most significant ("sign") bit of the input port, and a relay that controls the lamp is connected to the least significant bit of the output

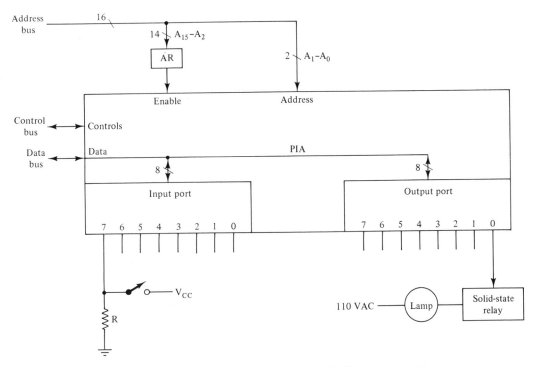

Figure 11–3. A switch operating a lamp, controlled by programmed input-output through a PIA.

port. (In Chapter 12 we discuss how to drive relays.) This exercise requires only these 2 bits.

We assign the address F008 (hexadecimal) to the input port and F00A to the output port. In response to either address our address recognizer asserts an enabling signal to the PIA. PIAs usually have several addressing inputs to distinguish between the chip's internal registers. We will attach the two least significant address bus lines A_1 and A_0 to the PIA's addressing inputs. Our address recognizer, working with the most significant 14 bits of the address bus, identifies the PIA as the target, and the remaining two address bus lines tell the PIA which register is being addressed.

In Fig. 11–4 we present a section of 6809 assembly-language code for the control algorithm. You can understand this simple program without knowledge of the 6809. LDA and STA are load and store instructions, which in this memory-mapped system are used to address the ports of the PIA. BPL and BMI are branches on plus and minus, respectively. JSR is a subroutine jump, and CLR is a clear or "store zero" instruction. EQU is the usual symbol-equivalencing assembly directive.

The code performs a software switch-debouncing action by pausing sufficiently long after detection of a switch transition. A valid change in the switch setting prompts the turning on or off of the light, and spurious switch signals evoke no response.

```
INPORT    EQU  F008     ; assign PIA input port address
OUTPORT   EQU  F00A     ; assign PIA output port address

INITPIA   —              ; initialize the PIA
          —              ; this may take several program steps
INITLITE  CLR  OUTPORT   ; begin with the light off

SWOFF     LDA  INPORT    ; get the switch signal
          BPL  SWOFF     ; repeat while switch is off (sign bit off)
          JSR  WAIT      ; switch is on. wait 10 msec for debouncing
          LDA  INPORT    ; get switch again to check validity
          BPL  SWOFF     ; if not the same, exit to beginning
LIGHTON   LDA  #1        ; set bit 0 on to turn on light
          STA  OUTPORT   ; store the bit in the PIA output port
SWON      LDA  INPORT    ; get switch and wait until it is turned off
          BMI  SWON      ; is it still on? if so, try again
          JSR  WAIT      ; off. wait 10 msec to be sure it's real
          LDA  INPORT    ; get switch again
          BMI  SWON      ; if false alarm, try again
LIGHTOFF  CLR  OUTPORT   ; clear bit 0 of output port to turn off
                         ; light
```

Figure 11–4 Programmed control of the architecture in Figure 11–3.

The architecture of the interface circuitry and the spirit of the programmed control of more complex designs are similar to this simple example.

Programmed Input-Output with Interrupts

Some types of peripheral interactions are sporadic and unpredictable. Keyboard entry from a low-speed terminal is an example: A user may depress a key the next second or not until next week. To control such a device with normal DMA or programmed input-output, we would execute a loop that tested the status of the keyboard device frequently, waiting for a character to come along. This straightforward approach is acceptable if the MPU does not have more pressing business, but in busy systems this may use up too much time.

Most microcomputers support external interrupts. This feature requires an additional line on the control bus: the *interrupt request* signal. When the MPU senses that the interrupt request line is true, then at the end of the next instruction the MPU forces a subroutine jump to a special memory location set aside for this type of interrupt and begins executing the instructions located there. The programmer arranges for a suitable *interrupt subprogram* to reside in memory, beginning at the special interrupt location.

The interrupt subprogram must determine the reason for the interruption of normal processing, and take appropriate action. For example, consider the processing of character data from a terminal keyboard, using interrupts to announce the availability of a character. The MPU program goes about its main business, for example, monitoring the status of temperature and pressure sensors in a chemical process and controlling a heater and a vacuum pump motor. The main MPU program takes no heed of the keyboard, which is present only for an occasional human intervention. When the experimenter presses a key on the terminal, the terminal's interface hardware asserts an interrupt request. This causes the MPU to suspend the execution of the program and make a subroutine call to the interrupt location. There, the interrupt subprogram determines that the reason for the interrupt is the availability of a character from the terminal. The interrupt subprogram reads the character, using programmed input, and deals appropriately with the character. The subprogram then does a subroutine return, which causes control to return to the next instruction following the original point of interruption. If the interrupt program has been properly written, all the relevant registers of the MPU will be unaffected by the interruption and the interrupted program will not be aware that the interrupt has occurred.

As you learned in Chapter 7, computers always provide a way to turn the interrupt recognition hardware off and on, so as to give the MPU program some control over its operating environment. There are two situations in which we want the interrupt recognition system to be turned off: when the algorithm is not designed to handle any interrupts, and when the interrupt subprogram is already dealing with an interrupt. This last case is interesting; it arises to avoid the possibility of one interrupt piling on top of another one.

In the simplest form, the entire interrupt detection structure would be inactive during the processing of an interrupt. A more sophisticated method is available in many microcomputers. The *priority interrupt* scheme provides hardware support for prioritized levels of interrupts. Here, the designer assigns each type of interrupt request a priority level, with interrupts requiring rapid processing having higher priorities. An interrupt occurring at a particular priority level will disable detection of all interrupt requests of equal or lower priority but will preserve the ability to detect higher-priority requests. In such a system, each priority level has its own interrupt location in the memory, which must contain the start of an interrupt subprogram capable of dealing with the interrupt. During the processing of an interrupt request at a certain level, if the MPU detects a higher-priority request, a new interrupt subroutine jump will occur.

It sounds tricky, and it is. Interrupts are potent tools for providing service on demand, but they cause a host of difficult problems.

Difficulties with interrupts. Interrupt programming is much more difficult than ordinary software development. The trouble is that interrupts are not reproducible. The *time* of an interrupt request is not under the direct control of the MPU; interrupts do not occur in lockstep with the MPU algorithm. It is an intrinsic property of the interrupt concept that we cannot predict exactly when an interrupt will occur. Most complex programs have weak, perhaps little used, sections that have not undergone rigorous testing. It is difficult to test an interruptable program thoroughly. An interrupt may occur during a section of the program that is not correctly written for interrupt processing. For instance, the main program must be careful in using memory data that an interrupt subprogram might also use. *Might* is the key word; an incorrect program may run for a long time without an interruption in the erroneous section. Such nonreproducible errors are nearly impossible to diagnose after the fact. And creating flawless programs in an interrupt environment is orders of magnitude more difficult than in simpler circumstances.

Since interrupts can occur between any two instructions of an executing program, the interrupt subprogram must be written with great care. Its execution must leave *no effects* that might disturb the correct execution of the interrupted program. In particular, the interrupt subprogram must not alter the MPU's operating registers, or at least must restore these registers to their original state before returning to the interrupted program.

Of course, the interrupt subprogram will make *some* change in the state of the computer system; otherwise, why bother detecting the interrupt? Usually, the subprogram will store some information in the memory, or initiate a new input-output activity. Programmers must design their programs with great care, so that they perform flawlessly even in the presence of unpredictable interrupts.

Interrupts are the software equivalent of asynchronous hardware circuit design. Both are powerful techniques loaded with opportunities for errors. In hardware, we found that synchronous design allows simpler handling of most problems, with little or no penalty. This was possible primarily because in hardware and in microprogramming we may perform many operations in parallel and can thereby achieve great overall speed whether or not the design is clocked. In conventional computer software, each instruction performs only a small task and activities are highly serial. Serial processing is slow. We need any facility that can help speed performance and enlarge the spectrum of problems amenable to conventional computer processing. We cannot forego interrupts as easily as we sidestepped asynchronous design.

So interrupts are often necessary, but you must always be aware of the design and debugging problems introduced by this powerful concept.

DESIGN EXAMPLE 1: A WIRE-WRAP CONTROLLER—PURE SOFTWARE CONTROL

Printed circuits, wire-wrap, and wire plugboards are common ways of fabricating digital circuits. Plugboards—special connector boards with holes into which the designer pushes wires—provide a manual method suitable for testing small prototypes. Printed circuits are best for high-volume production but require substantial efforts at circuit layout and fabrication. Wire-wrap serves a wide range of applications, including testing prototype of large circuits destined for printed circuits and production versions of circuits of varied complexity.

In wire-wrapped circuits, integrated circuit chips are plugged into mounted sockets having special wire-wrap pins. Electrical connections are made through wires strung between pins and wrapped to each pin by means of a wire-wrap tool or "gun." Wire-wrap is inexpensive, and commercial products such as insulation stripping tools, wrapping tools, and sockets are plentiful. Wire-wrapped circuits are easy to modify in the field.

Manual wire-wrapping is feasible, but in a large circuit the forest of pins and wires confuses the eye and leads to occasional errors. Locating the correct pins is tedious and slow, a likely candidate for automation. Fully automatic numerically controlled wire-wrap machines are commonplace in high-volume-production plants, but these machines are expensive and require elaborate setting up.

With a more modest investment, using commercial equipment, we can develop a semiautomatic wire-wrap facility. The system relies on a human operator to wrap the wires but provides for the automatic positioning of a guide for a wrapping tool so there is no doubt about a pin's proper location. The equipment consists of a wire-wrap table and its controller. In the following example we use a commercial wire-wrap table and a microcomputer for control.

The Architecture

The wire-wrap table is an upright X-Y positioner with a rigid rectangular frame that holds the board to be wrapped. On the frame is mounted a vertical beam that can move in the horizontal (X) direction along a precision screw. Attached to the vertical beam is a smaller beam that can move along another precision screw in the vertical (Y) direction. This smaller beam holds the guide for the wire-wrap tool. By a suitable combination of motions in the X and Y directions, the controller may place the guide at any location within the rectangular frame.

Each beam has a stepping motor that can position the beam precisely by rotating the screw. A stepping motor responds to a control pulse by moving one step in a specified direction. In our system, a motor step results in a beam movement of 0.005 in.; thus there are 200 beam steps per inch. To move a beam 5 in. would require 1000 properly administered steps.

The stepping motors are independent, so both beams may be in motion at the same time. Each stepping motor has a driver that responds to digital TTL control signals and converts them into high-power motor signals. In our example, each stepping motor control has a *direction signal* (specifying left or right; up or down) and a *motor signal*. To move a beam one step, we establish the proper value of the direction signal and then make the motor signal true for a specified time (about 1 msec). At other times, the motor signal must be false.

Starting from a resting position, a stepping motor may reliably begin motion at a rate of up to 200 steps per second; once in motion, the motor may be gradually speeded up to 600 steps per second. For reliable operation, the stepping must be decelerated to a maximum of 200 steps per second before the motor is brought to a halt. Stepping the motors too fast can result in lost steps. The manufacturer specifies the desirable rates of increase and decrease of the stepping speed.

Specifications for the Controller

The wire-wrap table is ideal for microcomputer control:

   a. It is slow. Almost any microprocessor can drive the stepping motors at full speed and can easily keep up with the human operator.

b. The control is a complex task, easily handled in software, yet difficult for a pure hardware controller.

c. The design is likely to change as our experience suggests minor modifications.

d. We can use the microcomputer's interface with the operator to good advantage. A display terminal, floppy disk, editor, and assembler will assist the development task and may later become part of the production system.

The wire list. The *wire list* contains the basic data for wire wrapping. Each wire connects two pins; the wire list gives the pins' location in terms of X and Y coordinates relative to an arbitrary origin. For convenience, each entry in the wire list contains the length of the wire and a commentary to assist the operator to identify the wire.

The designer prepares the wire list before wrapping a board. There are several devices commonly available in microcomputer systems that can retain the wire list: paper tape, floppy disk and hard disk. The floppy disk is the most versatile.

The operator's controls. The operator should have a footswitch to signal the controller that a wrap has been completed and the wire-wrap tool is removed from the guide.

A small keyboard with 10 or 12 buttons is a convenient way to enter information into the controller. The keyboard has two basic functions; the panic button, which will cause the immediate halting of the stepping motor, and a set of four buttons to permit manual movement of the table beams. The operator can press any one of the four buttons $-X$, $+X$, $-Y$, or $+Y$ to move the guide in the indicated direction. If the operator holds a button down longer than 0.2 sec, for example, the controller moves the beam at a rate of 200 steps per second until the button is released. Single steps result from depression of less than 0.2 sec.

Another keyboard button can cause the computer to position the wire-wrap guide at the point the controller believes to be the origin. This is useful to verify that the stepping motors have not missed steps during previous wrapping operations.

We may design additional operator controls to be entered either from the keyboard or from the microcomputer's main display terminal. For example, it is useful to allow the operator to search the wire list for a given wire which can serve as a starting point for wrapping operations. This need arises when an operator has suspended wrapping and wishes to begin again.

The controller. Any microcomputer can control the wire-wrap table. We might choose a Motorola MC6809 microprocessor with 32K RAM storage, a floppy disk, and a display terminal. This machine has memory-mapped input-output for communication, and supports external interrupts. A peripheral interface adapter (PIA) provides a convenient interface between the wire-wrap table's motor drivers and the microcomputer. The PIA contains registers for data written from the microprocessor, so the control algorithm's responsibility is to enter proper signal changes into the bits of the PIA registers. Figure 11–5 shows the

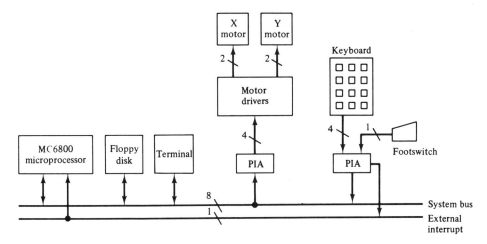

Figure 11–5. Signal paths for the wire-wrap controller.

logical interconnections of the microcomputer, its peripheral devices, and the wire-wrap components.

The Control Algorithm

The microcomputer manages a variety of operations:

    a. Positioning the wire-wrap tool guide, either from information in the wire list or in response to a pushbutton.

    b. Processing the panic button signal.

    c. Display of data pertinent to a wrap, including the length of the wire and an identification for the wire.

    d. Interpretation and execution of operator's commands, such as searching the wire list.

    e. Storing the wire list.

With the aid of the software accompanying the display terminal and the peripheral storage device, parts (c), (d), and (e) become routine programming tasks. The organization of the control program is shown in Fig. 11–6. In response to the operator's commands, the supervisor program selects a subprogram to perform a task, such as seeking the origin or wrapping wires. Upon completion of the task, the subprogram returns to the supervisor, where the controller will await further instructions from the operator.

Part (b), panic-button activity, is best handled by attaching the panic-button signal to the microcomputer's interrupt request line. Asserting the panic signal interrupts the program, causing immediate suspension of its normal activities. The interrupt subprogram must return control to a standard point in the supervisor program, to await further commands from the operator.

Part (a), moving the stepping motors, requires careful study.

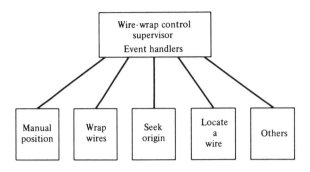

Figure 11–6. Organization of the wire-wrap control program.

The wire-wrap algorithm. Figure 11–7 shows the software control for wrapping the wires given in the wire list. The footswitch provides the basic sequencing information for this algorithm. Each depression of the footswitch is a "clear to proceed" signal; the algorithm responds by positioning the wrapping tool guide over the next pin. The treatment is similar for the two wraps of a wire; however, prior to beginning each wire, the algorithm provides an opportunity for the operator to intervene. Upon sensing the assertion of the signal from the footswitch, the microcomputer algorithm reads the next record from the wire list and displays appropriate information for the operator's use. The important information to display is the wire length, to allow the operator to select the proper wire, and the wire identification, to permit the operator to know the present point in the wire list.

To process a wrap, the algorithm must use the coordinates of the new pin and the present location of the guide to compute the number of X and Y steps required to reach the next pin. With this information, an X–Y positioning sub-program MOVE can direct the stepping of the two motors. The operator will then wrap the first pin.

Prompted by another depression of the footswitch, the algorithm will move the guide to the second pin of the wire and then return to the beginning of the loop to await further instructions. The operator will wrap the second pin and then press the footswitch or enter another command from the keyboard or the terminal.

The MOVE subprogram should move the guide into position over a pin in minimal time. In general, both motors are in motion during each move, but for different durations. Each motor must start slowly and, depending on the distance the beam must travel, may be accelerated to a maximum speed before deceleration. A motor engaged in a short movement may not reach full speed before deceleration must begin.

The optimal acceleration and deceleration are complex functions of the stepping speed, but in our application only a conservative approximation is necessary. We may change the stepping speed in discrete jumps, using a constant acceleration factor. Figure 11–8 is a typical pattern of software-generated X and Y motor speed changes. The X motor reaches the full speed of 600 steps

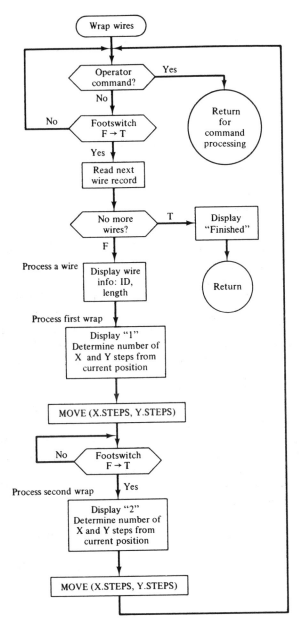

Figure 11-7. A software algorithm for wrapping wires in the wire list.

per second, whereas the Y motor must decelerate before reaching maximum speed.

Managing the independent motions of the X and Y motors with a single processor requires careful planning of the algorithm. We use a simplified version of the technique of discrete event simulation, which permits a single algorithm to manage several independent sequences of events. In our example, there are

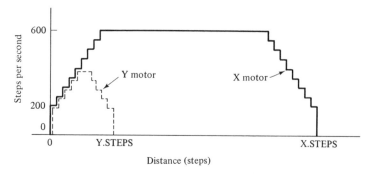

Figure 11–8. Acceleration and deceleration patterns of stepping motors.

two sequences of events: stepping the X motor and stepping the Y motor. The control algorithm must produce the proper sequence of true and false signals for each motor, for instance as shown in Fig. 11–9.

Managing one motor is relatively simple, requiring only that the algorithm delays the proper time between changes to the motor signals. Given a formula for accelerating and decelerating the motor, the sequence of signals depends only on the distance to be traveled and is completely defined when the stepping subprogram starts. Therefore, when each motor event is processed, the subprogram can determine the time and nature of the next event.

To control both motors, the algorithm keeps track of two intervals: for each motor, the computed time from the present event until the next event. Upon completing an event, the algorithm delays until the earlier of the two next events arrives, then executes that event and creates a new next event for that motor. This again allows one interval for each motor, allowing the stepping process to advance. The algorithm always processes the earlier event.

Figure 11–10 shows the flowchart of the motor-movement subprogram MOVE, the heart of the control. The time-delay subprogram DELAY, called in the main loop of MOVE, accepts a single argument D and pauses for the appropriate interval D before returning. The delays are on the order of milliseconds, whereas the microprocessor's instruction times are on the order of microseconds. Since in this application we do not require exact timings, we may create delays by

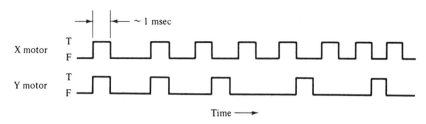

Figure 11–9. Typical signal patterns for the independent X and Y stepping motors.

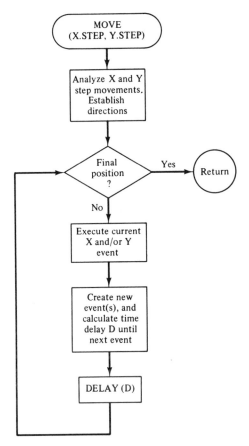

Figure 11–10. Flowchart of the sub-program MOVE for controlling the simultaneous movement of the X and Y stepping motors.

executing an appropriate number of microprocessor instructions in a loop in the subprogram DELAY.

Wrapping Up

This completes our treatment of the controller for the wire-wrap table. We leave as an exercise the development of the algorithm for manual control of the stepping motors. The wire-wrap processes are slow enough that we may perform all the control with microcomputer software. Such control is easier to debug and modify than the hardwired form and allows us to implement more complex designs. In the next example, we describe a system that requires combined hardware and software control.

DESIGN EXAMPLE 2: A TERMINAL MULTIPLEXER—HYBRID HARDWARE-SOFTWARE CONTROL

The device in our second design example is a terminal multiplexer, which will manage the transmission of data between a large number of low-speed computer terminals and a central computer system, merging the information to and from each terminal into a single path in each direction.

This example which arose in a university, represents a common problem in which large numbers of terminals must communicate with a distant point. One method is to connect each terminal to the computer by a pair of signal paths, one for moving data from the terminal to the computer, and another for moving data in the other direction. Typically, there are separate connections for each terminal. These wires may belong to the organization, or be private lines leased from the telephone company, or be a part of the normal telephone dial-up network.

System designers would like to minimize the overall cost of the communication system; one approach is to minimize the number of long connections to the computer. When only a single terminal exists at a location, there is little one can do except use a set of wires solely for that terminal, or rent a telephone dial-up connection. Where many terminals are close together—in the same room or the same building—there may be an opportunity to save on the cost of separate long communication links with each terminal. This university computing network has a switching system near its main computers. The switching system, purchased from a commercial manufacturer, provides a variety of complex functions to move information among the elements of the network. Among them, it provides access between the computers and individual low-speed terminals.

A more sophisticated service is to handle the high-speed lines over which several terminals communicate. The sharing of the high-speed line among several terminals occurs by *time multiplexing* the characters: The terminals take turns sending characters over one signal path and take turns receiving characters over another path. Such sharing of high-speed lines is beyond the capabilities of the terminals themselves. We need a *terminal multiplexer*—a controller such as shown in Fig. 11–11, located near the terminals, to manage the complex operations involved in the time multiplexing of the data.

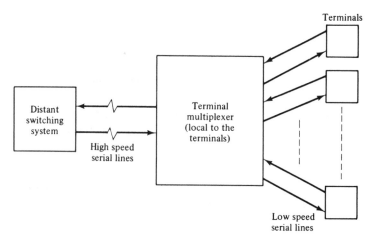

Figure 11–11. The environment of the terminal multiplexer.

With the aid of some circuits developed in Chapter 6, we can construct such a terminal multiplexer, controlled by a microcomputer.

Specifications

Here are the conditions under which the multiplexer must function. We have simplified the specifications somewhat.

Terminal communications. Each terminal communicates at a maximum rate of 300 bits per second in each direction, using the standard asynchronous character communication protocol described in Chapter 9. Each 8-bit ASCII character requires a total of 10 bits, including the asynchronous protocol's start and stop bits. The transmission is full duplex, since simultaneous communication in both directions is possible. All the terminal communication lines merge into the nearby terminal multiplexer, but so far as a user is concerned, each terminal is independent.

High-speed communications. The terminal multiplexer system is connected to a commercial switching system at a distant site. The transmissions travel over a single full duplex high-speed link, according to the protocols defined by the switching system. The speed of transmission is 9600 bits per second in each direction, although by consulting with the switching system's manufacturer we could raise or lower this value. In normal activity, transmission in each direction is continuous, without a break. The continuous streams of bits consist of 9-bit bytes organized into "frames." A frame begins with a special *SYNC* byte that marks its start and also permits the identification of byte boundaries in an incoming bit stream. Following the *SYNC* byte, the frame has one 9-bit byte for each terminal in the configuration. A given terminal's byte always occupies the same relative location in each frame. The byte for a given terminal may contain a terminal data character or, if there is no data character for this time slot, the byte will be a filler, or *NULL*, byte.

Each 9-bit byte begins with a type specification bit: 0 means that the remaining 8 bits are data, 1 means that the 8 bits are a control code. The control codes are defined by the switching system's manufacturer; the only codes we will consider are *SYNC* and *NULL*. Figure 11–12 shows the structure of a frame.

This high-speed protocol is an example of synchronous data transmission. (The terms "synchronous" and "asynchronous" in data communications do not have the same meaning as they do in digital design.)

Number of terminals. We may compute the maximum number of terminals as follows: For N terminals, a frame contains $N + 1$ bytes, or $(N + 1) \times 9$ bits. At 9600 bits per second, a frame requires $(N + 1) \times 9/9600$ sec. The frame contains a spot for each terminal, so each terminal receives processing once in each frame time period.

Each transmission of a character in the asynchronous communication protocol involves 10 bits, of which 8 bits are an ASCII character. At 300 bits per second,

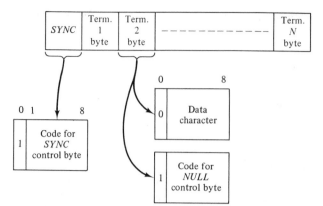

Figure 11–12. The frame structure for the terminal multiplexer's high-speed transmissions.

each terminal can process at most 30 characters per second in each direction. Therefore, each terminal requires $\frac{1}{30}$ sec to process a character.

Solving for the number of terminals N, we have

$$\frac{(N + 1) \times 9}{9600} = \frac{1}{30}$$

$$N \approx 34$$

We will use a maximum of 32 terminals in our example; all frames will contain 32 terminal bytes, regardless of the actual number of terminals attached to the multiplexer or turned on.

The Structure of the Terminal Multiplexer

At the terminal side of the multiplexer in Fig. 11–11, the appearance is of a large number of independent and sporadic transmissions using the asynchronous communication protocol. On the switching system side, the communications are synchronous over a pair of high-speed signal lines, using the specific protocol defined by the switching system manufacturer. Our task is to design something to fit between these two disparate activities. The work of transforming between the synchronous frames and the asynchronous terminal communications will be structured around characters, yet both ends of our system require bit-serial communications, of which the character itself is only a part. We are surely faced with serial-parallel data conversions at each end, to produce the characters that are the common element of each form of communication.

Our choices for dividing the responsibilities are:

a. To translate the high-speed synchronous serial bit stream into 9-bit bytes, and vice versa, using hardware to provide the necessary speed for these operations.

b. To translate serial asynchronous communications from terminals to 8-bit characters (and vice versa), using hardware if possible.

c. To perform the complex manipulations on the resulting characters and bytes using a microcomputer. The hardware in items (a) and (b) will be attached to the microcomputer system bus. We will assume that the data bus is 16 bits wide. Figure 11–13 shows the components of the terminal multiplexer. Let's see how each of these pieces will work.

Switching system communications. In Chapter 6, we developed serial-parallel conversion circuits that support the 9-bit byte format dictated by the switching system's protocol. These circuits include a serial-to-parallel (S $\rightarrow$ P) converter to produce bytes from the serial input line, and a parallel-to-serial (P $\rightarrow$ S) converter to serialize bytes onto the serial output line. The microcomputer is attached to the parallel side of both converters. The rules of operation require the microcomputer to process incoming and outgoing bytes fast enough to keep up with the continuous synchronous serial transmissions. The microcomputer obtains its basic algorithm timing through two status signals from the converters. A change in the value of *READIT* prompts the microcomputer to accept a 9-bit byte from the S $\rightarrow$ P converter; a change in *FILLIT* requires the microcomputer to write a 9-bit byte to the P $\rightarrow$ S converter. The only independent control the microcomputer can exert over the serial bit-stream processing is through the *RESET* line. Momentarily asserting *RESET* will cause the S $\rightarrow$ P converter to abandon its normal byte acquisition and search for the special *SYNC* bit pattern from among the bits of the incoming serial bit stream. This fixes byte acquisition on proper byte boundaries—a necessary operation upon startup, or if the microcomputer loses synchronization with the incoming bytes in a frame, or if there is a break in the serial communications.

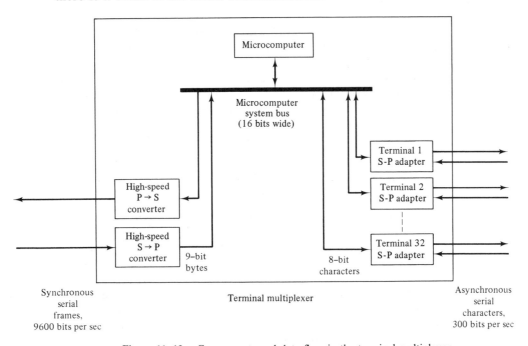

Figure 11–13. Components and data flow in the terminal multiplexer.

To attach the serial-parallel converters to the microcomputer system bus, we must give address and data bus assignments to the various device signals. We will make the following allocations:

```
Address ADDR.A (read only):    READIT              Data bus bit 10
                               FILLIT              Data bus bit 9
                               S → P byte buffer   Data bus bits 8-0
Address ADDR.B (write-only):   RESET               Data bus bit 0
Address ADDR.C (write-only):   P → S byte buffer   Data bus bits 8-0
```

(Read-only and write-only are from the viewpoint of the microprocessor.)

In a typical microcomputer memory-mapped input-output bus, the three addresses will each have a 16-bit code; a separated input-output bus will have a smaller code for the device addresses. We must provide our converters with bus interface logic to detect the presence of the three address patterns, and to assert the signals *ADDR.A*, *ADDR.B*, and *ADDR.C* appropriately. The bus interface must deliver the required control signals to the converters to load data from the bus at the proper time.

Since address *ADDR.A* is read-only (i.e., input *to* the microcomputer), and addresses *ADDR.B* and *ADDR.C* are write-only (output *from* the microcomputer), we do not need bus transceivers. Simple three-state buffers will serve as bus drivers for the elements of *ADDR.A*. The output-enable signal for this collection of 11 bus drivers is *ADDR.A•READ*, where *READ* is the read signal on the control bus. When the output enable signal is true, the *ADDR.A* data is available on the system data bus.

Managing *ADDR.B* and *ADDR.C* requires a knowledge of the type of elements used in the converter's interfaces with the microcomputer bus. In Chapter 6, we suggested using a controlled (JK) flip-flop to receive the *RESET* signal, and a 9-bit enabled D register for the P → S byte buffer. In order to load these elements under the bus's control, we must generate the correct clock and control signals from the information available on the system bus. The following equations are suitable, where *WRITE* is the write signal on the control bus:

$$RESET*.FLIP.FLOP(CLOCK) \quad = WRITE$$

$$RESET*.FLIP.FLOP(SET) \quad = RESET•ADDR.B$$

$$RESET*.FLIP.FLOP(CLR) \quad = \overline{RESET}•ADDR.B$$

$$P{\rightarrow}S.BYTE.BUFFER\ (CLOCK) \quad = WRITE$$

$$P{\rightarrow}S.BYTE.BUFFER(ENABLE) = ADDR.B$$

Terminal communications. You are already acquainted with an LSI device that performs conversions between parallel characters and serial asynchronous bit streams; the UART described in Chapter 9 does exactly that. We may use a UART for each terminal, and interface each UART to the microcomputer system bus. Various LSI chips perform UART-like functions. For a particular microcomputer, interfacing some chips to the bus is simpler than others. Nevertheless, for all UARTs, there are three vital types of data movement on the bus:

8-bit character data from a terminal into the microcomputer, 8-bit character data from the microcomputer into a terminal, and UART status information for the microcomputer. At the least, the status must include a character-ready signal to describe input from the terminal, and a transmit-buffer-empty signal to control output to the terminal.

We may manage the bus transactions by allocating two microcomputer addresses to each UART, one for data transfer and the other for status. The microcomputer will read a character from a particular terminal by first reading the UART's status, using the assigned bus address for the status port of the selected terminal. If the status indicates that a character is present in the UART, the microcomputer program will read the data portion of the UART. Writing proceeds by a similar sequence of two bus operations: a status read followed, if appropriate, by a data write.

Designing the detailed logic to interface the UART terminal controller with the system bus is not difficult; the pattern is much the same as in the high-speed serial-parallel converters discussed earlier.

The Control Algorithm for the Microcomputer

We have specified the hardware of our terminal multiplexer system: the high-speed synchronous serial-parallel converters, the terminal's asynchronous serial-parallel converters, and the microcomputer system bus's interfaces to these elements. Let's turn to the master controller for the multiplexer—the microcomputer program.

We require our microcomputer to process bytes to and from the high-speed synchronous system at a rate fast enough so that the algorithm *never* misses its cues. The high-speed serial activities go on all the time, and each frame has a specific position for each terminal. When a terminal's time slot arrives in either an incoming or an outgoing frame, we must process that terminal. It will do us no good to process a terminal at any other time! This important realization shows that the *READIT* and *FILLIT* signals are the master timing elements; a change in either of these flags must cause the microcomputer program to process a 9-bit byte. We can forget about the timings of the terminal characters; if we keep track of the position in the frames, we will know when to process each terminal. The S → P and P → S bit streams operate independently, so the *READIT* and *FILLIT* signals are not synchronized with each other. The microcomputer program should *poll* (look at) each flag as frequently as it is able, to detect the changes that occur periodically.

How much time is available for processing a byte? Bytes move at the rate of $9600/9 = 1067$ bytes per second in each direction. The system must handle $1067 \times 2 = 2134$ bytes per second, so there is about 470 μsec for each byte. This corresponds to 100 or more microcomputer instruction executions. Our intuition, borne out by experience, tells us that this is ample time.

Our initial model of the microcomputer software is in Fig. 11–14. (Note that this and the following diagrams are software flowcharts, not hardware ASMs.) The initialization occurs only during the startup of the system; thereafter, control

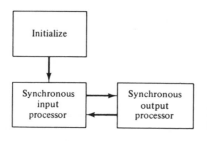

Figure 11–14. Algorithm for the terminal multiplexer: basic modules.

flips back and forth between the synchronous input (S → P) and the synchronous output (P → S) sections, as rapidly as the algorithm can manage.

In each of these two sections of the algorithm, we must poll the appropriate status flag. If we detect no change in the flag, we will go immediately to the other section to perform a similar test on the other flag. If a change occurred in the flag, the algorithm must process a byte before going to the other section.

Consider the synchronous output processor—the software that constructs the bytes of the P → S frames. Figure 11–15 shows the algorithm for synchronous output. Each time the software detects a change in *FILLIT*, our program must produce the correct 9-bit byte. To keep track of which byte position in the frame the algorithm is to fill, we use a counter PSCNT whose value ranges from 1 through MAX + 1, where MAX is the maximum number of terminals in the system. In our example, MAX is 32.

Values of PSCNT from 1 to MAX designate the frame slots for each terminal. When PSCNT exceeds MAX, we have completed a P → S frame, so we must transmit a *SYNC* byte to begin the next frame and reset PSCNT to 1. When PSCNT is in the range of 1 to MAX, we must acquire a character from the designated terminal, if one is available. We then form the 9-bit byte by adding a 0 bit to the front of the character. If the terminal has no new character, we transmit a 9-bit *NULL* byte, which begins with a 1 bit.

The role of the synchronous input processor is to manage the bytes in the incoming S → P frames, dispersing any data characters to appropriate terminals. Figure 11–16 shows the algorithm. If *READIT* has changed, the meaning of the current byte is governed by a counter SPCNT, similar in function to the PSCNT of the P → S algorithm. If SPCNT is greater than MAX, the current input byte must be a SYNC pattern, marking the beginning of a new frame. Not finding a *SYNC* pattern at this time is a calamity! The only solution is to start S → P operations over by issuing *RESET* to the converter. (Our protocol calls for the microcomputer to follow its assertion of *RESET* with a clearing of *RESET*, as shown in Fig. 11–16.) Once we initiate the reset process, *READIT* will not change until the S → P converter has detected a *SYNC* byte pattern among the incoming serial bits; therefore, the next byte to appear for processing by the MPU's synchronous input algorithm will be *SYNC*, to begin a new frame.

When SPCNT has a value of 1 to MAX, the input byte is from a terminal and may contain a data character or be *NULL*. We transmit data characters to the appropriate terminal through its UART, and ignore null bytes.

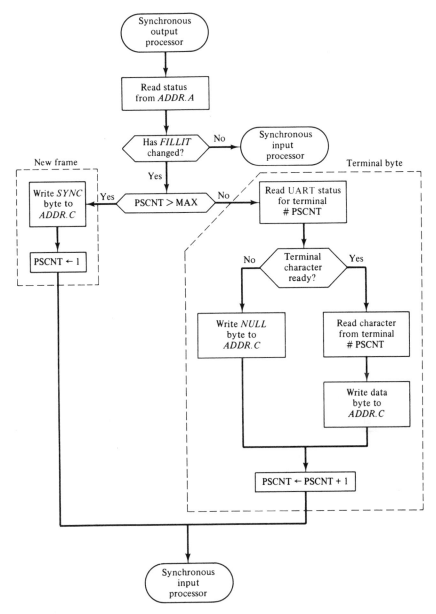

Figure 11–15. Algorithm for the terminal multiplexer: synchronous output (P → S) processing.

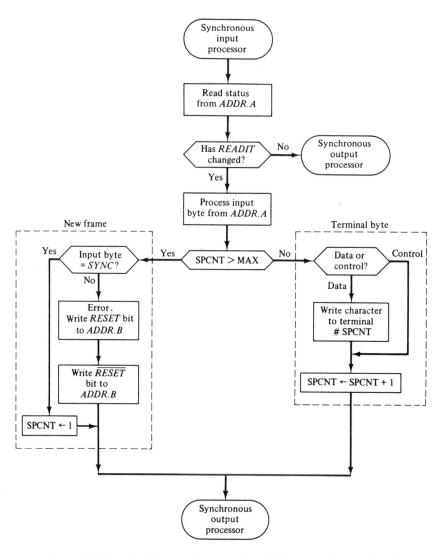

Figure 11-16. Algorithm for the terminal multiplexer: synchronous input (S → P) processing.

Figure 11-17 shows the activities of the initialization section of the microcomputer algorithm. In addition to other routine activities, initialization sets SPCNT and PSCNT and resets the S → P converter system, so that all operations will begin with a new frame.

Checking Out the System

The terminal multiplexer has several hardware components that we were careful to retain as separate modules: the S → P and P → S converters, the serial bit

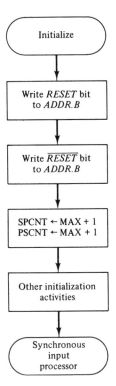

Figure 11–17. Algorithm for the terminal multiplexer: initialization.

clock, and the individual terminal controllers. The microcomputer software is also modular, with well-defined sections for processing incoming and outgoing bytes and for initializing the system. All this suggests that we can check out the system in a modular fashion.

The worst approach is to design and construct the system, plug in the parts, hook up the terminals and the high-speed lines, and turn on the power. It is a virtual certainty that the system will not work at this stage, and such an attack may cause considerable damage to valuable components, perhaps even to the switching system or the terminals. In hardware, and in software to a somewhat lesser extent, there is much opportunity for error. Top-down design techniques reduce the likelihood of error in the logical and implementation phases of the project; in Chapter 12 we discuss some techniques for avoiding trouble in the circuit-fabrication phase. Despite this, it would be unusual for a system to work the first time.

The proper checkout technique would be to test each hardware module separately, using manually produced signals to simulate the inputs that will eventually come from other modules. Here is an advantage of having a manual mode on the circuit clocks. After the hardware module passes this static debugging, we may subject the software modules to similar testing. Here, it may be necessary to use a single-instruction execution mode or to introduce breakpoints into the code. Again, we are trying to do as much debugging as possible under carefully controlled static conditions. Only when the modules pass these tests should you move toward more elaborate testing.

Next, hook two or more of the components together and do more static testing. The extent to which this is feasible depends on the details of the project, but time and effort spent at this level of testing will usually pay big dividends. When all systems respond satisfactorily, try running the local terminal multiplexer system up to speed. It is still unwise to attach the distance "innocent parties" to your unproven system. Instead, you may wish to consider building into the hardware or software some facilities for providing *local loops*. These are ways to simulate the effect of the distant communicator (the switching system or a terminal) by generating outgoing data and using this data as incoming data to test the system; we create a local loop to provide some input that would otherwise be difficult to provide during testing.

Next, hook up a terminal and test the system again. One test might be to use a local loop on the high-speed side of the terminal multiplexer, along with a live terminal on the low-speed side. Another stage might be to add a second terminal, to investigate whether the system will perform with more than one device attached. Then, at last, begin tests with the high-speed synchronous lines; this will not only exercise the hardware and software, but will help to determine if your understanding of the high-speed protocol is accurate.

Debugging and checkout methods vary with the individual project, but a systematic, cautious approach is necessary to avoid expensive and time-consuming errors.

Alternative Approaches

Was our separation of the terminal multiplexer problem into software and hardware components a wise choice? Let's investigate some other options. At one extreme, we might have built the whole system in hardware. At the other end, we might have implemented all the control logic in microcomputer software, using minimal hardware.

The algorithms in our example are uncomplicated enough so that a pure hardware solution is feasible. In the actual system on which this example is based, the switching system defines many more frame byte control codes than the *SYNC* and *NULL* used here. Although the basic control algorithms remain the same, the full algorithm is heavy with special coding to handle the control sequences. The extra frame byte codes deal with line protocols, terminal connect and disconnect operations, error handling, and other features that make the control algorithm messy with detail. Handling these controls with hardware would be an agonizing chore. The ability to write and correct the control algorithm easily is a major advantage of software control. Microcomputers are fast enough to keep up with the data, using the division of duties in our example, and so we have a strong bias against a pure hardware solution.

Suppose we do everything in software, including the serial-to-parallel and parallel-to-serial conversions. In other words, suppose our microcomputer deals directly with the various serial data streams associated with the terminals and the high-speed lines. Since a computer is a universal logic implementer, it is certainly possible in theory to handle the logic of any algorithm. There are 64 asynchronous bit streams and two high-speed synchronous bit streams to manage in this problem, and we must deal with every bit at just the right time. The software overhead to perform all the necessary chores would be ruinous. The code to interleave the bit processing for all the terminals and the synchronous lines would be very cumbersome. The computer might not be fast enough to do everything required at the proper rate. The effort required to write the software would be considerably greater than the energy expended to produce our hybrid system.

A more modest move toward software would be to use UARTs to reduce the terminal's asynchronous communications to parallel characters, as we did, but also to perform all the high-speed synchronous serial transfers with software. This approach has some merit, and is probably within the speed limitations of conventional microcomputers, at least for the data rates in our example. However, in developing such a system, we would find numerous places where hardware would be desirable and even necessary, such as in handling the system bus protocols and in deriving the serial bit clock. Furthermore, with a software-heavy approach we could not tolerate much increase in the terminal or high-speed data rates before the microcomputer is consumed with its duties. The method in our example will permit a considerable increase in the data rates before the system becomes saturated.

CONCLUSION

You have now completed Part III of this book, which concludes our presentation of the art of digital design. In Part I we developed the basic logic operations, circuit drafting methods, and building blocks. In Part II we applied this material to a study of digital design at the fundamental MSI level. In Part III we introduced powerful programming-oriented techniques used in handling many types of complex control problems.

In Parts I and II, we tried to include all the material you need to design at the hardwired level. To avoid distracting you with interesting but less relevant material, we have firmly omitted a number of topics usually found in older or traditional treatments of design.

The introduction to microprogramming and computer-based control in Part III is a foundation for further study. We have presented basic information without covering these subject areas in depth.

Whereas the building blocks available at the hardwired MSI level of design are well established and subject to only minor embellishments, the situation at the LSI and VLSI levels is quite different. New products appear weekly. To maintain excellence and currency as a designer, one must have a sound knowledge of the fundamental principles underlying digital design at all levels. MSI design, feasible and appropriate for complete small systems and for special high-performance systems, is usually not effective for large and complex systems. On the other hand, microprogramming and microcomputer-based control can succeed only when supported by MSI design to provide architectural structures and "glue."

In presenting the art of design, we have tried to give you fundamental knowledge that will not be outdated by new technological developments. As you continue to mature as a designer, always remember to seek systematic and fundamental principles and procedures amid the welter of new devices. Happy problem solving!

READINGS AND SOURCES

ARTWICK, BRUCE A., *Microcomputer Interfacing*. Prentice-Hall, Englewood Cliffs, N.J., 1980.

Bipolar/MOS Memories. Advanced Micro Devices, 901 Thompson Place, P.O. Box 3453, Sunnyvale, Calif. 94088. Data book.

BLAKESLEE, THOMAS R., *Digital Design with Standard MSI and LSI,* 2nd ed. John Wiley & Sons, New York, 1979.

HILL, FREDERICK J., and GERALD R. PETERSON, *Digital Logic and Microprocessors*. John Wiley & Sons, New York, 1984.

KLINGMAN, EDWIN E., *Microprocessor System Design*. Prentice-Hall, Englewood Cliffs, N.J., 1977.

KLINGMAN, EDWIN E., *Microprocessor System Design,* Vol. 2: *Microcoding, Array Logic, and Architectural Design*. Prentice-Hall, Englewood Cliffs, N.J., 1982.

MOS Microprocessors and Peripherals. Advanced Micro Devices, 901 Thompson Place, P.O. Box 3453, Sunnyvale, Calif. 94088. Data book.

OSBORNE, ADAM, *An Introduction to Microcomputers,* Vol. 1: *Basic Concepts,* 2nd ed. Osborne/McGraw-Hill, Berkeley, Calif., 1980. One of the best technical presentations on microcomputer principles and operations.

STONE, HAROLD S., *Microcomputer Interfacing*. Addison-Wesley Publishing Co., Reading, Mass., 1982. Good discussions of microcomputer structures, busses, handshaking, arbitration, and so forth.

WIATROWSKI, CLAUDE A., and CHARLES H. HOUSE, *Logic Circuits and Microcomputer Systems*. McGraw-Hill Book Co., New York, 1980.

EXERCISES

11–1. What is a minicomputer? A microprocessor? A microcomputer? In digital design, does a microcomputer offer support for control, for architecture, or for both?

11–2. Describe the methodology underlying each of these means of digital control:
(a) Hardwired.
(b) Microprogrammed.
(c) Microcomputer-based.

11–3. Describe the advantages and disadvantages of each of the methods in Exercise 11–2.

11-4. What are the basic components of a microcomputer system? Choose two popular microcomputers and investigate the major components of each. For each system, show how the components are interconnected.

11-5. Characterize the two forms of microcomputer system bussing methods—memory-mapped and separate input-output.

11-6. Design address detection circuits for the following three device controller addressing configurations:
(a) A fixed 16-bit address $2A00_{16}$.
(b) A switch-selectable 8-bit address.
(c) A 16-bit address with the upper 12 bits fixed at $F00_{16}$ and the lower 4 bits selectable by switch settings.

11-7. Describe bidirectional and unidirectional bus interfaces. One often hears of "bi-directional data bus" and "unidirectional data bus." Do these phrases have useful meaning? Why or why not?

11-8. What is a bus driver? A bus transceiver? Using discrete interface components, draw a circuit diagram of a bus interface for a controller with a unidirectional address bus interface and a bidirectional data bus interface.

11-9. Consult manufacturers' literature and investigate the characteristics of the following LSI chips:
(a) The Intel 8251 Programmable Communications Interface.
(b) The Motorola MC6820 Peripheral Interface Adapter.

11-10. Two popular microcomputer bus protocols are the VMEBUS and the MULTIBUS-II. Locate specifications for these two standards and study their properties.

11-11. Describe programmed input-output. Choose a large computer, minicomputer, and microprocessor, and determine whether programmed input-output exists. Where it is present, show how it works.

11-12. You must design a controller for a low-speed (15 bytes per second) paper tape reader, to be used in a microcomputer system with separate input-output. Design at the functional level two controllers for use with the following input-output modes of operation:
(a) Programmed input-output without interrupts.
(b) Programmed input-output with interrupts.

11-13. Describe the interrupt system of two microcomputer systems of your choice.

11-14. Develop a software flowchart for the manual-positioning event handler in Fig. 11-6.

11-15. Develop a software flowchart for the "seek origin" event handler in Fig. 11-6.

11-16. For a microprocessor of your choice, implement the algorithm in Fig. 11-7 for wrapping wires with the wire-wrap table.

11-17. Discuss the motivation behind the terminal multiplexer project. For the implementation, why did we choose a hybrid hardware-software approach instead of a pure hardware or a pure software solution?

11-18. In the terminal multiplexer, why is a reset capability necessary for the serial-to-parallel converter, but not for the parallel-to-serial?

11-19. Discuss the effect on the terminal multiplexer hardware and software of:
(a) Increasing the terminal data rate from 300 bits per second to 1200 bits per second, while retaining the synchronous transmission rate of 9600 bits per second.

(b) Increasing the synchronous transmission rate to 100,000 bits per second.

(c) Reducing the maximum number of terminals from 32 to 10.

11–20. Modify the terminal multiplexer hardware or software (or both) as required to accommodate a frame structure that begins with two SYNC bytes instead of one.

11–21. Why is the *NULL* control code a necessary part of the synchronous frame in the terminal multiplexer?

11–22. In the terminal multiplexer, how does the microprocessor know when to accept a new byte from the synchronous input stream? How does the microprocessor know when to seek a new character from a terminal?

11–23. Consider the bytes for a particular terminal arriving over the synchronous data line in the terminal multiplexer example. If every byte contains a proper terminal data character, the rate at which characters for this terminal arrive in the system is $9600/(9 \times 33) = 32.3$ characters per second. Since each character sent to a terminal requires 10 bits in the asynchronous communication protocol, the effective data rate at the terminal is $10 \times 32.3 = 323$ bits per second—faster than the rated speed of the terminal. To cope with this situation, we must modify the structure of the *incoming* frame. The switching system at the far end of the 9600-bit-per-second synchronous line must insert a certain number of dummy or "filler" bytes in each frame, preferably just after the *SYNC* byte or at the end of the frame, to lengthen the frame time.

(a) For the 32-terminal maximum used in the example, calculate the number of dummy bytes required to assure that the data rate to any terminal does not exceed 300 bits per second.

(b) Modify the software to accommodate the new incoming frame structure.

(c) Why is it not necessary to also modify the outgoing frame structure?

11–24. In the terminal multiplexer example we assumed a microcomputer data bus of 16 bits. Modify the hardware and software to accommodate an 8-bit data bus.

11–25. Design address detection circuits for the hardware of the terminal multiplexer to accommodate the following:

(a) The Motorola MC6809 system bus.

(b) The Intel 8080 input-output bus.

11–26. For the terminal multiplexer example, implement the interface between a terminal and a microcomputer of your choice. For example, you might choose the Motorola MC6809 with a MC6850 Asynchronous Communications Interface Adapter.

12

Meeting the Real World

In this chapter, we discuss a selection of topics pertaining to digital hardware technology. In the first three parts of this book we emphasized the logical, functional, and organizational aspects of digital design. The design approach was top-down, with emphasis on the problem and its algorithmic and architectural solution, rather than on the details of the hardware. Of course, hardware is important; without it there would be no implementations! In this chapter we present some basic information about the hardware used to support digital design. Many of the topics have a nondigital flavor to them—at least, a nonbinary flavor—being concerned with transistors, currents, power distribution, and so on. Part of our goal in this chapter is to give you sufficient insight into critical areas of hardware technology so that you can perceive the *limitations* of digital logic and can avoid wandering blindly in areas that are by nature nonlogical. This is not a thorough course in hardware technology. We have tried to keep the number of topics and their depth of treatment to a reasonable size, and there is inevitably a certain arbitrariness in our choice. The Readings and Sources give sources of additional information.

ON TRANSISTORS AND GATES

Semiconductors form the basis of the electronic revolution and are the primary components of our circuits, so we will sketch some of their properties as background for the topics discussed in this chapter.

The Flow of Electrical Charge

The conduction of electricity is the result of charged particles moving from one place to another. In a flashlight, electrons leave the negative battery terminal, flow through the flashlight case, through a wire (the light-bulb filament), and back into the positive battery terminal. Here the charged particles are negative electrons.

When experiments on electrical conduction were first performed, electrons had not been discovered. The early workers guessed that the carriers were positively charged, and the convention of positive current remains with us today. Therefore, we draw arrows for the flow of charge from the positive battery terminal to the negative, just the opposite of the electron flow. This causes no real difficulty. The gain in clarity is not worth the chaos that would result from changing conventions. Therefore, all current arrows on diodes, transistors, and other electronic devices refer to positive current.

Metallic Conduction

In metals the mobile particles are electrons that have been stripped from the metal's atoms and are free to roam throughout the metal. The electrons act much like a gas of negative electricity trapped within the metal by the positively charged metal ions formed when the electrons left the neutral atoms. Compared with the electrons, these positive ions are massive chunks of matter, locked into a regular geometric pattern called a *crystal lattice*. It is this locked pattern that gives metals their strength. The free electrons contribute to metallic properties such as high thermal and electrical conductivity, and cause the bright, shiny appearance characteristic of metals.

Insulators

Insulators have no free electrons; all the electrons are locked up in the chemical bonds tying the atoms of the insulator together. The firmer the bonds, the better the insulator. These bonds are also responsible for the characteristic brittleness of insulators.

Semiconductors

Semiconductors are fundamentally insulators with a tiny proportion of broken chemical bonds. The bonds are broken by thermal vibrations. The vibrations set free electrons that form a gas which can then conduct electricity just as in a metal. The resulting gas is billions of times more dilute than in a metal, and semiconductors are correspondingly poor conductors. The most common semiconductor is silicon, which has four electrons available for bonding to other silicon atoms. The resulting crystal structure is a tetrahedral arrangement with each silicon atom bonded to four others.

Interesting things happen when a phosphorus atom with five bonding electrons is placed in the silicon crystal. The phosphorus atom adapts as best it can by using four electrons to bond with surrounding silicon atoms, and releases the fifth electron to migrate throughout the crystal. This electron is *in addition* to those few that are set free by thermally broken silicon bonds. Now we have a way to control the number of free electrons in the crystal, by deliberately introducing phosphorus as an impurity. Since the resulting charge carriers are negative, we call the "doped" crystal an *n-type* conductor.

But we may play the game with boron atoms instead of phosphorus. Boron has only three electrons available for bonding. When placed into a silicon crystal, a boron atom tries to get four electrons so it can look like its silicon neighbors. It does this by robbing an electron from a neighboring silicon atom, which then has only three electrons. This electron-deficient silicon atom may then steal an electron from a neighbor. The deficiency is called a *hole*, and it acts like a mobile positive charge; the result is callled a *p-type* semiconductor.

Our present-day digital technology is built on the precise creation of adjacent chunks of p- and n-type silicon. The boundary between the p and n regions is called a *p,n junction*. In many cases, useful electronic properties arise at the p,n junction, so we will discuss some of the devices that can be built from junctions.

Diodes

The diode will pass current easily in one direction but block it in the opposite direction. The symbol has an arrow showing the direction of positive charge flow. The diode is constructed of a tiny chip of silicon with a single p,n junction as follows:

Diode symbol   Construction

A diode is *forward-biased* when the voltage at the output end (the direction of the arrow) is more negative than at the input, thus causing the diode to pass current. When the output voltage is positive with respect to the input, no charge flows, and the diode is *reverse-biased:*

Forward biased   Reverse biased
(charge flows →)   (no charge flow)

Bipolar Transistors

Bipolar devices are three-layer sandwiches that come in two forms—pnp and npn. As shown below, the pnp transistor has an n layer sandwiched between two p layers; npn has the reverse arrangement. The term *bipolar* arises because conduction involves both p- and n-type charge carriers. The following figure shows the construction and symbol of each type:

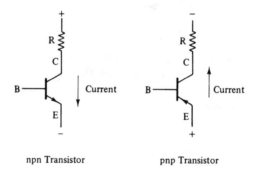

npn
transistor

pnp
transistor

As always, the arrow shows the direction of positive current. In both types of transistor, the major current is between the *emitter* (E) and the *collector* (C). In a pnp transistor, the charge carriers are p-type holes; electrons are the carriers in the npn transistor. In both types, the charge must pass through a thin layer called the *base* (B), of opposite semiconductor type. The base can modulate (control) the emitter-collector current by increasing or decreasing the density of its charges. Changing the voltage on the base lead accomplishes this modulation.

Supplying operation power. The arrow in the transistor symbol is the key to biasing the power supply properly. The arrow points in the direction of positive current, so we must have these arrangements:

npn Transistor

pnp Transistor

In each case the resistor R limits the current whenever the transistor is conducting; without the resistor, the transistor could be damaged. Most bipolar digital circuits are made with npn transistors. The positive supply voltage in these circuits is called V_{CC}.

Turning on and off. In digital work the conducting path between emitter and collector is in one of two states, conducting or nonconducting. In the nonconducting, or off, state no current is flowing, whereas in the conducting, or on, state there is almost a short circuit between emitter and collector. We may think of the transistor as a small switch that can be flipped between open and closed positions. The analogy with a switch gives rise to the common name for a transistor used in digital work: the *switching transistor*.

The base lead controls the opening and closing of the switch in the following way. If we bias the base to inject current in the same direction as the emitter arrow, the transistor switch will turn on. For example:

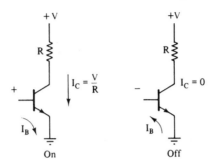

The transistor acts as an amplifier of current, in that a small base current can control a much larger current in the emitter-collector path. The term $\Delta I_C/\Delta I_B$ defines the *gain* of the transistor; in common switching transistors, the gain is from 20 to 100.

In an npn transistor, current will begin to flow in the base lead when it is 0.7 V more positive than the emitter. A pnp transistor will turn on when the base is 0.7 V more negative than the emitter.

MOS Transistors

The MOS (metal oxide semiconductor) transistor is a *unipolar* device—its conduction mechanism depends only on charge carriers of one type. The MOS transistor was proposed before the bipolar transistor, but technological difficulties slowed its development until the need for smaller cells in memory chips caused sufficient attention to be focused on the fabrication process.

Consider a bar of p-type silicon with two n-type regions, *source* and *drain*, diffused into its surface:

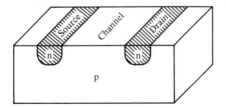

The p-type region between the n-type diffusions is called the *channel*. This arrangement appears similar to the bipolar npn transistor, except that the channel is several microns wide—much wider than the base region in a bipolar transistor.

Conductive electrodes, usually made of aluminum, are attached to the source and the drain, and are biased with the drain positive with respect to the source. The channel-drain junction is not forward-biased, so no current flows across the channel through the transistor. Now place a third electrode—the *gate*—over the channel, separated from the channel by a thin insulating layer of silicon dioxide.

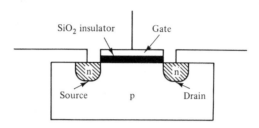

The gate and channel form a capacitor. The mobile carriers in the channel are positively charged holes. If the gate is unbiased or negatively biased with respect to the channel, the holes dominate in the channel, and no current can flow between the source and drain. If the gate is made positive with respect to the channel, electrons will be attracted to the channel region. With sufficient positive gate bias, enough electrons will be available to overcome the holes in a thin region of the channel adjacent to the insulating layer. The *induced* negative charge carriers form a conducting path from source to drain, allowing a current to flow across the transistor. A portion of the channel is temporarily converted from p-type to n-type. This is the principle of operation of the unipolar n-MOS transistor.

Such a device is an *enhancement-mode* transistor, since biasing the gate enhances the concentration of the (n-type) charge carriers. It is also possible to fabricate *depletion-mode* n-MOS transistors, in which biasing the gate depletes the charge carriers. During fabrication of a depletion-mode n-MOS transistor, a thin layer under the gate is diffused with n-type impurities. Like the enhancement-mode transistor, the drain is biased positive with respect to the source. When the gate is unbiased (or positively biased), the transistor conducts current from the source to the drain across the n-type channel. When the gate is biased negatively, positive charge carriers are attracted into the n-type channel region until the n carriers are neutralized. Conduction across the channel ceases, and the transistor turns off.

The n-MOS transistors have negative charge carriers. MOS transistors with positive charge carriers may be constructed by forming a p-type source and drain on the surface of an n-type substrate. Such p-MOS transistors exist in enhancement-mode and depletion-mode types.

Several notations are used for representing MOS transistors. The preferred notation is a circle on the gate input of a p-MOS transistor and no circle on the n-MOS transistor.

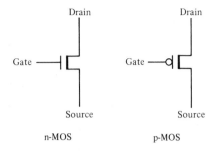

n-MOS   p-MOS

The drain voltage is V_{DD}; the source voltage is V_{SS}. Typical n-MOS supply voltages are 0 volts for V_{SS} and 5 volts for V_{DD}. In good mixed-logic style, the circle shows that a low gate-voltage turns on the p-MOS transistor. In circuits that are entirely of one type of transistor, the distinguishing notations are often omitted.

Gates from Transistors

An inverter. Consider the bipolar transistor circuit of Fig. 12–1. Resistors R_1 and R_2 limit the flow of current in the base and collector leads. As the input voltage *IN* rises above 0.7 V, current will begin to flow in the base, the switch will close, and the emitter-collector path will approach a short circuit. The output will then be at ground voltage, so we see that an H input will yield an L output. As we reduce the voltage at input *IN* to below 0.7 V, current will cease to flow in the base, and the emitter-collector path will open. With the switch open, little current flows through R_2, so the voltage at the output *OUT* is high. Thus an L input yields an H output, just the behavior of an inverter.

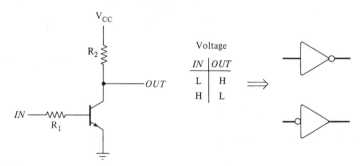

Figure 12–1. An npn transistor used as an inverter.

A two-input gate. Connect two transistors to a common collector resistor as shown in Fig. 12–2. If either or both transistors are on, the output will be shorted to ground, and the *OUT* voltage will be low. If both transistors are off (both switches open), the collector resistor R_2 can pull the output voltage high. The voltage table follows directly from this behavior, and the logic symbols

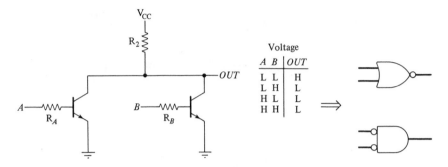

Figure 12–2. A 2-input gate made from two transistors.

result from routine applications of mixed logic. This circuit is the basis of bipolar gate construction. Actual commercial gates are more complex, in order to achieve various advantages of speed, power, and stability. Similar logic gates arise from MOS technology. For instance, Fig. 12–3 shows an inverter and a Nand gate made using n-MOS transistors.

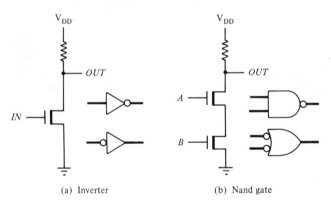

(a) Inverter   (b) Nand gate   Figure 12–3. n-Mos logic gates.

BIPOLAR LOGIC FAMILIES

RTL: Resistor-Transistor Logic

How transistors interconnect to form gates determines the type of integrated circuit logic, or logic family. The bipolar transistor arrangements in the previous section were used in an early family called RTL, which was simple to make but was sensitive to noise and had longer propagation delays on the L → H transition than on the H → L transition. When the output transistor turned on, there was a sudden low impedance to ground, which rapidly discharged stray capacitance, so the H → L transitions were fast. When the transistor switched off, resistor R_2 had to charge up the stray capacitance, which took much longer. RTL soon gave way to DTL (diode-transistor logic), which in turn was supplanted by TTL.

TTL: Transistor-Transistor Logic

To make propagation times more symmetric, the output driver must have a low impedance path to V_{CC} as well as to ground. This requirement resulted in the "totem pole" configuration of two output driver transistors, one to pull the output high and the other to pull it low; see Fig. 12-4. Here R is small, so it can charge and discharge stray capacitance on the output line rapidly. To produce an H output, transistor G must be on and G' off. An L output requires G to be off and G' on. The control signals G and G' are the complements of each other.

A totem pole output stage, providing solid pullup to V_{CC} and solid pulldown to ground, is common in modern logic families. However, the configuration has two problems:

a. *Current spikes.* It is hard to switch one of the totem pole transistors off before the other comes on. There is an overlap of a few nanoseconds when both are on, and during this time the power supply is shorted to ground, except for the small resistance R. This puts a sudden short load on the power supply—a spike—which in turn can affect other gates. TTL circuits require good power supply and distribution of power to alleviate this problem.

b. *Output fights.* One transistor in each totem pole is always on. Suppose you inadvertently tie two totem pole outputs together. If one circuit is trying to produce an L and the other an H, the two circuits provide a low impedance path from V_{CC} to ground through the two turned-on transistors. In this case, the output voltage is indeterminate, and things will get hot!

TTL logic families. TTL is a well-developed line of integrated circuits, with many subfamilies. Some of the subfamilies, such as the 74Lxx and 74Hxx series, are obsolete, and we will not discuss them. Here is a brief summary of the most important TTL subfamilies:

a. 74xx series: regular TTL. This is the oldest TTL line, still found in some old systems. It was fairly fast but drew considerable power.

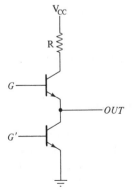

Figure 12-4. The TTL totem pole output configuration. The internal signals G and G' are complements of each other.

b. 74Sxx series: Schottky TTL. This is the fastest TTL line, but it requires high power and suffers from severe noise problems because of its fast switching speed. Its use requires close attention to power distribution, wire length, and power supply bypassing.

c. 74LSxx series: low-power Schottky TTL. This is a fairly fast series, with low power requirements. It was the predecessor to the improved series described below. Its speed is slow enough to permit standard wiring practices.

d. 74ALSxx series: advanced low-power Schottky TTL. Twice as fast as 74LSxx but consumes half the power. A good choice for new designs.

e. 74Fxx series: Fairchild's advanced low-power Schottky TTL ("FAST"). Three times as fast as 74LSxx (and therefore faster than 74ALSxx) but requires nearly three times the power of 74LSxx. A good choice when speed is important.

Schottky technology. Most digital circuits drive their switching transistors fully on and fully off, a switching mode called *saturation*. Such devices tend to be rugged, simple, and inexpensive, but slower than circuits that do not saturate their transistors. Consider an npn transistor: When the base voltage becomes sufficiently high, the transistor is conducting enough to establish a valid TTL output level. Further increases in the base voltage serve only to force more charge into the base until the transistor is saturated, without improving the logical capabilities of the device. When the base voltage drops, the transistor shuts off, but not before the accumulated charge on the base has bled off. The time required to bleed off the base charge is called *storage time*; to achieve high speed, the storage time must be reduced.

Saturation can be avoided by clamping the circuit's inputs with *Schottky diodes* to prevent unnecessary upward excursions of base voltages. The resulting speedup is substantial. Nearly all newer TTL devices use this technique. Chip designers have used the extra speed in two ways, illustrated by the 74S and 74LS series. In general, transistors will switch faster at higher power levels. In 74S, the original Schottky series, the power consumption is about double that of regular TTL, but the speed is three times as great. In 74LS, the designers reduced the power consumption to one-fifth that of 74S, to give a family with the same speed as regular TTL. Subsequent improvements of 74LS have resulted in several TTL series that consume even less power, and have greater speeds.

The internal structure of modern TTL devices is quite complex, but the newer series provide rugged, economical chips that are the workhorses of the digital industry.

ECL: Emitter-Coupled Logic

ECL is the fastest logic line in common use. It takes advantage of the increase in speed attained by operating transistors in the unsaturated mode. An ECL circuit contains a *differential amplifier* (an amplifier whose output is a function of the difference between two input voltages) that compares a logic input with a reference voltage. The result is used to steer a constant current into one of two paths without saturating the circuit's transistors. The output circuit always contains two paths that have complementary voltages, thus providing easy access to high- and low-active versions of the output logic variable.

ECL circuits have gate propagation delays of about 1 nsec, and ECL flip-flops can be toggled at a rate of 500 MHz—several times faster than their nearest competitor, 74S TTL.

ECL has several advantages for designers:

a. Since the current is nearly constant, even during switching, ECL circuits do not tend to cause current spikes on the power supply. This property simplifies the distribution of power to the ECL design—an important advantage since power distribution tends to become troublesome with faster logic lines. Yet ECL is the fastest of all.
b. ECL outputs are available in .H and .L forms, with equal propagation delays. This virtually eliminates the need for voltage inverters.
c. The complementary outputs are excellent for use as line drivers.
d. In the most convenient ECL family, the rise time of the gate outputs is purposely made slower than the internal propagation delay. This simplifies the clocking of synchronous circuits and greatly reduces electrical crosstalk.
e. The slowed rise times permit the use of ECL in properly designed wire-wrapped circuits.
f. ECL outputs may be directly connected, to achieve a wired-OR or wired-AND logic function.
g. Power supply current is independent of the system clock rate. This enhances ECL's role as a high-speed logic family. In most logic families, the power supply current increases with clock speed.

Offsetting these advantages are several serious disadvantages of ECL:

a. Although power consumption is nearly independent of clock speed, ECL is a "power hog."
b. The high speed and fast rise time of signals in ECL circuits makes the propagation delay along circuit interconnections a significant factor in the design. To achieve reliable performance, one must frequently use transmission-line techniques to inhibit signal reflections.

Manufacturers have made ingenious modifications and compromises in ECL products to ease the problems brought on by their blinding speed, and ECL is a reasonable choice when speed is essential. However, using ECL in a design requires much more care than a slower family such as 74LS TTL, so ECL is not the first choice for the design of many circuits.

UNIPOLAR LOGIC FAMILIES

MOS technology provides the basis for several unipolar logic lines. All have the advantages of a small unit area and low power consumption and the disadvantages of sensitivity to damage from static charge and of relatively slow speed.

p-MOS and n-MOS Logic

MOS technology is popular for large, complex digital systems because the basic MOS gate is much smaller than its bipolar counterpart. Since less power is consumed, the problem of heat dissipation is more manageable than in bipolar technology. Nearly all dynamic memories and most microprocessors are made with n-MOS technology. The p-MOS technology is little used today, except as a component of complementary MOS (CMOS) circuits.

CMOS: Complementary MOS Logic

CMOS is a logic family that uses both n-MOS and p-MOS transistors to produce a circuit that consumes much less power than either alone. Low power consumption is obviously desirable for such lightweight and portable devices as watches and pocket calculators, and it is also necessary if the packing density of transistors on chips is to increase. Power generates heat, which must be dissipated efficiently to keep the device's temperature within reliable working limits.

To understand CMOS operation, consider the simple bipolar and unipolar inverters shown in Figs. 12–1 and 12–3. These inverters consume power whenever the transistor switch is conducting, because current flows between power and ground through the main output pullup resistor. When the transistor switch is not conducting, only a small current flows across the resistor. The resistor provides a means of establishing the high voltage level when the transistor is not conducting, and limits the current flowing through the transistor when the transistor is conducting. In CMOS, both these functions of the resistor are provided by a p-MOS transistor whose operation is complementary to the usual n-MOS output transistor:

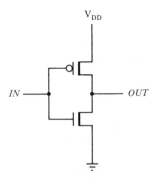

In this arrangement, one transistor is conducting whenever the other is shut off. When IN is a high voltage, the n-MOS transistor is on and the p-MOS transistor is off, producing a low voltage level at the output OUT. When IN is a low voltage, the n-MOS transistor is off and the p-MOS transistor is on, producing a high voltage level at OUT. Except during the brief instant of switching, no current flows between V_{DD} and V_{SS}. The input *IN* and the destinations for *OUT*, connected only to the capacitative gate of MOS transistors, have no steady-state currents. Therefore, the power consumption of the CMOS gate is extremely small. The CMOS cell is nearly as small as in n-MOS.

Whereas earlier CMOS integrated circuits were slow, the newer Hi-Speed CMOS family rivals TTL in speed while retaining the extremely low power consumption of the earlier CMOS devices. CMOS is popular in VLSI and is also available as pin-compatible counterparts to 74LS-series SSI, MSI, and LSI components. One may develop a circuit using the rugged 74LS chips and then substitute the less-robust CMOS equivalents to drastically reduce the power consumption.

THREE-STATE AND OPEN-COLLECTOR OUTPUTS

Open-Collector Outputs

The TTL open-collector circuit is essentially a switching transistor without an internal connection to V_{CC}: the collector is open. To complete the circuit, we must supply a connection from the open-collector output to V_{CC}, through a suitable external resistor. Here is the circuit for an open-collector inverter, with one commonly used circuit symbol:

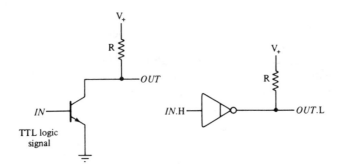

The supply voltage V_+ may vary over a wide range, and we must choose the resistor R to suit the requirements of the circuit. Since the open-collector circuit does not use a totem pole configuration, it is slower than its other TTL counterparts. Various chips in the 74 and 74LS lines are available with open-collector outputs, including inverters and various gates that perform AND and OR logic. We use the vertical line within the gate symbol to designate open-collector output; there is no universal standard notation.

There are two main uses of open-collector gates:

Buffer-Driver. Whereas the normal open-collector chip has current behavior similar to its parent 74 or 74LS family, it is possible to build an open-collector current buffer that can accept much higher currents than the ordinary variety. This, coupled with the ability to set V_+ at any reasonable level, makes the open collector useful for driving display lamps and other devices with non-TTL voltage requirements or with high current demands; see the later section on lamp driving for more details.

Wired OR. Suppose we tie the outputs of several open-collector gates together, and power them all from V_+ through a single resistor, as in Fig. 12–5. In this circuit, if any open-collector transistor is on, the output will be L. For a transistor to be on, its base must be H, so we observe the voltage behavior shown in the figure. This voltage table corresponds to an AND or OR logic operation, depending on the voltage convention we select. Since we achieve this logic operation without an actual AND or OR gate, we speak of a "wired OR" or a "wired AND." The rule is:

Wired OR   if T = L.
Wired AND if T = H.

The wired OR has some uses in bussing and interfacing, but it appears more attractive than it turns out to be in practice. The circuit is spread over several packages. Suppose it doesn't work: which package is bad? This alone is sufficient cause for caution. Another minor problem is that now we have to put an external resistor into the circuit, which is awkward packaging. Still a third objection is the open collector's slow speed. The open collector has its place, but as a logic tool it is not popular with designers; the three-state output has largely eliminated the need for it as an implementer of logic. Chapters 2 and 3 have additional discussions of open-collector circuits.

Open-collector resistor value. The open-collector pullup resistor R must be within a certain range for proper operation. We will demonstrate a computation of R for a wired OR circuit. In this case, where we must produce a TTL-compatible logic signal as output, V_+ will be equal to V_{CC}, namely, 5 V. Assume that we have six ordinary 74LS open-collector outputs wired together, with the wired output feeding four 74LS inputs. We must consult the data sheets for the information needed for the computations. A 74LS open-collector output can safely pass only a certain amount of current, at most 8 mA. This maximum current could occur when the transistor is on and the voltage at the output is close to zero. We must make certain that the total current that can flow through

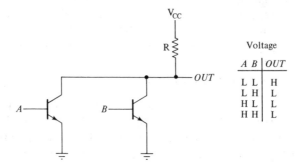

Figure 12–5. An open-collector wired OR.

the transistor is less than its maximum rating. This current comes from two sources: through the resistor, and from each input tied to the open-collector output. Each 74LS input may provide as much as 0.4 mA in the low state. The current allowed through R is the difference of the maximum open-collector current (8 mA) and the total current from the inputs. The computation is

$$R_{min} = \frac{E}{I} = \frac{5 - 0}{0.008 - 4 \times 0.0004} = 780 \ \Omega$$

We may determine the maximum R from the number of inputs and outputs attached to the output line. Low-power Schottky gates will treat an input voltage of 2.0 V or greater as a high level, and we must make sure that the voltage drop across R is not so great that the actual high-level voltage of the output drops below 2.0 V. Each 74LS input draws as much as 20 μA in the H state, and each 74LS open-collector output will draw 100 μA (into the output). All of this current must come from V_{CC} through R, so

$$R_{max} = \frac{5 - 2}{6 \times 0.0001 + 4 \times 0.00002} = 4400 \Omega$$

One more consideration: the resistor must be of sufficient wattage to dissipate the power generated in it. Power is P = E $\times$ I. Using Ohm's Law, we may write R = E^2/P. The maximum power loss in the resistor occurs when the voltage drop is greatest, about 5 V. If we wish to use a ¼-watt resistor, the minimum safe value of R is

$$R_{power \ min} = \frac{5^2}{0.25} = 100 \ \Omega$$

This value is much lower than the earlier minimum of 780 Ω, so our ¼-watt resistor will be quite sufficient for any proper value of R.

From these computations we see that a value of 1000 to 2000 Ω is safe. More details will be found in manufacturers' data books.

Three-State Outputs

Many chips in the major integrated circuit families have three-state outputs. In addition to the normal H and L output states, a three-state gate has a third state Z, called the *high-impedance state.* When the gate is in the high-impedance state, the chip's output is effectively disconnected from the circuit. In addition to its normal inputs, a three-state chip also has a control input called the *three-state enable,* whose purpose is to select between normal (enabled) and high-impedance (disabled) modes.

Using TTL As a model, we may show how the high-impedance state comes about. In the normal TTL totem pole output stage, one of the two totem pole transistors is always on, so there is always a relatively low-impedance path through the totem pole. The three-state circuit permits the control input to turn both totem pole transistors off when the control is false. Schematically, the output circuit is as in Fig. 12–6. When the three-state signal is false, the outputs of the ANDs are both low, and both totem pole transistors are off. In this case, the current in the output circuit has no low-impedance path and the output behaves as if the gate were not attached.

As long as we make sure that *at most* one three-state element is enabled at a time, we may tie many such outputs together. This property has had a great, almost revolutionary, effect on data bus design, where we wish to wire all data sources onto the bus and allow at most one to "talk." (See the discussion in Chapter 3.)

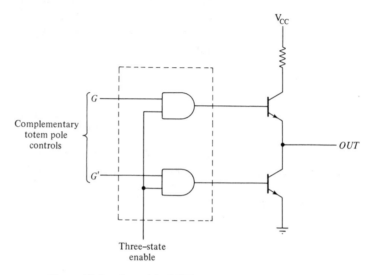

Figure 12–6. A model of TTL three-state output control.

INTEGRATED CIRCUIT DATA SHEETS

The integrated circuit data sheet is the document that defines a given device. It contains much important information, some unique to the device, and some fairly standard for the logic family. There is no standard format for data sheets, although some of the terminology is uniform among manufacturers. In a typical data sheet for a digital device, we expect to find:

a. The chip's name and number and a description of its major characteristics.

b. A logic symbol of the manufacturer's choosing, naming the functions of each input and output of the chip.

c. A pin diagram, showing the function associated with each pin.

d. A complete description of the use of the chip, often with a detailed voltage table. For SSI chips, this description can be quite brief; for MSI and LSI, it can be complex. For VLSI, the description may require dozens of pages.

e. Electrical and environmental specifications for the chip's operation.

f. For MSI, LSI, and VLSI circuits, perhaps some information about applications.

Electrical Data

The electrical information is the most difficult part of the data sheet to master, but it is important that you understand some of the details. Fortunately, the terminology is nearly standard, and for TTL many of the values of the parameters are standard, although the exact form of presenting the information varies. The following discussion applies to the low-power Schottky TTL family. Other logic families have different values of parameters, but the treatment is similar.

The data sheet will usually describe two versions of a chip, a military grade and a commercial grade. The military grade is designed to operate over a wider temperature range than the commercial grade and will have slightly different operating characteristics. Manufacturers differ in how they specify these grades, but most manufacturers use the prefix "74" for the commercial grade and "54" for the military grade. We will discuss the commercial grade.

Standard nomenclature is V for voltage, I for current, and subscripts I for input, O for output, H for high voltage level, and L for low voltage level. Thus V_{OL} stands for low-level output voltage. H and L represent the binary voltage *levels*; the electrical data will specify the actual range of voltages for these levels.

The electrical information falls into three categories: operating conditions, static operating characteristics, and switching data.

Operating conditions. These are the basic conditions under which the manufacturer states that the chip will perform as specified. The usual parameters are:

Temperature: 0° to 75°C for commercial-grade chips; −55° to 125°C for military grade.

V_{CC}: supply voltage. For all commercial TTL chips, the allowable range is 4.75 to 5.25 V, with a typical value of 5.0 V.

I_{OH}: high-level output current. I_{OH} must not exceed the stated value, which for 74LS is typically 400 μA, directed out of the output.

I_{OL}: low-level output current. This is the maximum allowable current through the output in the low level. For 74LS, the usual value is 8 mA, directed into the output.

Static characteristics. The static, or dc, electrical characteristics describe how the chip's outputs and inputs will react to specified stable operating conditions. The major parameters are:

V_{IH}: high-level input voltage. This is the minimum voltage that the chip guarantees to recognize as a high signal level. For 74LS, this is 2.0 V. The chip will treat any input voltage from 2 to 5 V as H.

V_{IL}: low-level input voltage. This is the maximum voltage that the chip guarantees to recognize as a low signal level. For 74LS, this is 0.8 V, so the chip accepts an input between 0 and 0.8 V as an L. An input voltage between V_{IL} and V_{IH} is in the forbidden region, and the manufacturer does not guarantee reliable performance.

V_{OH}: high-level output voltage. If the output-high current I_{OH} does not exceed its maximum rating (given in the operating conditions), then V_{OH} represents the voltage that the chip will produce for a high signal level. Usually, the data sheet will give a guaranteed minimum V_{OH} (2.7 V for 74LS) and a typical V_{OH} (usually 3.4 V for 74LS).

V_{OL}: low-level output voltage. This is the low-level counterpart of V_{OH}. If I_{OL} does not exceed its maximum rating, then V_{OL} represents the voltage that the chip will produce for a low signal level. The guaranteed maximum value of this voltage for 74LS is 0.5 V. The typical low-level output voltage for 74LS is usually 0.25 V.

I_{IH}: high-level input current. This is the current that an input will draw when presented with a high voltage level. The data sheet gives a maximum value, which for 74LS is 20 μA at 2.7 V. The direction of this current is into the input, and the value is a function of V_{IH}. The higher the voltage on the input, the higher is the current demanded by the input.

I_{IL}: low-level input current. This is the current that an input will require when presented with a low voltage level. For 74LS, the maximum value of I_{IL} is 0.4 mA at 0.4 V; the direction is out of the input. The lower the voltage on the input, the higher is the current.

I_{CC}: supply current. This is a measure of the chip's power consumption. The data sheet will specify the operating conditions under which the supply current is measured and will usually give a typical and a maximum value for I_{CC}. The supply current varies greatly with the chip type; the information is useful when estimating the size of the power supply needed for a given circuit.

Switching characteristics. The switching, or ac, characteristics show the time dependence of various transitions in the chip. The most fundamental switching parameter is the *propagation delay*, which is the elapsed time from the initiation of an action until the outputs reflect the result of the action.

In the usual notation, t_{pLH} is the propagation delay of an output going from L to H; t_{pHL} is the counterpart for an output going from H to L. In combinational circuits, the delay period starts with a change in an input. For complex circuits, either combinational or sequential, having several types of inputs and outputs, the data sheet may list several propagation delays. Propagation delay information usually has maximum and typical values. For simple 74LS gates, propagation delays are about 5 nsec; more complex circuits may have delays of a few tens of nanoseconds.

Sequential circuits have another class of switching information in the data sheet: the setup and hold times. *Setup* is the time prior to some event (such as a clock edge) during which an input must be stable to assure reliable device operation. *Hold* time is how long after an event the input must remain stable.

Setup and hold times are usually stated as minimum and typical times. Setup times for 74LS vary considerably but range up to about 30 nsec, with 15 or 20 nsec being typical. Generally, we like the hold times for inputs relative to the clock to be no greater than zero. The minimum setup and hold times listed in the data sheet are conservative values. If you meet these conditions, *all* chips of that type should work. The "typical" entries for setup and hold times are usually considerably smaller than the "minimum"; the typical chip will be faster than the slowest chip, to which the minimum values apply. It is not easy to tell, so it is best to stick with the conservative specifications.

PERFORMANCE PARAMETERS OF INTEGRATED-CIRCUIT FAMILIES

Loadings, noise margins, and propagation delays are given in integrated circuit data sheets or may be inferred from data sheet information. During the implementation phase of a project, the designer must take care to avoid violating the manufacturer's specifications. Here we discuss these design parameters and compare the TTL-compatible integrated circuit families.

Input and Output Loadings

Consider a wire representing a signal path from an output to one or more inputs. Every input on the wire places a current *load* on the signal source; the data sheet specifications I_{IH} and I_{IL} characterize the load. The output of an integrated circuit is capable of handling only a specified maximum current before the quality of its signal deteriorates; we measure the output *drive* capability by the data sheet specifications I_{OH} and I_{OL}. Each family of integrated circuits is designed around standard values of these four specifications of current; if the standard specifications of output current are not exceeded, the "standard gate" will perform correctly within the standard input current limits. The standard I_{IH} and I_{IL} values are called the *unit load* for the high and low logic levels. In most cases, inputs draw one unit load. Sometimes, when an input goes directly to several gates within the device, the input will draw two, three, or more unit loads. Familiarity with the concept of unit load will help you to interpret the specifications in the data sheet.

The ratio I_{OH}/I_{IH} of the standard values is called the *fanout* of the high voltage level; similarly, the ratio I_{OL}/I_{IL} is the fanout of the low voltage level. For a given family of integrated circuits, fanout describes how many standard inputs a standard output will drive. Designers use fanout as a rule of thumb to avoid overloading circuit outputs with too many inputs. For most families the fanout values for high and low voltage levels are the same; when they differ, we use the smaller value.

The TTL-compatible integrated circuit families have the current behavior shown in the accompanying table. A 74ALS standard output will drive 20 standard 74ALS loads, and a Hi-Speed CMOS standard output will drive 50 standard Hi-Speed CMOS loads. The output drive capability of integrated circuits is well standardized. However, since inputs may use several unit loads, you must do more than simply count inputs; you must check the data sheets for each chip.

Family	Maximum current (mA)				Fanout	Fanout to 74LS
	I_{IL}	I_{OL}	I_{IH}	I_{OH}		
74 TTL	1.6	16	0.04	0.4	10	40
74LS TTL	0.4	8	0.02	0.4	20	20
74ALS TTL	0.2	8	0.02	0.4	20	20
74F TTL ("FAST")	0.6	20	0.02	1.0	33	50
Hi-Speed CMOS (V_{CC} = 5 V)	~0	4	~0	4	50	10

When using chips from different families within a circuit, you must compute the loading of each output, using the actual inputs attached to the output. The table shows the fanout of each family into inputs of the same family, and also the fanout of each family into 74LS inputs. From the current data in the table you can compute the fanouts of any family into any other family by using the appropriate standard input and output currents. The mixing of "compatible" integrated-circuit families requires close attention to the loadings. In the course of constructing or debugging a circuit, it is easy to insert a 74F chip in place of a 74ALS chip without bothering about the possible consequences. You gain additional drive capability at the 74F output, but the inputs to the 74F chip require more current than the 74ALS chip and you may be overloading earlier outputs by making this exchange.

Handling overloaded outputs. The drive capability of modern logic chips is usually sufficient to handle the input loadings, but occasionally you will have to drive more loads than the output will handle. One solution to this problem is to split the inputs into groups and form an identical output to drive each group. For example:

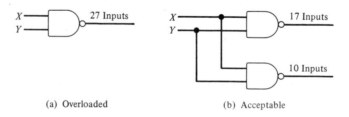

(a) Overloaded   (b) Acceptable

Another technique is to use a high-power amplifier called a *buffer* to drive the entire set of inputs. For instance, in a 74LS37 Quad Two-Input Nand Buffer I_{OH} = 1.2 mA and I_{OL} = 24 mA, three times the values of the standard 74LS gate output.

As you design a complex circuit, you must watch out for overloading. An undetected overload can pull your signal into the forbidden voltage region, producing unreliable performance which may be hard to diagnose.

Noise Margins

For each integrated circuit family, the allowable voltage ranges of valid high and low levels are rigidly defined. Inputs will accept a wider range of voltages than outputs produce. This is necessary because signals inevitably undergo deterioration between the source (an output) and the destinations (inputs). For each voltage level, the difference between what the input will tolerate and what the output is guaranteed to produce is called the *noise margin*. The low-level noise margin is

$$\Delta V_L = V_{IL} - V_{OL}$$

The high-level noise margin is

$$\Delta V_H = V_{OH} - V_{IH}$$

These figures represent how much deterioration a signal may suffer in going from an output to an input before the voltage enters the forbidden region between the levels, where it is no longer recognizable as a valid digital signal level. High noise margins are desirable. The accompanying table shows the standard voltage limits and derived noise margins of the TTL-compatible integrated circuit families.

Family	Guaranteed voltages (V)				Noise margins (V)	
	V_{IH}	V_{OH}	V_{IL}	V_{OL}	High	Low
74 TTL	2.0	2.4	0.8	0.4	0.4	0.4
74LS TTL	2.0	2.7	0.8	0.5	0.7	0.3
74ALS TTL	2.0	2.7	0.8	0.5	0.7	0.3
74F TTL ("FAST")	2.0	2.7	0.8	0.5	0.7	0.3
Hi-Speed CMOS ($V_{CC} = 5$ V)	3.5	4.9	1.0	0.1	1.4	0.9

CMOS can operate within a range of power supply voltages; our figures assume the use of the normal TTL power supply voltage of 5 V. Hi-Speed CMOS has excellent noise margins, so inputs are less affected by noise than with the other families. (On the other hand, CMOS is much more sensitive to damage from static electric charge than TTL, and must be handled with care.)

Noise margins are an important figure of merit for an integrated circuit family, since noise, whether introduced by circuit losses or from external sources, is a fact of life in digital equipment. Imagine the difficulty we would be in if the noise margins were zero!

UNUSED GATE INPUTS

Logic chips frequently have unused inputs. Gates often have too many inputs (e.g., when a 74LS00 Nand gate is used as an inverter) and so do flip-flops and registers having asynchronous set and clear inputs. Here are some procedures for neutralizing these unneeded inputs.

To force a *high voltage level* on an unused input: for 74 TTL, connect the input to V_{CC} through a 1-kΩ resistor; for 74LS, 74ALS, and 74F TTL and for CMOS, connect the input directly to V_{CC}.

To force a *low voltage level* on an unused input: connect the input directly to ground.

For unused AND and OR gate inputs, another option is available for standard 74 TTL: connect the unused input to a used input on the same gate. This is equivalent to an application of the Boolean algebraic identities $A \cdot A \equiv A + A \equiv A$. Because it makes the chips more sensitive to noise, *this method should not be used with low-power Schottky or CMOS integrated circuits.*

In the TTL families, an unattached (floating) input tends to assume a high voltage level, and it is tempting to just let the input float to achieve an H. *Don't do it!* The floating input acts like a small antenna and can easily pick up sufficient noise from the surroundings to cause spurious behavior. Always establish a firm value for unused inputs.

THE SCHMITT TRIGGER

The Schmitt trigger is an ordinary inverter with a little bit of internal feedback. The amount of feedback is chosen so that the circuit is not yet a flip-flop, but has some flip-flop characteristics superimposed on normal inverter behavior. Figuratively, it takes a big push to switch the output, but once switched it will stay there even if the push is partially relaxed.

Figure 12–7 describes the behavior of the Schmitt trigger's input and output voltages. If we start with $V_{in} = 0$, the output is H as with a normal inverter (point A). Now increase the input voltage from 0 toward Y. The output will remain H, on path A → B. Increase V_{in} a little more, and the output will switch suddenly to L at $V_{in} = Y$, and will remain L as the input voltage undergoes limited excursions around Y (between points E and D). Now decrease V_{in}. The output will not switch until V_{in} is considerably less than Y. In fact, we will have to lower V_{in} all the way to X (point F) before V_{out} will switch to H.

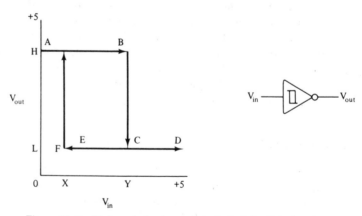

Figure 12–7. Voltage behavior and symbol of the Schmitt trigger.

It is this different switching point on increasing and decreasing voltage that gives the Schmitt trigger its utility. The action, called *hysteresis*, is measured by the difference between the X and Y voltages. The Schmitt trigger circuit symbol is a standard inverter with a hysteresis curve drawn inside the triangle, as shown in Fig. 12–7. For a TTL Schmitt trigger, typical values for X and Y are 0.8 and 1.6 V, respectively.

The Schmitt trigger is a potent device for combatting noise at the inputs to digital systems. Here is how it works. Suppose we have a poorly formed signal coming into our system, such as this:

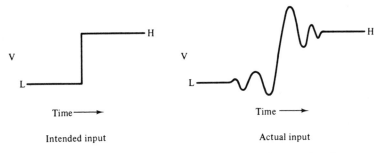

Let us put this waveform into a Schmitt trigger and see what happens. In Fig. 12–8, the sloppy input waveform with the Schmitt trigger switching voltages X and Y is shown. The input must rise to Y before the Schmitt trigger will switch. The hysteresis in the Schmitt trigger rejects all the noise in the lower part of the signal, and a clean switch occurs in the output at point S. Once switched,

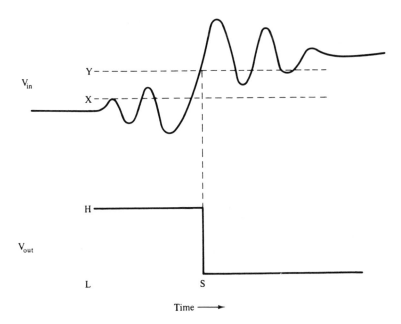

Figure 12–8. How the Schmitt trigger removes noise.

the Schmitt trigger will ignore voltage excursions above X. The hysteresis in the Schmitt trigger has cleaned up the terrible waveform and recaptured the original intent.

Schmitt triggers have another important application with slowly varying input voltages. A standard low-power Schottky gate interprets H as being > 2.0 V and an L as < 0.8 V. Voltages in the forbidden region between 0.8 V and 2.0 V can cause indeterminate or even oscillating outputs. The outputs of normal gates switch through the forbidden region so fast that subsequent inputs are not confused. Problems can arise if the inputs receive slowly changing signals, such as from a 60-Hz alternating current or a charging or discharging capacitor. To use such signals in logic circuits, you must first feed them into Schmitt triggers, whose "snap" action (hysteresis) produces clean, fast output transitions, no matter how slowly the inputs change.

A POWER-ON RESET CIRCUIT

Orderly startup of a digital system requires that certain crucial flip-flops be set to known values when power is applied. We need a reset signal that will come on for a short time as the power supply is coming up to voltage, and will then go off and remain off.

A resistor-capacitor combination provides a slowly increasing voltage across the capacitor in response to a step (rapidly increasing) input voltage. We made use of this property in the single-shot delay circuit discussed in Chapter 4. If we feed this slowly increasing voltage into a Schmitt trigger, we have a circuit that produces an immediate H signal as output, and then switches to L as the capacitor charges during power up. Figure 12–9 shows the circuit and the waveforms resulting from the power-up sequence. The duration of the reset signal is roughly the product of R and C: $t = R \times C$, where R is in ohms, C is farads, and t in seconds.

OSCILLATORS

Synchronous systems need a clock. There are many oscillators that can serve as the basis for a system clock. We mention a few here. Some of the criteria influencing the selection of an oscillator are cost, stability, and frequency range.

The Schmitt Trigger Oscillator

This is about the least expensive and simplest oscillator you can make. Figure 12–10 shows the circuit, which uses an R-C combination to cause the oscillation. This circuit will continue to oscillate reliably as long as power is applied. The Schmitt trigger oscillator suffers from a lack of stability—the clock frequency is not precisely constant. Another mild disadvantage is the 30 percent duty cycle of the square-wave output. If you want a perfectly symmetric square wave, you may run the output into the clock input of a JK flip-flop wired in toggle mode.

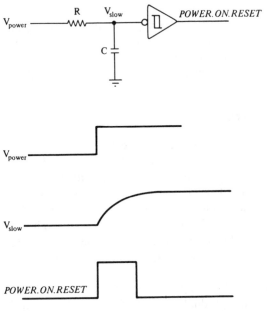

Figure 12–9. A power-on reset circuit.

This produces a flip-flop output transition at every rising edge of the oscillator. The flip-flop thus provides a clock with one-half the frequency of the oscillator.

In the Schmitt trigger oscillator, the capacitor continually charges and discharges between the hysteresis points. When the capacitor charges to the upper trip point, the output will switch to low. This will discharge the capacitor until the lower trip point is reached, which causes the output voltage to go high. The cycle repeats indefinitely.

The 555 Oscillator

The 555 Timer is a low-cost integrated circuit built for timing industrial devices. It can be configured as a single shot or as an oscillator. One of the virtues of

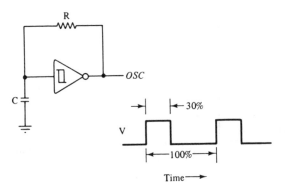

Figure 12–10. A simple Schmitt trigger oscillator. R and C control the frequency.

the 555 Timer is the slow oscillation frequency that can be achieved with capacitors of moderate size. It is also more stable than the Schmitt trigger oscillator. This is a popular chip—a fine choice for clocks with periods slower than several microseconds. Figure 12–11 shows the 555 Timer wired as an oscillator. The values of R and C determine the frequency of oscillation; consult the data sheet for timing computations.

Crystal-Controlled Oscillators

A variety of integrated circuit clock chips derive their frequency from an external quartz crystal. The Motorola MC12060 and MC12061 are examples of chips that permit a wide range of clock frequencies that are determined by the oscillation frequency of the quartz crystal. The MC14411 is useful for clocking data transmission, since the chip provides a large selection of standard clock frequencies, including such popular ones as 110, 300, 1200, and 9600 Hz. The chip uses a crystal of fixed frequency (1.8432 MHz) and has internal logic to reduce this frequency to the values selected by the control inputs. Another excellent wide-range oscillator is the RCA CD4047.

The disadvantage of these chips in general synchronous design is their fixed frequencies, which preclude any gradual varying of the clock rate in debugging.

The System Clock

Remember that in synchronous system design, we usually want more from a clock than just an oscillating logic signal. In Chapter 6 we designed a useful system clock that supports an automatic clock derived from an oscillator, and also a manual mode that permits the designer to issue manual clock transitions. The oscillator forms the first stage of such a system clock.

Also, remember that system clocks for synchronous circuits must often

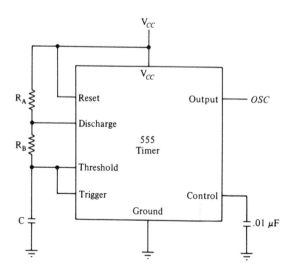

Figure 12–11. The 555 timer used as an oscillator. R_A, R_B, and C control the frequency.

drive many chips, so the clock's output must be well buffered to provide the necessary operating power.

SWITCH DEBOUNCING

Mechanical switches are important parts of digital devices, but they present signals that are unsuitable for use in digital logic circuits. The switch contacts do not open or close cleanly; they undergo a period of "bounce" in which the electrical signal from the switch will change noisily between its voltage extremes many times in the course of a few milliseconds. A switch debouncer removes the bounce by responding to only one of the voltage excursions during each switch operation. In Chapter 4 we analyzed this situation and offered a solution that used RS flip-flops.

Here we repeat these circuits, and discuss the component values and the electrical efficiency of the solutions. Figure 12–12 shows two switch-debouncing circuits, each using a single-pole, double-throw toggle switch and gates from the 74LS logic line.

Consider the circuit in Fig. 12–12a. Whenever the switch closes the On contact, the voltage on that line goes to ground (L) and causes the cross-coupled gate flip-flop to record truth on the *SWITCH.ON* outputs. Once the flip-flop responds to the On signal, further bounces on the line will have no effect on the outputs. The debouncer will register a change only when the Off contact closes, at which time the debouncer will respond to only one of the Off bounces.

The resistors have a pull-up or pull-down function to assure a proper default

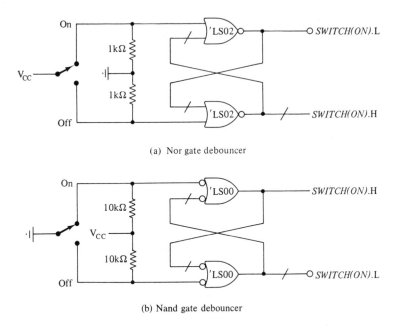

(a) Nor gate debouncer

(b) Nand gate debouncer

Figure 12–12. Switch debouncers made with Nor and Nand gates.

H or L voltage level at the gate inputs. We derive the values of the resistors in Fig. 12–12a by noting that when a contact is open the current path is from the gate input through a resistor to ground. The 74LS02 input requires a maximum of 0.4 mA current in the L state. We must make sure that the voltage at the gate input does not rise above the low-level logic threshold of 0.8 V; for safety, we will choose 0.4 V as our maximum allowable low-level voltage. Then

$$R_{max} = \frac{0.4 - 0}{0.0004} = 1000 \ \Omega$$

The maximum suitable resistance value is 1 kΩ.

To calculate the resistance in Fig. 12–12b, we must ensure that the voltage at a gate input is at a valid H level when the switch is open. Choosing 3 V as our lower limit for H, and using $I_{IH} = 20 \ \mu A$ for 74LS gates, we have

$$R_{max} = \frac{5 - 3}{0.00002} = 100 \ k\Omega$$

In our circuit, we showed a smaller resistance value, 10 kΩ.

There is a small advantage in the circuit in Fig. 12–12b, since it consumes less power. Except during switching transitions, one side of the switch is always closed, so there is a virtually constant 5 V drop across one of the resistors. The current drawn by this circuit is 5 mA for Fig. 12–12a, which is equivalent to 12 or more normal 74LS loads. In Fig. 12–12b, the circuit draws only one-tenth of this current, and so is about equivalent to one 74LS load. The power saved in Fig. 12–12b can be significant if the design has several debounced switches.

This illustrates the typical reasoning when choosing values for pull-up or pull-down resistors in digital circuits. In most cases, only Ohm's Law is needed, coupled with the ability to identify the important issues.

LAMP DRIVERS

It is often necessary to display critical signals in a digital system. Gates seldom have enough current-handling capability to drive lamps as well as other gates, so it is usually necessary to isolate the lamp from the logic signal by an open collector buffer. We may use standard buffers that will load the logic signal with one additional standard TTL current load; alternatively, Darlington buffers are available that require close to zero input current.

In digital display, the incandescent lamp has given way to a solid-state device called a *LED* (*light-emitting diode*). LEDs have the advantages of low cost, low power drain, and long service life. The brightness of a LED is a function of the current passing through it. Most LEDs are limited to currents of about 10 mA. Higher currents produce more light but shorten the life of the

LED. The typical LED driver circuit is:

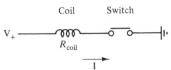

where the open collector gate might be a 7406 buffer or a 9667 Darlington buffer.

If the input signal X is H, the output of the open collector inverter is L, and the LED will light. The resistor R limits the current through the LED. We may determine a suitable value for R using Ohm's Law. The voltage at the output of the open-collector gate is close to zero; the voltage drop across the LED itself is about 1.6 V when it is conducting, regardless of the current. Thus for a 5 V power supply, the voltage drop across resistor R when the LED is lit is $(5 - 1.6 - 0) = 3.4$ V. If we wish to limit the current through the LED to 5 mA, then

$$R = \frac{E}{I} = \frac{3.4}{0.005} = 680 \ \Omega$$

A resistor in the range of 700 to 1000 Ω would be appropriate.

DRIVING INDUCTIVE LOADS

Mechanical relays, solenoids, and motors depend for their operation on coils of wire, which exhibit electrical inductance. Although inductance is sometimes useful, it can cause difficulties for the digital designer who wishes to switch an inductive load on and off. Even at low voltages, inductance can damage switches. Inductance tends to stabilize current by resisting changes in current through the coil. Consider a coil containing a sufficient length of wire to provide a safe current-limiting resistance when the coil is connected between its design voltage and ground:

With the switch closed, the current is given by Ohm's Law:

$$I = \frac{V_{CC}}{R_{coil}}$$

When the switch is opened, the resistance across it rapidly moves from zero toward infinity. Ohm's Law predicts that the current would rapidly drop to zero, but the inductance in the coil prevents the current from dropping immediately. Where can this current go? The inductance creates a large transient voltage across the switch, polarized to maintain the flow of the current across the switch. In many cases, the voltage becomes great enough to create a spark across the switch's contacts. If the switch is a switching transistor, the voltage surge will destroy it unless it is protected. Even mechanical switches will burn out unless they are robust or otherwise protected.

Protecting Switches with Diodes

In dc circuits, a diode across the coil will protect the switch from damage:

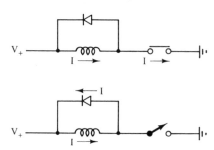

When the switch is closed, the diode is back-biased and no current flows through the diode. When the switch opens and the inductance tries to maintain the current, the voltage rises at the switch terminal, and the diode becomes forward-biased. Current can flow in a circular path through the diode until the inductive effect has dissipated. Therefore, no large voltage surge is developed across the switch.

Protected by diodes, switching transistors can safely switch many inductive loads. The 1N4000 series of diodes is useful for bypassing common inductances arising with relays and solenoids.

Solid-State Relays

For switching heavy dc or ac loads, we recommend a *solid-state relay*. These devices are carefully engineered to amplify weak logic signals to the level where they can drive a final switching transistor for dc loads or a TRIAC for ac loads. The logic switching component is often optically isolated from the relay, providing excellent protection to the digital circuit driving the relay. When using a solid-state relay to switch a dc load, you should still protect the relay from dc voltage surges by using an external diode, as described previously.

THE OPTICAL COUPLER

When interfacing distant or dissimilar devices, we frequently wish to assure that electrical problems at one end of the system will not damage components at the other end. As long as there is an electrical connection between elements of a system, we face the risk of one malfunctioning element damaging another. For instance, the Teletype produces its signals from electromechanical circuits. A mechanical short circuit within the Teletype might produce violent electrical disturbances on the serial signal lines.

Fortunately, there is a nondigital device called the *optical coupler* or *optical isolator* that uses light instead of electrons to pass signals. The optical coupler has a light-emitting diode as its input component and a phototransistor as the output, sealed together in one unit. You are already familiar with the light-emitting diode. In a *phototransistor*, the switching action is controlled by light shining on the base of the transistor rather than by electrical current flowing through the base. When current flows through the light-emitting diode, it lights up. The phototransistor, which is off (switch open) when the diode is dark, responds to the light by closing its transistor switch. Figure 12–13 shows the optical coupler circuit in its customary notation.

There is no electrical connection between the input and the output sections; within wide ranges of power supply voltages, the input and output stages are electrically independent of each other. It is this property that proves so valuable in protecting or isolating systems.

The principal disadvantage of the optical coupler is its relatively slow switching speed; 1 to 100 μsec is typical performance—much slower than digital logic circuits, but satisfactory in many data communication applications.

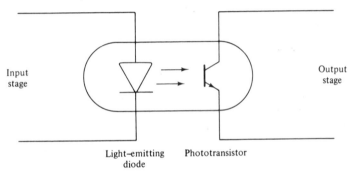

Input stage

Output stage

Light–emitting diode   Phototransistor

Figure 12–13. An optical coupler.

POWER SUPPLIES

Power supplies are nondigital devices, and the construction of a good power supply is something of a mystery. Although the elements of power-supply technology are easy to understand and primitive supplies appear simple to construct, it usually pays to purchase them unless you have a good understanding of analog design techniques. Usually, you get what you pay for, and usually you are paying for things other than volts and amperes—for example overload protection, service life, voltage regulation, efficient cooling methods, compactness, and ability to run at high ambient temperatures.

A typical power supply has four major components, shown in Fig. 12–14.

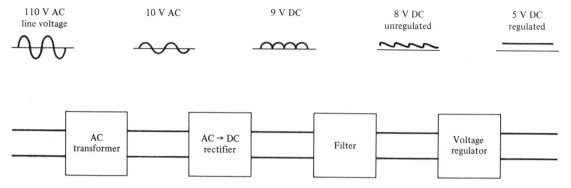

Figure 12-14. Components and waveforms of a dc power supply.

The raw power comes from the power company as 110 or 220 V ac. A transformer lowers the voltage to several volts above that required by the supply output. A full-wave rectifier changes this alternating voltage to a bumpy dc version. A filter smooths this waveform, producing a somewhat fluctuating voltage slightly above the desired one. Finally, a voltage regulator removes the last fluctuations and fixes the output voltage at the precise level desired.

Remote Sensing

Quality power supplies have a *remote sensing* feature to compensate for the small voltage drop generated in the conducting paths connecting the power supply with the circuit. Two small wires (sense wires) sample the voltage in the circuit itself and feed this voltage back to the power supply so that it can regulate the voltage into the digital circuit rather than the voltage produced at the power supply. The sense wires can be small, since they carry virtually no current.

Integrated-Circuit Voltage Regulators

Commercial power supplies often provide all the components in Fig. 12-14 (and other refinements) in one package, providing regulated power that must then travel to the digital systems. Another approach is to use regulators contained in an integrated circuit chip. These can supply a few amperes at fixed voltages and can be an economical way to supply power. They have the advantage that all the analog complexities of power regulation are wrapped up inside the package; troubleshooting is reduced to swapping components.

The main power supply, without the regulation section, is easier to build than the complete supply. This approach allows us to furnish unregulated power to each circuit card and to perform the voltage regulation right on the card, a design that tends to minimize ground loops and other undesirable power distribution phenomena discussed in the next section. However, on-board regulation places a major heat-producing element on each logic card, making efficient heat dissipation essential.

POWER DISTRIBUTION

In digital logic, well-developed formalisms exist for solving problems in terms of the pure binary concepts of truth and falsity. What happens when these binary concepts are emulated by voltages? If the voltages are clean and stable, digital hardware provides a reliable correspondence between the logic values and the voltage levels. If the voltages are noisy or unstable, we are no longer in the true digital domain and unpleasant things can happen.

Two common problems that remove us from a simple binary world are the generation and distribution of power. Chips require power, and a circuit with many chips may require much power. You should develop an intuitive under-standing of where current is flowing in digital circuits and what effect that flow has upon noise generation. Fortunately, it is not difficult to gain this understanding.

Every integrated circuit package has a pair of power pins. In the TTL logic family, these pins are V_{CC} and ground. We do not (and should not) show these pins on logic diagrams, since they are not logical in nature. V_{CC} is a source of current that flows into the chip and out through the ground pin. The chip accepts voltages representing *logical* signals as inputs and uses a portion of the supply power to produce voltages for logical output signals. The conversion of power is never 100 percent efficient. The lost power appears as heat, and the chip warms up. The circuit assembly must be able to dissipate this heat, or the chip's temperatures may rise to the point that the circuits behave improperly.

Losses in Power Distribution Systems

Resistive losses. Figure 12–15 shows two gates powered by the same power supply; the inevitable resistances of the power distribution system are lumped in the diagram for convenience. The R_V terms are resistances associated with V_{CC} distribution; R_G is ground distribution. For the sake of our exposition, let us assume that each gate requires 10 mA of operating current and that all resistances are zero except R_{G1}, which is 10 Ω. The voltage at the ground pin of gate 1 will be 100 mV above the power supply (system) ground. This comes from Ohm's Law:

$$E = I \times R = 0.01 \times 10 = 0.10 \text{ V}$$

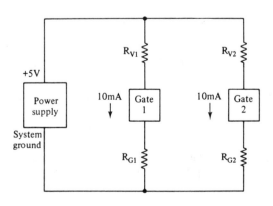

Figure 12–15. Two gates powered by one supply, with resistance in the power-distribution system. Line resist-ances are lumped for convenience. Logic inputs and outputs are omitted.

Each gate produces logical output signals referenced to the internal ground voltage of the integrated circuit chip. Suppose that each gate produces an output of l volts in the low voltage level and h volts in the high level. Gate 1 produces outputs of $l + 0.1$ V and $h + 0.1$ V with respect to the system ground. Now suppose that the output of gate 1 goes into the input of gate 2. Since R_{G2} is zero, gate 2 is operating with its internal ground equal to the system ground. The 0.1 V offset of gate 1's output reduces the noise margin at gate 2, since the gates have a maximum threshold with respect to their own internal grounds, above which they will not interpret a signal as a low voltage level. The offset at gate 1 is in the direction that moves gate 2's input closer to the threshold, thereby reducing gate 2's margin for error. In 74LS TTL, the low-level noise margin is 0.3 V. In our example, instead of a 0.5 V low-level signal, gate 2 would receive 0.6 V; only 0.2 V remains of the safety factor to handle other types of voltage disturbance.

Inductive losses.   Ten ohms is an intolerably high resistance in a power circuit, and almost any power-distribution system will have resistances much lower than this. Don't relax! The problem may still be with you. We used resistances in the example above because they provided us with a simple, static example. The real villain is inductance. Inductance exists in any conductor, independent of resistance, and it depends on the geometry of the conductor. An inductor will generate a voltage across itself if the current it carries changes. For a given inductance, the faster the rate of change of the current, the larger is the voltage across the conductor.

Now look again at Fig. 12–15. The power distribution resistances are typically a few milliohms, and we can usually neglect them. As long as each gate draws constant current, the rate of change of current is zero. The induced voltage due to the inductance of the power paths will also be zero, so we have no problem. Unfortunately, the power-supply current drawn by a gate surges up and down when its logic output changes. This phenomenon is particularly acute in TTL, the most popular logic family. Since any useful logic system will have changing internal logic signals, we have no alternative but to reduce stray inductances to a minimum. Circuit geometry is the only avenue available to us. We will discuss several fabrication techniques for combatting inductance.

Combatting Inductance in Power-Distribution Systems

Ground planes.   Solid-sheet conductors have the lowest possible inductance, and it is almost independent of the sheet's thickness. The best system is to mount integrated circuits on such a plane and connect the chip's ground pins to it. Piercing the plane with small holes to allow the other pins to pass through the ground plane to the other side of the board has virtually no harmful effect.

Strip distribution.   Many commercial integrated-circuit packaging systems have interdigitated (interleaved) strip busses for power distribution, as shown in Fig. 12–16. In the section on TTL integrated-circuit data sheets we showed that

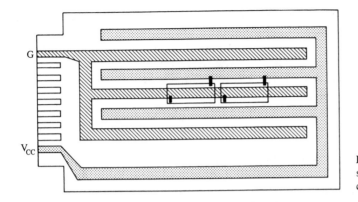

Figure 12-16. A circuit board with strip power busses. Two integrated-circuit chips are attached.

noise margins in the L state are often smaller than in the H state, so we want the lowest possible inductance on the ground distribution system. Inductance is proportional to length, so in Fig. 12-16 we should use the bus marked G for ground since it has the shortest run to the supply pin on the edge connector.

Wire distribution. A round wire has much more inductance per unit length than a strip bus; it is usually pointless to try to use wire for power distribution on a printed circuit board. Nonetheless, wires are usually necessary to carry power from the power supply to a digital system. There are two ways to reduce the wire inductance—stranding and changing the shape to a flat braid. Flat braid is stranded wire that has been rolled into strip form, thereby gaining the low inductance of strips while retaining the flexibility of woven wire.

Many commercial systems use laminated bus bars to distribute ground and V_{CC} in a system. These bus bars have low inductance and low resistance, and also represent a distributed capacitance useful for filtering high-frequency noise. For power distribution on printed circuit boards, bus bars often allow the use of two-sided rather than multilayered boards.

Recommendations. If you are new to digital logic systems, we strongly recommend that you use a commercial ground-plane system for your first few projects. This will keep you safely in the digital domain and make success more likely.

Some commercial ground-plane systems are expensive, but you can justify them if the system will see extensive use. Be prepared to pay more for power supplies, power distribution, and circuit cabinetry than you pay for the integrated circuits.

For small systems, printed circuit cards with a power bus on one side and a ground plane on the other are available at low cost. They are more desirable than a one-sided interdigitated bus card.

Bypassing the Power Supply

Even a distribution system with the least inductance will need help in order to deliver clean, stable power to the integrated circuits. The solution is a capacitor.

The function of the capacitor is to store charge that can power the surges of an integrated circuit when its outputs change. The ideal geometry is to place a capacitor directly across the V_{CC} and ground pins of each chip. In a sense, this capacitor is a local power supply that is inexpensive enough to dedicate to each chip and will thereby have short leads and low inductance. More properly, it is a power averager that supplies surge demands and is then recharged from the main power supply. A capacitor used in this way is called a *bypass capacitor*, since the surges of current are absorbed in the capacitor and bypass the power supply.

Most capacitors come with round wire leads that have a high inductance per unit length. A rule of thumb is to keep these leads less than 3 mm long. Only certain types of capacitors have low enough internal inductance to be useful for bypassing. Ceramic capacitors have low internal inductance and are suitable if in the range 0.01 to 0.1 μF. Ceramic dielectric disks are commonly used because of their small cost.

Tantalum dielectric capacitors also have low internal inductance and more capacitance per unit volume than ceramics, but with higher price. It is good practice to bypass the power supply leads with a 10 μF tantalum capacitor where the leads enter the circuit card, in order to handle any current surge remaining from the smaller bypass capacitors at the individual chips.

A common mistake is to mount one bypass capacitor per chip on interdigitated power strip cards, but across the wrong strips. The correct way is the shortest possible path through the capacitor between the V_{CC} and ground pins of a chip. The wrong way will force the surge current from a chip to go all the way to the end of a strip and back down an adjacent strip to reach the bypass capacitor. Short is beautiful in digital fabrication!

NOISE

So you have done a fine job on the algorithmic and architectural phases of your digital design, your logic implementations are flawless, and you have constructed your circuit neatly on a circuit board with a good ground plane and well-bypassed power supply. Is everything just fine? Not necessarily. Despite all your efforts, noise may still be a problem. Noise is unwanted voltage on the signal lines. Good digital logic design attacks noise at the logical level. For instance, in synchronous design we process information only at certain fixed times governed by a clock signal, so that we are sure that, in our design, all the vital signals are stable. After a clock edge, we expect a certain amount of turbulence on the signal lines as they adjust to their new status, but this will die out long before the next edge. Similarly, we avoid using asynchronous inputs as decision variables, since we cannot control when they change.

However, other sources of noise arise from the physical layout of the hardware. They are unrelated to the logic of our problem, and we must attack them on the hardware level. Fortunately, there are ways to lessen the effect of the noisemakers.

Crosstalk

One source of spurious signal information is *crosstalk*, a signal picked up from a changing voltage on another wire. The most common source of crosstalk is nearby logic signal wires. Each wire acts like a little antenna, capable of receiving "transmissions" from energy radiated by other wires. The antenna is sensitive to a range of frequencies that is a function of the length of the wire. The longer the wire, the longer the duration of the noise pulses picked up by the antenna.

Why is the length of a noise pulse important? Sequential logic elements such as flip-flops change their state upon being activated by their input signals. These inputs must remain in the active state long enough for the flip-flop circuitry to respond. This required stable period is similar to the setup time of the flip-flop. As a general rule, if the noise has a shorter duration than the setup time, the flip-flop will still behave correctly. Now you can see the significance of the length of the input wire. If the wire is long enough, crosstalk can create noise of sufficient duration to cause the flip-flop to act incorrectly.

For the TTL-compatible families of integrated circuits, typical setup times range from about 20 nsec (for regular 74, 74LS, and Hi-Speed CMOS) to about 3 nsec (for 74S and 74F). ECL setup times are on the order of 1 or 2 nsec. We may encounter problems with noise when the wires are longer than about $2\frac{1}{2}$ ft for chips with the longer setup times; with 74S we are in trouble when the interconnections are longer than about 6 in. ECL is even worse; limits are about 3 in. unless special precautions are taken.

Remedies. What do we do? First, use synchronous design. In this mode, the logic signals change only for a short time following the active clock edge. If we are not running our circuit too fast, the signals will settle down before the next clock pulse. Crosstalk in logic signals is thus primarily limited to an insensitive period of the clock cycle.

Second, do not run the clock so fast that the unsettled period of the logic signals can endure until the next clock edge. Remember that the crucial propagation delay is the sum of the delays in the slowest circuit path. Third, do not use the very fast integrated circuit families such as Schottky TTL and ECL unless the design definitely requires them.

Crosstalk onto the clock line during the signal's transition period can be serious, since noise on the clock line *at any time* can generate spurious clock pulses. To make further progress, we must look to the actual fabrication of the circuit. As in power distribution, short is beautiful. Keep interconnections as short as possible. In addition, there are other ways to reduce the antenna effect on a signal wire. The coupling of one signal to another (the crosstalk) is significant only when the two wires are roughly parallel and fairly close to each other. When wiring a circuit board, run your wires directly from point to point, making them as short as possible, so that the collection of wires tends to form a random arrangement. This reduces the likelihood that any two wires will be parallel for a long distance. *Do not* run neat channels of wires between the rows of chips. Such packets look pretty, but they greatly increase the chance of crosstalk.

On printed circuit boards, where parallel signal traces are inevitable, interposing a trace tied to either ground or V_{CC} will shield adjacent signals.

Reflections

Even a signal wire isolated from other wires can cause noise. Consider a signal injected into a wire; for example, let the voltage at the sending end of the wire change rapidly from 0 to 4 V. Launching this signal into the wire causes a change in current because of the change in voltage. The change in current (and with it a voltage step function) travels down the wire at a rate of about 2 nsec/ft until it reaches the other end. We would like the signal to be absorbed at the receiving end, and that would be the end of it. However, if the receiving end is an open circuit (a dangling wire, for example), the current has no place to go and simply turns around and reflects back toward the sending end. Part or all of the current may be reflected, causing the received voltage to be different from the intended value during the reflections. The reflections may continue to bounce back and forth, forming amazing and quite unacceptable patterns of voltage. Eventually, the reflections will die out and the receiving end will have the same voltage as the sending end (less the very small resistive loss in the wire).

The open circuit approximates a gate input attached to the receiving end of the wire, since the gate input draws little current and acts like a high-impedance load. Once again, the severity of this type of unwanted behavior is related to the speed of the integrated circuit logic family we are using. Our earliier warnings about maximum wire lengths for the various logic families apply to reflections as well as to other types of noise. In short wires, the reflections die out before they can do any harm. However, if the wire is long enough, reflections become a serious matter, even in an otherwise noise-free environment. In this case, we must take further action.

Synchronous design solves many of the problems of reflections by assuring that they will occur only in insensitive parts of the clock cycle. However, when we encounter crosstalk, the clock line and any *asynchronous* inputs are vulnerable to spurious voltages at any time. Reflections are an inherent property of the particular type of signal wire and its source and load. The method of combatting reflections in the hardware is more exact and definitive than the procedures for crosstalk.

The astonishing behavior of voltage reflections is the subject of transmission-line theory, whch predicts the observed waveforms quite accurately. Although we are hardly inclined to think of our circuit board signal wires as transmission lines, that is exactly the viewpoint we must take. The theory predicts that the reflections are a function of the relative impedance of the signal line and the receiving device. Impedance is the counterpart of resistance that includes the frequency-dependent capacitative and inductive effects introduced when the flow of charge undergoes a change.

Every wire has a *characteristic impedance* that is a function of the capacitance and inductance of the wire. These depend on the type of wire and its insulation, where the wire is positioned with respect to ground, and so on. The characteristic impedance is independent of the wire's length, and in the frequency range of interest in digital switching (above 100 kHz) the characteristic impedance of all practical conductors falls in a narrow range from about 30 to 600 Ω. For the type of conductors used in printed circuit boards and wire-wrap assemblies, the characteristic impedance is 150 Ω ± 50 Ω; 150 Ω is sufficiently accurate for our purposes.

Transmission-line theory shows that if the receiver has the same impedance as the line, there is no reflection; the entire current step is absorbed and the received voltage is correct on the first try. This is the key to dealing with long signal wires. The inputs of integrated circuits present a high impedance to the line, so we can match the line's characteristic impedance by placing a 150 Ω resistance in parallel with the path of the input current (for instance, between the input terminal and ground). This is called *terminating* the line in its characteristic impedance. It is sometimes desirable to terminate the sending end of the wire also, to damp out any reflections that arrive back at the sender. Figure 12–17 shows the typical behavior on an unterminated and a properly terminated line. The figure displays a common technique in which the termination is shared between ground and the high-voltage supply. For this purpose, which depends on ac rather than dc behavior, the two terminating resistors are in parallel. The net impedance is 132 Ω, well within the suitable range.

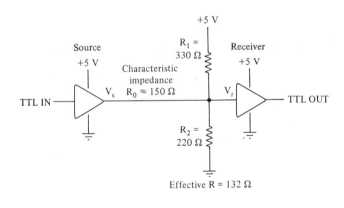

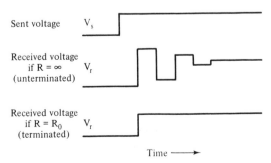

Figure 12–17. Signal transmissions on a long line. The unterminated line has reflections; proper termination solves the problem.

Line Drivers and Line Receivers

Line-termination resistors consume power. Five volts across a 150 Ω resistor requires 33 mA of current, far more than the typical logic gate output can supply. Therefore, when terminating long lines, we usually make use of special integrated circuits called *line drivers* and *line receivers*. These chips are designed to mesh with logic signals such as TTL, and transmit and receive the signals over the long lines according to one of several accepted methods. The driver, in addition to other characteristics desirable in this application, will have sufficient drive capability to handle the extra load created by the terminating resistor.

Since the long lines often run many feet, they are especially susceptible to external noise. The noise travels toward both ends of the wire from the point of pickup. Although in a properly terminated line the noise is not reflected, it nevertheless appears as a part of the received signal. The simple line-driving technique illustrated in Fig. 12–19 is called *single-ended* and is rather susceptible to external noise. Another method, much more effective at rejecting the noise, is called *differential* line driving and is shown in Fig. 12–18. Instead of sending the signal as a voltage on one wire, referenced to ground, the differential line driver and receiver communicate over two signal wires, carrying equal voltages opposite in sign. In the differential method, it is only the *signed* difference in the voltages on the two wires that carries signal information, a positive difference representing an H and a negative difference being an L. If one wire picks up noise, it is likely that the other wire will pick up the same noise, since the wires are exposed to almost identical environments. Noise added to each wire has no effect on recovering the signal.

Another advantage of differential line driving is its insensitivity to differences in the ground potential at the ends of the line. Because of variations in the earth ground potential and other factors, this *common-mode* voltage can sometimes be many volts. Since the transmitted and received signal voltages refer to the local ground level as the zero reference point, a voltage difference between the two ground points may result in loss of information in the single-ended method. The differential method relies solely on the voltage difference on its two wires, so differences in ground potential, which affect both wires equally, do not affect the detection of signals at the receiver.

Line driving and line termination are common in digital design, and often involve many signal and data lines. Groups of resistors of appropriate values are available in dual in-line packages that can be mounted in the same way as

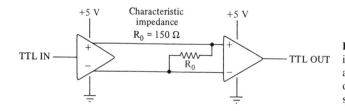

Figure 12–18. Differential line driving. The lines are properly terminated and the sign of the voltage difference on the two lines determines the logic signal information.

conventional SSI integrated circuits. These "resistor packs" greatly simplify the fabrication of circuits requiring terminators.

Summary

Combatting noise requires careful attention to design methods and fabrication practices. We have concentrated on noise arising within the digital system itself but, as indicated in the previous section, noise from external sources such as automobile engines, transmitters, and motors can also cause problems. Our recommendations for noise abatement serve to protect against external as well as internal noise.

Here are the basic steps to minimize noise in digital circuits:

a. Use synchronous design methods.

b. Keep wires short.

c. Don't use fast integrated-circuit families.

d. Use a well-designed power supply and power-distribution system.

e. Keep wires close to the ground plane.

f. Let the wires run in random directions whenever possible.

g. Take special care with the clock line.

h. Use line-termination procedures when necessary, but avoid them wherever possible by conforming to points (b) and (c).

Noise in digital systems is a complex subject. We have touched on only its major aspects. Your best bet is to stay well within the safe area described in the recommendations.

METASTABILITY IN SEQUENTIAL CIRCUITS

In Chapter 4 we saw how hazards could cause a physical version of a logic circuit to produce momentary signals that were different than those predicted by Boolean algebra. To nullify the effects of these hazards, we adopted synchronous design. In Chapter 5 we saw that the introduction of asynchronous test inputs into an otherwise synchronous ASM could result in transition and output races. To avoid these problems, we synchronized the asynchronous external inputs by using D flip-flops. Throughout these procedures we assumed that we are not violating the setup and hold times of the devices in the circuit. But when there are external signals that change asynchronously to our circuit, we cannot make such a guarantee. What happens when, despite our best efforts, an input transition occurs close to the active clock edge?

Realizing that the clock signal in a synchronous device can be viewed as just another input, we can broaden our question to ask what happens when two inputs to a sequential circuit change shortly before or after each other. The facts, which were not widely recognized nor even believed until recently, are disquieting.

Consider the standard D flip-flop, the 7474. Figure 12–19 is the observed behavior of this device when the D-input changes about 3 nsec before the rising clock edge. The outputs of the transistors in the main feedback circuit can assume voltages that do not correspond to either of the valid digital voltage levels. This behavior is called *metastability*, and flip-flops exhibiting these characteristics are said to be in a *metastable state*. Eventually, the metastability will resolve and the flip-flop outputs will assume valid (and complementary) voltage levels, but the duration of the metastable state is indeterminate and may be quite long compared with the propagation delay listed in the data sheet.

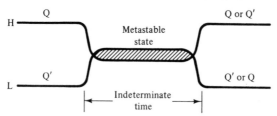

Figure 12–19. Metastability in the 7474 D flip-flop. The final stable state is unpredictable, but Q and Q' are inverses.

Other bizarre and awkward forms of metastability may occur. If the 7474's clock and D-input signals switch to high at nearly the same time, the outputs may display proper digital voltage levels with normal signal rise and fall times, but the transitions may be delayed for an indeterminate time. In an RS flip-flop constructed from crosscoupled 7400 Nand Gates, the outputs may behave as shown in Fig. 12–20 if the gates are set or cleared by a runt pulse. In this form of metastability, which occurs in circuit families such as TTL in which signal rise times are less than the gate propagation delay, the Q and Q' outputs oscillate in-phase for an indeterminate time before resolving into valid digital voltage levels.

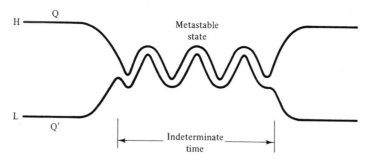

Figure 12–20. Metastability in cross-coupled 7400 Nand gates. When set or cleared with a runt pulse, the outputs may oscillate in phase.

Why do flip-flops behave this way? A gate is really an analog *amplifier*. In digital circuits, the gate inputs are overdriven so that the output transistor *saturates* (is fully conducting or nonconducting) in its high or low state. These states form the familiar digital voltage levels. Digital logic relies on inputs passing quickly through the transition region between the voltage levels. But by judiciously biasing the inputs we can force a gate's output to be in the linear region. For instance, consider a 7404 Inverter gate driving another inverter gate:

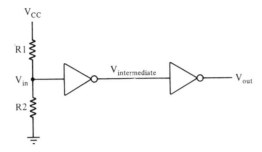

By proper choice of the resistors R1 and R2, we can set the input voltage V_{in} such that $V_{out} = V_{in}$ and $V_{intermediate} \approx V_{in}$. Both inverters are in their transition region. Now, if we carefully connect the output to the input and remove the source of V_{in}, we will have constructed a crosscoupled flip-flop circuit that is metastable:

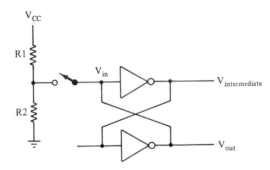

Eventually, electronic noise or the random thermal fluctuations of the atoms in the gates will upset the metastability and drive the outputs to a stable condition. The duration of the metastable interval is unpredictable. Real flip-flops always have feedback and can always be balanced at this metastable point. *There is no way to avoid this phenomenon!*

How serious is this problem? If a flip-flop is in a metastable state, the probability P of still being in the metastable state after time t can be approximated by

$$P = \kappa e^{-t/\tau}$$

where κ and τ are constants determined by experiment or calculation. The probability of remaining metastable decreases exponentially with time—a reassuring trend, but insufficient to guarantee the flip-flop's performance or to alleviate problems at fast clock rates.

To understand the effective treatment of metastability, we must gain a quantitative understanding of the timings and probabilities. At present, this information is available only from detailed experiments with devices in the various logic families. These complex experiments rely on circuits with both digital and nondigital elements to detect the existence of metastable behavior and record the frequency of metastability resulting from repeated attempts to induce the metastable condition. In the experiments, the system clock rate is fixed, and input transitions randomly distributed about the clock edge are produced at a selected rate. The usually complementary outputs of the flip-flop being tested are compared using analog comparators to detect metastability. At a specified time after the system clock transition, the result of the comparison is captured in a flip-flop. A true output, indicating metastability lingering at least as long as the sampling delay, increments a digital counter. From the count of the metastable errors that occur during the experiment, the mean time between failures (MTBF) under the given conditions can be determined.

The relationship among the experimental variables is given by

$$\frac{1}{\text{MTBF}} = f_{\text{cp}} \times f_{\text{data}} \times K_1 e^{-K_2 \Delta t}$$

where f_{cp} is the system clock frequency, f_{data} is the frequency of data transitions, Δt is the sampling delay, and K_1 and K_2 are constants describing the metastable characteristics of the device being tested. By determining the MTBF at two sampling delays, the experimenter may determine the constants K_1 and K_2 for the device under test.

T. J. Chaney[†], who has been a leader in the experimental study of metastability, has published extensive tables of K_1 and K_2 for many devices. For example, for the 74LS74 D Flip-flop, $K_1 = 0.4$ sec and $K_2 = 0.67$ nsec^{-1}. Consider an experiment in which the system clock rate is 10 MHz and the average rate of input transitions is 0.1 MHz. Assume we are interested in the probability of a metastable failure that endures until the next clock pulse—a measure of the reliability of a simple D flip-flop synchronizer. The calculated MTBF is several billion years. But doubling the clock frequency reduces the MTBF to only 3.7 sec! The results are extraordinarily sensitive to the clock frequency. By running the clock slowly enough, you can obtain any degree of reliability you wish. There is a wide difference between flip-flops of the same type produced by different manufacturers and even between different batches of the same flip-flop from one manufacturer. Experiments on several 74S74 flip-flops running at a frequency of 25 MHz, receiving 10^5 input transitions per second, show a variation in failure rates from 2 hours to 5×10^{31} sec. (The age of the universe is about 3×10^{17} sec.) Push the clock speed at your own risk!

What might we do about the metastability problem?

a. We might do nothing, aside from synchronizing our external input with a D flip-flop. This method is satisfactory if we wait long enough, since the probable duration of a metastable state decreases exponentially with time.

b. We might choose a flip-flop that resolves input conflicts faster. This requires a small K_2, not just a small propagation delay. When we wish to run the

[†] T. J. Chaney, "Measured flip-flop responses to marginal triggering," *IEEE Transactions on Computers C-32*, December 1983, p. 1207.

clock at maximum speed and can't afford special detection circuits, such as in VLSI design, this method can be beneficial. Chaney's tables of metastability constants for standard elements can be helpful here. The Fairchild FAST series, an extension of the 74ALS series, seems to have especially good resolving characteristics. Special flip-flops constructed with tunnel diode resolvers or with ECL or gallium arsenide technologies can work at clock frequencies up to five times those suitable for standard TTL elements†.

   c. We might run our clock at high speed until a detector senses metastability and then freeze the clock until the metastability resolves. This method requires two tricky circuits: a system clock that can be stopped synchronously and started asynchronously and a metastability detector.

The detection circuit used in Chaney's experiments can be adapted to our purpose. In Fig. 12–21 we show the basis of this detector. The reference voltages V_{OH} and V_{OL} for the comparators correspond to the boundaries of the voltage transition region for the D flip-flop's outputs. If the flip-flop's output voltage V_Q is lower than the high-level threshold and higher than the low-level threshold, the signal *METASTABLE* is true. This type of detector is useful with circuits in which the flip-flop output is directly available and the flip-flop does not suffer from the oscillating form of metastability. Standard metastability detectors are available for inclusion within VLSI designs.

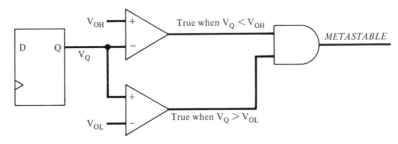

Figure 12–21. A simple metastability detector.

An accepted way to form a system clock that can be stopped synchronously (the false portion of the clock is extended indefinitely) and started asynchronously begins with a delay line and an inverter in a feedback loop:

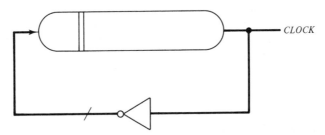

† G. Elineau and W. Wiesbeck, "A new JK flip-flop for synchronizers," *IEEE Transactions on Computers C-26*, December 1977, p. 1277.

Once the circuit is properly initialized upon power-up, the delay line processes two internal states, the true and false portions of the clock signal, that travel down the line. Whenever the output *CLOCK* is false, the delay line is processing the next true signal through its internal storage, to emerge after the appropriate delay. Whenever the output is true, the delay line processes the next false. The waveform repeats after two delay-line time periods. We may modify this circuit to include the control signal *METASTABLE* in the feedback loop:

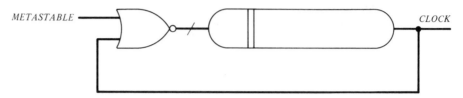

In using this circuit, the assumption is made that *METASTABLE* will become true only during the positive part of the clock cycle. This is reasonable for the common form of metastability detected by the circuit in Fig. 12–21 since, if metastability occurs, it arises soon after a clock transition. Then the positive portion of the clock cycle will run its normal course and the negative portion will be stretched by the assertion of *METASTABLE*. If *METASTABLE* becomes true during the negative phase of the clock, the impending positive phase traveling down the delay line will cause another rising edge on the clock line, defeating our intention to freeze or stretch the clock cycle in order to wait out the metastable period. Furthermore, the new positive phase will be foreshortened, and the delay line may begin to "multimode," operate at a multiple of the desired frequency.

An even number of inverters may be substituted for the delay line. Delay lines are quite stable, but they are sensitive to multimoding. Inverter chains do not have the stable delays of delay lines, but are less sensitive to multimoding.

Three potentially useful treatments of metastability were suggested above, but the technical literature is full of incorrect proposals for solving the problem. Many of these just move the source of potential metastability from one point in the circuit to another. Here are a few treatments that do not work:

a. Using Schmitt triggers in feedback loops to clip runt pulses. If a runt pulse causes metastability in a circuit with feedback, then that same runt pulse will not cause metastability in that circuit if a Schmitt trigger is inserted in the feedback loop. However, a runt pulse of different duration and magnitude can still cause metastability in the modified circuit. This has been verified mathematically by Marino[†] and experimentally by Chaney[‡].

b. Using unit distance codes so that only one state variable depends on an

[†] L. R. Marino, "The effect of asynchronous inputs on sequential network reliability," *IEEE Transactions on Computers C-26*, November 1977, p. 1082.

[‡] T. C. Chaney, "Comments on 'A note on synchronizer or interlock maloperation,' " *IEEE Transactions on Computers C-28*, October 1979, p. 802.

asynchronous qualifier. This technique can eliminate transition races in state machines, but does not address the metastability dilemma. The state flip-flop affected by the asynchronous qualifier can become metastable. The state flip-flops contribute to the production of the state machine outputs through combinational logic circuits, so metastability in a flip-flop output can have drastic consequences.

c. Using a chain of flip-flops. If a clocked flip-flop fed by an asynchronous input can become metastable, they why not feed the flip-flop output into another flip-flop to clean it up? Such a technique can indeed reduce the frequency of occurrence of metastability in the final output, but cannot eliminate it. With each flip-flop in the chain, the probability of failure decreases, yet at fast clock speeds the final probability may still be uncomfortably high, and is certainly not zero.

d. Starting and stopping crystal oscillators used for clock generation. Crystal oscillators and LC resonant circuits are often used to generate precise clock frequencies. Such oscillators do not start cleanly and instantly—many cycles will pass before the oscillator's amplitude stabilizes. Therefore, an attempt to stop and restart these oscillators to stretch the clock pulse when metastability occurs will not produce satisfactory performance.

e. Gating the output of continuously running clock oscillators. "Gating the clock" was discouraged in synchronous circuits because of the possibilities of clock skew, runt pulses, and hazards. With these drawbacks, this method has little to recommend it for the treatment of metastability.

f. Using an "asynchronous arbiter." Many such complicated circuits have been proposed; all fail to solve the problem of metastability, but instead move the sensitive points to other parts of the circuit.

What Should You Do?

Metastable behavior cannot be eliminated in real circuits. The proper defenses against metastability involve lowering the probability of its occurrence, shortening its duration, or sensing its presence and freezing the action of the system until the metastable state has passed.

The simplest defense is to run your synchronous systems at moderate speeds. If you avoid pressing the system clock to the limit of your circuit, the probability of inducing metastability can be made acceptably low. As you approach a crucially fast clock speed, your circuit's behavior can deteriorate dramatically.

If running the system clock at a slow speed is not acceptable, then the next defense against metastability is to use a well-designed interruptible clock and a metastability detector. The detector will catch metastable states that manifest themselves as voltages in the transition region but will not detect the delayed-transition form of metastability.

If your clock must run fast and stretching it is intolerable, then you must investigate using flip-flops that resolve metastable states faster. At this stage, you may have to do your own experiments on metastability.

A SELECTION OF INTEGRATED CIRCUITS

The low-power Schottky families of integrated circuits are our mainstays at the MSI level of design. The popular chips are produced by several manufacturers under the standard 74LS, 74ALS, and 74F nomenclatures. In Table 12–1 we list a selection of chips that we have found useful. The table is not exhaustive, nor is it a substitute for the manufacturers' data books. We list 74LS chips, if available; many of these are also available in the improved 74ALS and 74F lines. Some manufacturers' data books are cited at the end of the chapter.

TABLE 12-1 SELECTED INTEGRATED CIRCUITS

Number of pins	Chip designation	Description
AND and OR		
14	74LS00	Quad 2-input Nand gates
14	74LS02	Quad 2-input Nor gates
14	74LS08	Quad 2-input And gates
14	74LS32	Quad 2-input Or gates
14	74LS10	Triple 3-input Nand gates
14	74LS27	Triple 3-input Nor gates
14	74LS11	Triple 3-input And gates
14	74LS20	Dual 4-input Nand gates
14	74LS21	Dual 4-input And gates
14	74LS30	8-input Nand gate
16	74LS133	13-input Nand gate
20	74ALS804	Hex 2-input Nand gates
20	74ALS805	Hex 2-input Nor gates
20	74ALS808	Hex 2-input And gates
20	74ALS832	Hex 2-input Or gates
Inverter		
14	74LS04	Hex inverters
EOR and COINCIDENCE		
14	74LS86	Quad Exclusive Or gates
14	74ALS810	Quad 2-input Coincidence gates
Buffers, Drivers, and Bus Transceivers		
14	7406	Hex Open-collector inverting buffers
14	7407	Hex Open-collector noninverting buffers
14	74LS125	Quad 3-state buffers with independent enables
16	74LS367	Hex 3-state buffers
16	74LS368	Hex 3-state buffers, inverting
20	74LS240	Octal 3-state buffers, inverting
20	74LS241	Octal 3-state buffers
20	74ALS640	Octal 3-state transceivers, inverting
20	74ALS641	Octal open-collector transceivers
20	74ALS642	Octal open-collector transceivers, inverting
20	74ALS643	Octal 3-state transceivers, inverting and true
20	74ALS644	Octal open-collector transceivers, inverting and true
20	74ALS645	Octal 3-state transceivers
20	74ALS62x	Series of octal transceivers, similar to the 64x series

TABLE 12-1 (Continued)

Number of pins	Chip designation	Description
Multiplexers (with common selects and enable, unless specified)		
16	74LS157	Quad 2-input multiplexers
16	74LS158	Quad 2-input multiplexers, inverted outputs
16	74LS257	Quad 2-input multiplexers, 3-state outputs
16	74LS258	Quad 2-input multiplexers, 3-state inverted outputs
16	74LS153	Dual 4-input multiplexers, separate enables
16	74LS352	Dual 4-input multiplexers, inverted outputs, separate enables
16	74LS253	Dual 4-input multiplexers, 3-state outputs, separate enables
16	74LS353	Dual 4-input multiplexers, 3-state inverted outputs, separate enables
16	74LS151	8-input multiplexer, inverted and noninverted outputs
16	74LS251	8-input multiplexer, 3-state inverted and noninverted outputs
24	74150	16-input multiplexer, inverted output
Decoders		
16	74LS139	Dual 2–to–4 decoders, independent enables
16	74LS138	3–to–8 decoder, with 3 enables
16	74LS42	4–to–10 decoder (also functions as enabled 3–to–8)
24	74154	4–to–16 decoder, enabled
Comparators		
16	74LS85	4-bit arithmetic magnitude comparator: $<, =, >$
20	74ALS518-522	Series of 8-bit comparators
20	74ALS677-678	Series of 16-bit comparators against a fixed pattern
Flip-flops		
16	74LS109	Dual JK flip flops, with direct set and direct clear
14	74LS74A	Dual D flip flops, with direct set and direct clear
D Registers		
16	74LS175	4-bit D register with asynchronous clear; inverting and non-inverting outputs
16	74LS379	4-bit D register with enabled load; inverting and noninverting outputs
16	74LS173	4-bit D register with enabled load, 3-state outputs, and asynchronous clear
16	74LS174	6-bit D register with asynchronous clear
16	74LS378	6-bit D register with enable
20	74LS273	8-bit D register with asynchronous clear
20	74LS377	8-bit D register with enable
20	74LS374	8-bit D register with 3-state outputs
Shift Registers		
16	74LS194	4-bit parallel-in, parallel-out universal shift register
20	74LS299	8-bit parallel-in, parallel-out bus-oriented shift register with asynchronous clear

TABLE 12-1 (Continued)

Number of pins	Chip designation	Description
20	74LS323	8-bit parallel-in, parallel-out bus-oriented shift register with synchronous clear
16	74LS166	8-bit parallel-in, serial-out shift register with asynchronous clear
16	74LS164	8-bit serial-in, parallel-out shift register
24	74LS674	16-bit parallel-in, serial-out shift register
24	74LS673	16-bit serial-in, parallel-out shift register

Counters

16	74LS161	4-bit binary counter with asynchronous clear
16	74LS163	4-bit binary counter with synchronous clear
16	74LS669	4-bit binary up-down counter
24	74AS869	8-bit binary up-down counter

PALs

Many useful chips exist, too numerous to list here. Monolithic Memories pioneered the concept; many others also manufacture PALs. Some PALs are erasable.

Single Shots

16	96L02	Dual single shot
14	TTLPWG-x	Rhombus TTL pulse-width generator. x can be 5–100 nsec (see Dynamic RAM Support for Rhombus Industries' address).

ALU Bit Slices

24	74LS181	4-bit ALU, combinational, with look-ahead and ripple carry
20	74LS381	4-bit ALU, combinational, with look-ahead carry
20	74LS382	4-bit ALU, combinational, with ripple carry
48	74LS481	4-bit processor slice
68	74AS888	8-bit processor slice
40	AM2901	4-bit processor slice
48	AM2903	4-bit processor slice
84	MM677081	16-bit ALU, combinational

Microsequencers

68	74AS890	Sequencer, 14-bit address
40	AM2910	Sequencer, 12-bit address

Dynamic RAM Support

40	AM2963	Address multiplexer, refresh controller
20	AM2966	High-speed address line driver
14	DZTM1-300	Rhombus 300-nsec delay line
		Rhombus Industries, 15801 Chemical Lane, Huntington Beach, Calif. 92649, makes a wide series of delay lines with internal drivers and terminators. Available with standard ECL or TTL drivers; programmable versions with 8 or 16 steps also available.

READINGS AND SOURCES

BLAKESLEE, THOMAS R., *Digital Design with Standard MSI and LSI,* 2nd ed. John Wiley & Sons, New York, 1979. Sound advice about implementations. Good section on noise and transmission lines. Includes a thoughtful treatment of the social consequences of engineering.

BREUNINGER, ROBERT K., and KEVIN FRANK, *Metastable Characteristics of Texas Instruments Advanced Bipolar Logic Families.* Texas Instruments, 1985. Test procedure and results of metastability experiments.

CHANEY, THOMAS J., "Comment on 'A note on synchronizer or interlock maloperation,' " *IEEE Transactions on Computers C-28,* October 1979, p. 802. Metastability experiments.

CHANEY, THOMAS J., "Measured flip-flop responses to marginal triggering," *IEEE Transactions on Computers C-32,* December 1983, p. 1207. Metastability experiments.

ELINEAU, GERALD, and WERNER WIESBECK, "A new JK flip-flop for synchronizers," *IEEE Transactions on Computers C-26,* December 1977, p. 1277. Metastability.

FLETCHER, WILLIAM I., *An Engineering Approach to Digital Design.* Prentice-Hall, Englewood Cliffs, N.J., 1980. Practical engineering information throughout.

GLASSER, LANCE A., and DANIEL W. DOBBERPUHL, *The Design and Analysis of VLSI Circuits.* Addison-Wesley Publishing Co., Reading, Mass., 1985.

MARINO, LEONARD R., "The effect of asynchronous inputs on sequential network reliability," *IEEE Transactions on Computers C-26,* November 1977, p. 1082. Metastability.

MEAD, CARVER, and LYNN CONWAY, *Introduction to VLSI Systems.* Addison-Wesley Publishing Co., Reading, Mass., 1980. The first VLSI textbook.

MICK, JOHN, and JAMES BRICK, *Bit-Slice Microprocessor Design.* McGraw-Hill Book Co., New York, 1980. A collection of design notes for the Advanced Micro Devices 2900 bit-slice family. This book is useful far beyond the Am2900 chips.

STONE, H., *Microcomputer Interfacing.* Addison-Wesley Publishing Co., Reading, Mass., 1982. Good discussion of noise and shielding in Chapter 2.

WESTE, NEIL, and KAMRAN ESHRAGHIAN, *Principles of CMOS VLSI Design: A Systems Perspective.* Addison-Wesley Publishing Co., Reading, Mass., 1985.

WILLIAMS, GERALD E., *Digital Technology.* Science Research Associates, Chicago, 1977. Good nuts-and-bolts technology.

Manuals and Data Books

Am29300 Family Handbook. Advanced Micro Devices, 901 Thompson Place, P.O. Box 3453, Sunnyvale, Calif. 94088, 1985. High-performance 32-bit building blocks.

Bipolar Microprocessor Logic and Interface. Advanced Micro Devices, 901 Thompson Place, P.O. Box 3453, Sunnyvale, Calif. 94088. The Am2900 family and support devices. Good discussion of Schottky logic.

Bipolar/MOS Memories. Advanced Micro Devices, 901 Thompson Place, P.O. Box 3453, Sunnyvale, Calif. 94088. Data book.

FAST: Fairchild Advanced Schottky TTL. Fairchild Camera and Instrument Corporation, Digital Products Division, South Portland, Maine 04101. Data book.

FAST: High Speed Logic Systems and Solutions. Fairchild Camera and Instrument Corp., Digital Unit, 333 Western Avenue, South Portland, Maine 04101. Notes for a 1984 Fairchild seminar on FAST/LSI. Discusses metastability.

High-Speed CMOS Logic Data. Motorola Semiconductor Products, 3501 Ed Bluestein Blvd., Austin, Tex. 78721. Data Book.

LSI Databook. Monolithic Memories, 2175 Mission College Blvd., Santa Clara, Calif. 95954. PALS, memory products, arithmetic units, system building blocks.

MECL Device Data. Motorola Semiconductor Products, Box 20912, Phoenix, Ariz. 85036. Data book.

MECL System Design Handbook, 4th ed. Motorola Semiconductor Products, Box 20912, Phoenix, Ariz. 85036, 1983. A wealth of practical ECL design advice. Good general treatment of noise and transmission lines.

Memory Components Handbook. Intel Corp., Literature Department, 3065 Bowers Ave., Santa Clara, Calif. 95051. Data book and applications.

MOS Microprocessors and Peripherals. Advanced Micro Devices, 901 Thompson Place, P.O. Box 3453, Sunnyvale, Calif. 94088. Data book.

PAL Programmable Array Logic Handbook. Monolithic Memories, 2175 Mission College Blvd., Santa Clara, Calif. 95054.

Systems Design Handbook, 2nd ed. Monolithic Memories, 2175 Mission College Blvd., Santa Clara, Calif. 95054, 1985. Excellent manufacturer's design handbook. Good treatment of ECL (pp. 2–40).

The TTL Data Book. Texas Instruments, P.O. Box 225012, Dallas, Tex. 75265.

Index

Abstraction in digital design, 2, 167
Access time, RAM, 138
Accumulator, LD20, 272
Active clock edge, 120, 171
Adder, 75, 92–94
  full, 92
  half, 92
  ripple carry, 92
Addition, binary, 92–98
  speedup, 95–98
Address:
  microcomputer device, 436
  PDP-8, 262–66
  RAM, 135
  table lookup, 79
Address bus, microprocessor, 434
Address recognizer, 435, 439
Algorithm (*See* Control algorithm)
Algorithmic state machine, 170–88 (*See also*
  Controller)
  ASM chart, 172–75
  asynchronous, 192–93
  branch, 173–74
  conditional output, 175
  output, 172–73, 175
  qualifier, 375
  state, 172
  synchronous, 171

  synthesis, 176–88
  unconditional output, 175
ALU (*See* Arithmetic logic unit)
Amplifier, 506
  differential, 474
AND logic function:
  definition, 3, 6–7
  drafting symbol, 36
  implementation, 37–38, 40
  logic notation, 6
Arbiter, asynchronous, 511
Architecture, 166–70
Arithmetic, signed, 93–94
Arithmetic logic unit (ALU), 94–98
  in LD20, 273–74
  in LD30, 406–7
Arithmetic magnitude comparator, 87–89
Arithmetic shift, 162
ASCII character code, 352–53
ASM (*See* Algorithmic state machine;
  Controller)
Assembler, Logic Engine (*See* Logic Engine
  microassembler)
Associative law, 9, 11
Asynchronous arbiter, 511
Asynchronous ASM, 192–93, 359–64
Asynchronous circuit, 114–15, 152, 359–64,
  451

Asynchronous communication discipline, 352
Asynchronous design (*See* Asynchronous circuit)
Asynchronous input, 121, 129, 190–93, 244–45
Auto-indexing:
   LD20, 296–97
   PDP-8, 265–66

Bank, memory, 137
Base, bipolar transistor, 468
Bias, transistor, 469
Bipolar logic families (*See* Logic families)
Bipolar transistor, 467–69
Bistable device, 114
Bit slice, 95, 133–34
   AMD 2901, 134
   AMD 2903, 134
   TI 74AS888, 134
Black Jack Dealer, 237–53
   microprogrammed, 377–78, 384–85
Block carry, 97–98
Boolean algebra, 3, 8–19
   conventions, 9
   duality principle, 10, 40
   indentities, 10–11, 19
   manipulations, 10–11
   operator hierarchy, 9
Breakpoint, microprogram, 398
Buffer:
   open collector, 102, 477–78
   for power, 436, 484
   register for data, 218
   three-state, 103–4, 479–80
Building block:
   combinational, 75–104
   sequential, 120–34
Bus, 100–104, 136–37
   bidirectional, 436–37
   LD20 data, 274–79
   microcomputer, 434
   protocol, 437–38
   unidirectional, 436–37
Bus control methods, 100–104
Bypass capacitor, 500
Bypassing, power supply, 499–500

CA (*See* Contents of effective address)
Canonical product-of-sums form, 13–14
Canonical sum-of-products form, 12–13
Canonical truth table, 5, 19–20
Capacitance, 152–53, 470, 499–500
Carry, ripple, 92–93
Carry-generate function, 96–98
Carry look-ahead, 97–98
Carry-propagate function, 96–98

Channel, MOS transistor, 469
Characteristic impedance (*See* Impedance, characteristic)
Charge, electrical, 466
   induced, 470
Charge flow (*See* Current, electrical)
Chip (*See* Integrated circuit)
Circuit analysis, 35, 49–50
Circuit diagram, 35
Circuit synthesis, 35, 50–54
Clock, 118, 171–72 (*See also* Oscillator)
   serial bit, 212–19, 358, 366
   system, 208–12, 257, 490–91
Clocked circuit (*See* Synchronous circuit)
Clock frequency, 171
Clock period, 171
Clock skew, 188–90
CMOS (Complementary MOS) (*See* Logic families, CMOS)
Code:
   binary, 83–87, 161
   gray, 161
   unit-distance, 255
Code conversion:
   with ROM, 140–42
Coincidence logic function, 61–63, 91
Collector, bipolar transistor, 468
Column address, 135
Column address strobe, 139
Combinational circuit, 75
Combination lock, 229–37
Command output, 172–73, 175
Common-mode voltage, 504
Communications (*See* Asynchronous communication discipline)
Commutative law, 9, 11
Comparator (*See* Arithmetic magnitude comparator)
Comparison, equality, 88
Complementary MOS logic (CMOS) (*See* Logic families, CMOS)
Conditional output, 175
Conduction, 466
Conductor, 466
Conjunctive normal form, 13
Contents of effective address (*CA*):
   in PDP-8, 264
Control (*See* Controller)
Control algorithm, 166–70
Control bus, microprocessor, 434
Controller:
   ASM, 176–88
   binary-counter, 201
   hardwired, 176–88, 433
   microcomputer, 372, 430, 433, 460–61

microprogrammed, 372, 433
multiplexer, 178–83
one-hot, 183–86
ROM-based, 186–88
Control store, microprogram, 393
Counter, 108, 126–31
  asynchronous, 128
  binary, 126–31, 161
  gray-code, 161
  moebius, 162
  ripple, 128
  synchronous, 128
Coupler, optical (*See* Optical coupler)
CPU (*See* Microprocessor)
Crosstalk, 501
Crystal, quartz, 490
Crystal lattice, 466
Crystal oscillator, 490, 511
Current, electrical, 152–53, 466–71
Current spike, 473
Cycle Time, RAM, 138

Data bus, (*See* Bus)
Data sheet (*See* Integrated circuit data sheet)
Debouncer, switch (*See* Switch debouncing)
Decoder, 83–86
Delay:
  interconnection, 109
  propagation, 109, 482–83
Delay line, 153–54
DeMorgan's law, 10–11
Demultiplexer, 80–83, 86
Depletion mode, 470
Differential amplifier, 474
Differential line driving, 504–5
Digital drafting, 35
Diode, 373–75, 467
  light-emitting, 493–95
  for switch protection, 494
DIP (*See* Dual inline package)
Direct clear, 121
Direct set, 121
Disable (*See* Enable)
Disjunctive normal form, 12
Display, seven-segment, 31–32, 73, 140–41
Distributive law, 9, 11
Documentation, 2, 35, 194
Don't-care outputs, 18–19, 22–24
Doping, 467
Downloading, 397
Drain, MOS transistor, 469
Drive, output, 483
Driver (*See also* Buffer):
  lamp, 492–93

line, 504–5
open-collector, 102, 477–78
Dual inline package, 34
Duality principle, 10, 40
Dynamic RAM (*See* Random-access memory, dynamic)

EA (*See* Effective address)
ECL (Emitter-coupled logic) (*See* Logic Families, ECL)
Edge-driven circuit (*See* Edge-triggered circuit)
Edge-triggered circuit, 119
EEPROM, 143
Effective address (EA), 264
  contents of (CA), 264
Electrode, 470
Electrons, 466
Emitter, bipolar transistor, 468
Emitter-coupled logic (ECL) (*See* Logic Families, ECL)
Enable:
  decoder, 85–86
  interrupt, 271
  multiplexer, 77
  RAM, 136
  three-state-output, 103, 136, 479
Encoder (*See* Priority encoder)
Enhancement mode, 470
EOR logic function (*See* Exclusive OR logic function)
EPROM, 142–43
EPROM eraser, 143
Equality comparison (*See* Comparison, equality)
Equivalence logic function, 61–63, 91
Excitation table, 117, 120
Exclusive OR logic function, 61–63, 90
Execute phase, LD20, 281, 300–319

Fanout, 483–84
Feedback, 111–12, 486, 507
Fetch phase, LD20, 281–85, 291–97
Fight, output, 473
Filter, power-supply, 496
Firmware, 140
Flip-flop, 108, 112–26
  asynchronous RS, 113–17
  clocked RS, 118–19
  D, 124–26
  edge-driven, 119–20
  edge-triggered, 119–20
  enabled D, 125–26
  JK, 120–24, 127–28
  level-driven, 119
  master-slave, 119

Flip-flop (*cont.*)
pure edge-triggered, 119–20
RS, 113–18
SOC, 161
toggle, 161
Formalism in digital design, 3 (*See also* Boolean algebra; Algorithmic
state machine, ASM chart; Mixed logic)
FORTH Machine, 398–406
Forward-biased diode, 467
Frequency divider, 161
Full adder (*See* Adder)

Gain, transistor, 469
Gate, 35, 38, 76, 101–2, 471–72
MOS transistor, 470
Gating the clock, 125, 189–90
Glitch, 110 (*See also* Hazard)
Gray-code counter (*See* Counter, gray-code)
Ground, integrated circuit, 34, 497–98
Ground plane, 498

Handshake, synchronizing, 246, 255–56
Hardwired control, 176–88, 368
Hazard, 108–11, 208
Hierarchy of logical operators, 9
High-impedence mode, 103, 479–80
.H naming convention, 38–39
Hold time:
flip-flop, 120
RAM, 138
sequential circuit, 482–83
Hole, 467
Hysteresis, 487

IC (*See* Integrated circuit)
Idle state, LD20, 298–300
Impedance, characteristic, 502–5
Implication logic function, 67, 73, 91
Index, table-lookup, 79
Indirect addressing:
LD20, 296–97
PDP-8, 263–64
Indirect bit, PDP-8, 263
Inductance, 493–94, 498
Induction, charge, 470
Inductive load, 493–94
Input, unused, 485–86
Input load, 483–84
Input-output:
interface, in LD20, 355–66
interrupt-driven, 440–42
LD20, 304–8, 339–40, 353–55, 366–68
memory-mapped, 435

microcomputer, 438–42
PDP-8, 270, 353–55
port, 437
programmed, 438–42
separate, 435–36
Instruction register, LD20, 274
Insulator, 466
Integrated circuit, 34, 76
characteristics, 481–85
data sheet, 34, 74, 480–82
loadings, 483–84
names, 38
noise margins, 485
operating conditions, 481
static (DC) characteristics, 481–82
switching (AC) characteristics, 482–83
Integrated circuit devices, 512–14
12060 oscillator, 490
12061 oscillator, 490
1488 line driver, 366
1489 line receiver, 366
14411 baud rate generator, 490
2901 ALU bit-slice, 134
2903 ALU bit-slice, 134
2909 sequencer, 390, 427
2910 sequencer, 390–94
555 timer, 366, 489–90
6147 static RAM, 343–44
6809 microprocessor, 435, 438
68000-series microprocessor, 435
7400 nand gate, 506
7404 inverter, 507
7406 open-collector inverter, 64–66
7407 open-collector buffer, 102–3
7474 D flip-flop, 506
74154 decoder, 334
74AS888 ALU bit-slice, 134
74AS889 sequencer, 390
74LS00 nand gate, 38, 43–46, 211
74LS02 nor gate, 38, 43–46
74LS04 inverter, 42–43, 46–48
74LS08 and gate, 40, 43–46
74LS10 nand gate, 40
74LS25 nor gate, 88
74LS32 or gate, 43–46
74LS37 nand buffer, 484
74LS42 decoder, 83, 85–86, 105, 236
74LS74 D flip-flop, 158, 216
74LS85 arithmetic comparator, 88, 106, 236
74LS86 exclusive-or gate, 62–63, 88
74LS109 JK flip-flop, 121–24, 216, 221, 224, 345
74LS133 nand gate, 224
74LS139 decoder, 82
74LS147 priority encoder, 87

74LS151 multiplexer, 235–36, 328, 333, 339
74LS153 multiplexer, 91, 106
74LS157 multiplexer, 78, 215–16
74LS163 binary counter, 130–31, 215–16, 221, 224, 233, 235, 349, 383
74LS164 shift register, 224
74LS165 shift register, 221
74LS175 D register, 185, 236, 333, 335
74LS181 ALU, 95–98, 107, 273–74, 328–30
74LS182 ALU carry-generate-propagate, 98
74LS193 binary counter, 131
74LS194 shift register, 133, 331–32, 340
74LS243 bus transceiver, 437
74LS244 three-state buffer, 103, 133
74LS251 multiplexer, 105
74LS280 parity generator-checker, 106
74LS283 adder, 93, 106
74LS352 multiplexer, 79, 81
74LS378 D-register, 126, 331
74LS379 D-register, 126, 331
74LS669 binary counter, 131
8080 microprocessor, 435
96L02 single-shot, 153, 344–45, 350
CD4047 oscillator, 490
DM7575 PLA, 145
PAL16R4, 350
PAL16R8, 150, 163
PAL18H4, 149
PAL18L4, 147
PAL20L10, 350
Interrupt:
    input-output, 440–42
    LD20, 284–85, 292, 309, 340–41
    PDP-8, 270–72
    priority, 441
    terminal, in LD20, 355, 364
Interrupt-driven input-output, 440–42
Interrupt request, 440
Interrupt subprogram:
    microcomputer, 440–41
    PDP-8, 271–72
Inverter, voltage, 42–43, 46–47, 471

Jam transfer, 124, 126

Karnaugh map, 19–26
    construction, 21
    definition, 19–20
    errors, 24–25
    in hazard elimination, 110
    simplification, 22–25
K-map (See Karnaugh map)

Lamp driver, 492–93
Large-scale integration (LSI), 76

Latch, 112–13
Lattice, crystal (See Crystal lattice)
LD20 minicomputer, 261
    ALU controls, 329–31
    architecture, 272–79, 315–19
    ASM, 286–319
    ASM chart auxiliary variables, 324–25
    ASM chart labels, 326
    conditional output terms, 326–27
    control, 279–86
    data multiplexer inputs, 327–29
    execute phase, 281, 300–319
    fetch phase, 281–85, 291–97
    interrupt system controls, 340–41
    IOP signal enabler, 339–40
    link-bit control, 333–34
    manual operations, 298–300, 302–4, 341–42
    memory control, 278, 287–91, 342–46
    microprogrammed implementation, 406–24
    priority circuit, 337–39
    register-load signals, 331–32
    state assignment, 334–35
    state decoder, 334
    state generator, 334–36
    state multiplexer inputs, 335
    state transitions, 334–36
LD30 minicomputer, 406–24
    architecture, 406–7, 423–24
    command bits, 407–8
    execute phase, 413–16
    fetch phase, 412–13
    idle phase, 409–11
    instruction decoding, 417–19
    interrupt processing, 416–17
    manual phase, 411–12
    microcode, 408–23
    test inputs, 422
LEASMB (See Logic Engine microassembler)
LED (See Light-emitting diode)
Level-driven circuit, 119
Light-emitting diode, 492–95
Line driver, 504–5
Line receiver, 504–5
Line termination, 503
Link, LD20, 273
.L naming convention, 38–39
Load:
    inductive, 493–94
    input, 483–84
    integrated circuit, 483–84
Loadings, integrated circuit, 483–84
Lock (See Combination lock)
Logical constants, 4, 9
    implementation, 53–54, 66–67

Logical operators, 3, 6–8
  hierarchy, 9
  implementation, 35–38, 40–49, 55–58, 61–68
Logical variables, 4–5
Logic Engine, 393–98
  backpanel, 397
  base unit, 396–97
  microassembler, 401–5
  support software, 397–98
Logic families:
  bipolar, 472–75
  CMOS, 476–77
  ECL, 474–75
  MOS, 475–77
  n-MOS, 476
  p-MOS, 476
  RTL, 472
  TTL, 473–74
  unipolar, 475–77
Low-power Schottky TTL, 33–34, 474
LSI (*See* Large-scale integration)

Mark level, 352–53
Maxterm, 13–14
Medium-scale integration (MSI), 76
Memory, 3, 108, 134–43 (*See also* EEPROM,
  EPROM; Flip-flop; PAL; PLA; PLE;
  PROM; RAM; ROM)
  LD20, 273
  microcomputer, 434
Memory addressing:
  PDP-8, 262–66
  RAM, 136–37
Memory address register, LD20, 273
Memory bank, 137
Memory buffer register, LD20, 273
Memory-mapped input-output, 435
Metal-oxide semiconductor (MOS) (*See* Logic
  families, MOS)
Metastability, 117, 154–55, 505–11
Metastable state (*See* Metastability)
Microcomputer, 372, 432
  6809, 438
  68000, 435
  8080, 435
Microinstruction, 375 (*See also*
  Microprogramming)
  control store, 393
  LD20, 309–15
  PDP-8, 267–69
  pipeline register, 388
  writable control store, 393
Microprocessor, 432
Microprogramming, 374
  classical, 372–75

Microprogram sequencer, 388–90
  2909, 390
  2910, 390
  74AS889, 390
Minicomputer, 260, 371–72, 431
Minterm, 12–13
Mixed logic, 35–55, 61–68
  AND/OR duals, 40
  circuit analysis, 40–41, 49–50
  circuit synthesis, 50–55, 65–66
  drafting conventions, 35–40, 51–52, 55, 163
  NOT implementation, 46–48
  "oops," 48
  operator symbols, 36, 47, 61
  in PALs, 149, 163
  signal naming conventions, 38–40
  theory, 42–49
  voltage inversion, 42–43, 47–48
Mixed-logic convention, 37
Modulus counting, 126–27, 218
MOS (Metal-oxide semiconductor) (*See* Logic
  families, MOS)
MOS transistor, 469–71
  depletion-mode, 470
  enhancement-mode, 470
MPU (*See* Microprocessor)
MSI (*See* Medium-scale integration)
Multiple qualifier (*See* Qualifier, multiple)
Multiplexer, 75, 77–80, 91, 100–101
  terminal, 449–61
Multiplexer controller, 178–83
Multivibrator, monostable (*See* Single-shot)
Mux (*See* Multiplexer)

Nand gate, 38, 472
  as voltage inverter, 49
NAND logic function, 27, 38, 55–58, 67–68, 91
Negative clock edge, 119
Negative logic, 37
Negative-logic convention, 37
n-MOS (*See* Logic families, MOS)
n-MOS transistor, 470–71
Noise, 500–505 (*See also* Glitch; Hazard)
Noise margin, 485
Nor gate, 38, 472
  as voltage inverter, 49
NOR logic function, 27, 38, 55–58, 67–68, 91
NOT logic function:
  definition, 3, 6
  drafting symbol, 47–48
  implementation, 46–48
  logic notation, 6
npn transistor, 467–68
n-type silicon, 467
Numbered slash, 89

Ohm's Law, 479, 492–93, 497
One-hot controller, 183–86
1's catching, 113, 129
One's complement, 93
"Oops" function, 48
Open-collector:
    buffer, 102, 477–78
    circuit, 63–66, 102–3, 367–68, 477–79
    driver, 102, 477–78
    resistor, 478–79
Operate instruction:
    LD20, 309–15, 337–39
    PDP-8, 267–69
Optical coupler, 494–95
Optical isolator (*See* Optical coupler)
OR logic function:
    definition, 3, 7–8
    drafting symbol, 36–37
    implementation, 38, 40
    logic notation, 8
Oscillating output, 111
Oscillator, 488–91
    crystal-controlled, 490
    555 timer, 489–90
    Schmitt trigger, 485–89
    system clock, 490–91
Output drive, 483
Output fight, 473

Page bit, PDP-8, 263
Page offset, PDP-8, 262
PAL (*See* Programmable array logic)
Parallel-to-serial conversion, 217–21, 453
Parity, 106
PDP-8 minicomputer, 260, 371–72
    description, 261–72
    as device controller, 431
    input-output protocol, 270, 353–55
    instructions, 266–70
    interrupts, 270–72
    memory addressing, 262–66
    operation codes, 84, 261, 266–72
    single-chip, 272
PDP-8E minicomputer, 351
PDP-8I minicomputer (*See* PDP-8
    minicomputer)
Peripheral interface adapter, 437
Phototransistor, 495
PIA (*See* Peripheral interface adapter)
PLA (*See* Programmable Logic Array)
PLE (*See* Programmable Logic Element)
p-MOS logic family (*See* Logic families, MOS)
p-MOS transistor, 470–71
p,n junction, 467
pnp transistor, 467–68

Port, input-output, 437
Positive clock edge, 119
Positive logic, 37
Positive-logic convention, 37, 55–61
    analyzing circuits, 58–61
    converting to mixed logic, 59–60
    synthesizing circuits, 56–58
Power distribution, 497–500
    losses, 497–98
Power supply, 495–96
Precedence rules (*See* Hierarchy of logic
    operators)
Preclear (*See* Direct clear)
Preset (*See* Direct set)
Priority encoder, 86–87
Priority interrupt, 441
Priority system, LD20, 309–12, 337–39
Product-of-sums form, 13–14, 16, 25
Program counter, LD20, 274
Programmable Array Logic, 147–52
Programmable logic, 143–52
    programming, 150–52
Programmable Logic Array, 144–45
Programmable Logic Element, 147
Programmable Read-Only Memory, 142, 145–
    47
    electrically-erasable, 143
Programmed input-output, 438–42
PROM (*See* Programmable Read-Only
    Memory)
Propagation delay (*See* Delay, propagation)
Protocol, microcomputer bus, 437–38
Pseudo-clock, 359–64
p-type silicon, 467
Pullup resistor (*See* Resistor, pullup)

Qualifier, 374
    multiple, 376–79, 385
    single, 379–85
Queue, 258

Race, 190–92
    output, 191–92, 244–45
    transition, 190–91, 244–45
RAM (*See* Random access memory)
Random access memory, 108, 135–39, 376
    access time, 138
    dynamic, 139
    read cycle time, 138
    refresh, 139
    static, 139
    write cycle time, 138
Read-Only Memory, 108, 139–43, 375–76
    field-programmable, 142–43
Receiver, line, 504–5

Rectifier, 496
Reflection, voltage, 502–3
Refresh, 139
Register, 75, 108, 112, 126–34
    counter, 126–31
    D, 126
    enabled D, 126
    shift, 131–33
Register file, 133
Regulator, voltage, 496
Relay, solid-state, 494
Remote sensing, 496
Reset:
    asynchronous, 121, 186
    LD20, 299–300
    power-on, 488
Reset state, flip-flop, 114
Resistor:
    pulldown, 116, 491–92
    pullup, 63, 116, 491–92
Resistor-transistor logic (RTL) (*See* Logic families, RTL)
Reverse-biased diode, 467
ROM (*See* Read-Only Memory)
ROM-based controller, 186–88
Row address, 135
Row-address strobe, 139
RTL (Resistor-transistor logic) (*See* Logic families, RTL)

Saturation, transistor, 474, 506
Schmitt trigger, 486–88, 510
    as oscillator, 488–89
Schottky diode, 474
Schottky technology, 474
Select, multiplexer, 77
Semiconductor, 466–67
Separate input-output, 435–36
Sequential circuit, 75, 112–20
Serial data transmission, 212, 352–53, 451
Serial-to-parallel conversion, 217–19, 221–24, 453
Set state, flip-flop, 114
Setup time:
    flip-flop, 120
    RAM, 138
    sequential circuit, 482–83
74xx TTL, 473
74ALSxx TTL, 474, 509, 512
74Fxx TTL, 474, 509, 512
74LSxx TTL, 33–34, 474, 512–14
74Sxx TTL, 474
Shift register, 131–33
    LD20, 308–9
Signal, 38–40

Signed arithmetic (*See* Arithmetic, signed)
Silicon, 466–67
Single-ended line driving, 504–5
Single-pulser, 203–8, 247
    building block, 207
    generalized, 207
    one-state, 204–6
    two-state, 206–7
Single qualifier (*See* Qualifier, single)
Single-shot, 152–53
Slash (*See* NOT logic function, drafting symbol; Numbered slash)
Small-scale integration (SSI), 76
Solid-state relay, 494
Source, MOS transistor, 469
Space level, 352–53
Spike, current, 473
SSI (*See* Small-scale integration)
Stable state (*See* State, stable)
Stack, 258, 391, 399
Start bit, 353
State:
    ASM, 171–72
    flip-flop, 113–15
    stable, 114
State assignment, 176
State generator, 176
State machine (*See* Algorithmic state machine)
State variable, 176
Static design, 194
Static RAM (*See* Random Access Memory, static)
Stepping motor, 443–49
Stop bit, 353
Storage time, transistor, 474
Strobe (*See* Enable)
Structured design (*See* Top-down design)
Style, digital design, 1, 52–54, 194–95
Sum-of-products form, 12–13, 15, 19, 25
Switch debouncing, 115–16, 491–92
Switching transistor, 469
Switch register, LD20, 268–69, 273
Synchronization, process, 246
Synchronizer, 125, 219
Synchronous circuit, 111, 115, 118, 170, 193–94
Synchronous communications, 218–24, 451
Synchronous design (*See* Synchronous circuit)
System clock (*See* Clock, system)

Table lookup, 79–80, 179, 374, 418–19
Teletype, 494–95
Terminal:
    communications protocol, 352–53
    input-output with PDP-8, 352–55

LD20 interface, 355–66
multiplexer, 449–61
Termination, line, 503
Three-state output, 103–4, 133, 479–80
Time-multiplexing, 450
Timer, 225–26, 489–90
Timing devices, 152–54 (*See also* Clock; Oscillator)
Timing diagram, 109
Toggle, 121
Top-down design, 2, 166
Totem-pole output, 473, 477
Traffic-light controller, 224–29
Transceiver, bus, 437
Transistor, 467–72
    bipolar, 467–69
    depletion-mode, 470
    enhancement-mode, 470
    MOS, 469–71
    npn, 467–68
    pnp, 467–68
    switching, 469
    unipolar, 469
Transistor-transistor logic (TTL), 33–34, 474
    families, 473–74
    low-power Schottky, 33–34, 474
    Schottky technology, 474
Transmission line theory, 502–5
Tri-state output (*See* Three-state output)
True, false (*See* Logical constants)
Truth table, 5–19
    in Boolean algebra, 10–19
    canonical, 5
    condensed, 17–18
    definition, 5

don't-care output, 18–19
vector representation, 6
TTL (*See* Transistor-transistor logic)
Two's complement, 93–94

UART (*See* Universal asynchronous receiver-transmitter)
Unipolar logic families (*See* Logic families)
Unipolar transistor, 469
Unit-distance code (*See* Code, unit-distance)
Unit load, 483
Universal asynchronous receiver-transmitter, 357–58, 360, 362–63, 365–66, 370, 454
Universal logic element, 90–91
Unused gate inputs, 485–86

V_{CC}, 34, 468
V_{DD}, 471
V_{SS}, 471
Vector, truth table, 6
Very-large-scale integration (VLSI), 76, 509
Vietch diagram (*See* Karnaugh map)
VLSI (*See* Very-large-scale integration)
Voltage inverter (*See* Inverter, voltage)
Voltage regulator, 496

Waveform, 109
    clock, 171
Wilkes, Maurice, 371–75
Wired AND, 63, 478
Wired OR, 63, 102–3, 478
Wire list, 444
Wire-wrap, 442–43
Wire-wrap controller, 445–49
Writable control store, 393